Soil Science Americana

Alfred E. Hartemink

Soil Science Americana

Chronicles and Progressions 1860–1960

 Springer

Alfred E. Hartemink
Department of Soil Science, FD Hole Soils Lab
University of Wisconsin
Madison, WI, USA

ISBN 978-3-030-71137-5   ISBN 978-3-030-71135-1   (eBook)
https://doi.org/10.1007/978-3-030-71135-1

Cover photo: Participants of the Transcontinental Excursion observing a Cecil clay loam profile near
Athens, Georgia on 25th June 1927. On the ladder: David Hissink, at the bottom of the ladder Boris
Polynov. Photo by M.E. Diemer, Chemist and Photographer of the University of Wisconsin.

This Springer imprint is published by the registered company Springer Nature Switzerland AG
The registered company address is: Gewerbestrasse 11, 6330 Cham, Switzerland

To my parents Derk[†] and Thea
For their ground gave me wings.

"*In every bit of honest writing in the world there is a base theme. Try to understand men, if you understand each other you will be kind to each other. Knowing a man well never leads to hate and nearly always leads to love. There are shorter means, many of them. There is writing promoting social change, writing punishing injustice, writing in celebration of heroism, but always that base theme. Try to understand each other.*"

John Steinbeck, 1938

"*It is rather a knowledge of relationships that the general reader seeks, not facts per se.*"

"*…knowledge is, after all, one great body, not departmentalized, but a huge jewel with many facets.*"

Charles Kellogg, 1940

"*Soil Science Americana* is an intellectual biography, not of one individual but of a new scientific field from its emergence to its complete coming of age. A tour the force with engaging sketches of the key players and often complex international relations. A fascinating read for all those interested not just in soil, climate and agriculture but in the history of science."

—Louise O. Fresco, *President, Wageningen University and Research*

"In a lively, personal voice, Hartemink traces the roots of modern soil science in the United States. He tells us of the melding of the Russian and American schools; the lives of those who shaped the field; formation of the famous academic departments of soil science; and the small wars fought by the giants of the field. Hartemink offers keen insight into how individuals and world history influenced the field, creating a book that will engage both the expert and non-expert in the underappreciated field of soil science."

—Jo Handelsman, *Director, Wisconsin Institute for Discovery*

"*Soil Science Americana* by Alfred Hartemink vividly outlines the historic evolution of soil science, its disciplinary diversity, and relevance to addressing historic, current and emerging issues of national and global significance. Using reader-friendly language, Hartemink draws on intellectual biography of pioneer soil scientists of the U.S. and their interaction with peers from Europe, and explains how global events influenced the evolution of soil science. The intellectual master piece is of interest to soil scientists, general public and the policy makers, and will remain pertinent for generations to come."

—Rattan Lal, *World Food Prize Laureate 2020, The Ohio State University*

Foreword by Ron Amundson

The soil of the Earth's surface is a historical object, a planetary skin that reflects the journey, experiences, and lineage of the patron of interest. Thus, it is not surprising that pedologists, as a group of scientists, have had a long interest in the history of their field and the origin of the ideas and concepts on which it is built.

In this book, *Soil Science Americana*, Alfred Hartemink has harnessed his long-standing ability and enthusiasm for examining science history and has created a riveting tale of American pedology, one of wide-ranging scope and depth. I anticipate that every reader will find themselves, as I was for a full afternoon as I read about Roy Simonson, carried to different times and places, mesmerized as the participants come again to life, regain their youth, and embark on their life journeys once again. As I remarked many years ago in the Foreword to the reprinting of Hans Jenny's Factors of Soil Formation, we are commonly introduced to eminent scientists after their death, and we see them only through the lens of photographs taken late in life. Not here. The illustrations are novel, and we see people in their prime. We begin to understand them as humans, with all the associated humility, vanity, foibles, mistakes, successes, and brilliance that come with the bargain.

I chose to preview Chap. 5, a synopsis of the life of pedologist Roy Simonson. Simonson and I were born a bit more than a generation apart, though I shared with him the experience of an upbringing on a farm on the Dakota plains, and as a descendent of Scandinavian immigrants. Alfred Hartemink illuminates Simonson's skills and training in basic skills such as mathematics and in writing. Simonson used his skills—and keen criticism—as an editor when he was just a graduate student, on the monumental book *Soils & Men*. As I read about this period in Simonson's life, it jarred my memory to an incident that occurred early in my career at Berkeley, when I was submitting some of my first manuscripts as a young Assistant Professor. I, as possibly many did, had the experience of submitting manuscripts to the journal Geoderma during Simonson's long tenure as Editor-in-Chief. In the late 1980s, my group had worked diligently on a manuscript, and we finally submitted it to Geoderma. Some months later, a package appeared in the mail, loaded with a heavily edited copy of the manuscript and the accompanying reviews. In addition to the comments provided by three anonymous reviewers was a "personal" review by Roy Simonson. The letter from Roy began by noting that due to his immense interest in pedology, he had taken it upon himself to personally provide a review of my manuscript. The comments he provided went on for more than 11 pages. Additionally, Simonson wrote, though I am still uncertain what he thought as he penned this, "that I also include for you a document that I normally provide for our non-English speaking authors." The document—as I remember it—was entitled How to write in the English language. As I continued to read Alfred's chapter about Simonson, it provided a bit of consolation to learn that, many years before when he was a graduate student, Simonson had rejected a manuscript submitted to the book *Soils & Men* by the future Nobel laureate, Selman Waksman. Criticism, sometimes a bit liberally administered, can be helpful in the end, especially if judiciously diluted to aid in its digestion.

One advantage of good history is that we can stand back and survey the bend of its arc and contemplate the arrow of time as it intersects with the here and now. The history of pedology that follows on these pages illustrates the story of the demography of science up to the beginning of the latter half of the twentieth century. While the characters in this history are mostly dead, and (yes) white, and (yes) men, the immigrant backgrounds that many came from, and the economic and sometimes political hardships that many faced, are indeed a fundamental part of the fabric of Americana. And, of course, they were not always old. They are part of our soil science family, which now is becoming an ever-broader slice of the modern human landscape of America. This lively book is an important first installment on a youthful science embedded in an evolving cultural landscape, one in which we now are actively shaping and expanding.

Ron Amundson
Professor of Pedology
University of California Berkeley
Berkeley, CA, USA

Foreword by Maxine J. Levin

My formative years in Soils and Pedology were the 1970s. I was in the first wave of women to become interested in soil science in the agronomy and forestry departments of universities. As students, we were thinking about soil science's links to environmental science and ecology, but the textbooks were not making it that easy. Nyle Brady's textbook on soil science was a dry, pictureless text, relating mostly to the basics of soil chemistry, plant nutrition, and some soil physics. The relationships with landscape and land management interpretations were described in some lectures—mostly in response to our questions—and presented anecdotally… essentially as hearsay, not fully developed ideas. We had some contacts with the icons of Pedology, which was itself a term rarely used in soil science. One exception was when it was used in passing references to soil surveys and Guy Smith's budding *7th Approximation* in Soil Taxonomy. Most of the founding pedologists were already retired or gone, and for the most part, their second-generation successors were teaching without texts or references. Paradigms and foundational concepts were still being tested.

At the University of California Berkeley, undergraduates and graduates had the benefit of Hans Jenny, emeritus professor of soils who was still coming into his office most days. For us students, his small book, Factors of Soil Formation, was our bible and we treated him with respect and awe. I was charmed by his yearly lectures on soil in art and esthetics and fervently hoped to get a seat near him at the coffee hours where he sometimes entertained graduate students. As a lowly undergraduate, I hoped some of his knowledge would rub off on me and was delighted when he sorted my face from all his other devotees, or acknowledged a remark as reasonable, if not profound or provocative. Despite his relative celebrity, he was a kind man, and a humble student of the science himself.

This book, Alfred Hartemink's *Soil Science Americana,* is a definitive start at a synthesis of the first 100 years of Pedology in the United States. I hope that readers will use this book as a platform for further discussion. As a soil mapper from the early 1980s to the 2000s, I often felt that I needed a broader perspective in order to fully contribute to the *Soil Taxonomy* that was being built, using field input gathered through the US National Cooperative Soil Survey. When I was a student and field soil scientist, we lacked the synthesis documents and the train of thought used to build the systems. In the field, I often found myself questioning my correlators and oversight staff as to whether the western or urban landscapes that I was mapping really fit into the systems to which I was supposed to conform to my observations.

It is hard to know really whether you are seeing and measuring something new if it is not easy to retrieve the logic and assertions needed for comparisons. Pedology is still in a constant flux, even after 100 years of exploration and discovery. We are now redefining our observations with the help of digital technology. At times, the picture that the data presents is skewed, not for reasons of nature but because of the conventions and assumptions used to gather the data. I welcome the reevaluation of historical data and the perspective that can be gained.

There also is the story of personalities and characters in the science that has colored and enriched my life and career. I have found that the pursuit of science is often linked to the influence of friends and adversaries, and that the story of these linkages in this book is as interesting as the science itself. As program manager and a leader of the National Cooperative Soil Survey in Washington, D.C., I finally had access to many of the seminal papers that had shaped the Survey's policy and methods—the same policies and methodologies that my correlators and I had mentally sparred over early in my career. Also, I was able to see the internal memos and correspondence that Emil Truog, Charles Kellogg, Hugh Bennett, Curtis Marbut, Hans Jenny, Milton Whitney, Eugene Hilgard, and Roy Simonson as well as the many Russians and other characters, like Guy Smith, all shared.

These written arguments built the foundations of the science—based on close observation of nature, and then the practicality of interpretation—first for farming, and later for various other kinds of economic development. In addition, the literature of the *International Congresses of Soil Science* has brought another level of understanding of the development and dialogue of Pedology that Alfred Hartemink has clearly articulated in this book. All would have been helpful in my early mapping days to clarify my thoughts and put my observations in context. I encourage newcomers to the science to use this text as a jumping-off point for their own journey of discovery, particularly in the influence of digital enhancement and applications to the environment.

Maxine J. Levin
2020 Pedology Division Chair, SSSA
Adjunct Lecturer Soil Science, University of Maryland
National Leader Interpretations
USDA NRCS, retired

Preface

The world would have been different if soil science had not emerged in the nineteenth century. By nature, it applies to many—if not all—of the sciences, but the impact of soil science on society and the world is largely untold. Soil science is an earth science which findings are applied in the judicious management of the land, whether that is for growing food, filtering water, storing carbon, or building houses. Most of all, it is a science that aims to understand the clothes of the earth, those top few meters that we live on. The need for understanding soil equals the need for comprehending all animals and plants, rocks, and the ever-changing climate; they are partners in an earth system from which we, humans, and all other life derive being and survival.

Scientific soil studies were started in the mid-1800s, and since then, thousands of people have studied soils. Progress has been made through individual and group discoveries combined with the international exchange that progressed at the beginning of the twentieth century. Slowly, an independent science evolved, and a global soil science community formed that advanced common thinking and practices. The globalization of soil science was decelerated by two world wars, but a firm basis was laid in 1924 when the *International Society of Soil Science* was established. Chemistry, geology, microbiology, and agronomy had well-established professional organizations, but now soil science has its own global society to foster soil studies. In 1927, soil science had its foremost pinnacle with the *First International Congress of Soil Science* in Washington. Since then, international soil congresses were regularly organized, and in 1960, the *Seventh International Congress of Soil Science* was held in Madison, Wisconsin. The 1960 congress brought 1,260 soil scientists together from all over the world. That was more than any previous international soil congress. The theme of the congress was *To Promote Peace and Health by Alleviating Hunger through Soil Science*. The Second World War was long over, and a cold war was brewing, but in 1960 the soil science community sent a message of science devoted to peace and hunger alleviation across the world.

This book is about soil scientists, their connections, and the budding and early blossoming of American and international soil science in the period 1860 to 1960. Interwoven is a tale of two farm boys who grew up 900 km apart, in the Midwest USA in the late 1800s and early 1900s. Emil Truog and Charles Kellogg met in the late 1920s and shared a natural connection to the soil. Both were practical pioneers and believed that understanding soils was crucial to helping people on the land make a better living. They had a profound influence on the soil science discipline, and were leaders in the organization of the 1960 soil congress [1]. Also interwoven is the story of Roy Simonson who was a student of Charles Kellogg in North Dakota and Emil Truog in Wisconsin. As a matter of fact, this book is about a few hundred people who studied soil in the USA, and other parts of the world.

I read the unpublished diaries of Charles Kellogg and Roy Simonson, and the biographies of Justus von Liebig [2], Eugene Hilgard [3], Nathaniel Shaler [4], Curtis Marbut [5, 6], Hugh Bennett [7], Macy Lapham [8], John Russell [9], Bernard Tinker [10], Jacob Lipman [11], Selman Waksman [12], Sergei Vinogradskii [13], Dan Yaalon [14], A. P. A. Vink [15], John Arno [16], Marjory Stephenson [17], Leen Pons [18], Dennis Keeney [19], Anthony Young [20], René Dubos [21], Martinus Beijerinck [22], Gary Petersen [23], the published interviews with Hans Jenny [1], and Guy Smith [37], and I read obituaries and homages of soil scientists whose works and lives are chronicled in this book.

The geographic focus of this book is the USA, and I read the histories of soil and related departments in Illinois [24], Maine [25], Michigan [26], Minnesota [27], Missouri [28], New York [29], North Dakota [30], North Carolina [31], Ohio [32], Wisconsin [33, 34], Pennsylvania [35], and bits and pieces of the history of soil departments in California. Three studies on the history of the soil survey in the USA were reviewed [36–38], in addition to some of the main soil science history books [37, 39–44]. Given that soil science is a global science that has extensive international roots, contextualizing chronicles and progressions required a wide perspective. Therefore, several historical studies were used to frame American soil science in a global light including those written for Germany [45], France [41], Russia [39], the Netherlands [46], and more globally [47].

Soil Science Americana is an attempt to knit human interest into soil science developments. It is about people, human history, and soil science in the USA and across the world. Soil science in the USA has not suffered from abysmal isolationism, and international cooperation and exchange became more common after the first international congress in 1927. It is not possible to see progressions in soil science in the USA in the early twentieth century without taking into account the cradle of soil science in Russia, and the immigrants that came from Russia and its surrounding countries. I studied the publications of the Israel Program for Scientific Translations that, in return for American wheat, translated soil publications from Russian into English in the 1950s and 1960s [48].

Much of this book was written at odd hours in my office of the Department of Soil Science at the University of Wisconsin, Madison—the same office that Emil Truog occupied as Head of Department 80 years earlier. He stayed in that office for 15 years, and even though he was succeeded in the spring of 1954, he did not vacate the office for another 9 months. The office overlooks Lake Mendota, and on its walls hang the painted portraits of Franklin King, Andrew Whitson, and Emil Truog—the first three Heads of the Department. Below them hangs a portrait of Eugene Hilgard painted by Sergei Wilde in 1944. They all looked over my shoulder and ascertained that I 'came in with the birds' as Emil Truog recommended, if not directed, to his students.

Some of the chapters were written at Franklin King's desk that has a holder with eight of his pipes. In the drawer, two small notebooks with the finances of the first 15 years of the department starting in 1889, and one could only wish for the simplicity of such financial accounting. In other drawers, old photographs from the department, the 1927 and 1960 congresses, and letters from Emil Truog and his predecessors. The invitation was there: sit, mine the material, read, think a little, read some more, and write it up. While writing, I strived for what Bernard Tinker attempted when he obtained a postdoctoral appointment at Rothamsted in 1954: "I learned to think logically, to plan practically and to write sensibly [10]."

The Department of Soil Science and the University of Wisconsin, Madison, have an extensive Emil Truog archive. The Charles Kellogg archives are at the USDA National Agricultural Library and Library of Congress. The grandchildren of Emil Truog and Charles Kellogg, and the son of Roy Simonson provided information, and all three had extensive correspondence and kept detailed records of their activities. Charles Kellogg maintained a logbook of events that he had begun in 1950, and I have used his dairy up to 1934, and the years 1950 and 1960. I have cited from the diaries, letters, and papers, but realize that it is a snapshot of the events as they occurred and guided by my interpretation and framing of events. I am distanced both in time and space from the main people in this book, and all had died before I knew what's up and down. The exceptions are Roy Simonson with whom I corresponded for a while about soil science history and *Geoderma* [49], and Peter LeMare with whom I have corresponded for years and who died in 2018 at the age of 95.

In 2003, Henry Janzen from Agriculture and Agri-Food Canada reviewed my book *Publishing in Soil Science*, which was inspired by his paper *Is the Scientific Paper Obsolete?* [50, 51] He wrote a fair review about my book, but what caught my attention was the comment: "…I might also grouse about the 'administrative' flavour of the history. I learn about dates and commissions and congresses, when what I really seek is 'story.'" I took that comment at heart, had years to think, and figure out how to blend human history with soil science progressions. In this book, I have tried to let the story prevail, and often thought of Johan Bouma who had been an indelible promoter of soil stories [52]. *Soil Science Americana* narrates stories of people and soil science developments in the USA embedded in a global setting between 1860 and 1960. The stories became chronicles as they were shaped and experienced by a university professor and chief of the soil survey. I have aimed to rise above the level of a bundle of pedology trivia along a strict historic timeline but throughout the book, there is the intermittent highlight of a frivolous fact or event. There is some overlap between the chapters, and occasionally the same story is narrated in different chapters as it was experienced by different people.

As long as I have studied soils, I developed an interest in soil science history and in those for whom soil science is a profession: Who they were, where they came from, what they had in common—not that I ever found it, but conceivably, I was seeking my ground in the soil science world which had a new impetus after I moved to the USA in 2011. In the search for the ground and soul of American soil science, I had to think what Stephen Fry once said: "Everything you say about America is true, and so is the opposite." Admittedly, I am not a historian or a novelist, and so much of what I have written here is the outcome of seeing people, stories, and developments through my soil science and human lens. The structure of the book is fabricated, and I may have overlooked, overemphasized, or misinterpreted happenings. I do not claim that my judgment has been infallible in all chronicles and progressions, but except for some obvious wanderings, none of them are fictitious. It has been a joy writing this book and may that be reflected in what follows henceforward.

Alfred E. Hartemink
Madison, USA
January 2021

References

1. Maher D, Stuart K, eds. Hans Jenny - Soil scientist, teacher, and scholar. Berkeley: University of California; 1989.
2. Brock WH. Justus von Liebig. The Chemical Gatekeeper. Cambridge: Cambridge University Press; 1997.

3. Jenny H. E. W. Hilgard and the birth of modern soil science. Pisa: Collana della revista agrochemica; 1961.

4. Shaler NS. The autobiography of Nathaniel Southgate Shaler; with a supplementary memoir by his wife. Boston: Houghton Mifflin; 1909.

5. Krusekopf HH, ed Life and works of C.F. Marbut. Madison: Soil Science Society of America; 1936.

6. Marbut-Moomaw L. Curtis Fletcher Marbut. Life and works of C.F. Marbut. Madison: Soil Science Society of America;1936:11–35.

7. Brink W. Big Hugh. The father of soil conservation. With a preface by Louis Bromfield. New York: The MacMillan Company; 1951.

8. Lapham MC. Crisscross trails: narrative of a soil surveyor. Berkeley: W.E. Berg; 1949.

9. Russell EJ. The land called me. An autobiography. London: George Allen & Unwi; 1956.

10. Tinker PB. My life's adventures. Memoir by Bernard Tinker. Bloomington, In: AuthorHouse; 2017.

11. Waksman SA. Jacob G. Lipman. Agricultural scientist and humanitarian. New Brunswick, New Jersey: Rutgers University Press; 1966.

12. Waksman SA. My life with the microbes. New York: Simon and Schuster, Inc.; 1954.

13. Waksman SA. Sergei N. Winogradsky. His life and work. The story of a great bacteriologist. New Brunswick, N. J.: Rutgers University Press; 1953.

14. Yaalon DH. A passion for science and zion. Jerusalem: Maor Wallach Press; 2012.

15. Vink APA. Vijf en veertig jaar uit het blote hoofd. Bodemkartering, Landschapsecologie, Landevaluatie. Bussum: unpiblished biography; 1989.

16. Arno JR. Experiences in soil survey. Soil surveyor in Maine 1936–1976. unpublished biography; undated.

17. Štrbáňová S. Holding Hands with Bacteria. The Life and Work of Marjory Stephenson. Heidelberg: SpringerBriefs in Molecular Science; 2016.

18. Fanning DS, ed Leen Pons. Father of the International Acid Sulphate Soils Symposia/Conferences. University of Maryland; 2008; No. College Park MD.

19. Keeney DR. The Keeney Place: A Life in the Heartland. Levins Publishing; 2015.

20. Young A. Thin on the ground. Soil science in the tropics. Second edition. Norwich: Land Resources Books; 2017.

21. Moberg CL. René Dubos, Friend of the Good Earth: Microbiologist, Medical Scientist, Environmentalist. Washington DC: American Society for Microbiology; 2005.

22. van Iterson G, den Dooren de Jong LE, Kluyver AJ. Martinus Willem Beijerinck. His life and works. Delft: Alles Komt Teregt; 1940.

23. Petersen GW. The best of Luck. Lulu Press Inc.; 2018.

24. Moores RG. Field of rich toil. The development of the University of Illinois College of Agriculture. Urbana: University of Illinois Press; 1970.

25. Kalloch NR. The early soil survey in Maine. unpublished not dated.
26. Robertson LS, Whiteside EP, Lucas RE, Cook RL. The Soil Science Department 1909–1969 - A historical narrative. East Lansing: Michigan State University; 1988.
27. Anon. The Department of Soil, Water, and Climate: 100 years at the University of Minnesota. Minneapolis: Universty of Minnesota; 2013.
28. Woodruff CM. A History of the Department of Soils and Soil Science at the University of Missouri. Special Report 413 College of Agriculture. Columbia, Missouri: University of Missouri; 1990.
29. Cline MG. Agronomy at Cornell 1868 to 1980. Agronomy Mimeo No. 82–16. Ithaca, NY: Cornell University; 1982.
30. Thompson KW. A history of soil survey in North Dakota. Dickinson: King Speed Publishing; 1992.
31. Soil Science Society of North Carolina. Papers Commemorating a Century of Soil Science. Report by the SSS of NC. 2003.
32. Baver LD, Himes FL. History of the Department of Agronomy 1905–1970. Columbus: Ohio State University; 1970.
33. Beatty MT. Soil science at the University of Wisconsin-Madison. A history of the department 1889–1989. Madison: University of Wisconsin-Madison; 1991.
34. USDA-NRCS. History of Wisconsin soil survey. Madison: Natural Resource Conservation Service; 2007.
35. Eckenrode JJ, Ciolkosz EJ. Pennsylvania Soil Survey. The First 100 Years. Agronomy Series Number 144. The Pennsylvania State University: Agronomy Department; 1999.
36. Gardner DR. The National Cooperative Soil Survey of the United States (Thesis presented to Harvard Univsersity, May 1957). Washington DC: USDA; 1957.
37. Helms D, Effland ABW, Durana PJ, eds. Profiles in the History of the U.S. Soil Survey. Ames: Iowa State Press; 2002.
38. Weber GA. The Bureau of Chemistry and Soils. Its history, activities and organization. Baltimore, Maryland: The Johns Hopkins Press; 1928.
39. Krupenikov IA. History of Soil Science From its Inception to the Present. New Delhi: Oxonian Press; 1992.
40. Yaalon DH, Berkowicz S, eds. History of Soil Science - International Perspectives. Reiskirchen: Catena Verlag; 1997.
41. Boulaine J. Histoire des Pédologues et de la Science des Sols. Paris: INRA; 1989.
42. Simonson RW. Historical highlights of soil survey and soil classification with emphasis on the United States, 1899–1979. Wageningen: ISRIC; 1989.
43. Tandarich JP, Darmody RG, Follmer LR, Johnson DL. Historical development of soil and weathering profile concepts from Europe to the United States of America. Soil Sci Soc Am J. 2002;66:335–46.
44. Warkentin BP, ed Footprints in the soil. Amsterdam: Eslevier; 2006.
45. Giesecke F. Geschichtlicher Uberblick under die Entwicklung der Bodenkunde bis zur Wende des 20. Jahrhunderts. Berlin: Verlag von Julius Springer; 1929.
46. Bouma J, Hartemink AE. Soil science and society in the Dutch context. Neth J Agr Sci. 2002;50(2):133–40.

47. Yaalon DH. History of soil science in context: International perspective. In: Yaalon DH, Berkowicz S, eds. History of Soil Science International Perspectives. Reiskirchen: Catena Verlag; 1997:1–13.
48. Yaalon DH. V.A. Kovda - Meeting with a great and unique man. Newsletter IUSS Commission on the History, Philosophy and Sociology of Soil Science. 2004;11:4–9.
49. Hartemink AE, McBratney AB, Cattle JA. Developments and trends in soil science: 100 volumes of Geoderma (1967–2001). Geoderma. 2001;100(3–4):217–68.
50. Hartemink AE. Publishing in soil science - Historical developments and current trends. Vienna: International Union of Soil Sciences; 2002.
51. Janzen HH. Is the scientific paper obsolete? Can J Soil Sci. 1996;76(4):447–51.
52. Bouma J, Droogers P. Translating soil science into environmental policy: A case study on implementing the EU soil protection strategy in The Netherlands. Environ Sc Policy. 2007;10(5):454–63.
53. Beatty MT. Nature of the difficulty soluble inorganic phosphorus in acid soils. Madison, University of Wisconsin; 1955.

The original version of the book was revised: Chapter-wise abstracts have been included in the online version. The correction to the book is available at https://doi.org/10.1007/978-3-030-71135-1_16

Acknowledgements

On a cold January evening in 2019, I had dinner with Marv Beatty in Madison. There was a frost warning, it snowed with gusty northern winds, and roads were icy. All of that is nothing to worry about in Wisconsin. I had met Marv some years before I moved to the University of Wisconsin, and we have had a conversation about the history of soil science. Marv obtained his PhD in 1955 that was supervised by Emil Truog [53], and he retired when I started my first soil science position. In 1989, he wrote the history of the Department of Soil Science following its 100th anniversary. Marv was close with Emil Truog and Roy Simonson, and an assistant for the organizing committee of the *Seventh International Congress of Soil Science* in 1960, as were Dennis Keeney, Leo Walsh, and all students of the Department of Soil Science.

Some 50 years before that cold evening in 2019, Marv had dinner with Charles Kellogg at the annual meeting of the *Soil Science Society of America* in Cincinnati. They knew each other through Ken Ableiter who worked for Charles Kellogg and who came to inspect a soil survey that Marv had conducted. That evening in Cincinnati, Charles told a story about how Emil Truog had given him a new suit in 1929, and that it helped him in being representable for a job interview. Charles Kellogg got the job, and 27 years later, when he had become an influential soil scientist, he returned a favor to Emil Truog. That story sparked my interest and I pondered how seemingly tiny gestures have shaped the world of soil science. It became the start, if not *Leitmotiv*, for this book. While digging through other stories, papers, books, and diaries, it became apparent that the world of American soil science is small indeed, and that it takes a bird's eye view and a deep breath to see it that way. So, I have uplifted numerous small events and framed them in a broad and long perspective. A strong inner compulsion made *Soil Science Americana* as comprehensive as it is, but I had the help of many people.

Thank you, Marv, for that story and the pedology trip to Montana. I am grateful to Jim Bockheim, Leo Walsh, Robin Harris, and Dennis Keeney for various bits of information and encouragement. They are emeriti professors in the Department of Soil Science. I am grateful to Jim and Leo for the visits to Palo and Independence in the summer of 2019. Prior to my current position at the University of Wisconsin, I had the good fortune of working alongside Wim Sombroek and Hans van Baren who had leadership positions in the *International Society of Soil Science*. I learned from them how the world of soil science was affected by politics, strong individuals, and disparaging ambiances. At that time, I also inherited, which meant saving it from the recycling dumpster, archival material from Jules Mohr, David Hissink, and Cees Edelman when the Department of Soil Science & Geology moved out of its Duivendaal 10 building in Wageningen where it had nested in 1966. Mariette Edelman gave me part of her father's archive as no institution had shown interest in the material, and after Hans van Baren died in 2009, I inherited books and some correspondence from his uncle Ferdinand and grandfather Jan van Baren. These Dutch soil scientists played a role in international soil science and were well-connected with soil scientists in the USA.

In 2009, Alex McBratney and I boarded a train in Amsterdam to attend a meeting in Budapest to commemorate the centennial of the *First International Conference of Agrogeology*. During the 18-hour train ride, we discussed the world of soil science as far as our horizons permitted. It was a conversation interspersed with some work on the laptop, gazing out the train window, little naps, and beverages. Along the way, we realized that scientific directions sometimes were determined by a few individuals, and that is not always possible to judge in hindsight whether those directions were correct. Over the past 30 years, I have had similar conversations with Dan Yaalon, Johan Bouma, David Dent, Mary Beth Kirkham, Pedro Sanchez, Don Sparks, Dennis Greenland, Peter Buurman, Stan Buol, Jim Bockheim, John Tandarich, and many others. My understanding of progressions in soil science has been shaped by those conversations, and likely there are sprinkles of that in every chapter.

The following people provided publications, books, anecdotes, diaries, photographs, stories, clarifications, or suggestions, for which I am grateful: Jim Tiedje of Michigan State University, Pax Blamey from the University of Queensland, John Galbraith from Virginia Tech, Anne Efland of USDA, Diane Wunsche of the National Agricultural Library, David Hopkins of North Dakota State University, Marty Rabenhorst of University of Maryland, Mary Beth Kirkham of Kansas State University, Maxine Levin, Ron Amundson of University of California Berkeley, Eric van Ranst of Ghent University, Dennis Merkel from the Lake Superior State University, Stephen Anderson from the University of Missouri, Jim Robertson from the University of Alberta, Gary Peterson from Colorado State University, Yakov Pachepsky of ARS-Beltsville, Budiman Minasny from the University of Sydney, Mogens Greve from Aarhus University, Erika Micheli from Szent István University,

Carl Rosen from the University of Minnesota, Peter Jacobs from the University of Wisconsin-Whitewater, Lesley Robertson of Delft University of Technology, Stan Buol from North Carolina State University, Peter Schad from the Technische Universität München, Susan Chapman from the *Soil Science Society of America*, Horea Cacovean from the Pedological and Soil Chemistry Institute at Cluj-Napoca, Senator Mark Miller, Tony Young, Henk Edelman, Elise Frattura, Els de Jong, Yakun Zhang, Jingyi Huang, Carol Duffy, and the staff at the archives in the Steenbock Library.

Some of the Russian literature, interpretation, and portraits of Russian soil scientists were provided by Pavel Krasilnikov from Lomonosov Moscow State University. Thank you Pavel. I am grateful to Robert Doe, Executive Editor at Springer Nature for his assistance and suggestions, and to Ho Ying Fan and David Dent for careful editing and making my English understandable. Jim Bockheim commented on the draft of the book, and he knows more about soil than anyone else I know. Lastly, thank you Steven Kellogg and Pat Saiki for photos and information on your grandfathers Charles Kellogg and Emil Truog, and thank you Bruce Simonson for information on your father Roy Simonson.

About This Book

The first chapter reviews soil explorations in the nineteenth century, it sets the stage for the progressions in the 1900s. The second chapter, named Pochva Americana, unwraps differences in soil research between Russia and the USA in the period 1870 to the 1920s. Chapters 3 and 4 deal with the growing up, schooling, and working life of Emil Truog and Charles Kellogg. Chapter 5 narrates the life and works of Roy Simonson who was a student of Charles Kellogg and Emil Truog. The following chapter is on the University of Missouri in Columbia, and its contributions to soil science. Chapter 7 relates the start of the soil survey in the USA and the lives and works of Milton Whitney and Curtis Marbut. Chapter 8 reviews the soil survey as directed by Charles Kellogg, and Chap. 9 focuses on immigrants who came from Russia and profoundly influenced American soil science. Chapters 10–13 deal with the formation of *the International Society of Soil Science*, the first congress in 1927, the period 1927 to 1960, and the *Seventh International Congress of Soil Science* in 1960. The book ends with a chapter in which connections in space and time between people and events are endeavored, and in the epilogue the legacy of Emil Truog and Charles Kellogg is discussed.

The book is about soils and people and roughly covers the period 1860 to 1960. It is about human history, chronicles, and progressions in soil science, but it is not about everybody in soil science during that period nor about all progression in soil science in the USA. I have included numerous small biographies that provide some human insight and narrate specific developments. Both Emil Truog and Charles Kellogg interacted with many people and it was not possible to depict the whole web of their professional relations. Little attention is given to the early soil surveys in the western part of the country, which is described in the autobiography of Macy Lapham covering the period—late 1800s to 1940s [8]. Charles Kellogg had given him stenographic assistance to write his autobiography and named the book a "prize piece of Western Americana."

The web of human chronicles has been woven widely across the book, and the main characters are given under the chapter title in Table of Contents, and in Person indices in the back. Emil Truog and Charles Kellogg are central, whereas the American soil survey and the formation of the *International Society of Soil Science* in 1924, and the 1927 and 1960 soil congresses form national and international cornerstones in the book.

Now something about the names of people and places. All people have been named by their first and last names, like Emil Truog and Jadwiga Ziemięcka, so no initials are used nor just surnames. Some people like J. C. Russel from Nebraska always used his initials; his first name was Jouette. There were others like W. G. Ogg, the first name William, but mostly named Gammie. The Dutchman D. J. Hissink's first name was David, but friends named him Ko from his second name Jacobus. Charles Kellogg was named Chuck by intimate friends, but Charles Kellogg is used throughout the book. Some names have different spellings like Sergei or Sergey, and Winogradsky or Vinogradskii. I chose the name that was most recently used by, for example, Wikipedia or the US Library of Congress system. That may seem rather arbitrary criteria, but it has been used consistently. The names of towns, cities, and countries are all current, so it is St. Petersburg instead of Leningrad, Lubumbashi instead of Élisabethville, Ghana instead of the Goldcoast, and Russia instead of the USSR. Past names do not always match current political boundaries but since that is not the subject of this book, it does not affect the story.

Over 800 references have been listed by the Vancouver Style with notes in superscript in order to let the text flow. Not all letters from Charles Kellogg or Emil Truog have been referenced but they are in archival boxes in the Department of Soil Science. I have generously quoted from the letters, diaries, and papers for I sensed that the original wording contributed more to the story than my rephrasing and interpretation. Besides, I thoroughly enjoyed reading the way it was written reflecting much about the person. In all quotes, I have kept the original spelling and grammar including errors.

Considerable time was spent finding and selecting photos that have not been widely published, and as no paragraph headings are used, the illustrations and photographs are meant to illuminate the text. All photos have been acknowledged; most photos were from the archives of the Department of Soil Science and from the Truog archives at the University of Wisconsin—Madison, the Truog family album, Steven Kellogg family album, Bruce Simonson, the Charles Kellogg archives at USDA, the Library of Congress, *International Society of Soil Science* Proceedings, Missouri State Archive, Senator Mark Miller, and the Smithsonian Institution Archives.

About the Author

Alfred likes soil, soil science, and people who study the soil. What is there not to like about all of that. He is professor of soil science at the University of Wisconsin—Madison. Alfred was trained in pedology and soil fertility, and has an MS degree from Wageningen, the Netherlands, and a PhD from The Reading University in the UK. Since 2011, he works at the University of Wisconsin—Madison, where he teaches Pedology, and Earth's Soil, and his research focuses on novel ways to explore the soil profile, and the management of soil carbon in natural and agricultural ecosystems. Before his current position, he was for 12 years at ISRIC—World Soil Information in the Netherlands and was responsible for the World Soil Museum and the GlobalSoilMap project. Between 1987 and 1999, he worked at research institutes and universities in Tanzania, Congo, Indonesia, Kenya, Papua New Guinea, and Australia. From 2001 to 2014, Alfred served as (Deputy) Secretary General of the *International Union of Soil Sciences*. He is the founding Editor-in-Chief of *Geoderma Regional* and the *World Soils Book series*.

His other books are *The Soils of Wisconsin* (2017) with Jim Bockheim, *The Soils of the USA* (2017) edited with Larry West and Mike Singer, *Digital Soil Morphometrics* (2016) edited with Budiman Minasny, *Soil Carbon* (2014) edited with Kevin McSweeney, *Digital Soil Mapping: Bridging Research, Production, and Environmental Application* (2010) edited with Janis Boettinger *et al.*, 2010, *Profiel van de Nederlandse Bodemkunde* (2010) edited with Johan Bouma *et al.*, *Soil Science* (2009) edited with Alex McBratney and Bob White, and *Digital Soil Mapping with Limited Data* (2008) edited with Alex McBratney and Lou Mendonca. Alfred edited *The Future of Soil Science* (2006) and wrote the books: *Invasion of Piper Aduncum in the Shifting Cultivation Systems of Papua New Guinea* (2006), *Soil Fertility Decline in the Tropics* (2003), and *Publishing in Soil Science* (2002).

Contents

1

Prologue—The Roots of Soil Science

"I propose to regard the soil as a creature sui generis, sustaining living bodies whilst it is itself sustained by them."

Robert Wood, 1850

This book meanders through progressions in soil science and the institutionalization of the discipline over the period 1860 to 1960. The story is narrated through the activities, ideas, publications, and correspondence of people who influenced those progressions, and how they formed national and international learned societies. As soil science is a global science with a broad and deep root system, this first chapter sets the scene for stories that unfold in the subsequent chapters. Two lines of soil studies will be reviewed here: the search for the soil matter that makes plants grow and the formation and genetics of soil. Eventually, these lines intersected and then paralleled in a direction that became the discipline of soil science.

As long as people have cultivated the earth, they have observed the growth and fruits of what they had planted. As nature showed them, they distinguished which plants grew well on different soils. Observation did not bring much understanding but was needed for mimicking nature so as to avoid failure in crops grown for sustenance. While experience became tradition, the world underfoot was essentially unknown to the early cultivators. Some tried to manipulate nature and mend the soil so that plants grew better and yielded more; they dug ditches so excess water drained away, plowed up dense layers, and ashes or limestone were fed to their crops. There was little comprehension, but it generated practical knowledge that was passed on from generation to generation. The world could have been left in those spheres for millennia, but it was not; people figured out the scientific method to advance knowledge and that, of itself, is possibly humanity's most successful discovery.

With the progress in chemistry, theories were developed on the origin of substances in plants. An idea which gained popularity was that plants fed upon substances that were similar to them, or in other words, plants grew

© The Author(s), under exclusive license to Springer Nature
Switzerland AG 2021
A. E. Hartemink, *Soil Science Americana*
https://doi.org/10.1007/978-3-030-71135-1_1

out of substances that originated from plants that had grown before them. Humus, or decomposed plant material, was regarded as the main nutrient for plants, and it was taken up by roots and then converted to new plant material [1]. This humus theory made sense when a few of the chemical elements in nature had been identified. How could plants take up substances that were not similar to them? With time, the idea of how plants were feeding from the soil altered drastically. At the same time, the soil was studied independently from geology and chemistry—studies focused on the soil, its origin, the elements it contained, and its relation to climate and vegetation. It was postulated that soils were deposited with the great flood; some were convinced that stones grew in the soil; others thought that soils were merely broken-down rocks mixed with dead and alive plant materials so that, if the geology of an area were known, one would know what kind of soil there would be.

Discoveries in these two areas—plant nutrition, and the origin of soil—occurred at more or less the same time, but they were made by different groups of people. One group was trained in agricultural chemistry, and the other group was mostly trained as geologists. It was a period of searching and discoveries, some more inadvertent than others. Theory was immature or lacking, myths perpetuated, and communication and exchange of ideas and findings were sparse to absent. Some soil investigators had university education related to chemistry, plant physiology, mineralogy, or geology, but others were travelers, naturalists, skilled amateurs, and keen observers. Most studies took place in isolation and researchers were not conversant with what others were doing. Such was the situation in the eighteenth and nineteenth centuries when there was no obstruction to establish and nurture an individual soil scientific universe.

The prime discoveries in plant nutrition and on the origins of soil were made in Europe and Russia. The USA was not on the scientific frontier in the nineteenth century but became a living laboratory where the theories and discoveries of soil research developed elsewhere could be tested and refined. The country had vast areas of unfamiliar soil and a limited research tradition, but it had a willingness to learn, to step out of its European shadow, and lead some of the major soil progressions. The long and winding road to understanding soil crossed vast areas of mysterious land where it sought answers to down-to-earth questions and discovered problems that were far from practical.

The New World was an unknown land for the European travelers, colonizers, and freedom seekers who set foot in the early 1600s, but it was not a new world, nor was it uninhabited. The Clovis people had arrived some 24,000 years earlier from Beringia [2], and millions of American Indians lived all across the land. European colonists claimed the discovery of land that had been settled for thousands of years. For the colonizer, the land's boundaries, enormity, and nature were mysterious and of mythical proportions. Several explorers wrote up their impressions and reflections while they traversed along the east coast and into the Appalachians which had folded half a billion years earlier. These accounts were well-read in Europe and, among those who could read, there was an inquisitiveness about America and its natural wealth. The land drew people from Spain, France, the UK, Russia, and the Netherlands. Tenacious wars, poverty, feudal dread, hunger, religious persecution, crop failures, and land shortages drove people to America. There were the élite, the bourgeois and free-spirited, the convicts, and the adventurers who settled in the first 13 colonies.

Soil observations in the USA began with American Indians who had transformed from nomadic to sedentary lifestyles, started growing crops, and discovered the diversity and peculiarity of soil, but that empirical knowledge, gained over many centuries, was ignored by the European colonizers and settlers. Some observations on the soil came from the early explorers and travelers. In 1751, John Bartram traveled by foot, horse, and canoe from Philadelphia to Ontario, and he evaluated the land based on the kind of trees that were growing [3]. He wrote a detailed travel account: *Observations on the Inhabitants, Climate, Soil, Rivers, Productive Animals and other matters worthy of notice*, in which the growth of the trees was connected to the soil: "… the land and timber good, brown soils, and the stones flat and gritty."

In the late 1700s, the French traveler and philosopher C. F. Volney (Constantin François de Chasseboeuf) voyaged for 3 years through the central and eastern part of the colony. He also observed the trees and the forests, and published his report that was translated in 1804 as *A view of the soil and climate of the United States of America* [4]. His travels took him through the states east of the Mississippi, with the purpose of "…studying the climate, laws, people, and their manners, chiefly in the relations of social and domestic life." It was a wide-open country, sparsely populated; Louisville in Kentucky had only one hundred houses. Constantin Volney observed: "To a traveler from Europe, the most striking feature of America is the rugged and dreary prospect of an almost universal forest. This forest is first discerned on the coast but continues thickening and enlarging from thence to the heart of the country. During a long journey, which I made in 1796 from the mouth of Delaware, through Pennsylvania, Maryland, Virginia, and Kentucky, to the Wabash and thence northward, across the North-west territory, to Detroit, through Lake Erie to Niagara and Albany; and, in the following year, from Boston to Richmond, in Virginia, I scarcely passed, for three miles together, through a tract of unwooded or cleared land. I always found the roads, or rather the paths, bordered and obscured by forest, whose silence, uniformity, and stillness was wearisome. The ground beneath it was sterile and rough, or encumbered with the fallen and decaying trunks of ancient trees. Clouds of gnats, mosquitoes, and flies hovered beneath the shade, and continually infested my peace. Such is the real state of these Elysian field, of which, in the bosom of European cities, romancers entertain us with their charming dreams. In the rest of the country, especially among the inland mountains, trees are found in such numbers, and their prevalence is so little checked and circumscribed, that the United States, compared with such a country as France, may justly be denominated one vast forest."

The 1804 book by Constantin Volney contained an illustration that can be considered the first depiction of an American soil profile. Two profiles up to about 6 m (20 *pied royal*) were observed some 6 km east of Louisville in Ohio. It does not describe how the profile was observed, but it likely was in a quarry or a river-cut cliff. One profile showed 30 cm of mold over 75 cm clay over limestone over gravel with clay, whereas the other profile showed 75 cm of mould, over 4.5 m of dark sand over 30 cm of sand and shells, over coarse rounded gravel followed by water at about 6 m overlying calcareous stratum. The second profile seemed to be alluvial material over limestone. Mould was often used for the organic matter-rich top layer of the soil, a term that was later used by Albrecht Thaer and Charles Darwin [5, 6]. The profile with clay up to great depth had Constantin Volney wondering: "I shall be asked, perhaps on what foundation this clay rests; but I have had no means of ascertaining this point; and we may as well stop here, without the trouble of proceeding till we find, like the Hindoos, the tortoise which bears up the world upon its back" [4]. It was turtles all the way down.

A view of the soil and climate of the United States of America had chapters on the following: The face of the country, Earthquakes and volcanoes, Of the Winds, Prevailing diseases, and comparisons between the USA and Europe with regard to rain, wind, dryness, and electricity of the atmosphere [4]. Constantin Volney had an interest in the changing climate: "…an opinion has, of late years, gained ground in the United States, that partial changes have taken place in the climate of the country, which have shown themselves in proportion as the land has been cleared." Similar observations had been made in Canada in the late 1600s and mid-1700s where, over time, summers were perceived to be hotter and winters colder. Seasons were found to change, and the weather was more variable than before. Constantin Volney listed testimonies on the change in climate including observations made by Thomas Jefferson in Virginia. It was believed that the changes were brought about

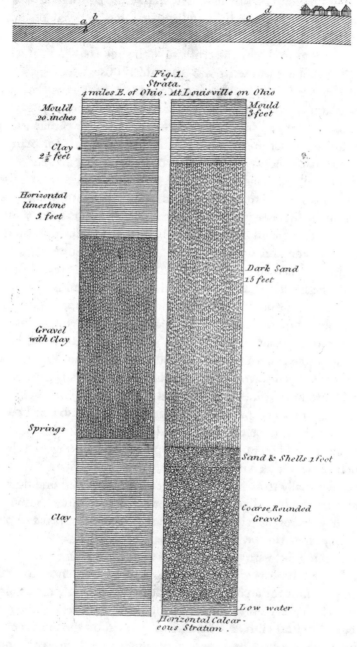

Soil profiles near Louisville on the Ohio River. From Constantin Volney (1757–1820): *A view of the soil and climate of the United States of America: with supplementary remarks upon Florida; on the French colonies of the Mississippi and Ohio, and in Canada; and on the Aboriginal tribes of America*, translated by C.B. Brown in 1804

by the felling of the trees which caused higher air temperatures. Data from measurements of 'heat in the field and in the forest' showed an increase of 4 and 13 °F between May and October, and it was concluded: "…there seems, therefore, no room to doubt the truth of a sensible change in the climate of the country" [4]. That was written in the late 1700s.

A view of the soil and climate of the United States of America was praised by the translator as the best and most complete on the physical condition of the country as far as it concerns the surface and the climate. Constantin Volney included a description of the Mississippi River that for many travelers was the western boundary of their journey: "The magnificent Mississippi, decorated with all the charms of a land of promise, is a most mischievous neighbor. It rolls along, a mass of yellow muddy water, a mile and a half wide, which it annually lifts twenty or twenty-five feet above its banks, and deluges with it a loose soil of sand and clay; forms islands and destroys them; throws trees upon one side and uproots them on the other; submitting its course to obstructions of its own creating; and at length overwhelms the spot which you thought the most secure. The sublimity of this stream is like that of most other grand agents in nature, to be admired safely only at a great distance." When Constantin Volney traveled, not many geologic investigations had been made and it took another 10 years before William Maclure published the first geologic map of the eastern part of the country, and it was 40 years before the geologist Charles Lyell traveled through the eastern states [7]. After 3 years of traveling, Constantin Volney was expelled by President John Adams. He was suspected of spying to prepare for the French re-occupation of Louisiana.

William Cobbett traveled in the 1810s and found that many of the previous travelers had completely false impressions: "Amongst all the publications which I have yet seen on the subject of the United States, as a country to live in, and especially to farm in, I have never yet observed one that conveyed to Englishmen anything like a correct notion of the matter. Some writers of Travels in these States have jolted along in the stages from place to place, have lounged away their time with the idle part of their own countrymen, and, taking every thing different from what they left at home for the effect of ignorance, and every thing not servile to the effect of insolence, have described the country as unfit for a civilized being to reside in. Others, coming with a resolution to find every thing better than at home, and weakly deeming themselves pledged to find climate, soil, and all blessed by the effects of freedom, have painted the country as a perfect paradise; they have seen nothing but blooming orchards and smiling faces" [8].

William Cobbett's book from 1818 was entitled *A year's residence in the United States of America* and had several observations on the soil and the way

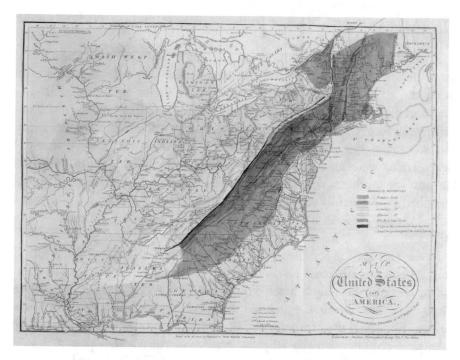

First geologic map of the USA produced in 1809 by William Maclure (1763–1840). The colors reflect rocks that were considered primitive, transition, secondary, alluvial, or old red sand stone. The thick black line showed "...to the westward of which has been found the greatest part of the Salt & Gypsum"

it was managed by the settlers. Locally, peat was cut in bricks, dried, and then burned. The ash was sold "...and were frequently carried 20 miles of more...but they are not equally potent upon every sort of soil." The ashes were used as a fertilizer for the poorer soils where they were applied to legumes such as clover and sainfoin. The settlers experimented with ash, but results were mixed, and "...there is no harm in making a trial. It is done with mere nothing of experience. A yard square in a garden is quite sufficient for the experiment" [8]. William Cobbett thought that most soils were of good quality for growing crops: "I know of no soil in the United States, in which rutabaga may not be cultivated with the greatest facility. A pure sand, or a very stiff clay, would not do well certainly; but I have never seen any of either in America." Like Constantin Volney in the late 1700s, William Cobbett found that the country was covered with forest, and he had some admiration for the settlers that farmed: "...for this is really and truly a country of farmers. Here, Governors, Legislators, Presidents, all are farmers. A farmer here is not the poor dependent wretch. A farmer here depends on nobody but *himself* and his own proper means; and if he be not at his ease, and even rich, it must be his own fault" [8].

There were several descriptions of soils in Florida, Georgia, and Kansas made in the late 1700s and early 1800s. Joshua White made observations along the coast of Georgia on the soils and climate, and related them to common diseases in the area [9]. He stated that the climate resembled that of a tropical country, although Georgia has extreme differences between winters and summers. He observed that sunburn and sunstrokes were absent in Georgia, and he considered drinking cold water needed but not dangerous. He saw no merit in draining the salt marshes along the coast that were described as 'unfriendly to health.' The clay lands were the most fertile and he named them the *mulatto soil*, a term that had been used by Thomas Jefferson [9].

In the 1850s, Sara Robinson described the soils of Kansas: "...the soils for richness can be surpassed in no country. It is of a black color, with a sub-soil of clay and limestone basis. Vegetation is most luxuriant. The soil and climate are most admirably adapted to the raising of grains of every known variety." She loved the prairies and found: "...the climate exceedingly lovely." Fires were common on the tallgrass prairies of the Eastern part of Kansas: "...the air is hazy from the many fires on the prairie, which are burning day and night. They are a grand and sublime sight when spreading over a large tract, the tall grass waving with every breeze, now fiercely blazing, and now with graceful undulating motion, looking indeed like a 'sea of flame,' when the fiery billows surge and dash fearfully; or when the winds are still like an unruffled, quiet burning lake" [10]. All these early travelers were in awe about the vastness of American nature but had no eye for the way American Indians managed their land and soils.

The European population in the USA grew rapidly. There were over 5 million Europeans in 1800, 23 million by 1850, and 50 million by 1880. The American Indian population decreased from several million in the 1500s to 600,000 by the early 1800s, and fewer than 250,000 by the late 1800s [11]. Smallpox, measles, influenza, as well as ruthless wars and murder killed 90% of the American Indians. Alarmed by this trend, the federal government changed its attitude from suppression and land cession to policies promoting acculturation and assimilation [11]. Reservation land was given to American Indians in addition to citizenship, and children were sent to boarding and reservation day schools where they were taught English. The amount of land owned by American Indians declined by 62% between 1887 and 1934 [11].

For Europeans, America was a magnet. Some of the attractions came from a popular book from 1869 entitled *Where to Emigrate and Why* [12]. It listed the climate, soils, price of farmland, and a range of conditions including railway and land law descriptions for all states that were part of the union. Of Missouri, *Where to Emigrate and Why* noted that the soil and climate were favorable to every staple of the temperate zone, and "...in every direction, there are unopened avenues leading to wealth." Wisconsin was described as having a cold but agreeable climate, and the soils were remarkably productive and would create "...extraordinarily wealth in production and quality in wheat, maize, etc." Michigan was described as having sterile soils in the north, and the peninsula was called the 'Siberia of Michigan.' Soils in the southern part of Michigan were assumed to be extremely fertile. The grasses on the plains and valleys of the Dakotas were described as most abundant and nutritious, but not much was written about their soils.

Settlers' experiences did not match those descriptions. Crop yields declined after some seasons, and the settlers developed the notion of soil exhaustion that was well-known in Europe where the chemist Humphry Davy had noted in 1815: "If land be unproductive, and a system of ameliorating it is to be attempted, the sure method of obtaining the object is to determine the cause of its sterility, which must necessarily depend upon some defect in the constitution of the soil" [13]. The cause for the soil exhaustion was assumed to be the continuous cropping, and George Ville, a professor of vegetable physiology in Paris, summarized it in 1866 as "To preserve to the earth its fertility, we must supply it periodically with substances in quantities equal to those removed by the crops" [14]. In the USA, the term soil exhaustion was used in the late 1700s and, by 1860, farmers in the east were well aware of the problem [15].

Soil exhaustion was used to describe a variety of conditions, including soils that were affected by soil erosion and depleted of their fertility [16]. Exhausted soils were also termed 'tired', 'sick', 'out of condition', or 'worn-out soils' [17]. They were not all worn out, and some of these soils had never been productive [18]. It was assumed that all soils were equally productive when brought under cultivation, but many soils were weathered and had low inherent soil fertility like in the eastern part where the settlers survived on abundant rainfall and a long growing season rather than the soil conditions [19, 20]. Soil exhaustion was caused by continuously cropping soils of low fertility, lack of knowledge how the soils should be made or kept productive, and some carelessness. As the historian Avery Craven, whose parents had left North Carolina because of slavery and racism, explained in the 1920s: "Frontier communities are, by their very nature, notorious exhausters of their soils. The wants and standards of living of such communities have been developed in older economic regions and they make demands upon the newer sections that cannot be met from normal returns. The abundance of land combined with a scarcity of capital and labor—a condition which characterizes all frontiers—throws the burden of intensified production upon the soil as the cheapest factor. The problem is one of rapid spending, not of conservation. Frontiers, like those who come to sudden wealth, are inclined to be spendthrifts" [16].

The weather was also more extreme than what the settlers knew from their homes in Europe. Settlers started growing crops they had never grown before or managed the soils like they would have done in the home country. It often led to the loss of soil and crop failures. It was believed that tobacco, which was an important crop for settlers, was the main cause for the exhaustion of the soil. Others put the blame on "...the cultivator...who was attributed to be easy-going, spendthrift and short sighted" [16]. With soil exhaustion came migration and experimentation: move west ('west is best') or improve the soil through amendments.

The early soil experimentation all took place in the east, as the western prairies, Rockies, and the Golden Bear State were *terra incognito* until Meriwether Lewis and William Clarke made their exploration in the early 1800s. Farmers, including George Washington, started using amendments and crop rotation. He ran an extensive farm with 300 slaves on the banks of the Potomac River in Virginia, where tobacco was grown and, later on, wheat. Further south in Virginia, Edmund Ruffin experimented with soil amendments. Throughout Virginia, emigration was seen as the only possibility following soil exhaustion, and "...abandoned mansions stood everywhere in silence and ruin while the former inmates made their melancholy way to the wilds" [16]. When Edmund Ruffin was 19 years old, he was in charge of a plantation [21]; like most farmers in the region, he owned over 30 slaves upon which he built his wealth. He knew some chemistry from reading Humphry Davy's *Elements of Agricultural Chemistry* that was published in 1813 [13]. Humphry Davy had discovered potassium, sodium, magnesium, calcium, boron, and barium as well as the elemental nature of chlorine and iodine. Plants could not grow, flower, produce seeds, and complete their life cycle without these essential elements, nor could they substitute the essential ones with others. Inspired by Humphry Davy's studies on calcium and from observations at his plantation, Edmund Ruffin started experimenting with what he called 'calcareous manures,' which was clayey soil with high amounts of calcium carbonate.

The soils at Edmund Ruffin's plantation were acidic and, in his view, it was the vegetable acid that caused sterility. The subsoil was viewed as "...barren because it contains very little putrescent matter, the only food for plants" [22].

Edmund Ruffin observed that certain grasses and trees grew on acid soils but were absent in soils where calcareous materials were present. His solution was to apply calcareous manure to his acid soils and to plow it under. His slaves dug up chalk-rich clay, which was named marl, from beds of fossilized shells that underlie coastal Virginia. The marl and shells were applied to a test plot that was planted with corn, and corn yields increased by 40%. He believed that the exhausted soils could return to their original fertility if the acid conditions were corrected by applying marl or calcareous manure [16]. According to Edmund Ruffin, his studies with marl were "…the first systematic attempt wherein a plain practical unpretending farmer has undertaken to examine into the real composition of the soils which he possesses, and had to cultivate" [22]. Applying marl to acid soils was not new. In Belgium and the Netherlands, farmers along the rivers applied 'white clay' to improve the soil long before the Roman occupation that started in 50 BC. In Belgium, the application of marl was termed 'mergeling' and in the Netherlands it was 'mergelen.' Likely, the Romans brought this practice back to Italy [23, 24].

In addition to marl, Edmund Ruffin recommended the application of manure, crop rotations, and the growing of the legume cowpea to restore the soil as it brought nitrogen in the soil and kept it covered. Cowpeas originated from West Africa and were brought in with the slave ships to Jamaica in the late 1600s. It spread across the West Indies, reached Florida in 1700, and was grown in Virginia in 1775. George Washington grew it at Mount Vernon. Edmund Ruffin named cowpeas the 'the clover of the south.' The growing of cowpeas was adopted so readily by American Indians that, by the end of the 1800s, it was thought to be indigenous [20].

Edmund Ruffin started presenting his findings at the agricultural society, and his *An Essay on Calcareous Manures* appeared in the *American Farmer*. He added more of his research findings and published it as a book running through five editions and growing from 116 pages in 1835 to almost 500 pages in 1852. In the 1850s, he was regarded as "…the great Nestor of an improved Southern Agriculture" [25]. Marl came to be in high demand and much of it was mined in New Jersey. In the USA, Edmund Ruffin was one of the first to experiment with marl and lime, but he had little notion of the soil itself and its origin: "…soils were formed as a result of rock decomposition and the action of lichen, mosses and other imperfect vegetables which are constantly floating in the atmosphere" [22]. Edmund Ruffin was convinced that exhausted soils could be restored and that "…the South was to be restored to national leadership or to become a prosperous independent nation" [16]. A pro-slavery confederate, he killed himself after the civil war had ended in 1865.

Not everyone was convinced that soil exhaustion could be repaired by applying manure, ashes, marl, or lime. William Goggin from Tennessee wrote an essay in the early 1880s asking: "Is it absolutely necessary for the cultivator of the soil to resort to the use of artificial fertilizers, or the natural fertilizers from the barnyard, either to restore fertility to an exhausted soil or to perpetuate its fertility when restored?" [26]. He was convinced that nature provided enough energy to adequately restore an exhausted soil: "…the farmer can stir his land deep or shallow, he can pulverize his soil fine or coarse, and open it minutely or partially deep or shallow to the penetration of the air, to have its primary elements vitalized by radiated light and heat from the sun according to the labor bestowed in preparing the seed-bed and cultivating the crop; and if the rain-fall and the dews are too small in quantity to be a sufficient solvent, he may supply that deficiency by irrigating his land; and in this way the cultivator of the soil can come to the aid of nature in the operation of her laws, directing them in the channels of fertility in the products necessary for the sustenance of man and beast" [26]. In his view, the energy from the sun and rain was sufficient to keep the soil productive. It did not become a widely shared view.

Some of the early soil experiments and findings were published in popular magazines such as the *American Farmer,* which was first published in 1819, and *The Cultivator—A Monthly Publication Designed to Improve the Soil and the Mind* published since 1834 by the New York State Agricultural Society. The magazine *The Cultivator* focused on all aspects of farming including cattle, crops, buildings, roads, new applications of electricity, the opinions of wise men, and investments in mental stock. It drew material from British agricultural journals [15] but, over time, American farmers sent in reports, as John Smith of Morristown, New Jersey, did in June 1836: "…applying lime to low land, the effect was to double the crops of buckwheat and corns, which were subsequently taken from the ground." The magazine noted that he had omitted to state the quality of the soil or the contents of the field on which he applied 50 bushels of lime, and so "…we content ourselves with this notice of his communication." Peat ash was used as fertilizer and considered "…a valuable manure for uplands." Agricultural chemistry became a favorite subject in *The Cultivator* [15], and the magazine published suggestions for sampling and soil analysis: "…take a small quantity of earth from different parts of the field, the soil of which you wish to ascertain, mix them well together and weigh them." Then the samples should be put in a bread oven, and after the sample was dry, it should be weighed again. The difference in weight showed "…the absorbent power of the earth." When it was over 50 of

400 grains, the power was considered great, below 20 it was low and "...the vegetable matter deficient."

Besides the *American Farmer* and *The Cultivator,* there were similar magazines that aimed to inform the settlers on any agricultural issue. They were early extension pamphlets, published before agricultural experiment stations were established and research became institutionalized. In addition to the pamphlets and magazines, numerous practical books were published on farming and soils [14, 27–29]. The sciences had not reached the art of farming that relied on European knowledge and tradition which, in many places, misaligned with the soils and climate. In the mid-1800s, American farming was backward and inefficient [15].

An unsolved problem in the growing of crops was the way they obtained nutrients. Early theories had stated that plants acquire their nutrients through the uptake of vegetative matter and, partly, through the atmosphere. This humus theory was asserted by the agronomist Albrecht Thaer who, in the early 1800s, established several schools and training centers in Germany, and published the four volumes *Grundsätze der rationellen Landwirthschaft*, which was translated into English as *The Principles of Agriculture* [5]. The humus or organic matter in the soil was thought to provide the carbon for the plants: "Although nature furnishes a number of substances which tend to quicken vegetation, either by augmenting the vital principles or by assisting in the decomposition of the mould, it is only the mould or humus (finely-divided organic matter) which is capable of the requisite degree of decomposition, or the vegeto-animal manures, which supply to plants the most essential particles, and those most necessary for their nourishment." Chemical techniques and concepts were too immature in the 1840s to understand the complexity of humus but the theory, although unfounded, can be seen as an early attempt to explain the relationship between soil fertility and soil organic matter [30]. In the USA, Samuel Dana expanded the humus theory; his *Muck Manual,* which ran through four editions in the 1840s, was popular with farmers but not with scientists [15, 31].

Jean-Baptiste Boussingault started the first agricultural field experiments in the 1830s, focusing on crop rotations and nitrogen uptake [32]. Nitrogen had been discovered as an element in 1772 by Daniel Rutherford who had removed oxygen and carbon dioxide from the air in a bottle and discovered that the residual gas would not support combustion and that a mouse could not live in it. The remaining gas had no oxygen and was called 'burnt' or 'dephlogisticated air.' That the atmosphere was rich in nitrogen was then known but how it ended up in plants was a mystery. In 1845, Jean-Baptiste Boussingault found that plants take up nitrogen as well as other nutrients from the soil:"…we see that the mineral substances which meet us in plants also exist in the soil independently of any addition from manure. We may therefore lay it down as a principle that the mineral substances encountered in vegetables are obtained in the soil, and that the whole of these substances come from rocks which form the solid crust of our earth." Jean-Baptiste Boussingault had demystified the humus theory.

The origin and importance of humus remained the subject of many studies in the mid-1800s. Robert Wood worked at the National Institute for the Promotion of Science that later became a part of the Smithsonian Institution. He studied humus, and wrote in 1850 a pamphlet entitled *The soil considered as a separate and distinct department of nature*, in which he viewed the soil with its 'vegeto-mineral character' as a connector between organized and unorganized matter [33]. The soil sustained living bodies and was sustained by the living bodies, but it would slowly weather away:"If the natural history of soil be studied, we find that although it may increase enormously under certain conditions, and although its term of maturity may be prolonged to an apparently indefinite extent, its ultimate dissolution, in whole or in part, is a matter of as much certainty as the lapse of ages." The soil could be maintained by applying manure which he considered more acceptable to vegetables than their own decaying leaves or the debris of other higher plants. He thought that plants raised with 'offensive manure' should be objected with: "…a very general repugnance." Like Constantin Volney in 1804 [4], Albrecht Thaer in the 1840s [5], and Charles Darwin in 1881 [6], Robert Wood named organic matter-rich topsoils: mold, vegetable mold, or humus, and he emphasized its importance: "…we do not propose adding compounds of nitrogen to worn out soil solely for the purpose of raising vegetable mould, although the improvement in the soil is the first step in the improvement of our vegetables, and consequently of our animal." Robert Wood concluded that humus in the soil should be preserved by judicially applying organic or mineral manures [33].

It was the German chemist Justus von Liebig who further demystified the humus theory, through a series of laboratory experiments and deductive reasoning [34]. Carbon in plants was not supplied by organic matter but taken up from the air. Justus von Liebig established that nitrogen, phosphorus, sulfur, potassium, calcium, magnesium, silica, and sodium were all essential for plant growth. Plants had no mystic power to synthesize elements from others, and Justus von Liebig broke with the alchemists who had speculated on the organic origin of plant nutrition [35, 36]. Much of this was based on observation and speculation rather than by experimentation, but he was correct. He was credited with discovering the mineral nutrition of plants, and from the 1840s onwards, the chemical school of thought on soils was led by Justus von Liebig. He became one of the most influential scientists, and the Duke of Hesse-Darmstadt made him a Baron [37]. Justus von Liebig and Albrecht Thaer were celebrated on German stamps and on Reichsmark banknotes—Albrecht Thaer in the 1920s (10 *RM*) and Justus von Liebig in the 1930s (100 *RM*).

Justus Liebig's laboratory findings did not hold up in the field. Crops did not respond to amendments as expected, and the nature and behavior of soils in the field were poorly understood. Much of Justus von Liebig's reasoning on plant nutrition was based on the idea of the soil as a balance sheet, in which he did not take into consideration biological nitrogen fixation, as it had not yet been discovered. Some changed his surname to 'Lie Big.' In his earlier writings, the soil was solely a medium for plant growth: "...the most careful examination of the chemical nature both of the soil in which a given plant grows, and of the plant itself, must be the foundation of all exact and economical methods of cultivation." The duty of the chemist was "...to explain the composition of a fertile soil, but the discovery of its proper state or condition belongs to the agriculturists." He appreciated that there were other factors relevant for the fertility of soil and the growing of crops: "The fertility of a soil is much influenced by its physical properties, such as its porosity, colour, attraction for moisture, or state of disintegration. But independently of these conditions, the fertility depends upon the chemical constituents of which the soil is composed" [38]. He was a good enough scientist to learn better in due course. In his later writings, Justus von Liebig noted: "The soil consists of disintegrated rocks, and either rests upon these same rocks or on others elsewhere; the transported soil may, nevertheless, have remained the same and corresponds at least to the rocks from which it has its origin" [39].

John Lawes was a wealthy and practical farmer and founded an experimental farm at his Rothamsted Manor in England. In the 1830s, he conducted a series of pot experiments showing that when mineral phosphate treated with sulfuric acid was applied to the soil, plants grew better. In 1842, he took out a patent for manufacturing superphosphate, becoming the 'superphosphate king' [15, 40]. Successive experiments were conducted with acidulated phosphate rock and the excrement of seabirds and bats, which was named guano. In 1843, John Lawes hired a chemist, Joseph Gilbert, who had studied with Justus von Liebig in Giessen. His wife, Maria Gilbert, had translated papers and books from Justus von Liebig into English. Joseph Gilbert and John Lawes favored field experiments over theoretical expansion and speculation [40]. In the 1840s, it was still assumed that organic manures were essential and that the manures provided more than just plant nutrients, but Joseph Gilbert and John Lawes showed that crop yields could be sustained solely by applying inorganic fertilizers. They also found that high crop yields could be achieved under continuous cropping and without rotation when inorganic fertilizers or organic manures were applied [40].

In the first half of the 1800s, there was little understanding of how soils were formed [41]. For example, it was thought that the presence of stones and gravel in the soils was the result of chemical precipitation, and that peat was formed by algae [42]. Some soil investigations in the early 1800s introduced biblical terms such as *diluvium* deposited during *le deluge*—the great floods. In the Netherlands, soils were considered to be *diluvial* or *alluvial*, and a similar distinction was used in the UK [23, 43, 44]. The mining and construction industries required a systematic study of the earth, and geology, as a scientific discipline, was born at the beginning in the early 1800s with James Hutton and Charles Lyell as pioneers [7, 45].

Scientific studies on the origin of soils had a geologic focus, and soils were seen as the weathering products of rocks [31, 46]. It took some time before it was realized that soils could be mixtures and derived from transported materials. The chemist Samuel Johnson distinguished in 1861 two types of soil based on their mode of formation: *sedentary* soils or soils in place that cover the rock from whose integration they originate, and *transported* soils that are subdivided into drift, alluvial, and colluvial [46]. Despite this distinction, he held a geologic view on soils: "...soils are broken and decomposed rocks. We find in nearly all soils fragments of rock, recognizable as such by the eye, and by help of the microscope it is often easy to perceive that those portions of the soil which are impalpable to the feel chiefly consist of minuter grains of the same rock" [46]. The conversion of rocks to soils was performed by changes in temperature, moving water or ice, the chemical action of water and air, and the influence of vegetable and animal life.

Not everyone in the 1800s had a geological view on the origin of soils. In 1861, John Nash, an instructor in agriculture at Amherst College, published a book entitled *Progressive Farmer* that was "...an effort to render science available to practical farmers." He had some original ideas on the soil: "All soils, whether alluvial, drift, or tertiary in their origin, are derived from rocks, broken down, ground to a greater or less degree of fineness, and so disseminated that the ruins of one rock may be supposed to be mixed, in most cases, with those of a great many others. The idea that soils have originated from the rock immediately under them is an error. The soil on nearly every foot of land in our country—and the same is true of Europe, at least, if not of the whole world—has come from many and widespread localities. Every soil may be considered as a mixture of many soils" [28].

Fewer people studied soils in the field than in the laboratory [47]. Those who studied soils in the field were mostly trained as geologists, and their findings were used to assess land for agricultural development or for tax assessment. They became known in the late 1800s as *agrogeologists* [48]. Geology dealt with the dead part of the globe while *agrogeology* dealt with the ever-changing, living earth cover. The agricultural geologists or agrogeologists considered the study of soil to be a branch of geology, and aimed to add an agricultural component to geology, in part to help secure funding for geology [49]. Agrogeology research had a regional focus, and there was little international cooperation [50]. The study of plant nutrition and agricultural chemistry had a broader international exchange.

Friedrich Fallou had studied jurisprudence at the University of Leipzig in Germany. He was a land tax assessor and developed an interest in soils although he had no formal training in geology or agricultural chemistry [51]. At that time, there were very few books that solely focused on soil, and most were compilations of geology, geography, agricultural chemistry, and plant physiology. In 1862, Friedrich Fallou published the book *Pedologie oder allgemeine und besondere Bodenkunde* [52], in which soil was the exclusive subject. Friedrich Fallou treated the study of soils independently and as a separate topic from geology, and he defined soil "…as loose masses of mineralic and organic components, which arise from weathering and reorganization of the earth's surface." He emphasized that soils were not formed at the same time nor in the same way [51]. Friedrich Fallou recognized the importance of the soil profile and its layers. The geologist and botanist Ferdinand Senft studied soils of the forests and had also distinguished soil layers in 1857 [53, 54]. Friedrich Fallou coined the terms *pedology* and *solum*, as well as *soil quality* [52]. These terms were ignored for years as jargon, or because Friedrich Fallou was not recognized as a scientist and had no student followers [51]. He was never featured on a German post stamp or the Reichsmark.

The Englishman John Morton was a contemporary of Friedrich Fallou, but it is unlikely that they ever met. John Morton was a farmer and "…employed his leisure periods in walking most over most of the counties noting their geology and farm practice" [55]. He emphasized that local variations of rock were reflected in the soil and that rocks with thin layers of alternating texture and finer material determined the nature of the soil. Local and distant agents of transport caused the mixing of mineral materials which differentiated the upper part of the soil. His book published in 1843, *The Nature and Property of Soils*, included bedrock geology maps that can be seen as early soil maps [42].

For John Morton, soils and the underlying rocks were inextricable: "The surface of the earth partakes of the nature and colour of the subsoil or rock on which it rests. The principal mineral of any district is that of the geological formation under it; hence, we find argillaceous soil resting on the various clay formations—calcareous soil over the chalk and oolitic rocks; and silicious soils over the various sandstones. On the chalk the soil is white; on the red sandstone, it is red; and on the sands and clays, the surface has nearly the same shade of colour as the subsoil" [43]. He stated that "...the surface is composed of the same materials as the subsoil, with the addition of vegetable and animal matter, in every state of decay, intimately mixed with it; and we perceive a change in the external appearance of the surface, whenever there is a change in the subsoil below." John Morton, a pioneer of soil science, has somehow been overlooked in soil science historiography [55–57].

The soil was seen as a mineral substance that originated from the weathering products of rocks and included plants and animals ('as the rock, so the soil') or as an inert mass of mineral debris [55, 58]. Those views persisted well into the 1900s. In 1927, Hermann Stremme worded this as follows: "Originally there is rock, a mixture of minerals in contact with the atmosphere, which has experienced a certain crushing and decomposition, depending on the intensity of atmospheric forces of physical and chemical nature. The result would be debris rather than soil. Soil only forms when plants and animals settle on the debris. They uptake directly and indirectly matter from the debris and return metabolites like humus, solutions and gases" [59].

There were numerous studies on isolated properties of soils, but there was no agreement on what the important properties were, or how they should be analyzed. Textbooks focused on the analyses of soils, limestones, and manures. The chemist James Johnston provided instructions for soil analyses in 1856, as he found that "...the benefits to be derived from the chemical examination and analysis of a soil are by many misunderstood. Some have represented it as the only sure guide to successful cultivation; while others have not scrupled to pronounce the analysis of soils to be entirely useless, and unfitted to lead to any profitable practical result" [60]. James Johnston provided instructions for the analysis of physical properties of the soil, organic matter, saline matter, earthy matters of the soil, and the analyses of limestone, marl, saline manure, bone manure, guano, and oil cake. Advice was given on how to sample a field. He discouraged sampling from across the field and then mixing; instead "...one or two pounds should be taken from each of four or five parts of the fields where the soil appears nearly alike" [60]. The development and standardization of soil analytical methods was an important subject in the

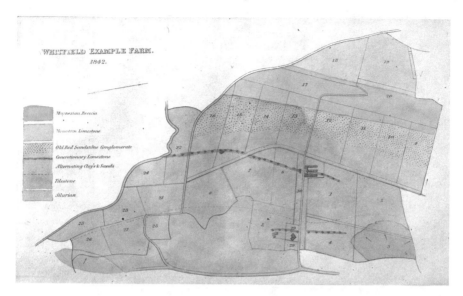

Bedrock geology map of the Whitfield farm in the Parish of Cromwall in the county of Gloucestershire in the UK. Green is magnesian breccia, blue is mountain limestone, brown is sandstone conglomerate, alternating clays and sands and tilestone, dark blue is Silurian. From John Morton, *The Nature and Property of Soils; their Connection with the Geological Formation on which They Rest*, 1843

mid-1800s. It helped in understanding the composition of soils and how they were formed.

In the 1870s, Peter Müller studied organic matter in soils under forest in Jutland and the eastern part of Denmark. With a microscope, he looked at humus forms and observed connections between fungi and roots [61]. Peter Müller made drawings of what appear to be Podzols, which is not unlikely as such soils cover about one-third of Denmark [62]. He made descriptions of the soil profiles and presented analytical data with depth. Letters were used for the different layers, whereby *a* was turf, *b* was bleached sand, *b'* was reddish earth, and *c* was underground. Peter Müller was familiar with the studies by Emil Ramann, who worked on soils of the forest at the Prussian Forestry Experiment Station and who followed in the tradition of Ferdinand Senft [54]. Emil Ramann later introduced the soil type *Braunerde* or brown forest soils [63]. Both Emil Ramann and Peter Müller worked on the relationship between soils and forest vegetation. Peter Müller's findings were published in a book that was translated into German in 1887. By that time Vasily Dokuchaev had published his genetic soil profile concept, but it had not reached Denmark [64].

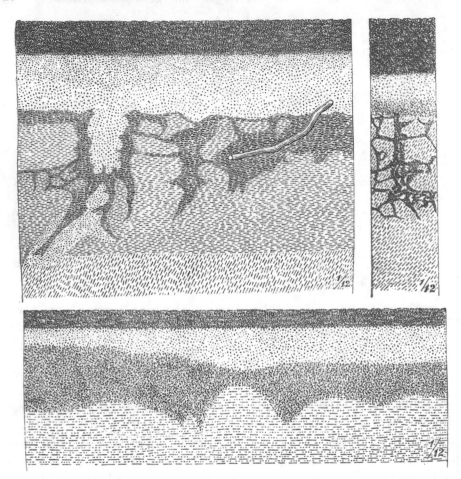

Soil profile drawing from the 1870s by Peter Erasmus Müller, sand under forest Jutland and eastern Denmark. Peter Müller made drawings of Podzols, had detailed descriptions of soil profiles and used letters for the different layers. From *Studien über die natürlichen Humusformen und der Einwirking auf Vegetation und Boden* published in 1887

Nathaniel Shaler was a professor of geology at Harvard University and had grown up on a farm in Kentucky in the 1840s. He had been a student of Louis Agassiz and was of the same generation as the geologist Thomas Chamberlin. In 1892, Nathaniel Shaler wrote a monograph entitled *The Origin and Nature of Soils* [65]. His view on soils was geological, and soils were derived from cliff talus, glaciation, volcanics, or elevated ocean bottoms. Peculiar soils included swamp soils, marine marshes, tule lands, ancient soils, prairie soils, and windblown soils. Soils influenced humans and animals in three ways: through the water in the soil that influenced the humidity of the surface, the

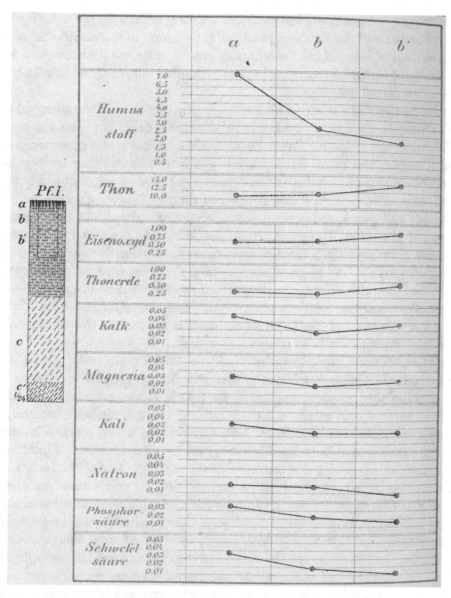

Soil profile (Pf. I.) from a forest with letters indicating different layers: a. mull, b. topsoil, c. subsoil. Profile consisted of loamy sandy over sandy loam, and gives depth functions for humus, clay (Thon), iron-oxide (Eisenoxyd.), Aluminite (Thonerde), lime (Kalk), M = magnesium (Magnesia), potassium (Kali), sodium (Natron), phosphoric acid (Phosphorsäure), sulfuric acid (Schwefelsäure). From *Studien über die natürlichen Humusformen und der Einwirking auf Vegetation und Boden* by Peter Müller published in 1887

bacteria in soil water, and the quality of drinking water from the soil. He was not familiar with Russian work, and had little notion of soil studies that were conducted in Europe. His most notable contribution to soil science was the training of Curtis Marbut, who in the 1910s lifted soil survey in America out of its geological trappings.

Nathaniel Shaler wrote in his autobiography how the soil was managed in Kentucky in the 1840s: "I never knew of manure being put upon the land. When, about 1855, my father began the use of it, he was much laughed at. The plan was to till a field until it was worn out and then let it go to grass or bushes of a kindly nature, helped by chance sowing; commonly the soil washed away until the lava rock was exposed. The crops were mainly tobacco and grains, and as there was no system of rotation, the fields rapidly became exhausted. The more careful landlords required that their tenants should plant tobacco, a most exhausting crop, only for three or four years, and then set the land in grass; but generally, there was no adequate enforcement of the rules, so that the cleared land rapidly became worthless. In the first 60 years of this atrocious process nearly one half of the arable soil of the northern counties of Kentucky, where most of the surface steeply inclined, became unremunerative to plough tillage" [66].

The childhood observations of Nathaniel Shaler came back in his 1892 essay on *The Origin and Nature of Soils* that he wrote with the purpose:"It is also intended to show that this slight superficial and inconstant covering of the earth should receive a measure of care which is rarely devoted to it; that even more than the deeper mineral resources it is a precious inheritance which should be guarded by every possible means against the insidious degradation to which the processes of tillage ordinarily lead" [65]. Nathaniel Shaler was convinced that humans had the duty to take care of the earth and that future generations would look back "…with great amazement and disgust" [65]. Likely, he was familiar with George Marsh and his influential book *Man and Nature, or physical geography as modified by human actions,* published in 1863. George Marsh showed the consequences of overexploitation on flooding and soil erosion. Most examples were from Europe and included a biblical warning: "Let us be wise in time, and profit by the errors of our older brethren" [67].

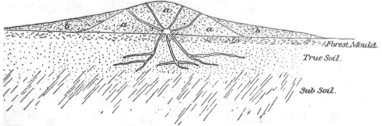

Soil profiles from *The origin and nature of soils* published in 1892 by the Harvard geologist Nathaniel Shaler (1841–1906). Original figure captions: "Effect of trees on the formation of soil; and Effects of ant hills on soil, with a—sand accumulated in hill' b—material washed form hill mingled with vegetable mold"

Samuel Johnson and Eugene Hilgard were pioneering soil researchers in the USA and stood for the primary two branches from which soil science emerged as an independent discipline: agricultural chemistry and geology. They were born in the 1830s and both obtained university degrees in Germany. Samuel Johnson was born on a farm in Fulton County, New York, in 1830, in a family that were early colonial settlers in Connecticut. He spent a year in the laboratories of Justus von Liebig, learned to read and write German and translated some of Justus von Liebig's works. While in Europe, he visited the Rothamsted laboratory of Joseph Gilbert, who had also studied with Justus von Liebig and whose wife, Maria Gilbert, had also translated works of Justus von Liebig. The translations of those works and Humphry Davy's book on *Elements of Agricultural Chemistry* were an impetus for agricultural chemistry in the USA [15]. When Samuel Johnson returned to the USA in 1855 and joined Yale University, he introduced the German practice of graduate training and the scientific seminar [68].

At Yale University, Samuel Johnson became the first professor of analytical and agricultural chemistry. Initially, he studied the absorption properties of soils but soon he moved his attention to field studies. The theories of Justus von Liebig and the contributions of agricultural chemistry to agricultural production had created high expectations. Field experiments in Rothamsted had shown the limitations, and Samuel Johnson became convinced that more science was needed. Although he was at the forefront of bringing chemistry to agriculture, he doubted its usefulness. In 1861, he noted that he would rather trust an old farmer for the judgment of land than the best chemist alive. Meaningful results could only come from field-based experiments and in 1875, Samuel Johnson requested state funding to establish an agricultural experiment station in Connecticut [69]. The aim of the experimental station was to bring science into farming, and Samuel Johnson emphasized that agricultural science had to move beyond well-meaning amateurs and laymen [15]. The funding request was approved and the State Agricultural Experimental Station in Connecticut became the first in the nation and included an agricultural chemistry laboratory [70]. After the establishment of the agricultural experiment station in Connecticut, similar stations were established by other Land Grant colleges following the Hatch Act of 1887. Some of these were modeled after experimental stations in Europe.

The first agricultural experiment station in the world was established in Pechelbronn in France in 1836 by Jean-Baptiste Boussingault. Rothamsted in the UK was established in the 1840s; and experimental stations were established in Möckern, in Germany, in 1852; in Sapporo, Japan, in 1871; in Gembloux, in Belgium, in 1872, and in Wageningen, the Netherlands, in 1877. All these stations started small and focused on research subjects that aimed to help agricultural development. Some stations had a double role: agricultural research, and analysis of agricultural products for regulation purposes. Over time, many agricultural experiment stations became leading scientific institutes, with renowned soil directors like Eugene Hilgard in California, Charles Dabney in North Carolina, Jacob Lipman in New Jersey, and Cyril Hopkins in Illinois. The agricultural experiment stations held their *First International Congress of Experiment Stations* in Versailles in 1881, and in 1889 in Paris where a monolith of the Chernozem or black soil was on display in the Russian pavilion. The monolith showed "…the depth to which the humus or black soil extends, with the yellow subsoil just showing at the bottom of the sample" [71]. The world, and in particular the USA, still had to learn about these soils and the principles that determined their formation.

In 1883, Samuel Johnson hired the 23-year-old Milton Whitney as an assistant chemist for the State Agricultural Experimental Station in Connecticut [70]. Milton Whitney had just finished a special course in chemistry at Johns Hopkins University and, in Connecticut, his task was to develop shade culture for the growing of tobacco [72]. Milton Whitney did not stay long in Connecticut and after 3 years, he joined the North Carolina Experiment Station which was directed by Charles Dabney. He then moved to his home state Maryland, where he conducted soil studies at the Maryland Agricultural Experimental Station that was established in 1888. On the land between the Chesapeake Bay and the Appalachians, he studied the relationships between soil texture and crop productivity, and developed instruments to measure soil moisture, temperature, and salt content in the soil [73, 74]. He perceived that the agricultural potential was closely related to the soil texture; the texture regulated soil temperature, moisture content, and air supply, and he termed those the 'climate of the soil' [75]. But, in his view, the chemical composition of the soil did not determine its fertility. He conceived the idea that all soils have solutions of the same concentration, and he was convinced that fertilizers did not feed the plant but acted in some other way [76]. These convictions brought him in conflict with several of the leading soil researchers at that time, including Eugene Hilgard.

Samuel Johnson (1830–1909) and Eugene Hilgard (1833–1916) were pioneering soil researchers in the USA. They stood for the two principal branches from which soil science as an independent discipline arose: agricultural chemistry and geology. Both had obtained university degrees in Germany. Photos Library of Congress

Eugene Hilgard was born in Germany in 1833, 3 years before Samuel Johnson, and 13 years before Vasily Dokuchaev. He was the youngest of nine children. The Hilgard family emigrated to the USA when Eugene was 2 years old, and they settled on a farm in Illinois. Eugene taught himself botany, physics, and chemistry, and at the age of 17 returned to Germany to study at the University of Heidelberg, followed by studies at the University of Zürich in Switzerland. For his PhD research, he investigated the constituent parts of a candle flame, and he obtained his degree when he was 20 years old. After graduation, he lived in Spain and Portugal for 2 years where he met his wife, Jesusa Alexandrina Bello. In 1855, Eugene Hilgard returned to the USA and became the State geologist in Mississippi, and professor of chemistry at the University of Mississippi in Oxford in 1866. After 20 years in Mississippi, he worked for 2 years at the University of Michigan in Ann Arbor, where he was professor of mineralogy, geology, zoology, and botany. In 1874, the regents at the University of California Berkeley, invited him to give a series of lectures. He had a national reputation as a geologist, soil researcher, and chemist, and had been elected to the University of California Berkeley. He was appointed at Berkeley as professor of agricultural chemistry and director of the Agricultural Experiment Station.

The work of Eugene Hilgard in Mississippi was influenced by the German agrogeological view, and he studied geology and soils in relation to agriculture [49]. Over time, he developed a more genetic approach to his soil studies and, in the 1890s, he investigated the relationship between climate and soil [77]. He introduced the term 'alkali soils' for soils that were common in the drier parts of California. The alkali soils had a high clay content, a high pH, and were characterized by a poor soil structure that reduced the infiltration of water. Eugene Hilgard showed that the properties of alkali soils were caused by high concentrations of soluble salts and, in particular, sodium salts. Some of these soils had a dense and hard calcareous layer at depth, for which Eugene Hilgard coined the term 'hardpan'. He worked with soil researchers from Hungary, where some of the seminal studies on alkali soils were conducted by Alex. de'Sigmond. Those soils were later studied by David Hissink in the Netherlands, and Konstantin Gedroiz in Russia.

In 1904, Eugene Hilgard retired at the age of 73. Two years later, he published his *magnum opus: Soils, their formation, properties, composition, and relation to climate, and plant growth in the humid and arid regions* [78]. The agencies of soil formation, rocks and minerals, and soil physical conditions were discussed, followed by the soil and subsoil in relation to soil organic matter, organisms, and vegetation. He reviewed how climatic factors affected the soil, and how soil and plant growth were related. In the last chapters, he synthesized the information on soils of the arid and humid regions including the effects of climate on inorganic matter. Eugene Hilgard did not conceptualize a framework that related soils, their properties, and climatic conditions [79]. His book provided much of what was known about soils, and focused on the USA, in particular the western part of the country, and provided elements for a systematic analysis of soil distribution and laid the foundation for understanding the factors that form a soil.

A nine-step approach was described for taking soil samples and directions for soil examination in the field or on the farm. These sampling guidelines were the first of their kind and showed that he was also a practical scientist. The book was meant for students at the University of California Berkeley, but also for "…the fast increasing class of farmers who are willing and even anxious to avail themselves of the results and principles of scientific investigations, without 'shying off' from the new or unfamiliar words necessary to embody new ideas." For Eugene Hilgard, farmers should learn to understand and appreciate both the terms and methods of scientific reasoning, and he thought such reasoning ought to be taught in public schools.

Eugene Hilgard was broadly trained and had worked and traveled across the country. He was well-known in Europe and Russia. Nikolai Sibirtsev cited his work, and Konstantin Glinka had reviewed his research on alkali soils [80, 81]. He was one of the first Americans to be receptive to Russian soil science, wrote in five languages but did not speak Russian [79]. In the 1880s, he contacted Vasily Dokuchaev, who spoke some French and German, but the correspondence has been lost. In his 1906 book, he cited Russian studies, and his own work was known in Russia [82]. Despite Eugene Hilgard's breadth of view and scholarly work, his influence on soil research in the USA was limited [69]. Some of that had to do with his isolation on the west coast, but most of it was caused by the relationship he had with one of the most influential soil scientists of his time, Milton Whitney. They did not get along.

In 1916, Eugene Hilgard died, 83 years old. Charles Lipman, who worked with him in Berkeley, wrote his obituary: "He was the first among soil scientists systematically to correlate soils with plant growth through a study of vegetation as it has established itself on virgin land" [83]. His brother, Jacob Lipman, dedicated the first issue of the journal *Soil Science* to Eugene Hilgard, which was published in 1916. Nine years after Eugene Hilgard's death, the California Agriculture Experiment Station started publishing *Hilgardia*, a journal devoted to agricultural science. It was published for 70 years and contained numerous soil articles from soil scientists who came after him in Berkeley: Charles Shaw, Walter Kelley, Dennis Hoagland, and Hans Jenny.

George Merrill's book, *A treatise on rocks, rock-weathering and soils,* was published in the same year as Eugene Hilgard's book. It focused on rock weathering and soil development, and he had some help from Milton Whitney, with whom he shared a geological view of the origin of soils. George Merrill considered volcanic dust and dune sands *aeolian rocks*, and the soil a superficial portion of the *regolith*—a term that he coined. For him, soils and the underlying rock were closely connected. There is not much on soils in his book; George Merrill has one reference to the Chernozems of Russia, and he saw their formation as "…a local phase of the loess, the color being due to the prevalence of organic matter" [84].

George Merrill (1854–1929), head curator of the National Museum (now the National Museum of Natural History of the Smithsonian Institution), pictured in 1925 with the largest perfect crystal globe in the world. He was the author of the 1906 book *A treatise on rocks, rock-weathering and soils*, and envisaged that "...the earth is fast becoming an unfit home for its noblest inhabitant, and another era of equal human crime and human improvidence, and of like duration with that through which traces of that crime and that improvidence extend, would reduce it to such a condition of impoverished productiveness, of shattered surface, of climatic excess, as to threaten the depravation, barbarism, and perhaps even extinction of the species." Photo Library of Congress

George Merrill's geological emphasis on the soil came naturally, as he was born in a family of geologists and mineralogists in Maine in 1854. He was 6 years younger than Vasily Dokuchaev. Trained as a chemist and geologist, he worked at the Columbian University in Washington D.C., then became head curator of the Department of Geology at the National Museum. While Eugene Hilgard had broadened the view on what characterized and formed the soil, George Merrill probed its geological origin. Konstantin Glinka was familiar with his work [81], and George Merrill introduced the work of Nicolai Sibirtsev to his student George Coffey, who conducted his MS and PhD research in 1908 and 1911 [70, 85]. George Coffey became an early adaptor of the Russian principles of soil that he attempted to introduce into the *Bureau of Soils* [85, 86].

In contrast to early travelers in the USA who were struck by the immense forests that blanketed the country, a 100 years later, George Merrill was concerned about its widespread deforestation: "When the forest is gone, the great reservoir of moisture stored up in its vegetable mould is evaporated, and returns only in deluges of rain to wash away the parched dust into which that mould has been converted. The well-wooded and humid hills are turned to ridges of dry rock, which encumbers the low grounds and chokes the watercourses with its débris, and except in countries favored with an equable distribution of rain through the seasons, and a moderate and regular inclination of surface—the whole earth, unless rescued by human art from the physical degradation to which it tends, becomes an assemblage of bald mountains, of barren, turfless hills, and of swampy and malarious plains" [84]. George Merrill ended his book with a gloomy outlook: "...the earth is fast becoming an unfit home for its noblest inhabitant, and another era of equal human crime and human improvidence, and of like duration with that through which traces of that crime and that improvidence extend, would reduce it to such a condition of impoverished productiveness, of shattered surface, of climatic excess, as to threaten the depravation, barbarism, and perhaps even extinction of the species" [84]. Some of that proved to be a thoughtful foresight; he had a good look into the crystal ball.

References

1. Feller CL, Thuriès LJM, Manlay RJ, Robin P, Frossard E. "The principles of rational agriculture" by Albrecht Daniel Thaer (1752–1828). An approach to the sustainability of cropping systems at the beginning of the 19th century. J Plant Nutr Soil Sci. 2003;166(6):687–98.

2. Bourgeon L, Burke A, Higham T. Earliest human presence in North America dated to the last glacial maximum: new radiocarbon dates from Bluefish Caves, Canada. Plos One. 2017;12(1):e0169486.
3. Bartram J. Observations on the inhabitants, climate, soil, rivers, productions, animals, and other matters worthy of notice made by Mr. John Bartram, in his travels from Pensilvania [sic] to Onondago, Oswego and the Lake Ontario, in Canada to which is annex'd a curious account of the cataracts at Niagara by Mr. Peter Kalm, a Swedish gentleman who travelled there. London: J. Whiston & B. White; 1751.
4. Volney CF. A view of the soil and climate of the United States of America: with supplementary remarks upon Florida; on the Frensch colonies of the Mississippi and Ohio, and in Canada; and on the Aboriginal tribes of America (translated by CB Brown) Philadelphia: J. Conrad & Co.; 1804.
5. Thaer A. The principles of agricuture (translated by W Shaw, CW Johnson). New York: Bangs, Brother, & Co.; 1852.
6. Darwin C. The formation of vegetable mould through the actions of worms with observations on their habits. London: John Murray; 1881.
7. Dott RH. Lyell in America—his lectures, field work, and mutual influences, 1841–1853. Earth Sci Hist. 1996;15(2):101–40.
8. Cobbett W. A year's residence in the United States of America. In three parts. Paternoster-Row: Sherwood, Neely, and Jones; 1818.
9. White JE. Cursory observations on the soil, climate, diseases of the State of Georgia. Medial Repository. 1806; III.
10. Robinson STL. Kansas: interior and exterior life including a full view of its settlement, political history, social life, climate, soil, productions, scenery, etc. 8th ed. Boston: Crosby, Nichols and Company; 1857.
11. Hacker JD, Haines MR. American Indian Mortality in the Late Nineteenth Century: the Impact of Federal Assimilation Policies on a Vulnerable Population. Annales de démographie historique. 2005;110(2):17–29.
12. Goddard FB. Where to emigrate and why. New York: Frederick B. Goddard; 1869.
13. Davy H. Elements of agricultural chemistry. New York: Eastburn, Kirk & Co; 1815.
14. Ville MG. Six lectures on agriculture. High farming without manure. Boston: Press of Geo. C. Rand & Avery; 1866.
15. Rossiter MW. The mergence of agricultural science. Justus von Liebig and the Americans, 1840–1880. New Haven and London: Yale University Press; 1975.
16. Craven AO. Soil exhaustionas as a factor in the agricultural history of Virginia and Maryland, 1606–1860. Urbana: University of Illinois; 1926.
17. Beal WH. The new science of the soil. Sci Am. 1911;104(7):168–87.
18. Kellogg CE. Conflicting doctrines about soils. Sci Mon. 1948;66:475–87.
19. Marbut CF. Soils of the United States. Part III. In: Baker OE, editor. Atlas of American Agriculture. Washington: United Sates Department of Agriculture. Bureau of Chemistry and Soils; 1935. p. 1–98.

20. Helms D. Soil and southern history. Agric Hist. 2000;74:723–58.
21. Truog E. Putting soil science to work. J Am Soc Agron. 1938;30:973–85.
22. Ruffin E. An essay on calcareous manures. Petersburg VA: J.W. Campbell; 1832.
23. Staring WCH. Natuurlijke Historie van Nederland. De Bodem van Nedereland. Haarlem: A.C. Kruseman; 1856.
24. Moll ML. Handboek voor den Landbouwer. Brussel: T. Parent; 1847.
25. Chambers WH. The soil of the south. A Monthly journal devoted to agriculture, horticulture and rural and domestic economy. Columbus, Georgia: Lomax & Ellis publishers; 1853.
26. Goggin WM. Artificial fertilizers, not a necessity. A treatise on the recuperating energies of nature in agricultural science. Selbyville, Tennessee: Printed for the author; 1882.
27. Allen RL. The American farm book. New York: A.O. Moore; 1858.
28. Nash JA. The progressive farmer: a scientific treatise on agriculture chemistry, the geology of agriculture. New York: C.M. Saxton, Barker & Co.; 1861.
29. Lawes JB, Morton JC, Morton J, Scott J, Thurber G. The soil of the farm. Washington: Orange Judd; 1883.
30. Kleber M, Johnson MG. Advances in understanding the molecular structure of soil organic matter: Implications for interactions in the environment. Adv Agron. 2010;106:77–142.
31. Dana SL. Muck manual for farmers. Boston: James P. Walker; 1842.
32. Boussingault JB. Rural economy in its relations with chemistry, physics, and meteorology or, chemistry applied to agriculture (translated by G Law). New York, PA: Appleton & Co; 1845.
33. Wood RS. The soil considered as a separate and distinct department of nature. Washington: Corresponding Member of the National Institute; 1850.
34. van der Ploeg RR, Bohm W, Kirkham MB. On the origin of the theory of mineral nutrition of plants and the law of the minimum. Soil Sci Soc Am J. 1999;63(5):1055–62.
35. Kellogg CE. Russian contributions to soil science. Land Policy Rev. 1946;9:9–14.
36. Marschner H. Mineral nutrition of higher plants. 2nd ed. San Diego: Academic Press; 1995.
37. Brock WH. Justus von Liebig. The chemical gatekeeper. Cambridge: Cambridge University Press; 1997.
38. von Liebig J. Die organische Chemie in ihrer Anwendung auf Agrikultur und Physiologie. Braunschweig: Verlag von Friedrich Vieweg und Sohn; 1840.
39. von Liebig J. Letters on modern agriculture. London: Walton and Maberly; 1859.
40. Waksman SA. The men who made Rothamsted. Proc Soil Sci Soc Am. 1944;8:5–5.
41. Vil'yams VR. V.V. Dokuchaev's role in the development of soil science. Selected works of V.V. Dokuchaev. Washington DC: Israel Program for Scientific Transaltions, U.S. Department of Agriculture; 1967.

42. Hartemink AE. The depiction of soil profiles since the late 1700s. Catena. 2009;79:113–27.
43. Morton J. The nature and property of soils: Their connexion with the geological formation on which they rest. London: James Ridgway; 1843.
44. Donaldson J. Rudimentary treatise on clay lands and loamy soils. London: John Weale; 1852.
45. Wm HH. James Hutton, the pioneer of modern geology. Science. 1926;64(1655):261–5.
46. Johnson SW. How crops feed—a treatise of the atmosphere and the soil as related to the nutrition of agricultural plants. New York: Orange Judd and Company; 1870.
47. Kellogg CE. Soil genesis, classification, and cartography: 1924–1974. Geoderma. 1974;12:347–62.
48. van Baren J. Agrogeology as a science. Mededeelingen van de Land-bouwhogeschool en van de daaraan verbonden instituten. In: Wulff A, editor. Bibliographia Agrogeologica, vol. XX. Wageningen; 1921. p. 5–9.
49. Krupenikov IA. History of soil science from its inception to the present. New Delhi: Oxonian Press; 1992.
50. Hartemink AE. On global soil science and regional solutions. Geoderma Regional. 2015;5:1–3.
51. Asio VB. Comments on "Historical development of soil and weathering profile concepts from Europe to the United States of America" . Soil Sci Soc Am J. 2005;69(2):571–2.
52. Fallou FA. Pedologie oder Allgemeine und Besondere Bodenkunde. Dresden: Schönfeld Buchhandlung; 1862.
53. Blume HP. Some aspects of the history of German soil science. J Plant Nutr Soil Sci-Zeitschrift fur Pflanzenernahrung und Bodenkunde. 2002;165(4):377–81.
54. Senft F. Lehrbuch der forstlichen Geognosie, Bodenkunde und Chemie. Jena: Manke; 1857.
55. Bunting BT. John Morton (1781–1864): a neglected pioneer of soil science. Geogr J. 1964;130:116–9.
56. Yaalon DH. History of soil science in context: international perspective. In: Yaalon DH, Berkowicz S, editors. History of soil science international perspectives. Reiskirchen: Catena Verlag; 1997. p. 1–13.
57. Boulaine J. Histoire des Pédologues et de la Science des Sols. Paris: INRA; 1989.
58. Johnston JFW. Elements of agricultural chemistry and geology. Edinburgh and London: William Blackwood and Sons; 1845.
59. Stremme H. Grundzüge der praktische Bodenkunde. Berlin: Verlag von Gebrüder Borntraeger; 1927.
60. Johnston JFW. Instructions for the analysis of soils, limestones and manures. Cleveland, Ohio: S.B. Shaw Publishers; 1856.
61. Müller PE. Studien über die natürlichen Humusformen und der Einwirking auf Vegetation und Boden. Berlin: Verlag von Julius Springer; 1887.

62. Adhikari K, Hartemink AE, Minasny B, Kheir RB, Greve MB, Greve MH. Digital mapping of soil organic carbon contents and stocks in Denmark. Plos One. 2014;9(8).

63. Vernier RT, Smith GD. The concept of Braunerde (Brown Forest Soil) in Europe and the United States. Adv Agron. 1957;9:217–89.

64. Tandarich JP, Darmody RG, Follmer LR, Johnson DL. Historical development of soil and weathering profile concepts from Europe to the United States of America. Soil Sci Soc Am J. 2002;66(2):335–46.

65. Shaler NS. The origin and nature of soils. Washington: Department of the Interior, U.S. Geological Survey, Government Printing Office; 1892.

66. Shaler NS. The autobiography of Nathaniel Southgate Shaler: with a supplementary memoir by his wife. Boston: Houghton Mifflin; 1909.

67. Marsh GP. Man and nature, or physical geography as modified by human actions. New York: Scribner, Armstrong & Co; 1863.

68. Kellogg CE, Knapp DC. The college of agriculture. Science in the public service. New York: McGraw-Hill Book Company; 1966.

69. Gardner DR. The National Cooperative Soil Survey of the United States. Thesis presented to Harvard Univsersity, Washington DC: USDA; 1957.

70. Helms D. Early leaders of the soil survey. In: Helms D, Effland ABW, Durana PJ, editors. Profiles in the History of the U.S. Soil Survey. Ames: Iowa State Press; 2002:19–64.

71. U.S. Department of Agriculture. Experiment Station Record, vol. XII. Washington: Government Printing Office; 1901. p. 1900–1.

72. Fanning DS, Fanning MCB. Milton Whitney: soil survey pioneer. Soil Horizons. 2001;42:83–9.

73. Whitney M, Briggs LJ. An electrical method of determining the temperature of soils. Bulletin no. 7. Washington: U.S. Department of Agriculture, Divisions of Soils; 1897.

74. Whitney M, Gardner FD, Briggs LJ. An electrical method of determining the moisture content of arable soils. Bulletin no. 6. Washington: U.S. Department of Agriculture, Divisions of Soils; 1897.

75. Whitney M. Some physical properties of soils in relation to moisture and crop distribution. Bulletin no. 4. U.S. Dept. Agric. Weather Bureau; 1892. p. 80.

76. Russell EJ. Obituary Prof. Milton Whitney. Nature. 1928;121(3036):27–27.

77. Hilgard EW. A report on the relations of soil to climate. Washington; 1892.

78. Hilgard EW. Soils, their formation, properties, composition, and relation to climate, and plant growth. New York: The Macmillan Company; 1906.

79. Maher D, Stuart K, editors. Hans Jenny—soil scientist, teacher, and scholar. Berkeley: University of California; 1989.

80. The MD, Steppes A. The unexpected Russian roots of great plains agriculture, 1870s–1930s. Cambridge: Cambridge University Press; 2020.

81. Glinka K. Die Typen der Bodenbildung—Ihre Klassifikation und Geographsiche Verbreitung. Berlin: Verlag von Gebrüder Borntraeger; 1914.

82. Moon D. The Russian academy of sciences expeditions to the steppes in the late eighteenth century. Slavon E Eur Rev. 2010;88(1–2):204

83. Lipman CB. Eugene Woldemar Hilgard. J Am Soc Agron. 1916;8:160–2.

84. Merrill GP. A treatise on rocks, rock-weathering and soils. New York: The MacMillan Company; 1906.

85. Brevik EC. George Nelson Coffey, early American pedologist. Soil Sci Soc Am J. 1999;63(6):1485–93.

86. Brevik EC. George Nelson Coffey, early soil surveyor. Soil Horizons. 2001;42.

2

Pochva Americana I

"Many American agricultural scientists were trained primarily in chemistry and plant physiology in the German tradition of Liebig and had little experience in the earth sciences, especially geography."

Charles Kellogg, 1946

Soil research in Russia in the 1870s revolutionized the understanding of the origin and formation of soil. The new insights came from studies across a large area that combined soil observations with laboratory data of soil profiles. The discovery was made that soils had inherited their features and properties from the weathered rock, the climate, and the vegetation, or in other words, the soil had a genetic base. Soils could be studied independently and be understood by a set of principles that made them unique natural bodies. It took decades before that discovery found followers in the USA, although the Russian work was exhibited at the World's Fair in Chicago in 1893, published in a Bulletin of the Department of Agriculture in 1901, and in several other publications in the 1910s. This chapter reviews some of the early soil research in Russia, how it evolved, and how it was eventually adopted in America.

The Russian Empire in the 1800s was vast and sparsely populated. There were enormous areas of wilderness that attracted naturalist travelers who were interested in the land, forests, and waters. Ivan Lepekhin passed through the Transvolgian steppe in 1768 and noted that it was characterized by great quantities of salt 'distributed throughout the steppe' [1]. Other travelers crossed the Caspian steppe into Turkmenistan and studied the composition of saline lakes, ashes in saline plants, as well as its soil. The origin of the salt sparked the interest of several researchers. The zoologist Alexander von Middendorff, professor of zoology at Kiev University, studied the saline soils of Kazakhstan, and in the summer of 1840, he joined the second expedition to Novaya Zemlya to study permafrost [2]. The nobility ruling the country started to appreciate science and technology in the 1860s. Slavery was abolished in 1861 which required new methods to work the land and

© The Author(s), under exclusive license to Springer Nature Switzerland AG 2021
A. E. Hartemink, *Soil Science Americana*
https://doi.org/10.1007/978-3-030-71135-1_2

increased understanding of the soil. Scientists turned their attention to the land, and a young Dmitri Mendeleev carried out fertilizer experiments and conducted soil chemical tests [3, 4]. However, he became disappointed with the results and turned his attention to chemistry, and eventually developed the periodic table of elements. Sergei Vinogradskii investigated soil microbes and found that nitrifying bacteria oxidized inorganic compounds. He discovered lithotrophy in which organisms used minerals for energy production and respiration [5].

The 1870s was a period of bourgeoning scientific and cultural growth, and it bore fruit in books by Ivan Turgenev, Leo Tolstoy, and Fyodor Dostoevsky, and the heavenly music of Nikolai Rimsky-Korsakov, Pyotr Tchaikovsky, and Igor Stravinsky. Despite severe political repression, there was a climate of innovation, creation, land, and labor reform, and the population grew to 85 million. Between 1850 and 1900, Russia's population doubled but remained a rural society. Wheat replaced rye as the principal grain crop and by the 1900s, Russia was the largest producer and exporter of cereals in the world.

The first soil mapping was done in the mid-1800s for the Ministry of Imperial Domain in cooperation with the Commission on Land Surveys. Existing knowledge was used to construct a soil map, but the black soils of the steppe were not shown, and those soils, although striking in appearance, were a mystery [6]. They were assumed to be a geological formation as they had originated from the weathering of black Jurassic shales, marine silts, or lake sediments. Others thought that the black soils had developed in the peats from the northern regions that were deposited by ancient floods [7]. The Austrian botanist Franz Joseph Ruprecht traveled across the Russian steppe in the early 1860s and became fascinated by the black soils. He studied soil samples under the microscope and found tiny structures made of silica that had been discovered in the 1830s. These microscopic silica structures, named phytoliths, were the remains of the roots of the steppe grasses and he concluded that the organic matter and the deep black color in the soils was from decomposed steppe grasses [7]. In 1866, he made a map showing the distribution of the black soils that were then named Chernozems [7].

Much of the agriculture in Russia was on the steppes that extended for thousands of kilometers from the mouth of the Danube River all the way east to the Pacific Ocean. The steppes, named prairies in the USA, were characterized by a continental climate with hot summers and cold winters. The land north of the steppe was under forests but in the south where the climate was drier and hotter, the steppe boundary was unclear. Across large areas of the Russian Empire were Zemstva or elected bodies of local self-government, and in some of these Zemstva, agricultural land was surveyed, maps were made, and the soils were described and classified on the basis of some soil properties. The surveys were paid for by local authorities and generated some awareness on soil distribution and the ways the soils could be managed. The field crew came from the Institute for Physical Sciences, and the Physico-Mathematical Institute in St. Petersburg, and the surveys were practical but lacked a systematic and theoretical approach to soil studies [1, 8].

In 1876, a research program was started for a steppe area that covered about 100 million hectares—roughly the size of six mid-western states in the USA. The impetus for the soil study was the Imperial Free Society of St. Petersburg that had Dmitri Mendeleev among its members [9]. Vasily Dokuchaev, 30 years old and trained as a mineralogist and geologist, was asked to lead the survey, and he had been a crew member in some other surveys [5]. He traveled about 10,000 kilometers through the area and conducted a detailed study in the Nizhny Novgorod region some 400 km west of Moscow on the Volga River. The aim of the study was to develop a system of land taxation and it was here that he wondered: "What, generally, is to be termed as soil?"

The study in Nizhny Novgorod turned Vasily Dokuchaev's geological experience and knowledge to the study of the soil. By observation and inference, he discovered that the same rock may give rise to different soil characteristics, and that the soil at any location was influenced by the climate and vegetation. The Nizhny Novgorod region was divided into geographical morphologic units, and for each unit he collected information on the soils such as the color and thickness, and information on geology, topography, vegetation, climate, and crop yields. He used letters to denote soil horizons, and three types of horizons were distinguished and labeled by the letters A, B, and C [10]. For each horizon, he wrote a definition. The A horizon was the first horizon, and its main character was its uniformity throughout the whole horizon. The color from the A horizon originated from humus. The C horizon was parent material that was not modified by soil-forming processes and contained no organic matter. The B horizon was the second from the surface and transitional from A to C. Sometime later when Podzols were studied, the A horizon became the eluvial horizon from which compounds were lost, and the B horizon became the illuvial horizon in which it accumulated. No distinction was made between the topsoil and subsoil, and Vasily Dokuchaev considered both to be part of the soil.

After the soils and the horizons were described, they were sampled and analyzed for chemical and mechanical properties. Vasily Dokuchaev synthesized the information, and he disentangled the effects of geologic, vegetation,

Anna Dokuchaev (née Sinkler) and Vasily Dokuchaev in the 1880s. Anna was born in 1848, and was a teacher when she met Vasily in 1880. He was invited to teach some classes in mineralogy for students at her college. They married and she became interested in the natural sciences and introduced Earth sciences at the curriculum in her college. Anna accompanied Vasily Dokuchaev for 20 years, mentored some of his students, and Vasily Dokuchaev found that she deserved the title of the first Russian female pedologist

climate, and topographic factors on soil properties [11]. He studied the inter-actions among the different factors and what determined a particular soil in a particular location. For example, he wondered whether Chernozems could develop under forest, whether loess was needed for the development of Chernozems, and whether the climate had influenced the formation of Cher-nozems [6]. It was known that the organic matter was derived from long tall grass roots, but the other factors were not well understood. Based on the different factors, Vasily Dokuchaev developed a soil classification scheme that included *Normal soils,* which were soils that had not been changed by dynamic processes other than soil making processes, and *Extra normal* soils such as poorly drained and re-deposited soils. Some years later, he added a third group named *Transition soils.* It also included the use of Chernozems and Podzols; Chernozem was Russian for black and soil, earth or land (*chorny + zemlya*) and described soils with a thick mineral topsoil high in organic matter. Podzol was Russian for 'under-ash soils' and referred to a leached, light-colored subsoil that appeared liked ash [12].

The exceptional aspects of Vasily Dokuchaev's approach were its aims, systematic methods, comprehensiveness, and his collection, combination, and synthesis of data. He was not the first to describe and study soil profiles as there had been similar efforts in Denmark, France, and Germany [13], but his research led to a unique description and understanding of soils, especially the black soils of the steppe, the Chernozems. In his book *Ruskii Chernozem*, he introduced an integrated and novel way of viewing soils and their genetic origin in the natural world [14]. From then on, pedology became a science and Chernozems became the most iconic soils in the world. That all happened in the earth-shaking year of 1883—Napoleon Bonaparte died, the Brooklyn Bridge opened in New York, and Krakatoa's eruption, which killed over 36,000 people, glowed the sky red.

Vasily Dokuchaev showed that soils were not equivalent to weathered rock, and that soil was the result of a series of processes in addition to weathering, and soil was more than a chemical medium or a mixture of organics and broken-down rock particles. The soil was a natural body and part of the landscape [15]. His work brought the study of soils out of the confusion of the geologic, chemical, and agronomic points of view [16]. Vasily Dokuchaev combined data and insights and formulated discoveries into a theoretical framework that could be tested elsewhere. Climate, parent material, organisms, and relief were factors that formed the soil, and he considered the action of time and humans less important. As he wrote in 1883: "Soils were the products of extremely complex interactions of the effects of local climate, plant and animal organisms, composition and structure of the parent rocks, topography and, finally the age of the country, and therefore it is only natural that the investigator must constantly excurse into various branches of science" [14].

Vasily Dokuchaev's research led to a new view of soils: "Soil is a natural independent body which like any other natural body or organism, has a specific origin, history of development, and external appearance." Soils were natural bodies, and each soil was characterized by a distinct morphology and they developed with the natural landscape, and reflected, at any moment, the combined influence of the living matter and climate acting upon the parent material as conditioned by the relief over a period of time. From that followed the theory that soils were distributed over the earth according to orderly geographic principles [17]. His fundamental principles included *geographicity* which he noted as the dependence of the character of a soil on its geographical position [18]. Vasily Dokuchaev founded pedology and wrote in 1899: "This branch of science, while still very young, is nevertheless full of extraordinarily high scientific interest and significance" [19]. His research influenced archaeology, ecology, geomorphology, paleopedology, and quaternary geology [20].

The need for soil understanding and information that could be used to tax the user or improve the land to obtain higher yields was brought to a new level. While the work was of practical value, Vasily Dokuchaev often stressed that soil distribution had scientific significance, and that there should be a scientific approach to soil studies [11, 14]. Having defined the soil as an entity of its own different from others in the natural world, soil formation was the focus of several new studies. The complexity of soil across the landscape, with depth and changes over time, led to some overgeneralizations and misconceptions. While soil studies in the 1800s were driven by the idea that geology and parent material were the main forming factors, the Russian school showed the effects of climate and introduced a zonal theory of soil distribution. In other words, soil distribution followed a climatic gradient, and climate was regarded as the key soil forming factor. It resulted in the fundamental law of pedology: "The law of the adaptability of soil types of the globe to definite natural (primarily climatic) conditions."

The work of Vasily Dokuchaev did not go unnoticed in Russia and Dmitri Mendeleev wrote to him in 1895 and expressed interest in the new field of pedology, and he thought there was much work to do: "...the soil may be described as a corpse in legends, but for us it is a living provider. I regard it as very useful and timely to begin teaching this science at universities. I have no doubt of the success, I have certain doubt as regards bacteria, but not for one moment regards soil" [19]. Vasily Dokuchaev thought that micro-organisms and animals were not as important for soil as plants [5]. Later on, he changed his view, mainly because of the studies by Sergei Vinogradskii who showed the importance of soil microbiology in the transformation of mineral and organic matter.

Vasily Dokuchaev's approach of relating soils to the climate and vegetation was novel and earthshaking, and it followed the biogeographical zoning of Alexander von Humboldt and studies by Friedrich Fallou, who had coined pedology and viewed it as an independent science separate from geology [21, 22]. Vasily Dokuchaev was aware of Friedrich Fallou's book of the 1860s and cited him once, and he was familiar with the work of Franz Joseph Ruprecht on the origin of the organic matter in Chernozems [7]. There were others who worked on the origin of soil and had developed different theories. For example, Rafail Rizpolozhenski, of Kazan University at the Volga River, viewed soil as the product of only two primary conditions: organisms and rock; organisms changed rocks into soil and thus created an environment in which they could survive [5]. Rafail Rizpolozhenski considered all other factors external conditions. The botanist Peter Kostichev from St. Petersburg disagreed with almost everything that Vasily Dokuchaev wrote and said about

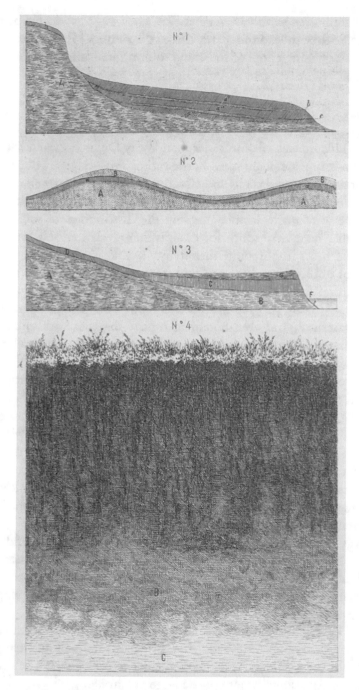

Chernozem soil profiles, with horizon designations. From the French translation *Tchernozéme (terre noire) de la Russie d'Europe*, by Vasily Dokuchaev published in 1879

soils. The animosity started when Peter Kostichev was not selected by the Free Economic Society to lead the survey across the steppes [5]. Over time, Vasily Dokuchaev's genetic school of soil came to dominate, and none of the other Russian soil scientists were as influential. His genetic school of soil lasted from 1875 to early 1900, and he trained some excellent followers. Vasily Dokuchaev's findings and theory spread slowly outside of Russia, and early attempts to introduce his ideas in the USA were unsuccesful [23–25].

Eugene Hilgard was 13 years older than Vasily Dokuchaev and had studied soils in Mississippi and California. In the early 1890s, Eugene Hilgard synthesized the relation between soils and climate and wrote *A report on the relations of soil to climate* for the precursor of the *Bureau of Soils* [26]. Rainfall and temperature were major factors affecting the distribution of soil types. But he noted that data were scarce for establishing a relation between soils and climate, and most importantly, that theory was lacking: "…the remarkable looseness of observations bearing upon it, their usually sporadic and unsystematic nature, and the fact that they are so thinly scattered in various kinds of literature—travels, geography, statistics, reports on geological, meteorological, topographical, and even botanical topics, together with encyclopedias and miscellaneous technical works of reference, renders their collation and elaboration very laborious and thus far rather unsatisfactory" [26]. Eugene Hilgard also arrived at the conclusion that the soil was independently formed, but he did not come up with a systematic framing and synthesis.

Rainfall and temperature in the Russian steppe followed more or less a north–south gradient. Georg Wiegner from ETH in Switzerland was familiar with the Russian work, and in 1928, he summarized how the rainfall decreased traveling east from Europe into Asia [27]. He tried to unravel a soil–climaterelationship across Europe, but the relationships were not clear because many of the soils were relatively young, shallow, eroded, mixed with various geological deposits, or altered by cultivation [28]. There was not a large area within any of the European countries that could be used to explore and unravel soil–climate relations. International cooperation was minimal and even the larger countries in Europe such as Germany, France, and England were too small to cover a wide range of soils to develop theory and principles of soil formation and distribution [29].

In the USA, a vast area existed between the Appalachians in the east and the Rocky Mountains in the west that also had a rainfall and temperature gradient, but the gradients were perpendicular to each other. The discoveries in Russia were possible because of the large extent along which soil could be studied, and the USA was the only country that had similar conditions and vastness where possibly a grand pedological theory could have been tested and

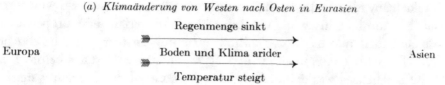

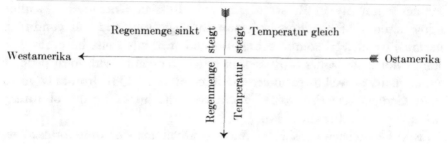

Climate gradients from west to east in Eurasia, and from east to west and north to south in North America. Regenmenge = rainfall; boden = soil; klima = climate; arider = more arid; gleich = the same; steigt = increases, sinkt = decreases. From George Wiegner's North American travel impression published in 1928

developed, but it did not happen. According to Curtis Marbut, the USA was too focused on agricultural development and suffered from societal individualism [30]. He also provided some other reasons why the scientific study of soil was neither started in Europe nor in the eastern part of the USA: "…because the soils of these regions have no striking characteristics. They are tame in their appearance. There is nothing unusual about them, nothing to arrest attention" [29]. Chernozems attracted attention because their black topsoils were regarded as extraordinarily fertile.

Curtis Marbut considered two soil regions in the world that provided a source of inspiration: the black, prairie, or grassland soils, and the red soils of the tropics. Red soils had been observed by Europeans before the black soils, and Curtis Marbut noted: "…they could have furnished the inspiration for the development of soil science if the right kind of people had seen them." English colonizers knew the red soils of the tropics, but they were focused on soil management and how to produce more crops. For Curtis Marbut, the work done in Rothamsted by John Russell and colleagues was characteristic for English soil scientists: they look at the soil from a practical or geological, rather than a scientific standpoint, and it explained to him why they failed to become pioneers in soil science [31]. The black soils of the Russian steppes were the source of inspiration leading to the foundational development of soil science. The red soils of the tropics had been observed earlier but were not studied in detail until well into the twentieth century [32].

Like many nineteenth-century scientists, Vasily Dokuchaev had wide interests, but most of his work was focused on the origin and properties of soils and their role in the agricultural economy. In that sense, he did not differ from his contemporaries or soil researchers who came before him. Vasily Dokuchaev was opposed to the application of the German system of agronomy to study Russian soils: "We should be ashamed of having applied German agronomy in Russia to the true Russian Chernozems, without taking account of conditions of climate, vegetation and soil conditions. Specific Russian agronomic techniques and methods must be evolved for the individual soil zones of Russia, in strict accordance with local pedological, climatic, as well as socio-economic conditions" [33]. Ironically, in the 1930s German scientists voiced alarm over the increasing use of Russian terminology in soil science [34].

Vasily Dokuchaev considered making a soil map of the world to test whether the principles of soil geography would hold across the globe. He sketched a soil map and distinguished Glacial soils, Podzolic soils, Gray transitional soils, and Chernozems [35]. In the south, soil names were mixed with climatic names: Kastanozems in Mexico, Subtropical soils across the Eurasian continent, Krasnozems or red soils, and Zheltozems or yellow soils in China. The Southern Hemisphere had the names of the climatic vegetative zones such as tropical areas, pampas, or coniferous forests. He was not sure if the distribution of soils was mirrored on the Earth, and the amount of data from the Southern Hemisphere was insufficient for good extrapolation but he drafted a zonal soil map which was shown at the *Exposition Universelle* in Paris in 1899 and 1900. Some 50 million visitors visited the *Exposition Universelle*.

For 20 years, Vasily Dokuchaev had help from his wife Anna Dokuchaev (née Sinkler) who was a teacher and became interested in the earth sciences, particularly in mineralogy. She died in 1897, at the age of 51, and Vasily Dokuchaev wrote the obituary in which he acknowledged her assistance for 20 years. She had encouraged many young pedologists, and he gave her much credit for the development and the existence of an independent Russian school of pedology: "…she, more than any other woman, deserves the title of the first Russian woman pedologist" [19]. Anna Dokuchaev died from cancer, but unofficial sources stated that Vasily Dokuchaev had syphilis and infected his wife. Vasily Dokuchaev, feeling guilt over her death and suffering from depression, attempted suicide in 1897. He tried to shoot himself, but the bullet went along the skull and he survived. To cover the scar on the head, he then started wearing a cap [36]. Vasily Dokuchaev died in 1903, 59 years old.

Vladimir Vernadskii, Nikolai Sibirtsev, and Konstantin Glinka were three disciples of Vasily Dokuchaev. Nikolai Sibirtsev and Konstantin Glinka expanded his pedological approach, whereas Vladimir Vernadskii advanced geochemical investigations of the soil. Vasily Dokuchaev had listed the five factors of soil formation in 1899: "Soils are functions of (a) climate (water, temperature, oxygen, atmospheric carbon dioxide, etc.), (b) parent rocks, (c) plants and animals, especially lower forms, (d) topography and elevation and finally (e) soil and partly also the geological age of the country" [19]. He emphasized that it was the interaction that led to a definite condition of soil formation [37].

Nikolai Sibirtsev focused on the role of climate and introduced the zonal, intrazonal, and azonal concepts of soil distribution based on temperature and rainfall. It was similar in approach to what Eugene Hilgard had attempted in 1892 [26]. The emphasis on the role of climate in soil formation was at the expense of the other soil forming factors [37]. Climate was not the sole factor determining the soil, but climatic gradients were a key to understanding soils and its geographic distribution. Nikolai Sibirtsev developed the first soil taxonomic classification, and he had broad interests that included soil ecology and farmers' soil knowledge, but died suddenly in 1900, only 40 years old [28]. His work was outshone by his contemporary Konstantin Glinka, whose books were translated earlier than that of Nikolai Sibirtsev. Konstantin Glinka also traveled widely and lived 27 years longer.

Konstantin Glinka continued the work of Vasily Dokuchaev on the development of the soil-geographic concepts. He contributed to advances in soil mapping and prepared the first soil map of the world at a scale of 1–80 million in 1908. Konstantin Glinka had extensive knowledge on the distribution of soils, and was, since 1906, the Head of Soil Survey of the Resettlement Administration, which had been created for the study of land resources of Siberia and Far East of Russia. He organized more than 100 field expeditions to those remote areas and participated in several of the expeditions. Konstantin Glinka was a member of the early international soil science community and attended the *First Agrogeological Congress* in Budapest in 1909. After the studies by Vasily Dokuchaev and Nikola Sibirtsev, there was a need to move from soil in climatic zones to particular soils [35]. A more refined system of soil geography was required to explain the many different soils for which the soil–climate relationship was unclear. In 1914, Konstantin Glinka published *Die Typen der Bodenbildung* [15]—11 years after Vasily Dokuchaev had died. *Die Typen der Bodenbildung* was rich in observations and data from the forested regions of northern Europe, the steppes of Eastern Europe, and the semi-arid regions of Southern Europe and Asia. *Die Typen*

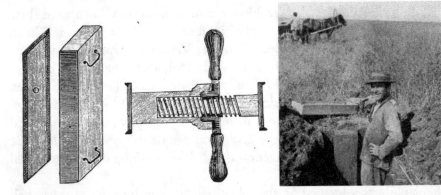

The Rizpolozhenski soil monolith sampler. The metal box on the left was pressed in the soil profile wall and then slowly pushed in by turning the handles. Rafail Rizpolozhenski developed the methods in the Perm Ural region in the late 1800s. For him, soil was the product of two primary conditions, organisms and rock, and monolith sampling allowed detailed studies in the laboratory on soil structure and soil color. He was opposed to much of the work by Vasily Dokuchaev. Another way of taking soil monoliths is illustrated in the right photo: digging the shape of the monolith, putting a box over it and carefully remove the soil behind; then transport the monolith in a horse cart to the laboratory. From: Konstantin Glinka, *Die Typen Der Bodenbildung* published in 1914

der Bodenbildung contained a 1–20 million soil map of Russia showing the north to south zones with typical soils for each zone [15].

Works from Alexander von Middendorff, Vasily Dokuchaev, Nikolai Sibirtsev, Sergei Neustreuv, Boris Polynov, and numerous Russian researchers was reviewed, and Konstantin Glinka synthesized studies by Albrecht Thaer, Friedrich Fallou, Alexander von Humboldt, Carl Sprengel, Ferdinand von Richthofen, Max Blanckenhorn, and Adolf Mayer, and the more recent works of Emil Ramann and Herman Stremme. He reviewed the mineralogical studies by Charles Van Hise, and the work on the soils of tropics by Francis Buchanan, Paul Vageler, and that of Jakob van Bemmelen and Jules Mohr in Indonesia. Konstantin Glinka had read some of the early 1900 reports of the *Bureau of Soils* by Milton Whitney, the 1906 book from George Merrill [38], and he was familiar with the research of Eugene Hilgard on alkali soils that he discussed over several pages. Konstantin Glinka traveled widely throughout Europe which brought him in contact with Alex. de'Sigmond and Peter Treitz from Hungary and Georghe Murgoci from Romania.

Konstantin Glinka listed three types of soil horizons distinguished by letters, and each had sub-horizons that were numbered. The A or eluvial horizon was material that had been removed by chemical or mechanical means. The B or illuvial horizon consisted of material that was chemically or mechanically deposited, and the C horizons were defined as the parent rock. The term horizon was preferred above stratum, which was used in geology and stratigraphy, and the horizon terms helped in distancing pedology from geology. *Die Typen der Bodenbildung* pursued independence from geology and agriculture, and advocated for studying soils just like plants, minerals, or rocks. The soil required an independent way of study: "…it possesses characteristics so peculiarly and exclusively its own that the methods by which it is investigated must necessarily differ from those adapted to petrographic or stratigraphic investigation. We may investigate the soil without interesting ourselves in its relation to man or his agricultural activity" [6].

Die Typen der Bodenbildung was a scholarly synthesis of soil information across the world and in search of the principles of universality: "…if the soil is looked upon merely as an ordinary geological formation, as any sort of a surface deposit, as a crop producer or as a nourishing layer for plants, we have no uniform basis or series of bases on which to establish any systematic scheme of grouping" [6]. The search for universality and principles was needed as he had seen similar soils on different parent materials, different soils on similar parent materials, different soils within one climatic zone, and similar soils in different climatic zones. The combination appeared to be endless but there was universality and further theory was needed that would explain these observations. The most logical way to develop such theory was through a scheme of soil classification that was inclusive of all soils, but rigidly maintained throughout the system. He reviewed the soil classification systems of Albrecht Thaer, Friedrich Fallou, Vasily Dokuchaev, Ferdinand von Richthofen, and Nikolai Sibirtsev, and dissected the problems with these systems: too climatic (Sibirtsev), or too petrographic (Fallou, von Richthofen). He considered Vasily Dokuchaev's system of genetic soil classification the correct base for soil investigations, although Konstantin Glinka realized that all systems of classification were provisional and must change with increasing knowledge [6].

Vasily Dokuchaev had introduced Normal soils, Transitional soils, and Extra normal soils [14]. Nikolai Sibirtsev did not find those classes useful, and he emphasized the climatic factors and introduced the zonal, azonal, and intrazonal concepts of the soil [28]. The zonal concept was objected to by Konstantin Glinka, who in 1914 suggested a system with two classes at the highest level—endodynamic and ectodynamorphic soils, and the classes were based on parent material and soil moisture. Most soils belonged to the ectodynamorphic class, which had six moisture conditions (optimum, average, moderate, insufficient, excessive, and temporarily excessive). In endodynamorphic soils, the parent rock was the key factor in determining the character of the soils, whereas in ectodynamorphic soils the others factors were important [39].

Based on the work of Nikolai Sibirtsev, Konstantin Glinka explained the variation in soil to be determined by the parent rock, organisms, physiographic conditions of the land, and the changes brought about by the soil forming processes and the climate. Soils were considered Mature or immature which represented the time factor of soil formation. In 1928, Curtis Marbut reviewed that approach and thought the classification in the two main groups was poorly balanced and undesirable [29]. However, in 1921, Konstantin Glinka had developed a different soil classification scheme and stressed that all systems were provisory [15]. He discarded the concept of endo- and ectodynamorphic soils, and his new system had five principal soil types (Lateritic, Podzolic, Steppe, Marsh, and Solonets) [40]. To Konstantin Glinka's surprise, Europeans, Russian pedologists, and even Curtis Marbut were still quoting his earlier classification of endo- and ectodynamorphic soils [41]. In 1926, one year before Konstantin Glinka died, he published the third edition of his book *Treatise on Soil Science* that did not get translated into English until 1963 [41].

Konstantin Glinka had chosen to have his 1914 book published in German, as it was commonly used for scientific publications and more widely read than publications in French or English [42]. His book was partly translated by Hermann Stremme who was the director of the Institut für Mineralogie und Geologie at the Technischen Hochschule in Danzig [15]. Konstantin Glinka and Herman Stremme knew that there was a language barrier affecting international discussion and dissemination of the Russian principles of soil science [34]. Herman Stremme found a publishing house in Berlin that was willing to publish the translation, and thankful for his help, Konstantin Glinka wrote in the preface: "My warmest thanks are due to Professor Dr H Stremme who undertook the laborious work, in single minded devotion to science and without profit to himself, of improving my imperfect German and who has extended to me at all times the most valuable assistance and the wisest counsel."

Die Typen der Bodenbildung was published in 1914 and 3 years later, Curtis Marbut translated the book into English. He only translated the first 218 pages of the total 358 pages and skipped the soil overview of the European, Asiatic, and mountain regions of Russia, nor did the translated version include the graphs and photographs of soil profiles and landscapes. Konstantin Glinka's *Die Typen der Bodenbildung* transformed Curtis Marbut's thinking about soils, and it eventually turned the geological emphasis of the American soil survey into a pedological approach. Only a few mimeographed

Farbenerklärung:

Tundrazone		Braune Böden mit Bodenkomplexen (Ssolonetz, Ssolontschak usw.)	
Podsolige Böden auf dem weichen Muttergestein		Braune Böden ohne Bodenkomplexe (turkestanischer Typus)	
Podsolige Böden auf vorwiegend harten Gesteinen		Grauerden der Ebenen	
Nördl. Teil der podsolig. Zone (Abschwächung des Podsolprozesses, Übergang z. waldig. Tundra)		Grauerden der Vorgebirge	
Podsolige Böden mit großer Menge der überwiegend befeuchteten Böden		Übergangsgebiet von braunen Bodenkomplexen zu den Grauerden	
Podsolige Böden mit großer Menge von Moorböden (»Marj«)		Sand	
Tschernosem, degradierter Tschernosem, graue Waldböden und sekundäre podsolige Böden		Vertikale Bodenzonen von Turkestan (kastanienfarbige, Tschernosem, Bergwiesenböden)	
Kastanienfarbige Böden		Ssolonetz- und Ssolontschakböden in der Podsolzone	

Soil map from Russia, showing a zonal soil pattern with tundra zone in the north (pink), several podzols below (greyish map units), followed by Chernozems (dark brown), Kastonozems (light brown), and Grauerde (yellow). From: Konstantin Glinka, *Die Typen Der Bodenbildung* published in 1914

copies of the translation by Curtis Marbut were available in the 1910s and early 1920s. The translated version was sought after by those working in the American soil survey. When Charles Kellogg started as student assistant in soil survey in the late 1920s in Michigan, he tried to obtain one of the mimeographed copies, but to no avail. Milton Whitney, the chief of the *Bureau of Soils*, likely banned the distribution of the translation, as he was opposed to the Russian work and considered Russian science to be 'backward' [42]. The English translation was officially published 1928, after Milton Whitney's death.

Although Curtis Marbut's conversion to the Russian school started somewhere in 1914, he was careful not to acknowledge that influence [43]. But in

1923, he openly admitted: "Through the opportunity offered to the Russian investigators by the unified political control of an immense area of country with a wide range of geologic, climatic, and topographic conditions they worked out many important laws of soil development and soil distribution and showed that these were intimately related to climatic and vegetational conditions. American workers have widened the field thus opened by showing the intimate relation of soil type character to topographic form and to stage in development" [44].

Besides field studies and theoretical development, there were advances in soil chemical analysis that contributed to the understanding of soil formation. Konstantin Gedroiz was one of the first to classify soils on the basis of the absorbing complexes and cations, which was a new approach in pedology [17, 45]. He used the climate classification of soils, and the principal climatic types, Podzols, Laterites, and Chernozems, were described in terms of their absorbing complexes and cations. The system of classification worked well for mature soils in which pedogenic processes had proceeded to such an extent that the profile characteristics reflected a climatic region. It worked less well in immature soils such as the alluvial soils and soils of regions where the surface was eroded. Konstantin Gedroiz had worked on base exchange and absorption in soils since 1912, but his work was not well-known outside Russia. The work was pioneering, and some of his papers were translated into English by Selman Waksman, who had emigrated from Ukraine to the USA in 1910. Copies of these translations were distributed by the United States Department of Agriculture.

After the studies by Vasily Dokuchaev and Konstantin Glinka, the integrated approach to soil studies and the factors of soil formation were well-established by the 1920s. For the *First International Congress of Soil Science* in 1927, Sergei Neustreuv wrote a bulletin on the genesis of soils that contained a discussion on the five factors or conditions of soil formation: "…in order to attain an understanding of the soil, it is indispensable to give one's consideration to the so called conditions of soil formation" [46]. He spelled it out in more detail in his book, *Elements of the geography of soils,* posthumously published in 1930, discussing the factors of soil formation: the role of climate, the effect of relief, the importance of geological conditions, the role of organisms, age of the soil, and time as a factor of soil formation [47]. This was followed by the types of soil formation that included the soil formation processes, characteristics of soil types, and the effects of cultivation on the soil.

Since its inception in the late 1800s, soil survey and research in the USA had progressed slowly. The geologic view on the origin of soils dominated,

Halbwüstenboden (Aridisol) from Turkestan in Kazakhstan, probably one of the first soil profiles that was pictured with a measuring tape (top photo). Sampling soils of the steppe in the Transbaikalien far east of Russia. From: Konstantin Glinka, *Die Typen Der Bodenbildung* published in 1914

and most soil studies had an agrarian and practical view, and according to Curtis Marbut: "…the greater part of the energy spent in the investigation of soils has been directed to an attempt to find a means by which the soil can be made to produce another bushel of corn or another bushel of wheat rather than to a determination of its characteristics. Our soil investigation has been empirical therefore rather fundamental." He noted that little work was done on soil processes or soil characteristics, and the fundamental character was ignored and shortcuts were used for soil studies [29]. Yet, there had been several opportunities to learn about the Russian soil principles long before Curtis Marbut's translation of *Die Typen der Bodenbildung*.

The 1901 report of the Agricultural Experiment Stations published by the Department of Agriculture contained a chapter named *Russian Soil Investigations* which was devoted to the studies by Vasily Dokuchaev, Nikolai Sibirtsev, and Rafail Rizpolozhenskii [48]. The zonal concept of soil classification was reviewed, and it was stated that the classification system of Nikolai Sibirtsev "…differed fundamentally from the petrographic and physico-chemical classifications commonly followed by investigators who have dealt with soils which have been profoundly modified under culture, rather than with those in a largely virgin conditions, as in Russia and in the western United States. As a comprehensive, systematic, and thorough piece of work which has been fruitful of remarkable results, these Russian soil investigations are worthy of the careful study of all interested in the subject. They should be of especial interest to American investigators, since the soil conditions of Russia are to a considerable extent duplicated on this continent, a fact which has been recognized by Hilgard and others, particularly in the study of the virgin soils of America" [48]. The 1901 report is probably the earliest account of introducing Russian work in the USA, and it came with a clear recommendation: "…especial interest to American investigators." It fell on barren ground, but America was no exception.

In the UK, agricultural chemistry dominated the way soils were studied ever since the work by Humphrey Davy in the early 1800s which over time morphed into agronomic studies by John Lawes and Joseph Gilbert, followed by the soil fertility and microbiological studies of Daniel Hall and John Russell. Soil research was directed toward the study of soil as a medium for plant growth based on the fertilizer experiments at Rothamsted [4]. Pedology had a late start, and early pioneers were Gammie Ogg in Scotland and Gilbert Robinson in Wales. Gilbert Robinson's book *Soils—Their origin, constitution and classification. An introduction to pedology* from 1932 was the first book in the UK with a sole pedology focus, and when Edward Crowther reviewed it for *Nature,* he noted: "It is free from those elementary expositions of plant physiology, microbiology, and biochemistry which often form a large part of soil textbooks intended primarily for agricultural students. The treatment is philosophical rather than technical, and it may therefore be recommended with every confidence to workers in many other branches of natural and applied science."

Instruments for pedology, and sampling a soil monolith, from Albert Demolon *La Dynamique du Sol* published in 1932. For Albert Demolon, the pedologist's tool were a hammer, spatula, tape measure, and bottle of hydrochloric acid, and he had adopted the sampling technique of a monolith from Konstantin Glinka's book

The UK was not large enough to investigate a soil–climate connection, and Edward Crowther viewed this in 1930 as follows: "…it has been apparent that the British and other workers familiar with small highly cultivated areas of irregular topography and varied geology attach much less importance to the climatic factors than do the Russians whose experience is largely of vast plains of fairly uniform loess material extending over well defined climatic zones" [49]. He became convinced that within the British Isles, a soil map would show the influence of climate and that the mapping approach developed in Russia would allow for modifications of the climatic soil type by variations in geology and topography.

In Germany, there were well-established schools in chemistry and geology, and soils were investigated in both schools. The 1893 textbook by Emil Ramann had a geologic focus although he had studied the relationship between the soil and forest vegetation [50]. Eilhard Mitscherlich used soil texture to classify soils [51]. After the *First International Congress of Soil Science* in 1927, soil science books in German such as by Friedrich Schucht [52], and books by Fritz Scheffer and Paul Schachtschabel emphasized the geologic origin of the soil but made little to no reference to Russian studies [53]. A German bibliography from 1931 had over a thousand pages and listed some of the key studies that originated from Russia [54]. Herman Stremme had helped translate the Konstantin book into German, but the Russian genetic principles of soils were not adopted in Germany. German soil science became also nationalistic in the 1930s banning influence from elsewhere, Russia in particular [34].

The Russian soil principles were adopted in Romania and Hungary but not in other European countries [8]. The geologic origin of soils remained dominant in some countries and soil communities, and some of this had to do with language, institutional barriers, the novelty of the Russian concept, differences in research focus, and the paradox that conventional thinking slowed the emerging field of the new soil science discipline. Language barriers hampered the exchange of findings even though in the early years of soil science, most scientists spoke more than one language. For Example, Herman Stremme spoke Russian, Curtis Marbut had taught himself German, Eugene Hilgard was fluent in German and English, as was David Hissink [55]. Studies in Russian, Hungarian, Japanese, Spanish, Danish, Swedish, Dutch, or French were not as widespread as publications in English or German and sometimes received minimal attention. As the Hungarian soil scientist Alex. de'Sigmond wrote in 1928: "The English literature of soil science is so rich that it may seem presumptuous on the part of a foreigner to suggest that his own work can contribute something more to the work of English or English reading soil experts" [56]. The book *La Dynamique du Sol* by Albert Demolon from 1932 [57], was not translated and only briefly reviewed in the journal *Soil Science* with the comment: "The book is especially recommended to graduate students in soils who are in need of improving their reading knowledge of French. Some refreshingly new points of view are presented for their consideration." Dissemination of soil knowledge was hampered by the slow rate of translation, and for example, Vasily Dokuchaev's monograph on the Russian Chernozem of 1883 was not translated in English until 1967 [58]. The French version *Tchernozéme (terre noire) de la Russie d'Europe* had been published in 1879 as it was a custom of the Free Economic Society to publish both in French and Russia.

Developments in soil science were affected by politics and personal opinions. Between the Russian Revolution of 1917 and 1933, there were no diplomatic relationships between the USA and Russia, and no exchange or visits were possible. Milton Whitney who directed the soil survey in the USA had never been to Europe and developed a somewhat isolationist view of how soils should be studied. He saw no reasons to adopt any of the Russian findings and disregarded the works of Eugene Hilgard. The isolationist view trickled down in some educational materials. Generations of students grew up with the *Nature and Properties of Soils* by Thomas Lyon and Harry Buckman from Cornell that was first published in 1915, and they used a geological classification of soils, until the fifth edition in 1952 [59]. Even in the 1969 edition, there was no acknowledgment of genetic soil science, and that edition included the zonal soils as well as the Great Soil Groups [60]. Other aspects that slowed down the acceptance of the Russian principles was that besides the USA, few countries had a large climatic spread. In other words, for relatively smaller countries such as the UK, Germany, and even France, there was no extensive climatic gradient across their landmass, and the Russian approach was perhaps seen as irrelevant. So different approaches for mapping soils were developed based on just one or two soil formation factors like, for example, the effect of drainage and water (hydromorphics) or terrain (physiography).

Some international collaboration in soil science was started in 1909, but the First World War stopped progress in scientific exchange. The world slowly recovered from the Great War and Europe was divided. International cooperation started again in the early 1920s and soil scientists from Europe, the USA, and Russia founded the *International Society of Soil Science* in 1924. The first activity was the organization of the *First International Congress of Soil Science* in 1927. At the meeting, the Russians, and in particular Konstantin Glinka, were welcomed as celebrities by those who admired their work, among them the most influential soil scientist in the USA being Curtis Marbut. By that time, the ideas of Vasily Dokuchaev had been altered to an emphasis on the climatic factor in soil formation. Some years later, the USA adopted a zonal system of soil classification in which soil distribution and climate were strongly linked. From then on, soil climatic relationships and the Russian genetic principles of soils were acknowledged in the USA, like for example in the textbook *Soil Management* by Firman Bear in 1927 [61]. In *Soil Science—Its Principles and Practice,* Wilbert Weir listed the following factors in the development of the soil from accumulated materials by rock weathering: climatic, edaphic, topographic, and biotic factors included plants and animals [62]. Climate was not regarded as a soil forming agent but heat and precipitation were, and he stressed that climate forces were dominant over extensive areas. He noted that the soil forming factors did not act separately but conjointly, and their action gave rise to four soil processes: podzolization, laterization, alkalization, and carbonation; he preferred carbonation over calcification as elements other than calcium were involved. Podzolization and laterization were dominant in humid regions, whereas alkalization and carbonation in the drier regions.

The *First International Congress of Soil Science* took place 10 years after the Russian Revolution and the rise of Communism. Russians were not allowed to cite foreign publications, and scientific exchange was limited. Russian soil science became isolated, doctrines were monopolized, and repressive actions were undertaken on those who opposed the regime [63]. The oldest soil journal, *Pochvovedenie,* stopped publication in 1917. Soil science stagnated in Russia [8]. Americans, however, were on an upward trajectory, and the *First International Congress of Soil Science* boosted soil research and soil survey, but severe droughts and economic depression halted the momentum. Then came another World War, more deadly, destructive, and widespread than two decades earlier. The USA came out of the Second World War with confidence and an increased sense of responsibility. American soil science drifted away from the Russian influence, which was facilitated by the widening political divergence between the two countries. It took until the 1950s for the Americans to realize they had ignored Eugene Hilgard's important and pioneering work, first in Mississippi, then in the Cotton Belt, and finally in California [64]. There were other American soil pioneers whose ideas had also been ignored, including George Coffey [65], and Charles Vanderford, who had noted that in the late 1800s: "The greatest difficulty encountered in studying the soils from an agricultural standpoint was the very limited number of intelligent observers" [66]. Lack of knowledge, wars, language and institutional barriers, and some form of traditionalism slowed progress in soil science worldwide in the early 1900s while it was rooting as a scientific discipline. Personal contacts, a few individuals, and events changed the course of soil science.

Russian work in the USA was introduced in 1901 in an Agricultural Bulletin of the Department of Agriculture [48], but it had been exhibited at the World's Columbian Exposition held in Chicago in 1893. The exposition was held for the first time in Chicago some 400 years after Columbus had sighted North America. The Exposition was a social and cultural event and reflected architecture, the arts, Chicago's self-image, and industrial optimism. The focus was on science, industry, and commerce, and the Agricultural building covered 6 ha of floor space. The center of the 280-ha fair was a large water pool that represented the long voyage of Columbus. The World's Columbian Exposition in Chicago attracted 27 million people—half the inhabitants of the USA at that time. A bewildering number of neo-classical and Renaissance buildings were spread across the shores of Lake Michigan. They were intended to be temporary and their facades were made of a mixture of plaster, cement, and jute fiber which was painted white. It was called 'The White City' and was impressive, but air pollution darkened the facades and plans were made to refinish the exteriors in marble or some other material. These plans were abandoned in July 1894, after the exposition had suffered four massive fires within 1 year, and 17 people were killed. Only the Museum of Science and Industry remains as a reminder of the 1893 World's Columbian Exposition.

The Russians had a pavilion in the Manufactures and Liberal Arts Buildings. It was a large carved and polished log cabin. The building was made in Russia and shipped to Chicago where it was assembled. The one-story building stood over twelve meters high, with a low sloping roof, and had a combination of Byzantine style with Slavic detail that the designers hoped

The agricultural building at the 1893 World Exposition in Chicago. Facades were not made of stone, but finished of a mixture of plaster, cement, and jute fiber that was painted white. Most of it burned down in 1894. Interior of the agricultural building at the World's Columbian Exposition in 1893 where the Russian and USA soils exhibit was located would become the property of an American museum [67]. For the exhibition, the Russian government had appointed a commission that included Leo Tolstoy [68]. There were exhibits of Russian products and arts: "It may surprise many to learn that she has a flourishing beet-sugar industry, for its product is displayed in great cones of snow-white sweetness. The importance of candles in Russia is indicated by a large exhibit. There are also displays of tobacco, wool, silk, astrakhan, flax, tow, cigars, and dairy machinery." There was grain in great bowls, and three court costumes that had been worn by the Empress.

Russia had a severe drought in 1891 and suffered a famine in 1892 that began along the Volga River, then spread as far as the Urals and the Black Sea. Most of the fields were cultivated with wooden plows and the grain was harvested with sickles. There was merciless taxation, exports of grain, and a rapidly growing population, and: "...the winter of 1890–1891 was one of little snow, and the unprotected frozen soil drank less than the usual moisture from that source. There was, indeed, an accumulation of all the plagues" [69, 70]. It affected 15 million people and caused 500,000 deaths. The USA sent shiploads of flour and cornmeal that were received with flying flags and the stirring music of 'God save the Tsar' [69]. The public anger at the Tsarist government's handling of the disaster led to the rise of Marxism and the 1917 revolution. In the end, God did not save the Tsar.

At the 1893 World's Columbian Exposition, there was an exhibit on soils from the USA and Russia. Eugene Hilgard had prepared the American exhibition on soils that consisted of 256 monoliths. Each monolith was about 90 cm long and 10 cm wide, and he had asked for chemical and physical analyses but only a few states provided such data [71]. The soil collection from Eugene Hilgard was from 24 states, and duplicate samples were sent to the Department of Agriculture [72]. Milton Whitney, who became the chief of the *Bureau of Soils* some years later, had prepared an exhibit that contained a draft map depicting the area and distribution of the soils in Maryland [73, 74]. The Division of Chemistry of the Department of Agriculture showcased a complete agricultural laboratory in which demonstrations of food adulteration, the saccharine value of sugar plants, and soil analyses were shown. A 17-year-old Merritt Miller from a farm in Grove City, Ohio, visited the fair; he would become a leading soil scientist in Missouri.

The soil exhibit from Russia was prepared by Vasily Dokuchaev and included soil samples, seven monoliths, maps, and essays [75]. A similar exhibition had been presented at the World Exposition in Paris in 1889 that was prepared by Vasily Dokuchaev with the help of Vladimir Vernadskii. In Paris they had presented several maps, a collection of soil monoliths, a large

Chernozem block from the Nizhny Novgorod region, and a schematic soil map of the Northern Hemisphere [35]. Explanatory texts were in French and Russian. There was a collection of publications by Vasily Dokuchaev and his students, including the 1883 Chernozem monograph in Russian, and articles published in French in 1879 [58]. The organizers placed the soil exhibit in a location where it attracted little attention, but the Russian exhibit won the gold medal, and there was some interest among French and Belgian scientists [58]. The Russian monoliths were left at the Sorbonne University but were destroyed during the student riots of 1968.

The 1893 Russian exhibition in Chicago included seven monoliths of major soil types that were arranged on a geographical basis. The soil monoliths were large glass tubes containing soil profiles with different horizons [58]. The exhibition had 139 soil and subsoil samples, 12 maps, tables, and 74 papers with special reports and catalogs printed in English. The report *The Russian steppes and study of the soil in Russia, its past and present* was prepared by Vasily Dokuchaev and translated into English by the American consul in St. Petersburg. Years later, Roy Simonson found this report in the National Agricultural Library in Beltsville where it had come from the Library of the University of Wisconsin in Madison and likely, Franklin King had deposited it there. In the report, there was a reference to the factors of soil formation: "It is completely demonstrated that normal soils are the result of very complex interactions of the following soil formers: ground, climate, vegetative and animal organisms, the age of the country, and the contour of the locality." Ground (Russian *grunt*) is equivalent to parent material or parent rock [71]. The report had details on the factors: "It may now be considered to be clearly established that all soils, clothing the earth's surface with a more or less thin film of half a foot to six feet in thickness, must be divided into normal, lying on the spot where they were formed and appearing as far as possible with their primitive properties: and abnormal, which are either excessively washed or even moved bodily to other situations. It is completely demonstrated that the first of these, the normal soils, are the result of the very complex interaction of the following soil-formers: ground, climate, vegetable and animal organisms, the age of the country and of the contour of the locality. In places where these variables are the same, the soils are identical; where they are different, the results of their activity cannot be the same" [76].

It seemed that neither the 1893 Russian report nor the exhibit attracted much attention [71], and the Russian agricultural display was overshadowed by the American exhibit. The soil exhibit did little to facilitate the dissemination of Russian soil science in the USA [58]. The *Chicago Tribune* described the Russian exhibit and praised the furs, lacquer boxes, and:

The Russians pavilion in the Manufactures and Liberal Arts Buildings next to Harvard University at the World Exposition in Chicago in 1893

"....embroideries, weapons, articles of dress and household ornamentation." The agricultural exhibit was noted but the soil exhibit was not mentioned [42]. Charles Vanderford's study of *The Soils of Tennessee* from 1897 contained photos on the collection of soil profiles, and likely he got the ideas from the Russian soil exhibit at Chicago [58, 66, 77]. The Russian influence on Charles Vanderford's work was a single case and received little attention [58].

In 1904, the World's Fair was held in St. Louis, and it was named the Louisiana Purchase Exposition. Curtis Marbut prepared a large map of Missouri soils made of plaster. The Missouri soil map exhibit won a gold medal [78]. The 1904 World's Fair also included a 'Human Zoo' where people from all over the world were 'exhibited' and competed in martial sports. The self-taught geologist William McGee was responsible for the anthropological exhibit at the World's Fair in St. Louis. Some years later, he was hired by Milton Whitney at the *Bureau of Soils* [79]. Almost 20 million people visited the fair. A passenger train with visitors from Kansas crashed at full speed into a freight train on its way to Columbia, Missouri. The train cars telescoped onto each other and killed 29 people.

Department A. Group 15. Class 83.

255. DOKUCHAIEV, B., Professor of the Imperial University of St. Petersburg, and SIBIRTSEV, N.

1. Samples of the soil of different parts of Russia.
2. Soil charts.
3. Sections of soil.
4. Pamphlets.

• Analysis according to Professor Dokuchaiev's method have been conducted since 1876. The black soil zone of Russia and the land bordering on it, as also different localities in Northern and Southern Russia have been studied. Detailed examinations of the soils have been made in the gov.'s of Nizhni-Novgorod and Poltava, and to a certain extent also in the gov.'s of Voronezh, Saratov, Kharkov, Ekaterinoslav, Smolensk, St. Petersburg, Vladimir etc. The following is a statement of the nature of work accomplished: genesis or origin of the soil; natural standards of the soil and its classification; relations of the soils to the geology ot the country, to the vegetable and animal kingdoms, to the formation of the surface, to the climate and to the water; chemical and physical analysis of the soils; the soils and agriculture; soils, forestry and water exploitation; valuation and taxation of grounds, according to the natural resources of the soil.

From the Catalogue of the Russian section, World's Columbian exposition in 1893 in Chicago. Neither the report nor the exhibit attracted much attention, and the soil exhibit did little to facilitate the dissemination of Russian soil science in the USA

The World's Fair was held for the second time in Chicago in 1933, and the fair was held at the height of the Great Depression and President Herbert Hoover maintained that "Any lack of confidence in the economic future of the basic strength of business in the United States is foolish" [80]. But Chicago had over 600,000 unemployed people, and homeless people built 'Hoovervilles' in empty lots and parks. The roaring 1920s were long over and Chicago suffered from gang warfare between mobs dealing in illegal liquor, gambling, and prostitution [80]. The city organized the World's Fair without

The agricultural building at the 1933 World's Exposition in Chicago, that had the motto *Science Finds, Industry Applies, Man Adapts.* Charles Kellogg and Roy Simonson drove the thousand kilometers from Fargo to Chicago to visit the exposition in August 1933

taxpayer money and collected money through bonds, shares, and contributions. The fair created revenue and jobs, and it became such a success that after President Franklin Roosevelt was elected, he encouraged the city to continue the fair for another year. By that time, prohibition, enacted in 1920, had ended, and the Mayor of Chicago arranged for 2,000 barrels of beer to celebrate the fair. The city came back alive. The Savoy Ballroom and Regal Theater in Bronzeville hosted the Mozarts of jazz: Louis Armstrong, Duke Ellington, and Count Basie—they were open 7 days per week [80], and who could resist the shuffle of Duke's 'Stormy Weather.'

The 1933 fair was named *A Century of Progress—International Exposition* and carried a theme of technological innovation. The motto was *Science Finds, Industry Applies, Man Adapts.* The fair aimed to offer a glimpse of happier futures driven by innovation in science and technology, but there was no Russian pavilion. There was light in the darkness, and tens of millions of people visited the fair, including President Roosevelt who was in office since March 1933 [80]. The thematic concentration on the services of society and humanity showcased the President's New Deal plan. It created the impression that the *Agricultural Adjustment Act* would restore the land of plenty [75]. Charles Kellogg and Roy Simonson visited the World's Fair in Chicago in August 1933.

References

1. Vilenskii DG. Saline and alkali soils of the Union of Socialist Soviet Republics. Pedology. 1930;4:32–86.
2. Tammiksaar E, Stone I. Alexander von Middendorff and his expedition to Siberia (1842–1845). Polar Rec. 2007;43:193–216.
3. Dmitriev IS, Sarkisov PD, Moiseev II. Dmitry Ivanovich Mendeleev: scientist, citizen and personality. Rend Fis Acc Lincei. 2010;21:111–30.
4. Russell EJ. Obituary. Prof. K.D. Glinka. Nature. 1927;120:887–8.
5. Ackert L. Sergei Vinogradskii and the cycle of life. From the thermodynamics of life to ecological microbiology, 1850–1950. Dordrecht: Springer; 2013.
6. Glinka KD. The great soil groups of the world and their development (translated from German by C.F. Marbut in 1917). Ann Arbor, Michigan: Mimeographed and Printed by Edward Brothers; 1928.
7. Fedotova AA. The origins of the Russian chernozem soil (Black Earth): Franz Joseph Ruprecht's 'Geo-botanical researches into the chernozem' of 1866. Environ Hist. 2010;16:271–93.
8. Joffe JS, editor. Russian contribution to soil science. Washington: American Association for the Advancement of Science; 1952. Christman RC, editor. Soviet science; a symposium presented on December 27, 1951, at the Philadelphia meeting of the American Association for the Advancement of Science.
9. Tandarich JP, Sprecher SW. The intellectual background for the factors of soil formation. Factors of soil formation: a fiftieth anniversary retrospective. Madison, WI: Soil Science Society of America; 1994. p. 1–13.
10. Nikiforoff CC. History of A, B and C. Bull Am Soil Surv Assoc. 1931;12:67–70.
11. Vil'yams VR. V.V. Dokuchaev's role in the development of soil science. Selected works of V.V. Dokuchaev. Washington DC: Israel Program for Scientific Translations. U.S. Department of Agriculture; 1967.
12. Bridges EM, Batjes NH, Nachtergaele FO, editors. World reference base for soil resources—Atlas. Leuven: Acco; 1998.
13. Feller C, Blanchart E, Yaalon DH. Some major scientists (Palissy, Buffon, Thaer, Darwin and Muller) have described soil profiles and developed soil survey techniques before 1883. In: Warkentin BP, editor. Footprints in the soils. People and ideas in soil history. Amsterdam: Elsevier; 2006. p. 85–106.

14. Dokuchaev VV. Russian Chernozem (Ruskii Chenozem). Selected works of V.V. Dokuchaev. Volume I (translated in 1967). Washington DC: Israel Program for Scientific Translations. U.S. Department of Agriculture; 1883.
15. Glinka K. Die Typen der Bodenbildung - Ihre Klassifikation und Geographsiche Verbreitung. Berlin: Verlag von Gebrüder Borntraeger; 1914.
16. Marbut CF. Introduction. In: Joffe JS, editor. Pedology. New Brunswick: Rutgers University Press; 1936. p. vii–xiii.
17. Kellogg CE. Russian contributions to soil science. Land Policy Rev. 1946;9:9–14.
18. Glinka KD. Dokuchaeiv's ideas in the development of pedology and cognate sciences. Vol Part I. Proceedings. Washington D.C.: The American Organizing Committee of the First International Congress of Soil Science; 1928.
19. Dokuchaev VV. The place and significance of modern pedology in science and life. (Translated from Russia by the Israel Program for Scientific Translations). Izbrannye Sochineniya. 1949;III:330–8.
20. Johnson DL, Schaetzl RJ. Differing views of soil and pedogenesis by two masters: Darwin and Dokuchaev. Geoderma. 2015;237–238:176–89.
21. Asio VB. Comments on "Historical development of soil and weathering profile concepts from Europe to the United States of America". Soil Sci Soc Am J. 2005;69(2):571–2.
22. Fallou FA. Pedologie oder Allgemeine und Besondere Bodenkunde. Dresden: Schönfeld Buchhandlung; 1862.
23. Joffe JS. Pedology (with a foreword by C.F. Marbut). New Brunswick: Rutgers University Press; 1936.
24. Tandarich JP, Darmody RG, Follmer LR, Johnson DL. Historical development of soil and weathering profile concepts from Europe to the United States of America. Soil Sci Soc Am J. 2002;66(4):1407.
25. Paton TR, Humphreys GS. A critical evaluation of the zonalistic foundations of soil science in the United States. Part I: The beginning of soil classification. Geoderma. 2007;139(3–4):257–67.
26. Hilgard EW. A report on the relations of soil to climate. Washington; 1892.
27. Wiegner G. Reiseeindrücke aus Nordamerika. In: Waksman S, Deemer R, editors. Proceedings and Paper First International Congress of Soil Science. Transcontinental Exursion and Impressions of the Congress and of America. Washington D.C.: The American Organizing Committee of the First International Congress of Soil Science; 1928. p. 89–120.
28. Sibirtsev NM. Pochvovedenie. St. Petersburg: Y.N. Skorokhodov; 1900.
29. Marbut CF. Soils: their genesis and classification (Lecture notes from 1928). Madison, Wi: Soil Science Society of America; 1951.
30. Marbut CF. Translator's preface. In: Glinka KD, editor. The great soil groups of the world and their development. Ann Arbor, Michigan: Edward Brothers; 1928.

31. Marbut CF. The excursion. In: Waksman S, Deemer R, editors. Proceedings and Paper First International Congress of Soil Science. Transcontinental Exursion and Impressions of the Congress and of America. Washington D.C.: The American Organizing Committee of the First International Congress of Soil Science; 1928. p. 40–88.

32. Hartemink AE. Soil science in tropical and temperate regions—some differences and similarities. Adv Agron. 2002;77:269–92.

33. Anon. Publisher's preface. Selected works of V.V. Dokuchaev. Washington DC: Israel Program for Scientific Translations. U.S. Department of Agriculture; 1967.

34. Arend J. Russian science in translation. How *Pochvovedenie* ws brought to the west c. 1875–1945. Kritika: Explor Rus Eurasian Hist. 2017;18:683–708.

35. Hartemink AE, Krasilnikov P, Bockheim JG. Soil maps of the world. Geoderma. 2013;207:256–67.

36. Zonn SV. Little-known facts of biography of V.V. Dokuchaev (in Russian). Pochvovedenie. 1991;6:106–12.

37. Whitson AR, Geib WJ, Dunnewald TJ, Truog E, Lounsbury C. Soil survey of Iowa County, Wisconsin. Madison, Wisconsin: Wisconsin Geological and Natural History Survey; 1914.

38. Merrill GP. A treatise on rocks, rock-weathering and soils. New York: The MacMillan Company; 1906.

39. Ogg WG. The contributions of Glinka and the Russian school to the study of soils. Scott Geogr Mag. 1928;44:100–6.

40. Wilde SA. Glinka's later ideas on soil classification. Soil Sci. 1949;67(5):411–4.

41. Glinka K. Treatise on soil science (Pochvovedenie), fourth posthumous edition. Translated from Russian in 1963 by the Israel Program for Scientific Translations. Washington DC: National Science Foundation; 1931.

42. Moon D. The American Steppes. The unexpected Russian Roots of Great Plains Agriculture, 1870s–1930s. Cambridge: Cambridge University Press; 2020.

43. Marbut-Moomaw L. Curtis Fletcher Marbut. Life and works of C.F. Marbut. Madison: Soil Science Society of America; 1936. p. 11–35.

44. Shantz HL, Marbut CF. The vegetation and soils of Africa. New York: National Research Council and the American Geographical Society; 1923.

45. Moon D. The international dissemination of Russian genetic soil science (pochvovedenie), 1870s–1914. J Reg Hist. 2018;2:75–91.

46. Keen BA, Hall D. The physicist in agriculture with special reference to soil problems. A lecture delivered in the Rooms of the Chemical Society, London, on 25th November 1925. 1925.

47. Robinson GW. Mother earth. Being letters on soil addressed to Sir R. George Stapledon C.B.E., M.A., F.R.S. London: Thomas Murby & Co.; 1937.

48. U.S. Department of Agriculture. Experiment Station Record. Volume XII. 1900–1901. Washington: Government Printing Office; 1901.

49. Crowther EM. The relationship of climatic and geological factors to the composition of soil clay and the distribution of soil types. Proc R Soc B. 1930;107:1–30.

50. Ramann E. Bodenkunde (Dritte, umgearbeitete und verbesserte Auflage). Berlin: Verlag von Julius Springer; 1911.

51. Mitscherlich EA. Bodenkunde für Land - und Forstwirte. Berlin: Verlagsbuchhandlung Paul Parey; 1905.

52. Schucht F. Grundzüge der Bodenkunde. Berlin: Verlagsbuchhandlung Paul Parey; 1930.

53. Scheffer F, Schachtschabel P. Lehrbuch der Agrikulturchemie und Bodenkunde. Stuttgart: Ferdinand Enke Verlag; 1956.

54. Niklas H, Czibulka F, Hock A. Literautirsammlung aus dem Gesamtgebiet der Agrikulturchemie. I. Bodenkunde. München: Verlag des Agrikultuchemischen Instituts Weihenstephan der Technische Hochschule; 1931.

55. Hartemink AE, editor. D.J Hissink (1874–1956). Wageningen: NBV; 2010. Bouma J, Hartemink AE, Jellema HWF, Grinsven JJMv, Verbauwen EC, editors. Profiel van de Nederlandse bodemkunde. 75 jaar Nederlandse Bodemkundige Vereniging (1935–2010).

56. de'Sigmond AAJ. The principles of soil science. London: Thomas Murby & Co.; 1938.

57. Demolon A. La Dynamique du Sol. Paris: Dunod; 1932.

58. Moon D. The Russian Academy of Sciences Expeditions to the Steppes in the Late Eighteenth Century. Slavon E Eur Rev. 2010;88(1–2):204–+.

59. Lyon TL, Buckman HO. The nature and properties of soils. A college text of edaphology. New York: The MacMillan Company; 1948.

60. Buckman HO, Brady NC. The nature and properties of soils. 7th ed. New York: MacMillan; 1969.

61. Bear FE. Soil management. 2nd ed. New York: Wiley; 1927.

62. Weir WW. Soil science. Its principles and practice. Chicago: J.B. Lipincott Company; 1936.

63. Yaalon DHVA. Kovda—Meeting with a great and unique man. Newslett IUSS Comm Hist Philos Sociol Soil Sci. 2004;11:4–9.

64. Soil Survey Staff. Soil survey manual. Washington DC: USDA; 1951.

65. Brevik EC. George Nelson Coffey, early American pedologist. Soil Sci Soc Am J. 1999;63(6):1485–93.

66. Vanderford CF. The soils of Tennessee. Knoxville: University of Tennessee, The Agricultural Experiment Station; 1897.

67. Walker JB. A world's fair. Introductory: the world's college of democracy. The Cosmopolitan. 1893;XV(5):517.

68. Pierce JW. Photographic history of the World's Fair and sketch of the city of the Chicago. Chicago: R.H. Woodward Company; 1893.

69. Queen GS. American relief in the Russian famine of 1891–1892. Rus Rev. 1955;14:140–50.

70. Smith CE. The famine in Russia. N Am Rev. 1892;154:541–51.

71. Simonson RW. Soil science at the World's Columbian Exposition, 1893. Soil Surv Horiz. 1989;30:41–2.

72. Whitney M. First four thousand samples in the soil collection of the Division of Soils. Washington: U.S. Department of Agriculture. Bulletin no. 16; 1899.

73. Fanning DS, Fanning MCB. Milton Whitney: soil survey pioneer. Soil Horiz. 2001;42:83–9.

74. Whitney M. The soils of Maryland. College Park, MD: Mary Land Agricultural Experiment Station. Bulletin No. 21; 1893.

75. White AF. Performing the promise of plenty in the USDA's 1933–34 World's fair exhibits. Text Perform Q. 2009;29:22–43.

76. Simonson RW. Concept of soil. Adv Agron. 1968;20:1–47.

77. Kellogg CE. Soil genesis, classification, and cartography: 1924–1974. Geoderma. 1974;12:347–62.

78. Woodruff CM. A history of the Department of Soils and Soil Science at the University of Missouri. Special Report 413 College of Agriculture. Columbia, Missouri: University of Missouri; 1990.

79. Helms D. Early leaders of the soil survey. In: Helms D, Effland ABW, Durana PJ, editors. Profiles in the history of the U.S. soil survey. Ames: Iowa State Press; 2002. p. 19–64.

80. Schonauer JR, Schonauer KG. Chicago in the Great Depression. Charleston: Arcadia Publishing; 2014.

3

From a Farm on Loess—Emil Truog

"While the metal of our minds is still hot and pliable, let us forge or hammer out something that is both good and acceptable."

Emil Truog

The glacial ice that concealed all the land in the north left a chunk of Wisconsin untouched. It was polar cold, but the ice did not rub over the landscape, crush and flatten its rocks, nor did it deposit till or drift—that unsorted angular and rounded rock rubble. The unglaciated land became known as the Driftless Area and covers about one-quarter of Wisconsin. Its bare landmass stood as uncrushed Sedimentary rocks which were formed tens of millions of years earlier when Wisconsin was situated in a tropical lowland. Strong winds blasted the barren land after the glacial period, and storms picked up fine silty deposits from the Missouri and Mississippi riverbeds, and carried them eastwards blanketing the old sedimentary rocks of the Driftless Area [1]. On these winds, a new landscape was borne.

The rugged terrain had some steep bluffs and broad valleys that were covered by silt, and this cover was thicker closer to the Mississippi River. Soils slowly grew in the silty loess and from the weathering of the underlying rock that was rich in calcium and magnesium and fossilized organisms. When the earth warmed up, plants carefully colonized the new ground. Bitter coldness returned annually, laying waste to all greenness, and for many months the land would return to its glacial state. Then came spring, and the warmth and rain revived green lushness, and the soils graciously harbored roots and decaying plants. Colorless carbon that had been chemically fixed by the plants from the air was brought into the soil where it decayed into a black organic matter. Minerals absorbed water, broke apart, and released nutrients that were taken up by the plants, and what was not taken up was held onto the finest soil particles.

The soils were alive; an invisible lively web of bacteria and fungi orchestrated the rearranging of every particle in the soil whether it was alive, dead,

A. E. Hartemink, *Soil Science Americana*
https://doi.org/10.1007/978-3-030-71135-1_3

Glaciers brought stones from up north and they became indicators that something had moved them. This particular example is from Southern Wisconsin, the Tinguely machine in the background is a stone-puller. From the book *Productive soils*, by Wilbert Weir published in 1920. He was born in 1882 on a farm in Southern Wisconsin

or had never lived. With the melting of snow in the spring and the rains in summer, calcium and magnesium washed down into the subsoil. The organic matter colored the surface while the iron from the weathered minerals gave the deeper soil a brownish color. In some soils, clay seeped down with the flow of water, and after the water had been taken up by plants or evaporated, the clay was left behind, cloaked up pores, and polished the peds. Soils covered in tall grass formed thick surface layers full of organic matter. On soils that had grown trees, such layers did not develop, but they had a cover of partly rotten leaves and twigs, and below that a bleached layer from which clay had seeped down overlaying a subsoil with clay-coated peds. Soils formed differently on the hills than in the valleys and were dissimilar under grass or trees, but they were all young, and their distinctness grew with the passing of each season.

Humans have lived in what is now called Wisconsin since the last ice age had ended. The Ho-Chunk, also known as Hoocągra or Winnebago, lived in the area along the Mississippi that formed the western boundary of the state. They were hunters and gatherers and ate rabbits, raccoons, squirrels, deer, and elk but with the warming climate, they settled in villages and grew corn, beans, and squash [2]. The Ho-Chunk maintained strong family ties and cultural traditions and connected to the land in a way that respected its past and future. There were about 100,000 people in the state of Wisconsin by the early 1600s [2].

The first French traveled through Wisconsin in the 1630s, arriving at the Mississippi River in 1673. They came from Canada and traded fur and some were missionaries. They cooperated with the American Indians. That changed when other Europeans came in who were not interested in trade but arrived to settle on the land. With land shortages and soil exhaustion in the eastern part of the country, settlers moved west in search of suitable land. Others came for the timber or the mining of lead and zinc, and the new Wisconsin government opened up land at a cost of $1.25 per acre [3]. This all led to inevitable disagreements between the oldest inhabitants and the European newcomers. The Ho-Chunk of western Wisconsin were reluctant to give up the land which had sustained them and where they had buried their dead for millennia, but they were forced from their land and had to move west of the Mississippi River. Several deadly wars were fought, and the Ho-Chunk lost many lives and practically all their land.

Some of the land in the Driftless Area was covered with a mixture of oak forests and prairie savanna [4, 5]. In other parts, settlers found little timber as the land had been regularly burned to promote the growth of berry crops and grazing. Timber grew along watercourses that the new settlers used for building and firewood. The land was cleared by hand and with plows pulled by oxen and horses. Wheat was the main crop sown, which was harvested by a scythe or hand sickle when it was ripe in August. Threshing was done with a flail. The first harvest often yielded more than 1.5 tonnes of wheat per hectare [2]. The 1860s saw a large increase in wheat production in Wisconsin. Trempealeau County in the Driftless Area was in 1876 a waving field of wheat, and became the largest wheat market of western Wisconsin [3]. After the harvest, caravans of oxen came to unload at the warehouse from which wheat was poured into the holds of steamboats. Wheat prices were high, and as there were frequent robberies, farmers traveled in groups as they returned home after selling their wheat.

By the early 1900s, all natural forests in Wisconsin had been logged [6]. As the land was sloping, rain was intense, and farmers plowed in the spring, erosion was common, and it did not take long before more than half of the cropped land had lost 15 cm of its topsoil [7]. The loss of soil by erosion, outbreaks of the wheat clinch bug, and low wheat prices brought hardship to the settlers. Several of them expanded their cow herds and started dairy farming [8]. There was no government help in the transition from wheat growing to establishing a dairy farm, and many farmers experimented or relied on the popular magazine *Hoard's Dairyman*, which was started in Fort Atkinson in 1885. It promoted the growing and feeding of alfalfa as opposed to clover, and the magazine created awareness on tuberculosis in dairy cattle.

There was no agricultural research until the College of Agriculture at the University of Wisconsin was established in the 1880s.

The village of Independence in Trempealeau County sprouted around a railway depot where the train stopped, loaded coal, and delivered mail and supplies. The depot drove economic activity, leading to the construction of houses, a flour mill, and grain elevators [3]. The village was registered in 1876 and named in honor of the nation's centennial. It was built on an old floodplain surrounded by wooded hills at the confluence of Elk Creek and the Trempealeau River, which was a tributary of the Mississippi River. The east–west streets were designated as Washington, Adams, Jefferson, and Madison, and the north–south streets were numbered First to Sixth. A post office was opened in 1876, followed by a lumber yard, a furniture store, and a drug store that occupied the first floor of a building on First street; the second floor housed the six classes of a primary school. A Presbyterian church was built in 1879, which was bought by Evangelical and Methodist groups in 1893. In 1895, the township assessment roll listed 12 bicycles and 33 gold and silver watches—all these items were taxable [3]. In 1895, a municipal well was drilled, and in 1903 Independence constructed an electric plant with a coal-fired steam boiler. A library opened in 1908, and the main roads were paved with brick in 1914.

In Wisconsin, many towns and villages have French names. The area of Trempealeau County was derived from the French 'soaked in water' (*Trempé a l'eau*) or from the phrase *La Montagne qui tremp a l'eau*, which translates as 'mountain in the water.' Immigrants from Europe and the eastern part of the country settled in Trempealeau County in the 1860s. They came from all over Europe, but quite a few were from Switzerland and settled in the Driftless Area which rugged terrain may have remotely reminded them of sierra Helvetia. The rest of Wisconsin, some three-quarters of the state, had become an undulating plain following millennia of flattening by colossal glaciers.

The Swiss farmers in Independence had a few cows for milk and meat. The excess milk was used for making butter, but it was of low quality and storekeepers took butter in trade for other goods. The butter was only suitable for lubricating purposes and was shipped to makers of axle grease in large cities [9]. Some farmers made cheese in the 1860s and soon after that, several small cheese factories were started. A failure of the wheat crop came in 1878. As the wheat was in the milk stage when the kernel began forming, there were periods of rainstorm and of intensive heat, which caused lodging and baking of the kernel. Only a small bucket of wheat was harvested from a hectare, and the catastrophe prompted a rush of farmers to the Dakotas, including many

The village of Independence, Wisconsin, in the late 1880s, and Washington Street in in 1900. Emil Truog was born in Independence in 1884

farmers from Trempealeau County. Those that stayed became dairy farmers [3].

In 1850, the Truog family immigrated from Chur, a thirteenth-century village in eastern Switzerland to the USA. There was an influx of settlers into Wisconsin, both from the eastern states and from Europe. The Truog family left Switzerland after they heard the stories from families that had emigrated some years before. They sailed across the Atlantic, landed at New Orleans, and took the steamboat up the Mississippi River. They settled in a Swiss colony in the western part of Wisconsin, and in the hills they could raise and milk cattle and make 'Swissconsin' cheese [10]. Thomas Truog married Magdalena Keller in 1856, and in 1884, they bought a farm in Independence some 40 km east of the Mississippi. The farm was 90 ha; 72 ha were cleared and brought under the plow and the remainder was grassland and forest [11].

Emil Truog's parents: Thomas Truog (1834–1906) and Magdalena Keller (1845–1936). Both were born in Switzerland and Thomas came to the USA in 1850, Magdalena in 1854. They married in 1865, and had ten children; Emil was their youngest. Photos from the Truog family

Within a year after they had settled in Independence, Thomas and Magdalena Truog had their tenth baby. They named him Emil, and they now had 4 boys and 6 girls. Thomas was 50 years old when Emil was born. The Truogs lived in a small house, and downstairs there was one room in which they cooked, lived, and ate. Upstairs there were bedrooms that were divided by curtains. Thomas Truog bred and milked Guernsey cows that gave milk rich in fat and protein, and the milk was a little golden-yellow because of its high β-carotene content. Like all farm children, Emil had to work on the farm, beginning when he was 6 years old. His tasks included filling the wood-box at night, carrying water, and taking lunch to the workers in the field. Emil often worked until darkness, after which cows had to be fed and milked. Rural life in the late 1800s was not easy; hygiene was poor, there were few medical doctors, and cholera, typhus, smallpox, and tuberculosis were common causes of death. Emil was 8 when he lost his brother Henry, and 20 years old when his brother John died. The Truogs lived frugally, and milk prices were fair, which gave them a steady income. During the long winter evenings, they played checkers and cards. A new house was built by his father when Emil was 15, and the house was not far from Plum creek that streamed south of Independence.

Emil worked 6 months each year on the farm and went to school for the other 6 months. He liked fishing, hunting, playing baseball, and horseshoes and saved pennies and nickels to buy his first pair of skates ($1.25). His school was about 2.5 km away and he walked every day. At school, he learned reading, writing, arithmetic, grammar, history, and geography. There were about a dozen books in the library, and the school meant much to him. His first teacher was Eva Mary Reid, who was born in 1873 on the farm that was homesteaded by her Scotch emigrant parents. She lived a long life, and in the 1950s Emil reflected on her as his first teacher: "She is a kindly, vibrant person to whom I owe much and to whom I wish here to pay a sincere tribute. It was my good fortune that, along the way, I received from my teachers the kind of encouragement and inspiration that led, step by step, to a long and fruitful career. Eva May Reid gave me the needed beginning impetus" [12]. They exchanged Christmas every year, and in 1965 at the age of 92, she wrote to Emil Truog in a letter: "Words fail to express my joy in hearing from you. Your description of bygone days was so perfect that the tears just rolled right down. I did not try to restrain them—nature's relief you know. I remember you as if it were yesterday, standing in the class like a statue (you were a trifle undersize) and as serious as a judge. I never saw you smile once, during the whole nine months' term of school. I would say to you: Emil what are you going to read about this morning? And without opening your book, you would repeat without faltering, the whole lesson. Your behaviour was such that I never even looked in the direction of your desk during the whole day."

At the age of 14, Emil Truog went to the high school in Arcadia, which was 4 km from the farm. He went to school every day and no longer worked on the farm for half the year. In 1893, his older brothers Thomas Jr. and John had rented the farm from his father. It became clear to Emil that there was no future on the farm, and his teacher suggested that he continued his studies. At home and in the library, he started reading farm magazines that focused on how cropping exhausts soils of lime, phosphorus, and nitrogen. There were articles on the restoration of nitrogen by legumes, and that phosphorus and lime would have to be applied even though those nutrients were returned in the animal manure. He had studied some chemistry in high school and was fascinated by these articles. Before and after school, he still had to milk the Guernsey cows, but by that time he had discovered that plants interested him more than cows.

His older brother John became sick, and Emil took a year off from high school to work at the farm. He returned to school, and although the principal recommended the study of wireless telegraphy which was new at the time, Emil fostered his interest in the growing of crops and in particular the use of fertilizers. The Truog land had been cultivated for 25 years and showed signs of soil depletion. They looked for land in Iowa but thought that much

Science laboratory of physics at Arcadia high school in March 1904, instructor Herbert Snowdon (with moustache). In the center front row: Emil Truog, 20 years old. He started his studies at the University of Wisconsin in 1905. Photo from the Truog family

of the state was a swamp, which was indeed the case before tile drains were installed. Thomas Truog thought that Emil should attend the short course in agriculture at the University of Wisconsin that taught Elementary agriculture during the winter months. Emil preferred to enroll in the 4-year course that would lead to a university degree. His father agreed, and Thomas Truog had, after all, always taught Emil that work should be done really well even if takes 'all summer.'

In 1905, Emil Truog enrolled in the Agriculture Program at the University of Wisconsin in Madison [12]. He was 21 years old and had saved enough to pay for some years in college; the remaining money was borrowed from his sister, who was a school teacher. Thomas Truog died at 72, during the second year of Emil's studies in Madison. His father, the Swiss upbringing, the land in Independence, his schools, and farming were deeply engrained in the life of Emil. Thomas Truog had often told his son that production of good crops never failed on new land, but "...with cropping the luxuriant growth of new virgin land declined rapidly" [12]. It was clear to Emil Truog that he should find answers to this problem.

The University of Wisconsin was founded in 1848. At the time, most people in Wisconsin were farmers, but the College of Agriculture was not established until 1889. The idea for such a college was met with resistance, as the study of agriculture was not considered an intellectual pursuit and therefore should not be part of a university. Some research was done at the Wisconsin Agricultural Experiment Station, which had been established years earlier. The College of Agriculture offered courses for farmers in the winter, and a traveling school of agriculture met with farmers throughout the state. Also in 1889, the Department of Agricultural Physics was formed. The geologist Thomas Chamberlin was President of the University, and he appointed his former assistant Franklin King as professor of agricultural physics and head of the new department. It was the first position of its kind in the country [13].

Franklin King had grown up on a farm on the Kettle Moraine near Whitewater in Wisconsin [14]. Soils on the moraine were coarse and gravelly, and at the edge of the moraine there were artesian wells and springs that sparked his interest in water. Franklin King attended the Whitewater Normal College and graduated in 1872. The college had prepared him to become a teacher, and for some years he taught at a high school, worked in the summer for the Wisconsin geological survey, and spent time at Cornell University [13]. As a professor of agricultural physics at the University of Wisconsin, he studied soil water, drainage, tillage, irrigation, swamps, nitrate movement, and erosion by wind on sandy soils. He was also interested in the construction of silos, maintenance of country roads, making of silage, and the comparative value of warm and cold water for dairy cows. Franklin King had worked with Thomas Chamberlin for the Wisconsin geological survey.

Thomas Chamberlin was born in Illinois in 1843 and raised on a farm in southern Wisconsin [15]. He studied geology at the University of Michigan a few years before Eugene Hilgard became professor of mineralogy in Michigan. After his studies, Thomas Chamberlin became the chief geologist for the Wisconsin geological survey, head of the glacial division of the Geological Survey, and in 1893 he founded the *Journal of Geology*. Thomas Chamberlin was president of the University of Wisconsin when in 1892 he was invited to establish a Department of Geology at the newly founded University of Chicago. He studied the relationship between atmospheric carbon dioxide and shifts in the climate, which was first explored by Svante Arrhenius in the 1890s [16].

Thomas Chamberlin was 5 years older than Franklin King and the two had their disagreements. In the beginning, Thomas Chamberlin did not recognize Franklin King's capabilities: "...my inability to judge human nature. I consider F. H. King to be Whitewater's greatest alumnus yet I once called him into my office on the campus and said: King, you're wasting your time; go back and follow the plow" [14]. Franklin King did not go back to the plow. He became the 'Benjamin Franklin of soil science' with an array of investigations and inventions, including the round barn that later became the silo, and numerous ingenious ventilation schemes [17]. Franklin King was a pioneering soil physicist, and wrote one of the first American soil textbooks, *The Soil*, which was published in 1895 [18, 19]. When Thomas Chamberlin joined the University of Chicago, Franklin King started a petition to keep him in Wisconsin to no avail. Thomas Chamberlin died in 1928, 85 years old. He had a crater named after him on the moon, as well as one on Mars.

As head of the Department of Agricultural Physics, Franklin King realized the importance of soil studies and in 1895: "...the spirit and results of investigation, which have grown so rapidly during our century, have already so widened our horizon of knowledge, and so changed the attitude of mind toward the phenomena of nature about us, that we are coming to study, in the spirit of science, many of those things which lie nearest to us, and with great moral, intellectual, and pecuniary profit; and since soil, air, and water are indispensable to all forms of life, we must know more and more of them as the demands for food and homes increase" [19]. More research was needed on the chemical composition of the soil, and in 1899 he hired Andrew Whitson to conduct research in soil fertility. At that time, soil fertility was defined as the overall capacity of the land to produce crops.

Born on a farm in Minnesota, Andrew Whitson was the sixth of seven children, and when he was 7 years old, he was put behind the plow and assigned to plow a 30-ha field [13]. He was trained in geology at the University of Chicago by Thomas Chamberlin, and became Head of Department when Franklin King moved to the *Bureau of Soils* [15]. Andrew Whitson remained in that position for 38 years [14]. In 1905, he changed the *Department of Agricultural Physics* to the *Department of Soils* because soils also encompassed chemistry and soil fertility, while agricultural physics mostly dealt with mechanization and engineering. The new department was one of the first Departments of Soils in the USA.

Andrew Whitson was also in charge of Soil Survey Division of *Wisconsin Geological and Natural History Survey* from 1909 to 1933. During that time, general soil maps were made for the northern half of Wisconsin, as well as detailed reconnaissance maps of the entire state. The Wisconsin soil survey was proposed in 1893, when a committee chaired by the geologist Charles Van Hise secured legislation for a geological and natural history survey. The survey studied mineral resources, soils, plants, animals, physical geography, natural history, and also conducted topographic mapping. The soil mapping program followed the legislative directive: "...to cause a soil survey and a soil map of the state," and began in 1899, in the same year as the national soil survey program directed by Milton Whitney.

Emil Truog participated with Andrew Whitson in a soil survey in Iowa County of Wisconsin in 1910, which was one of the early detailed soil surveys in the state [20]. Emil Truog, who had more interest in soil chemistry than soil survey, summarized his experiences as follows: "For the most part, we got along with such commonly understood descriptive words as dark or light, tight or open, top-soil or subsoil, level or hilly, and sand, silt, or clay. This had the advantage that both farmers and lawyers on asking us about our purpose or objective usually got the impression that we were engaged in a perfectly legitimate activity" [21].

In 1901, Milton Whitney of the *Bureau of Soils* hired Franklin King to investigate soil climatology, as part of a soon to be formed Department of Soil Management at his bureau [14]. In Washington, Franklin King studied the level of plant nutrients in soil solution and their relation to crop yields. Chemical analyses were made of soils and plants, and he wrote six manuscripts that were to be published by the *Bureau of Soils*. Milton Whitney rejected three of them as they were contrary to his own ideas that supposed that soil solution concentrations were the same in all soils. They could not agree [14]. Milton Whitney then asked him to resign, and Franklin King returned to the University of Wisconsin in 1904.

A few years later, Franklin King used money from his life insurance policies to travel for 9 months through Japan, Korea, and China. He was interested to learn how agriculture had been maintained in those countries for thousands of years. The maintenance of the soil for sustained production was not well understood, and it was perceived that many soils in the USA had been degraded. Franklin King calculated that there were about 16 ha to support every man, woman, and child in the USA, while the people in Japan have less than 1 ha per capita. He foresaw that the use of inorganic fertilizers in the USA and Western Europe could not be continued indefinitely, whereas its use to maintain soil fertility has never been possible in China, Korea, or Japan [22]. Based on his travels in Asia, he wrote the book *Farmers of Forty Centuries*, which was almost finished before his untimely death in 1911 [14].

Franklin Hiram King (1848–1911) and Andrew Whitson (1870–1945)—the first two Heads of the Department of Soils in Madison, both raised on the farm. Franklin King was born in LaGrange near Whitewater Wisconsin, Andrew Whitson was born near Stanton in Minnesota. They were student from the geologist Thomas Chamberlin. Franklin King and Andrew Whitson were buried about 100 meters from each other on Forest Hill cemetery in Madison, Wisconsin. Photos from Department of Soil Science

Franklin Hiram King was 63 when he died. He had laid the foundation for agricultural and soil physics and had written seven books. Cyril Hopkins from Illinois wrote that Franklin King founded soil physics, and that he knew more of chemistry than most chemists [14]. According to John Russell from Rothamsted in the UK, who had visited Franklin King in 1910, he was America's greatest soil physicist [14]. *Farmers of Forty Centuries* was privately published in 1911, published in the UK in 1927, and the book was re-published in 1939 by Jerome Rodale [23]. After that, Franklin King became, unintentionally, one of the heroes of the Organic agriculture movement.

Emil Truog took several soil courses that were taught by Franklin King and Andrew Whitson. During his first year at the University of Wisconsin, he studied soil chemistry and learned how to use the litmus paper test for soil acidity. He tried the test on the soils of the home farm, but it was unclear whether the soils needed lime or how much was needed. After his first year, the Dean of the College called in all students and said: "Now you boys are rich in hope, rich in expectation, and rich in the possibilities for the future, but awfully poor in ability, experience, judgement, and knowledge, especially of the ways of the world. So don't go home for the summer's work but go and work on a different farm in a different fraction of the state and get some new ideas and learn some new methods. Don't expect too much wages because you green boys will probably not be worth much. Experience will largely be your pay." Emil Truog followed the Dean's advice and worked on a large dairy farm where they had 60 cows. Up each day at four in the morning, he milked 15 cows, by hand.

Emil Truog received his BS degree in Agriculture at the University of Wisconsin in 1909 on a thesis entitled *The loss of phosphorus in heavily manured tobacco soils*. He had conducted a laboratory experiment on the leaching of phosphorus using soil columns that he named cylinders. These had been filled with soil and subjected to different rates of manure and acid phosphate and he found: "In the case of sandy soils the phosphorus may even be leached in considerable quantities entirely through the first four feet and is then probably lost in the groundwater." After his BS degree, he obtained an MS in chemistry at the University of Wisconsin in 1912 under the supervision of Victor Lenher. The research focused on the determination of silica in rocks, minerals, and soils, and the results were published in the *Journal of the American Chemical Society* with Victor Lenher as the first author [24]. Emil Truog finished all his course requirements in chemistry for a PhD, but was lacking a thesis.

Victor Lenher was an expert in research on gas and chemical warfare, and studied the chemistry of gold, selenium, and tellurium [25]. He was ready to accept the PhD thesis from Emil Truog, but became afflicted with a strange sickness, possibly as a result of some poisonous chemical that he studied. As Victor Lenher was seriously ill, Emil Truog became involved in teaching and research, and did not find the time to go through the formalities of submitting the thesis. He never obtained a PhD degree, and as he would say later in life: "…there was no one who could supervise my work" and when asked why, he would answer: "From whom?" Victor Lenher's ailment lasted for many years, and he died one day before the opening of the *First International Congress of Soil Science* in 1927.

Emil Truog was hired in 1912 by Andrew Whitson as an instructor in the Department of Soils after he had finished his MS degree in chemistry. At that time, the geologist Charles Van Hise was the president of the University of Wisconsin. The university grew rapidly, and the number of professors quadrupled in 15 years. Charles Van Hise founded the extension division as he felt that the work at the university should benefit the entire state. As part of that growth and investment, Emil Truog was hired. The Department of

Emil Truog at the age of 21, fresh at the University of Wisconsin in Madison. Photo from the Truog family

English also wanted to hire him, but he took the position in the Department of Soils, as he would say later: "…had I grown up in an industrial area, I quite probably would have moved my life inquiring into matters far removed from the soil, for from the beginning the basic sciences held a great appeal to me. It was when I began to see how the use of sciences could help solving of the problems of general agriculture, of which I had a first-hand knowledge, that my career in soil science was forged" [12]. Naturally, he focused his research on the decline in crop production in relation to the depletion of nutrients and how they should be measured and replenished by fertilizers.

In the 1910s, the Department of Soils taught three courses: *Elementary soils—Principles of soil fertility* had three lectures per week and a 4-h laboratory that included work in soil physics, simple chemical tests, and exercises that could be used in secondary schools. The second course, *Laboratory Course in Soils*, focused on quantitative chemistry, and the third course, *Soil Management and Mapping*, had three lectures per week and fieldwork on Saturday

Group of famers gathered to test soil for acidity and growing alfalfa ("Essential points in growing alfalfa: well drained sweet soil"). Original photo caption: "This picture was taken in Dupage County during the Alfalfa Campaign in 1913, showing the speaker testing the soil for acidity. It is necessary to have good sweet soil if your alfalfa is to do well. That is one of the essential points in growing alfalfa." Photo Department of Soil Science

mornings. The types of agriculture were related to soil types and the maintenance of soil fertility. Fieldwork included classification and mapping of soils in the vicinity of Madison, and at locations which could be easily reached by train. These courses were mostly taught by Andrew Whitson, with help from students. Andrew Whitson developed one of the first extension courses in soils, which was meant for 'self-instructed classes in moveable schools of agriculture' [26]. Besides classroom instruction, soil was being taught in short courses that lasted 12 weeks during the winter months, and on special trains that went around the state. There was little demand for college training by farmers and there was also no demand for extension but over time, the short courses and the extension trains became successful and were also used in other mid-western states.

General outline for a course on plant nutrition at the University of Wisconsin by Emil Truog in 1912. Photos from Department of Soil Science

Emil Truog's first task at the Department of Soils was to develop a course that focused on plant nutrition. He developed lectures on the structure of plants, the function of the essential mineral elements, the replaceability of mineral elements, and how to conduct 'plant house' and field experiments. His research focused on soil acidity, its measurement, and lime requirements. Emil Truog had read how George Washington's clover grew poorly in acid soils, and how Edmund Ruffin applied marl and lime to reduce soil acidity [27, 28]. Edmund Ruffin had experimented with marl and lime in the early 1800s, and recommended manure, improved crop rotations, and the growing of cowpeas to restore the soil. In the 1850s, he was regarded as "....the great Nestor of an improved Southern Agriculture" [27, 29], but by the early 1900s, he was forgotten. In 1938, some 100 years after the marl experiments, Emil Truog dedicated his presidential address to Edmund Ruffin at the *American Society of Agronomy* meeting [30].

Emil Truog's main goal was to develop a soil test that all farmers could use, and particularly a soil pH test that was an indication for how much lime should be applied. He called an acid soil 'chronically sick' and lime was the cure, and the work pleased the Head of the Department of Soils, Andrew Whitson, who had advocated for years that the application of lime and fertilizer was essential for successful and permanent agriculture [31, 32]. Emil Truog was convinced that soil testing was essential for managing the fertility of the soil and together with Andrew Whitson, he convinced in 1913 Wisconsin state lawmakers to provide funding for the establishment of the State Soils Laboratory. The laboratory was written into state law: "It shall be the purpose and the duty of the State Soils Laboratory to make field examinations and laboratory analyses of the soil of any land in this state, and to certify to the results of such examinations and analyses upon the request of the owner or the occupant of the land, and the payment by him of the fee or fees hereinafter prescribed." The costs of field examination, sampling, and chemical analyses were ten dollars.

One of the first tests by the state laboratory was a litmus paper test for soil acidity that Emil Truog had developed. With his graduate student, Tsic Tang, he developed a zinc sulfide test whereby moist filter paper collected hydrogen sulfide liberated from a boiling soil suspension [33]. The filter paper had been soaked in lead acetate, and it was placed over the steam that came from the flask. The paper darkened according to how acid the soil was, which meant more protons, more hydrogen sulfide, and the darker the paper. The color was compared to a paper of known acidity or pH. His MS degree in chemistry was put to use for solving a practical problem: how to test the pH of a soil in a rapid and relatively inexpensive way. It stimulated interest in soil tests at a time of diverse opinions about the nature of soil acidity. Some thought that soil acidity was not caused by acids but by a phenomenon of physical adsorption brought about by the soil colloids. Emil Truog was able to commercialize the zinc sulfide test [31]. Because of the test, Wisconsin farmers started using lime and by the 1940s, Wisconsin used almost two times more lime than the neighboring state of Minnesota.

Wisconsin State Soil Test Laboratory in 1915, grinding up dolostone. Note Truog Acidity tester on the shelf. Photo Department of Soil Science

Emil Truog observed that liming was highly beneficial for legume crops like alfalfa, as the nitrogen fixing bacteria, rhizobia, are severely affected by high acidity and lime improves nodulation. It delighted him to work on an issue that was his first name spelled backwards. The zinc sulfide test was in use until the 1940s, after which measuring soil pH was done by the glass electrode. After the zinc sulfide test, Emil Truog developed colorimetric tests to estimate the nutrient status of the soil and procedures for estimating available phosphorus, potassium, calcium, nitrogen, and boron based on chemical extractions. In the 1950s, phosphorus and potassium were determined by the flame photometer, fertilizer recommendations were made for the main crops, and by the early 1960s, the state laboratory used computerized soil test reports. Emil Truog also developed and patented a chemical treatment of clay that was used in brick manufacture. It drastically reduced production costs and improved brick quality. Successful patents brought revenue to him, and the university and was used to hire graduate students.

Emil Truog was interested in the mineral matter of the soil that he called the 'foundation material.' Part of the mineral matter could be identified by a petrographic microscope, but the colloidal fraction was too small and an X-ray was used to identify minerals [34]. The X-ray of a soil sample created a diffraction pattern from the constructive interference of scattered X-rays passing through a crystal. In the late 1920s, Emil Truog used the X-ray diffractometer of the Department of Geology and started investigating what phosphates were formed when phosphate fertilizers were applied. There was a poor response to phosphate fertilizers on some soils in Wisconsin, and phosphate combined with a hydrated iron oxide formed a basic phosphate which was unavailable to plants. Emil Truog and his students also used X-rays for assessing the nature of inorganic substances that caused soil acidity. They had great optimism in their research: "The X-ray now used to unlock the secrets of the soil." That was in 1930. Some 30 years later, X-ray fluorescence spectroscopy was presented as a new method for investigating soils [35].

The relationship between soil pH, liming, and the availability of phosphorus and other plant nutrients was a life-long interest of Emil Truog. In 40 years, and with many graduate students, he worked his way to unravel the complex relationship between the concentrations of hydrogen and how that affected the availability of the 17 elements needed by plants to come to fruition. If fertilizers or manure were applied, the balance of nutrients changed, and this dynamic, combined with the variation in soils, its three phases (gas, water, and solids), and the changing weather and cropping systems, made the assessment of nutrient availability and fertilizer recommendation a demanding task.

Emil Truog had an orderly and pragmatic approach to studying soils in relation to plants. He was well-trained in soil chemistry and had a down-to-earth farming background. Hans Jenny considered him more of an agronomist than a soil scientist [36], and Emil Truog might have agreed. In 1916, Emil Truog joined the *American Society of Agronomy* which had been established in 1907. He felt at home in the society, became its President in 1938, and was a society member for the rest of his life. In his 1938 presidential address, he stated: "We agronomists must all adopt a more incisive and positive attitude in promoting sound and practical programs of soil management" [30]. In the same speech, he discussed how soil depletion brought him to the university: "It is the duty of all agronomists to preach the doctrine that the major plant nutrient elements removed from the soil by crops must be returned, pound for pound, in the form of crop residues, animal manure, or commercial fertilizers, if soil fertility is to be maintained. Any other policy has in the past led to and will in the future lead to, first, soil depletion, then, soil destruction by erosion, and finally, economic ruin. To take more out of a soil than is returned is as certain to deplete a soil in time as the removal of money from a checking account at a greater rate than its return is of putting the checking account 'in the red.' Will this cycle of soil depletion and agricultural desolation be repeated, over and over, indefinitely? If the answer is to be no, then it will be necessary for the agronomist not only to take soil science from the sequestered cloister of the laboratory and hitch it to the plow, but also to go forth preaching a positive and realistic program of soil management and conservation as Edmund Ruffin did over 100 years ago" [30].

In 1946, Emil Truog published a diagram illustrating the relation of the soil pH to plant nutrients in which the width of the band at any pH value indicates the relative availability of plant nutrients [37]. It did not present the actual amount, as that was affected by other factors such as the type of crop, soil, and fertilization. The diagram had been designed in 1936 by Nicholas Pettinger and published in a *Bulletin of the Virginia Experiment Station* [38]. Born in Iowa, Nicholas Pettinger had obtained his PhD at the University of Illinois in 1927. In the bulletin and diagram that came with it, he discussed the range of soil pH in relation to the availability of nitrogen, phosphates, potassium, calcium and magnesium, manganese, iron, and the toxicity of aluminum. Suitable pH ranges for field and garden crops were provided, along with lime recommendations based on soil texture and type of lime. Nicholas Pettinger died at the age 34 from heart trouble, the same year the diagram was published [39].

Emil Truog liked this approach and expanded the chart to 11 nutrients and made it: "…more simple in form but more complete in several aspects" [37]. It was presented as conceptual in 1946, but contained several assumptions. The pH model assumed that the availability of nutrients was the same to all plants in all soils, and the diagram suggested that it was best to have the soil around pH 7 [40]. However, many acid soils were highly productive, as were some soils in the alkaline range. The diagram also suggested that deficiencies of metal micronutrients did not occur at low pH, and there were no problems with the availability of potassium or sulfur at high pH [40]. The 1946 paper provided no data and no references and stands among Emil Truog's scientific legacy. The diagram ended up in *Our Garden Soils,* the popular book by Charles Kellogg aimed to educate gardeners about soil [41]. The diagram continues to be used in textbooks and encyclopedia and numerous papers [42–45].

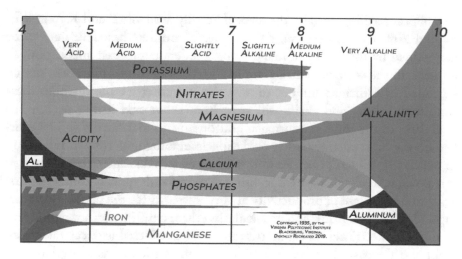

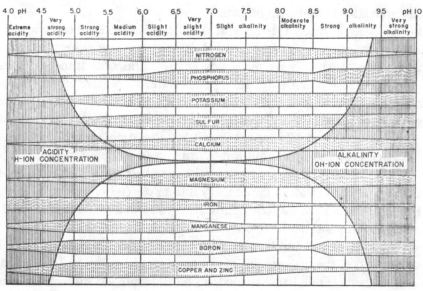

Diagram illustrating the trend of soil reaction (pH) to the availability of plant nutrients, original diagram from *A useful chart for teaching the relation of soil reaction to the availability of plant nutrients to crops* published in 1935 by Nicolas Pettinger, and the much reproduced version of Emil Truog that was published in the *Soil Science Society of America Proceedings* in 1946

The relationships between soil pH, liming, base exchange, and fertilizers were reasonably well-understood by the mid-1940s: "By rational application of fertilizer to depleted soils for general farming under soil and climatic conditions like those existing in Wisconsin is meant the initial application and thorough incorporation through the whole plow layer of sufficient phosphate and potash fertilizer to raise the level of readily available P and K to at least 50 and 200 pounds per acre, respectively. Thereafter these levels are to be maintained by the return of manure and moderate localized applications of fertilizer to the corn and small grain in the rotation. Lime, if needed, should be added in sufficient amount to raise the pH to at least 6.5. Under these conditions, clover and alfalfa will grow luxuriantly, and if they occupy the land 50% of the time, the nitrogen and organic matter factors will be satisfied for the most part. Extensive field experiments are underway in Wisconsin in which this system of fertilization is being tested. Results to date show that it is possible under this system to bring depleted soils quickly (often the first year) to a high state of productivity at very low costs for the marked increases in yields obtained. A method of calculating these costs in which the total cost of the initial fertilizer and lime treatments is considered as a capital investment and the cost of subsequent additions are considered as depreciation or depletion factors is explained in detail" [46].

In 1952, Emil Truog assisted with the Corn Pacemaker Program, which aimed at obtaining yields of 100 bushels of corn per acre rather than the 45 bushels per acre which were the yields at that time. One hundred bushels of corn per acre is about 6.7 tonnes per hectare. They selected 167 farmers who were instructed to follow the recommendations of soil testing, planting of corn hybrids, and proper weed control on a 4-ha field. The average yield of all those fields was 124 bushels per acre. The program received considerable attention, and in the following years, over 600 farmers participated. Wisconsin was not considered to be in the corn belt, but the state had some of the highest corn yields in the country during *the Corn Pacemaker Program.*

Having been born on a farm, Emil Truog was well aware of the fruits of a frugal, patient, and hard-working life. For him, there was no substitute for patience, perseverance and for the seeing eye, and inquiring mind [12]. He spent 42 years at the Department of Soils and was a mentor to 103 PhD and 72 MS students. Emil Truog trained no female graduate students. The first female that graduated with a PhD from the Department of Soil Science at the University of Wisconsin was Jaya Iyer in 1968, followed by Mary Beth Kirkham in 1971. All in all, Emil Truog has an extensive pedigree in the USA and across the world, although most of his former students have now retired or have passed away. He mentored Champ Tanner and Marion Jackson, the first Wisconsin soil scientists to become members of the National Academy— one of the highest honors for a scientist in the USA.

Every Saturday morning, he held seminars in which the students had to report on the progress in their research, often without warning. His questions were piercing and, according to Dennis Keeney, "…the questions would strike the fear of God into you." Emil Truog insisted on excellent writing and precise use of language. As one of his students would say later: "Prof emphasized clarity, directness, and simplicity in writing. Concise, clear writing can only come from clear thinking, he pointed out. To think and write clearly, he advised, get up with the birds, a time when distractions would be few" [47]. The PhD theses at the time were slimmer, and some theses resulted in one or two papers in a soil science or agronomy journal.

Emil Truog was a supportive mentor to many of his students, and his support went beyond the classroom, continuing after a student had graduated. Toward the end of 1929, Charles Kellogg visited Emil Truog to discuss a soil's position at *North Dakota Agricultural College*. He was on his way to Fargo for the interview. Emil Truog saw that Charles' suit was a bit worn, and that he probably had no money for a better one. He reached in his wallet and gave Charles money to buy a new suit, as well as a shirt and shoes [13]. Well-dressed and full of hope, Charles Kellogg drove to Fargo where he was interviewed for the position by Dean Harlow Walster. Shortly thereafter, he was appointed at *North Dakota Agricultural College*. Emil Truog never accepted financial repayment for the suit, shirt, and shoes; it was repaid some decades later, by alternate means.

Possibly because of his Swiss heritage, Emil Truog was an orderly man who expected the same of his students. He insisted on cleanliness and disliked open doors, which on occasion he would violently kick. When in 1952 it had come to his attention that some students were: "…monkeying with the delicate and expensive balances and getting them out of order," he wrote the following memo: "No matter how good your grades may be and they are very important, you cannot hope to get a good recommendation for a responsible and well-paying position if you do not exhibit a satisfactory attitude in promoting the general welfare of all by giving due care and attention to all of the equipment and facilities which are provided by the taxpayer for your use. If you do not wish to accept this responsibility, it pleases us immensely if you go elsewhere immediately."

Emil Truog in Room 270 of the Soils building at the University of Wisconsin teaching on the Miami Silt Loam and Spencer Silt Loam (both Alfisols) that cover about 186,000 ha and were widely used for growing alfalfa in Wisconsin; and with professors and students after a graduation ceremony. Photos from Department of Soil Science

During his teaching, he would talk for over an hour, stock still, with his eyes closed. He attended all seminars in the department and asked pertinent questions with such regularity that students became fearful of giving presentations in the department. His office contained publications stacked to considerable heights, but he was always able to find the right paper. After his retirement, he would continue working and visiting the department. Graduate students would have lunch at Francis Hole's office, and on several occasions, Emil Truog would appear at the door, and as he cleared his throat, they would all give him their attention. His speech was always the same and went something like: "I see the soil classifiers are having lunch." As they all agreed, he would go on: "How many new Series have you named today?" Francis Hole would usually reply, "Well Professor, we have not come up with any new Series." Emil Truog would then shake his head and say: "I am having coffee with the chancellor this afternoon and if you are not naming new Series I will not be able to say any good words about your work." He would then laugh and leave.

One winter Emil Truog was visiting Florida. At a meeting, he was asked to compare extemporaneously the merits of Wisconsin and Florida. He rose and said "I really have no competence for this kind of comparison except for agriculture. Now I know your citrus is your most important crop and grosses $100 million annually. I could compare that with our milk, cheese, our butter, but I don't want to take unfair advantage of you. However, if I compare this with the dollar value of the N, P and K of our animal manures, they are worth four times your citrus." The host from Florida remarked later that it sounded like a lot of B.S. [48].

Emil Truog was the Head of the Department of Soils from 1939 to 1954, during which time the department expanded considerably, as he was able to attract the best soil scientists and funding. After his retirement, he managed the organization of the *Seventh International Congress of Soil Science* in Madison in 1960. He became involved in the selection of a site for the new headquarters of the *Soil Science Society of America.* Preference was given to a city in the central part of the country and Columbus, Ohio, St. Paul, Minnesota, and Madison were on the ballot [49]. Madison received almost 80% of the votes. Emil Truog had purchased land close to the airport in Madison and had hoped the plot would be selected by the committee that included several of his former graduates [13]. The committee purchased half a hectare in March 1961 for $43,719, on 677 South Segoe Road in Madison. Much to his chagrin, it was not Emil Truog's land.

Shortly after the decision was made to build the new headquarters in Madison, Charles Kellogg wrote in his diary: "I was terribly disappointed

Emil Truog at the construction site of the Headquarters of the *American Society of Agronomy* on 677 South Segoe Road in Madison in 1962, and the building shortly after it was opened. The building was demolished in 2011. Photos from the Truog family

to learn that the Agronomy Society, already an anachronism, had decided to build an official building in Madison. This was a serious error. Apparently, the decision was made at the Purdue meeting when I was in the Soviet Union. I explained that any society that hoped to influence the Congress must have an able staff in Washington with an executive secretary eligible for membership in the Cosmos Club because that's where the real decisions on scientific plans were made. I convinced Truog of this but it was too late." The *Soil Science Society of America* never moved to Washington. It took another 38 years before the society hired Karl Glasener as a science policy director based in Washington.

Emil Truog officially retired in 1954, and by that time, the Department of Soils had graduated over 700 students, of which 320 had obtained an MS or a PhD degree in soil science. He had supervised the research of over half of the MS and PhD students. Upon his retirement, he donated all his journals and books to the department. The collection included complete runs of the journals *Soil Science* and the *Proceedings of the American Society of Agronomy* from their first issues. He had a sticker put in each issue: "Presented to the Department by Emil Truog for the furtherance of Soil Science." Some years after he had officially retired, Emil Truog was nominated for an Honorary Degree at the University of Wisconsin where he had worked since 1912, and where he had largely built the Department of Soils. It was not awarded.

Emil Truog married Lucy Price Rayne in June 1925; he was 41 and she was 10 years younger. Lucy was born in 1896 in Schell City, Missouri—some 160 km south of Kansas City. Her father was in the lumber business, and 2 years after Lucy was born, the family moved to Madison. She obtained a BS degree in 1913, made extensive travels across Europe and the USA, and worked at the Wisconsin State Historical Library, where she wrote a genealogical study on her ancestry—the Scottish Curd family who had emigrated to the USA in 1669 [50]. Emil and Lucy moved into what had been the Rayne family residence at Grant Street in Madison. It was a big house, and on the third floor, they provided guest rooms for international graduate students and visitors. Emil was 43 when his daughter was born, and 45 when they had twins. Lucy died in the summer of 1969, and Emil 5 months later, 85 years old.

The Truog residence in Madison in the 1960, and Lucille (left) and Emil Truog (right) with a visitor (center) in 1954. Photos from the Truog family

References

1. Bockheim JG, Hartemink AE. The soils of Wisconsin. Dordrecht: Springer; 2017.
2. Apps J. Wisconsin agriculture—a history. Madison: Wisconsin Historical Society Press; 2015.
3. Anon. 100 Years Independence 1876–1976. Historical album. Independence: City of Independence; 1976.
4. Campbell HC. Wisconsin in three centuries, 1684–1905. New York: Century History Company; 1906.
5. Hole FD. Soils of Wisconsin. Madison: The University of Wisconsin Press; 1976.
6. Whitson AR. Soils of Wisconsin. Bulletin no. 68, Soil Series no. 49. Madison: Wisconsin Geological and Natural History Survey; 1927.
7. Clark N. Soil erosion. Farmers and government together can whip it. Madison: Extension Service of College of Agriculture, the University of Wisconsin; 1940.
8. Hartemink AE. Some noteworthy soil science in Wisconsin. Soil Horiz. 2012. https://doi.org/10.2136/ssh2012-53-1-3hartemink.
9. Pierce EB. History of Trempealeau county. Chicago: H. C. Cooper, Jr., & Co.; 1917.
10. Wilde SA. Dr. Werner's facts of life. New Delhi: Oxford & IBH Publishing Co.; 1973.
11. Curtiss-Wedge F, Pierce EB. History of Trempeleau County, Wisconsin. Chicago and Winona: H.C. Cooper Jr. & Co.; 1917.
12. Truog E. Reflections of a professor of soil science. Soil Sci. 1965;99:143–6.
13. Beatty MT. Soil science at the University of Wisconsin-Madison. A history of the department 1889–1989. Madison: University of Wisconsin-Madison; 1991.
14. Tanner CB, Simonson RW. King Franklin Hiram—pioneer scientist. Soil Sci Soc Am J. 1993;57(1):286–92.
15. Tandarich JP. Wisconsin agricultural geologists: ahead of their time. Geosci Wis. 2001;18:21–6.
16. Chamberlin TC. An attempt to frame a working hypothesis of the case of glacial periods on an atmospheric basis. J Geol. 1899;7:545–84.
17. King FH. Ventilation for dwellings, rural schools and stables. Madison, Wi: Published by the Author; 1908.
18. Gardner WH. Early soil physics into the mid-20th century. New York, NY; 1986.
19. King FH. The soil: its nature, relations, and fundamental principles of management. New York: The MacMillan Company; 1895.
20. Whitson AR, Geib WJ, Dunnewald TJ, Truog E, Lounsbury C. Soil survey of Iowa County, Wisconsin. Madison, Wisconsin: Wisconsin Geological and Natural History Survey; 1914.

21. Truog E. Enlarging the use of soil survey maps and reports. Soil Sci Soc Am Proc. 1950;14:5–7.
22. King FH. Farmers of forty centuries or permanent agriculture in China, Korea and Japan. Emmaus, Pennsylvanaia: Republished by Rodale Press, Inc.; 1911.
23. Paull J. The making of an agricultural classic: farmers of forty centuries or permanent agriculture in China, Korea and Japan, 1911–2011. Agric Sci. 2011;2:175–80.
24. Lenher V, Truog E. The quantitative determination of silica. J Am Chem Soc. 1916;38:1050–63.
25. Anon. Resolution in memory of Victor Lenher. Science. 1927;LXVI:76.
26. Whitson A, Hendrick H. Extension course in soils. Bulletin no. 355. Washington: Government Printing Office; 1916.
27. Ruffin E. An essay on calcareous manures. Petersburg Va.: J.W. Campbell; 1832.
28. Truog E. The first American to make chemical soil tests. Soil Surv Horiz. 1963;4:4–5.
29. Chambers WH. The soil of the south. A Monthly journal devoted to agriculture, horticulture and rural and domestic economy. Columbus, Georgia: Lomax & Ellis Publishers; 1853.
30. Truog E. Putting soil science to work. J Am Soc Agron. 1938;30:973–85.
31. Truog E. Testing soils for acidity. Madison: Agricultural Experiment Station, University of Wisconsin; 1920.
32. Truog E. Andrew Robeson Whitson 1870–1945. Soil Sci. 1946;60:272–4.
33. Jackson ML, Attoe OJ. In Memoriam Emil Truog 1884–1969. Soil Sci. 1971;112(379–80).
34. Truog E. The X-ray is now used to unlock the secrets of the soils. Madison; 1930.
35. Fripiat JJ. New methods of investigation in soil chemistry. Paper presented at: 8th International Congress of Soil Science, Bucharest, Romania; 1967.
36. Maher D, Stuart K, eds. Hans Jenny—soil scientist, teacher, and scholar. Berkeley: University of California; 1989.
37. Truog E. Soil reaction influence on availability of plant nutrients. Soil Sci Soc Am Proc. 1946;11:305–8.
38. Pettinger NA. A useful chart for teaching the relation of soil reaction to the availability of plant nutrients to crops. Blacksburg: Virginia Agricultural and Mechanical College and Polytechnic Institute and the United States Department of Agriculture, Cooperating; 1936.
39. Food and Agriculture Organization of the United Nations and Intergovernmental Technical Panel on Soils. Status of the World's Soil Resources—Main Report (FAO & ITPS, 2015).
40. Blamey FPC. Comments on a figure in "Australian soils and landscapes: an illustrated compendium". ASSSI Newslett. 2005;142.
41. Kellogg CE. Our garden soils. New York: The MacMillan Company; 1952.
42. Larcher W. Physiological plant ecology. Berlin-Heidelberg: Springer; 2001.

43. McGrath JM, Spargo J, Penn CJ. Soil fertility and plant nutrition. In: van Alfen N, editor. Encyclopedia of agriculture and food systems, vol. 5. San Diego: Elsevier; 2014. p. 166–84.
44. Fernández-Martínez M, Sardans J, Chevallier F, et al. Global trends in carbon sinks and their relationships with CO_2 and temperature. Nat Clim Change. 2019;9(1):73–9.
45. Fine AK, van Es H, Schindelbeck R. Statistics, scoring functions, and regional analysis of a comprehensive soil health database. Soil Sci Soc Am J. 2017;81:589–601.
46. Truog E, Attoe O, Jackson M. Fertilizer application rationalized. Soil Sci Soc Am Proc. 1945;10:219–23.
47. Rich CI. Professor Emil Truog. Soil Sci. 1965;99:140–2.
48. Tanner CB. Letter to Walter Gardner. Department of Soil Science Archive. 1976.
49. Smith D. Building the ASA headquarters in Madison. Unpublished report ASA. 1970.
50. Curd WB, Rayne Truog LP. The Curd and allied families. Madison: Wisconsin State Historical Library; 1927.

4

From a Farm on Till—Charles Kellogg

"What is needed is boldness of imagination—minds that seek new combinations of relationships more appropriate to our needs."

Charles Kellogg

Not so long ago, the lower peninsula of Michigan was covered by glaciers. The ice was thick, several kilometers in some places, and laid on the land for tens of thousands of years. As the glaciers moved, ever so slowly, they scraped and crushed the earth, depositing stones and rocks originating from high up the northern reaches of Canada. When the ice finally melted, it left copious amounts of broken-up rocks, sand, and gravel over ridges and moraines. Roaring streams and rivers, full of water that had been frozen for millennia, ran west, and lakes formed from the gigantic ice blocks that were left behind across the land. Fierce winds blew over the barren landscape, and the land remained deeply frozen. Unhurriedly, it woke up.

It was cold when people moved in. Some of them had followed buffalo herds that crossed the Bering Strait, making an icy trek from Siberia into the Americas. The first people in Michigan belonged to the Sauk and Ottawa tribes. They moved around, fished, and gathered nuts, berries, and roots. Chipped chert was used for spears and arrowheads to hunt large animals. As bears were deeply revered, they were killed only with a special ceremony and apology. Mammoths roamed the land, leaving big footsteps in the sand. With the warming of the earth, trees colonized the land, and a dense forest eventually grew on supportive soils. There were scattered swamps and bogs, sand that blew up to form dunes, and land so gravelly and rocky that plants could not take root. Those lands stayed barren until the gravel and rocks had fallen apart by the weather and formed a soil thick enough that plants could feed and anchor.

Over time, people gathered in permanent settings, built little villages and started growing corn, squash, beans, and tobacco, and harvested wild rice that grew luxuriously in the calm waters of soft mucky lakes and streams. They

A. E. Hartemink, *Soil Science Americana*
https://doi.org/10.1007/978-3-030-71135-1_4

discovered copper deposits that had crystallized in lava layers a billion years earlier. The ice had ground up the edges of the basalt lava beds, exposed the copper, and deposited it further south. Some large and pure copper nuggets were found, but most of the copper was within volcanic rocks. To extract the copper, people burned wood over the copper vein which heated the rock and then poured cold water to crack it. The copper was pounded with rock hammers and stone chisels and shaped into spear points, arrowheads, knives, harpoons, and jewelry, and some copper was shipped across Lake Michigan [1]. All of that, from the retreat and melting of the ice to the making of copper jewelry, took thousands of years. The land transformed as slowly as the earth warmed up.

It also took thousands of years before there were words for the land that came from underneath the ice. No one described it more richly than Jethro Veatch who conducted soil surveys across Michigan: "The configuration features of glacial origin are generally curved and rounded in form, rather than straight-lined or angular, but they comprise a multitude of variants which have never been clearly defined or given distinctive names by geomorphologists; they are variously: knobs, domes, hillocks, mounds, oval or lenticular ridges, low, sinuous beach ridges, ice shove embankments, breasts, rolls, swells, insular hills, bulbous forms, fans, flaring U-shaped valleys, troughs, sags, dips, coves, saddles, swales, cupules, pot-holes, bowls and other calathiform depressions; lake basin swamps, linear or valley swamps, marshes, muskegs, bogs in pot-holes, lakes of great diversity in size and shape, buttonbush and cattail ponds, coastal lagoons, and estuaries" [2]. He who knew the land, the anatomy of the soil, created words to describe its spacious portrait and intimate aspects.

French colonizers moved through Michigan in the 1700s, trading fur, but they never settled and only built trading posts near the American Indian villages. They learned to live with the Indians, ate their food, and paddled their birchbark canoes. Most aristocratic French families had at least one member in North America during the seventeenth century, as they sought freedom from the binding ties of family [3]. Many of them ended up in the fur trade which was America's oldest industry and had begun in 1615 when a syndicate of Dutch fur traders and ship owners colonized Manhattan Island [4]. These fur traders had difficulty settling and their farms produced few crops. The French and French-Canadians were more adept with the axe than the plow, but some early Michigan settlers planted apple orchards that were brought from Normandy by way of Montreal [3].

After the French, numerous other Europeans settled in Michigan in the 1830s. In the east, and in particular, in New York, Michigan was perceived to be a good place to settle as it had metals, wood and farm products, and the navigable waters of the Great Lakes. There were jobs in mining, forestry, and agriculture. From the tin and copper mines of Cornwall in England to the vast forests of Finland and the steep hillsides of Croatia, many sailed to America and moved to Michigan. Some 40 nationalities settled [3]. If the Cornish Immigrants had an 'eye for ore,' the German settlers had 'eyes for the soil' and grew crops, and the Dutch came in the 1840s and settled along the shoreline of Lake Michigan in an area that became known as the 'Dutch coast.' In 1845, German publications, printed in Dutch, listed the advantages for new settlers in America [3].

1. Ordinary people are as good as rich folks.
2. He need not take off his hat to anyone.
3. Rich people honor the poor because they work for them.
4. There are good churches here and many of God's people.
5. The schools are free.
6. You need fear neither wild animals nor bad men.
7. You do not need to lock your door.
8. There are no poor people. If one is well he can earn good wages.
9. Women need to do nothing more than milk and prepare the food.

One of the first European families that settled was that of Joel Guild who came with his family from Utica, some 80 km east of Syracuse in New York. They traveled by horse cart from Buffalo to Detroit, and at each point where they stopped Joel Guild would sing a composition of his own [4]:

Come, all ye Yankee farmers
Who's like to change your lot,
Who've spunk enough to travel
Beyond your native spot,
And leave behind the village
Where pa and ma do stay,
Come, follow me and settle
In Michigania

The journey of the first settlers in Ionia was written by Prudence Tower in the 1890s [5]. Her father visited Michigan in 1832 and, a year later, the family packed up from Frankfort Village in New York. They bought a canal boat that fitted as many of their household goods as possible which they named 'Michigan caravan.' As she wrote: "The deck of the boat was piled with wagons and bound on every conceivable thing that could be taken to use in such country where there was nothing to be bought" [5]. From Buffalo to Detroit, they took the steamer and everyone was seasick. In Detroit, they bought oxen and cows and cooked provisions. Roads were rough and they camped out at night. Some 150 km north of Detroit, they had to cut roads as there had never been a wagon through and they had trouble crossing marshes and fording streams. Children were sick with 'canker rash or scarlet fever' and when they were in heavy timberland 50 km east of Ionia, one of the young boys died.

They arrived in Ionia and met two European families and an American Indian who was married to a French trader. When they finally reached their destination, Prudence Tower wrote: "In Ionia, there was a large population of American Indians. They have planted corn, melon, and squashes and did not like to leave; but through the aid of our interpreter father was able to pay them for their improvements and left peaceably." Such colonial narratives were common, probably far from the truth, and conveyed attempts to comfort the colonizers' culpability. The first Europeans took over the wigwams built from bark that were some ten square feet in size, after which they build small wooden houses. Transportation was by pole boat. The corn was pounded in a large mortar that the American Indians had dug out in a large hollow stump. Later on, they used a coffee mill. Prudence Tower ended her account with: "I want to pay this tribute to the Indians: They were very kind and peaceable, and seldom have us any trouble, never any serious trouble" [5]. She did not mention that they had lost their land.

The settlers built houses, hunted, cut the trees, sold the timber, and farmed the land like they had done in Europe or the eastern part of the country. Michigan changed rapidly, and counties were established to administer the settlements—they followed straight lines that had little connection to natural features such as waterways or landforms. One such county, Ionia, was established between the towns of Grand Rapids and Lansing. Initially, many counties had names that were derived from American Indians, but during a classical revival, new names were adopted using words from Latin, Arabic, Greek, and American Indian languages. Ionia County was named at that time, and this landlocked county ended up being named after a region in ancient Greece with inhabitants known for their love of philosophy, art, democracy, and pleasure. In the 1870s, some 40 years after the first European colonizers had settled, Ionia County had over 27,000 Europeans [6]. They lived on homesteads, in villages and small towns. Ionia County was founded at a time when the idea of free public schools was struggling for acceptance, and until the 1840s, education was seen as a luxury, not a necessity. That attitude slowly changed over time, and by the 1920s, the rural schools of Ionia County improved [4].

Once a year, the cold returned for a few months when the earth leaned away from the sun. Like the winter of 1912 when, in Michigan, the temperatures remained below minus 15 degree Celsius for eight consecutive days and nights. The ground remembered those days and deeply swallowed the cold, and after the frost, soils in Ionia remained wet and peats and swamps developed in hollows and around the lakes. Along the rivers and streams that thundered in the early days of spring, sediments were deposited, layer after layer. The deposits were coarse and formed broad riverbanks. When plants started growing along these waters, their roots penetrated through years of deposition, tapping water and nutrients and anchored against the prevailing western winds. In the sandier soils, evergreen trees produced a litter layer that could not be decomposed but leaked acidity. Rain and snowmelt dripped through the litter and picked up loose protons that were moved down by percolating water. The acidity carried iron and organic matter, coloring the subsoil. Such colorful soils became common and ended up as the state soil of Michigan. In the fine and dense till that was left behind by the glaciers, water stagnated close to the surface and in these soils, orange spots of iron oxides brightened up the matrix of dull colors. Far away, in the wetter parts of the Russian and Ukrainian steppe, such colored spots were named *glei*. [7, 8] Ionian soils were full of these, but the term glei had not yet surfaced.

The Grand River was the highway to reach the east of Ionia. The first steamboat plied the river waters in 1837, and smaller steamboats carried produce up and down the river. These boats were the region's lifeline until the Detroit and Milwaukee railroad were built in the mid-1800s. All of Ionia County drained westward: its rivers and their branches flowed into Lake Michigan. The European settlers cut trees to sell the timber and cleared the land so that crops could be planted, and pastures established. They threw the oak and pine logs into the Grand River and downstream they were picked up for milling. The cleared land was planted with wheat, barley, oats, and rye. Potatoes were grown on the sandier soils, and mint, celery, onions, and vegetables were planted on peat soils. Dairy and beef cattle roamed on little paddocks, and the manure was used to fertilize the cropped fields. Crops grew well during long summer days, and the short nights prevented late frosts in spring and early frosts in autumn. The rainfall was gentle, but freshly tilled soils washed downslope with the spring rain, and the settlers knew it was wrong, but there were not enough hands on the farm to halt it. Ionia was cold in winter and laid outside the moderating climatic effect of Lake Michigan, some 100 km to the west. Some settlers planted peach trees, but the trees died in winter. Apple trees did better, and apple blossoms became Michigan's state flower.

Palo was a small town in the northeast corner of Ionia County. The first road was opened in the 1840s and a trading post was built in 1849. The little village was named in commemoration of the 1846 victory at Palo Alto in the Mexican American War, and its naming aimed to bring history into the lives of the settlers. A post office was opened in the summer of 1857. Leander Millard built a tavern and Curtis Brook set up a blacksmith shop, followed some years later by a steam-saw mill. In 1868, a firm from Toledo, Ohio, put up a steam-grist mill upon a donation of $2,000 from the villagers and its surrounding farmers. The mill was to replace the hand-milling and offered the settlers more time to clear the forest and plant crops. It proved a poor investment and a local historian noted: "The Toledo men did not, however, do what was esteemed the fair thing in the premises, for they provided old and worn-out machinery, which failed to do satisfactory work and fell far short of fulfilling the expectations awakened at the beginning" [4]. All the money was lost. In 1868, a foundry was added followed by a grist and sawmill.

Little shops were built that sold everything that could not be grown, such as matches, calico, and sugar. A schoolhouse and church were erected. The Methodist Episcopal church was built in 1870, the church measured 10 by 18 m and cost $4,500, fully furnished. The congregation experienced steady growth and had 140 members in 1916 [4]. Roads were paved, electricity lines reached Palo, and a cemetery was established on the southside of the town. Palo grew slowly, and then growth ceased: "The village has striven earnestly for supremacy and recognition, but its efforts have failed in great measure. Palo of today is only a small village, which is maintained by the country trade for the several stores which exists there" [4]. It was in Palo that Charles Edwin Kellogg was born on 2nd of August 1902 [9].

Charles was named after his grandfather, an early settler in the Ionia region. They originated from Newfield in Tomkins County in upstate New York and homesteaded near the little Prairie Creek that meandered south into the Grand River. That same Grand River was somewhat serene when the Kelloggs settled, but its broad riverbed was a remainder of the enormous volumes of water that it carried when the glacial ice had melted. An only child, Charles was born on his grandfather's farm, and at his birth, his father Herbert was 23 and his mother 26 years old. When Charles was two, his grandfather died from gangrene and neglected diabetes, at the age of 64. One of Charles' early recollections was standing beside his grandfather's chair.

Street scenes of Palo in Michigan in the early 1900s with tree stumps, timber piles and the newly built Methodist church (top photo) and the main street heading north (below). Charles Kellogg was born in Palo in 1902, he was buried in Palo in 1980. Postcards Palo

Charles grew slowly, had many periods of sickness, and was cared for by his paternal grandmother, Mary Elizabeth (née Faussett) Kellogg. He had great curls that hung down to his shoulders until they were cut when he went to school at the age of five. Each day, he walked to school, which was three kilometers south of the farm. There were shorter routes through the woods and fields, but when deep snow blanketed the land, he followed the roads where horse carriages had left an easier walking path. The Palo school had one classroom, where Charles learned to read and write. The school had a library with five books, that he read and re-read, in particular *Children's stories of the Trojan War,* and *Arthur and his Knights.* The world in these books was far removed from the farm in Palo, but the heroism, fight for justice, and dominance appealed to him. Before and after school, he had to work at the farm. They grew grain, beans, apples, and most of their own vegetables. Pigs were butchered in autumn, some of the meat was sold, but ham, bacon, and salt pork were stored for the winter. Charles helped with all of that, but his father, Herbert, also hired men for the heavy work. Herbert Kellogg was critical and ill-tempered, so several of the farm workers ran away shortly after they had started. This meant there was more work for young Charles.

Charles' father and mother had not finished high school. Herbert worked for some time as an electrician in Lansing, but he was called back to the farm in Palo. The lack of schooling bothered Herbert, and he read as much as he could, particularly during the long winter evenings. Charles had a difficult relationship with his father, and it worsened when he became a teenager. His father was easily angered, and Charles found Herbert's hatred for labor unions and admiration for the wealthy irritating. The situation was so tense that Charles became depressed when he was 12. Every idea he had was ridiculed by his father, who often said that Charles was stupid and heedless. Charles started to believe that, and in his darkest moments, he considered suicide.

Charles read his father's books and all the books he could obtain from the library in Ionia. Books helped him through his period, slowly growing his confidence and self-esteem. He read *Self Reliance* from the philosopher and essayist Charles Emerson, who advocated following individual will instead of conforming to social expectations. Emerson advocated that a person should develop one's own culture and be focused on individual, rather than societal progress. Reading *Self-Reliance* helped Charles through those dark years. It dawned on him that he could be his own man and develop according to his own beliefs and knowledge gained. He did not have to follow the path of his parents, which for his father had led to bitterness and anger.

When Charles was 13, he went to Palo high school. Sometimes he rode a bicycle but most of the time he went by horse. He would like to have played basketball and tennis, but he was small for his age and lacked the time, as he had to work at the farm after school. His embarrassment was not that he was small and could not play basketball, but that they lacked money, and that he wore shabby clothes. Sometimes he took part in debating teams at school, where his class had seven students. One winter, he suffered from a severe attack of war-time influenza that had killed many soldiers. He had only barely recovered when he contracted measles. Though his mother and

Palo school, all students up to grade 12 with their two teachers. The photo taken in 1925, some years after Charles Kellogg had graduated. Postcards Palo

Charles Kellogg's parents: Herbert Kellogg (1879–1945) and Eunice Stocken Kellogg (1876–1953) in 1940. Photos from the Kellogg family

A thirteen years old Charles Kellogg in the barn door at the farm in Palo, and for a school photo. Photos from the Kellogg family

grandmother took good care of him, his poor health meant he grew slowly. Upon graduating high school, his father promised that he could go to college if he worked on the farm for another year. That summer, he had asthma after the grain harvest, and his father hired him out to a highway contractor to build graded and gravel-surfaced roads. It was rough 8 weeks for a boy who weighed only 50 kg. His father received all his earnings and the injustice of this angered Charles so much that he thought of killing his father. They argued over money and his future, over work on the farm, and the arguments culminated in a fight in the barn. Charles' anger had finally burst forth, as he was tired of years of his father's belligerent and belittling behaviors. It was a bad fight, but their relationship became less tense. The son had won over the father.

Charles worked another year at the farm and grew more independent. In the winter, he had to go to Ionia with team-drawn loads of grain or other produce, returning with coals and supplies. It was 16 km each way. Sometimes after work was done, he found an opportunity to feed his growing intellectual appetite. Judge Davis addressed the annual meeting of the historical society of Ionia County, remarking on the pioneers of the county and the many changes that have taken place. He ended his speech with: "We have so much to thank the pioneer for. We should keep alive their memories and great work by keeping up this annual meeting. We, as children of pioneers, must not rest content with the work of our fathers. We can't sit down and watch the cars go by filled with envy, nor be discouraged, but add to progress and the richness of the world by our energy and push. The world moves, and to be happy, we must move with it. This is an age of automobiles, electric cars, telephones, wireless telegraphy – and if we don't get too lazy to walk, too tired to talk, but ever remain true to our best instincts given through education, pure religion, honest politics and sound and healthy government, then we will be truly with all our other works the conquerors of Nature" [4]. These words engrained in young Charles a wish to conquer nature and an enthusiasm for individualism.

The Kelloggs were Methodists—hard-working and compassionate Calvinists with an underdeveloped ability for self-pity. They went to the Methodist church in Palo two or three times a month, but less as they grew older. Herbert Kellogg became senior deacon in Palo lodge 203 of the Free and Accepted Masons [4].Charles Kellogg found no comfort in the Methodist church, and like his father, he became a Freemason later in life. He started to date Lucille Reasoner, who was 14 years old and grew up on a nearby farm, and they had known each other since elementary school. They fell in love in the summer of 1920, visited each other's parents, and Charles was particularly glad that his parents liked her. When he turned 19 and she 16, they promised to get married.

Charles had no interest in taking over the farm and would not be forced into it, which was hard for his father to accept. The future of farming was uncertain and at the time, the cost of land was close to $100 per hectare. During the First World War, high commodity prices had benefited farmers, but after the war, prices fell sharply [10]. In addition, Charles' health was not strong, and as somewhat of a sullen bookworm, he was keen to get more education. He had the urge to get away from his father, the farm, and Palo—away from the restrictions of a small village built on dense drift.

Palo had one school, and Charles Kellogg and six others graduated from its high school in 1919. In October, he enrolled in the 2-year short course on

Charles Kellogg in 1921 whe he was 19 years old, and as cadet cavalry at the Michigan Agricultural College homecoming in November 1921 (nearest right). Photos from the Kellogg family

the basic principles of farming for farm boys at *Michigan Agricultural College* in East Lansing. The college was 80 km from Palo, which meant he could still work at the farm on weekends and holidays. He saved all his money and came home in the winter driving a little Model-T pickup. After the short course, he was interested to return to study for a college degree, but his father wanted him to run the farm. With money saved from selling sheep, trapping, and financial help from his grandmother, he returned in 1921 and started as freshmen at *Michigan Agricultural College*. He hoped to finish in 3 years and had a job in the college's greenhouse, as he wanted to become a horticulturist. Military was a mandatory subject, and he took cavalry and had a group of students to look after. It suited him. He knew horses ever since he rode them to school in Palo [9]. In his second year, his father had a kidney infection and Charles was called back to the farm and he stayed until his father was back on his feet, after which he promptly returned to East Lansing. There, he attended his first soil science course. It was disappointing, and not what he liked.

For the summer, he sought paid positions to avoid farm work back at home. He found a soil survey position with the State Department of Conservation that paid $80 per month plus expenses. The survey was 240 km to the north, and he learned to live in a tent, tolerate mosquitoes, and find his way in the woods. Gradually, he got insight into soils and soil mapping when he started working for Lee Roy Schoenmann, who had been worked for Wisconsin Soil Survey and had obtained a BS degree in soil science in 1911 from the University of Wisconsin [11] Lee Schoenmann had also worked for the *Bureau of Soils* conducting soil surveys in the summer in Michigan and Wisconsin, and in the summer in Alabama and Texas. In 1922, he had started a land economic survey in Michigan which was a cooperative survey program between the Michigan State Agencies, the Geological Survey, and the *Bureau of Soils* [12]. Lee Schoenmann surveyed 13 of the 83 counties in Michigan between 1924 and 1931. A kind and knowledgeable soil surveyor, he became a mentor to Charles.

The *Michigan Agricultural College*—M.A.C. as it was abbreviated—was formed in 1855 and served as a model for the land-grant universities created by the Morrill Act in 1862. The soils department was established in 1909, and Joseph Jeffrey, who had obtained his BS degree in Agriculture in 1896 under the supervision of Franklin King, was appointed as its Head and only member. In 1910, Charles Spurway was appointed as an instructor of soil physics, and a year later, George Bouyoucos was the first staff member with a PhD degree. Born in 1888 in a small village in a family of ten children, George Bouyoucos left Greece at the age of 11 and entered the University of Illinois when he was 15 years old. His PhD research, performed at Cornell University under Thomas Lyon, focused on transpiration and plant growth [13]. After obtaining his degree, he joined M.A.C. and worked at the college for 53 years. He studied soil moisture and its measurements, invented the hydrometer, and published his first paper in *Science* in 1915. Like everyone else at the time, these *Michigan Agricultural College* soil scientists studied a wide range of topics with little specialization; for example, Joseph Jeffery worked on corn selecting and seed testing, soil moisture, and its management, as well as the construction of cement silos. Most studies were practical but over time, the need for more fundamental understanding of soil followed the practical work which benefits were successful in some places but lacked universal applicability and understanding.

The soils department at the *Michigan Agricultural College* grew slowly. The telephone numbers listed contained only three digits and the numbers of the female students were listed separately. The department was concerned about the welfare and prosperity of its students, and the telephone directory included the following concerns and advice by the faculty [14]:

For students to remember

Be loyal.
Don't knock,
Don't start anything you can't finish.
Your instructors are your friends.
They are paid to help you.
Make them earn their salaries.

Joseph Jeffery was succeeded by Merris McCool as head of the Department of Soils. Merris McCool was born on a farm in the village of Amity in Missouri, and he attended the University of Missouri where he took geology courses from Curtis Marbut, and he obtained his BS degree in 1908. Part of his training included a soil survey of Dekalb County that was led by Henry Krusekopf [15]. After obtaining his MS and PhD degrees, Merris McCool worked in Oregon where he was one of the first to work on soils that had been developed from pumice. Volcanic soils had been reviewed by the geologist Nathaniel Shaler in the 1890s, who noted that: "...the soils are light and fertile" [16]. Oregon and its neighboring states had several volcanoes, but little was known about the soils, and Merris McCool called these volcanic soils 'unusual soils' because of their high porosity and water storage capacity [17]. After his work in Oregon, he was hired in Michigan where volcanoes had last erupted 2.5 billion years ago. One year after Merris McCool had left Oregon, Lassen Peak erupted, and volcanic ash rose more than nine kilometers into the air.

Inspired by the geological teaching of Curtis Marbut in Missouri, Merris McCool reorganized the soil courses at *Michigan Agricultural College* and he wanted to put more emphasis on scientific principles. He contacted Curtis Marbut for advice on how to prepare students for work in the *Bureau of Soils*, and it started a friendship and Merris McCool became an advocate for the changes that Curtis Marbut envisioned for the soil survey [18].Merris McCool was an early adopter of the concept of soil based on genetic and geographic relationships. Like his predecessor Joseph Jeffery, Merris McCool's research was diverse and included soil survey, plant nutrition, liming as well as soil freezing. In 1921, Merris McCool organized the *American Soil Survey Workers Association* meeting in East Lansing. At that meeting, Curtis Marbut presented the Russian concepts of classification and the study of soil profiles.

A handful of counties in Michigan had been mapped by the *Bureau of Soils* in the early 1900s by Elmer Fippin, Warren Geib, Thomas Rice, among others [19]. The majority of soil surveys were started after 1920 when Michigan signed a cooperative agreement with the *Bureau of Soils* [18]. Michigan had the advantage that it was not set in the system of soil provinces and types from Milton Whitney, and could thus adopt the new approaches that Curtis Marbut was promoting [18]. The Russian concept of soil was known among soil scientists in the early 1920s, had met resistance, but by the 1930s Curtis Marbut wrote confidently: "Every American pedologist should familiarize himself with at least the broad lines of the history of the development of pedological science in Russia. Russian work established it firmly as an independent science with criteria, point of view, method of approach, processes of development applicable to the soil alone and inapplicable to any other series of natural bodies. That work determined a definite relationship between the soil and the environment in which the soil is found, thus showing the soil to be related, on the plane of development, to biological bodies and not wholly physical" [20].

Jethro Otto Veatch was hired in 1921 as a soil surveyor in Michigan. For the *Bureau of Soils*, he had mapped soils in Pennsylvania and Iowa in summer, and in Texas in winter [19].Jethro Veatch was also born in Missouri, attended the University of Missouri, took geology courses from Curtis Marbut and worked for him as an assistant in geology. He spent a semester with Andrew Whitson in Wisconsin, and then worked in Georgia as an assistant geologist. Curtis Marbut hired him to conduct soil surveys in Texas and Pennsylvania, and found him to be one of the best soil surveyors [18].Jethro Veatch was hired in Michigan as the state leader for soil survey work, when the state signed its agreement with the *Bureau of Soils* [18]. He had an interest in organic or peat soils, and in 1936 proposed to study 'water soils,' or those of marsh, lake, pond, and rivers, and thought it was an excellent opportunity for pedologists to do pioneering work as those soils were not studied by others.

In 1925, Jethro Veatch spent a year in Scotland, where he was appointed by the International Board of Education of America. The purpose of the visit was to conduct studies in soil classification in cooperation with the agricultural colleges in Edinburgh, Glasgow, and Aberdeen. There was not much known about the soils in Scotland. The geographic distribution of soils was based on geologic formations or the till that covered the rock, and soils were named heather, moor soils, carse (alluvial) soils, and lowland or highland soils. Jethro Veatch definition of soil included all media capable of carrying vegetation such as peat, and in Scotland, he distinguished four classes of soils: common mineral soils (Communisols), organic soils (Plantasols), rock soils (Lithosols), and water soils (Hydrosols) [21]. He met with Gammie Ogg, an advisory officer in Soils at the Edinburgh, and they were of the same age and interested in pedology and soil survey. Some years later, Gammie Ogg accompanied Konstantin Glinka in an excursion across Hungary and America [22, 23]. He adopted the Russian methods of classification based on soil profile descriptions and found Konstantin Glinka the master in the field study of soils. Vasily Dokuchaev had founded pedology, but according to Gammie Ogg, it was Konstantin Glinka who brought Russian soil science to the world.

Merris McCool and Jethro Veatch understood the importance of the entire soil profile, and were among the first to demonstrate that horizons differentiated by the profile method was a meaningful method [24]. They used the Russian origin of the ABC soil profile, to which Jethro Veatch publicly referred as 'the Glinka scheme.' In 1923 Merris McCool suggested the term *podology* for the science of the entire soil profile, and years later, Roy Simonson wondered if that was an error and it should have been *pedology.* [24–26]. But an earlier version of the paper also contains the term *podologist* so it was likely not a misprint [18]. Podology never rooted. As his former students, Merris McCool and Jethro Veatch were important in promoting and testing Curtis Marbut's ideas [18].

In Michigan, land pressure was high, particular at the beginning of the First World War, and there was a high demand for agricultural products. The government encouraged farmers to produce more, and a loan act was created for farm expansion. Farmers used the loans to buy land, tractors, and equipment, but these loans came with high interest rates. Over 18 million hectares of land went under the plow in the wheat- and corn-producing states of the Midwest. When the war was over, relief efforts kept the demand for agricultural products high, and much of it was exported to allied countries. But prices dropped as Europe began to recover from the war, and the farm economy began a downward trend that spiraled toward the Great Depression. The price of land slumped, farms were hard to sell, and large areas were returned to the state when farmers and in particular, farms in marginal and cutover areas were abandoned, as homesteaders could not pay the taxes or high interest rates. Even the farms that had been profitable when commodity prices were high went bankrupt. Millions of hectares were laid fallow and returned to bush [11].

Michigan was also in a need of a land-use policy as timber removal left large areas barren, and logging was accompanied by forest fires. There was little land-use planning beyond the idea that the worst land could always be planted with trees whereas the rest of the land could be used for agriculture. The need for a land economic survey was summarized by Lee Schoenmann in 1922: "What is required are policies based on facts as obtained from two localized sources of information: First, an inventory by counties or regional units of the present status of the State's resources and the industries arising out of, and dependent upon their development and use; and second, a study of the economic conditions and factors which are accountable for their present existing state of use or lack of use. Such an inventory and study will reveal that certain regions of the State have problems peculiar to themselves; and that these can only be successfully handled when recognized as local" [27]. It was within this framework and economic setting that Charles Kellogg made his first steps in soil survey.

Charles Kellogg became interested in soils after working a summer with Lee Schoenmann in the field. Before attending college, he had seen soils on the farm, in roadcuts, and when they buried a dead horse or a cow. While the soil profile with its different layers fascinated him, his interest was wider; he saw how poor land was sold to people, and how houses and farms were abandoned after they went bankrupt and returned to the state. As he wrote in his diary: "Someone, must do something about this, and a soil survey will be needed." The need for change, and the field experiences with his mentor Lee Schoenmann drove him to soil science, and he started to study physical chemistry and geology. The combination of a good mentor, his hardscrabble upbringing on the farm, and the poverty and potential of the surrounding lands put Charles Kellogg on a path that he followed his entire life, a path carefully carved across the USA and eventually, the globe.

During his summer job, he noticed that soil surveying was more of an art than a science; they had no manuals and he learned by observing the senior surveyors. There was no standard vocabulary and different words were used for describing soil horizons. Initially, descriptions were being prepared of 'soil sections' which consisted of three layers defined by fixed depths. The layers were labeled surface soil, subsoil, and substratum [28]. Soil color terms were confusing, and the field crew debated whether the name 'chestnut soils' referred to the color of the nuts or to the bark of the tree. Letter designation for soil horizons was not used and they mapped soil series based on some soil physical observations. There was much to be learned, and although he had taken soil courses at M.A.C., he had not heard of the term pH nor had the more senior soil surveyors. By that time, the *Bureau of Soils* had already put out field books for soil survey and mapping, but they were not used in Michigan [29].

The lack of standardization in the Michigan soil surveys in the early 1920s was no exception. Andrew Whitson was in charge of the soil survey in Wisconsin, and Emil Truog participated in a county survey in 1910, and he summarized his experiences as follows: "...it was not too difficult. Our soil survey vocabulary at that time was not encumbered with such words as podzol, chernozem, wiesenboden, solonetz, horizon, profile, catena, eluviation, ortstein, etc. For the most part, we got along with such commonly understood descriptive words as dark or light, tight or open, top-soil or sub-soil, level or hilly, and sand, silt, or clay" [30].

The next summer, Charles Kellogg again joined Lee Schoenmann and Jethro Veatch for a soil survey in northern Michigan. He surveyed a township and produced a map that showed different soil series. Much of Charles' early field knowledge came from working with Lee Schoenmann, who was 21 years old when he had become a soil surveyor in 1910. They remained friends for many years. When Charles wrote his first book in 1941, it was dedicated to him: *To L.R. Schoenmann—who first showed me the soil* [31].Lee Schoenmann was a devoted pipe smoker, a custom that Charles Kellogg merrily adopted and he became convinced that there were no good soil surveyors who did not smoke.

The field crew of the surveys consisted of a surface geologist, an expert in peat investigation, two botanists, two experts in water power, drainage and water supplies, and an economic geographer who collected data on tax conditions, land occupation and ownership, and assessed the economic situations [27]. There were students from the University of Michigan, *Michigan Agricultural College*, University of Wisconsin, and the University of Chicago, all of whom had been trained in Forestry, Soils, Geology, or Geography. The field crew was housed in camp with six sleeping tents, one office tent, and a cook tent. Meals were prepared on army field ranges, and the order of the day was:

Roll out 6.00 am
Cakes 6.30 am
Start for field 7.15 am
In camp 5.30 pm
Supper 6.00 pm
Pipe Down 10.00 pm.

An area was assigned to pairs of assistants by sections, one square mile or 260 hectares, and by strips or transects in unsettled areas. The mapping was based on quarter corners or control lines, and the standard interval between lines was one-half mile. The forester had the compass, and the soil assistant depended on the forester for the location and distances. During the survey season, the evening was spent discussing the field conditions encountered during the day. Charles Kellogg found that the forestry schools stressed fieldwork more strongly than what was common in the soil courses. During the first part of the season, the Division Chiefs accompanied a different pair of field assistants each day and demonstrated the methods of classification, mapping, and description. As the teams gained experience, the chiefs spent more time on checking and revising the early work, and in scouting new field crew members. Two detailed maps were made in the field, each on a scale of eight inches to the mile, or about 1 to 8,000. Mark Baldwin from the *Bureau of Soils* came to inspect the mapping [27].

In 1925, M.A.C. renamed itself as the *Michigan State College of Agriculture and Applied Science* as its curriculum expanded beyond agriculture. That same year, Charles Kellogg received his BS degree, and he spent another

McIntire Kellogg Gillis Humsden Porter Wildermuth Moon VerWiebe Enbury
Foster Winters Stahl Booth Sayre Schoenmann McLaren.
Webb

Soil crew and their two cooks in front of their tents Michigan 1921. Charles Kellogg is second from the left in the back row. The crew was led by Lee Schoenmann, sitting second from the right. Photo from the Kellogg family

summer on the soil survey with Lee Schoenmann; it now paid $150 per month. One of the survey team members promised him a copy of Curtis Marbut's translation of *Die Typen der Bodenbildung* by Konstantin Glinka [32]. The work had been mimeographed by the publisher Edwards Brothers in Ann Arbor, Michigan, but very few copies were available, and Charles never received his copy. They continued soil surveying through the autumn and early winter of 1925, and Merris McCool came to visit the field crew several times. By December, the soil was solidly frozen, and the soil survey work stopped. On a cold Christmas day, Charles Kellogg married the love of his life, Lucille Jeanette Reasoner, and they made a home in a small apartment in East Lansing.

Over the winter, Charles worked in the soils laboratory at the college, and Lee Schoenmann recommended him for a position at Purdue University [33].Merris McCool, however, had negotiated a PhD fellowship with the engineer V.R. Burton of the Michigan State Highway Department, and had Charles in mind for that position. Merris McCool and V.R. Burton had

The newlywed couple Charles and Lucille Kellogg, a day before Christmas 1925 in Palo, Michigan. Herbert Kellogg closing the door of the Model T-Ford. Photo from the Kellogg family

met at the *First International Congress of Soil Science* in 1927, where V.R. Burton had presented some of his work in Michigan on highway engineering, declaring that: "Building of roads is a very old art, but a very new science" [34]. It was studied in Russia since the early 1920s, where it was noted that: "…the time has now come for us to find the means of putting the work of highway construction on a scientific basis. Our colleagues, soil scientists, who for the present consider their practical problems as being connected with agriculture alone, could in this matter be of the greatest help to us. They could also find in highway construction many interesting problems; hitherto ignored, such as engineering soil science. Pedology, in particular, in connection with highway construction, is still awaiting its creator" [33].

In 1925, a combined pavement condition and soil survey was started by the Michigan State Highway Department, and a year later, over 3,000 km of the highway system were surveyed. Subsoil conditions and the behavior of different pavements were not well-understood, particularly in soils that were not clayey or sandy. The *Bureau of Soils* classification of series and types of profiles was used, and about 100 types of soils had been mapped in Michigan by the mid-1920s. A problem in road construction was peat that in some areas was more than 20 m thick [34]. Upon drainage, the peat soils subsided and roads became bumpy as the peat layers and their thickness were irregular.

Charles Kellogg declined the position at Purdue University and took the position at the Michigan State Highway Department, where he was given the title of Research Engineer. He was in the field from April to October and

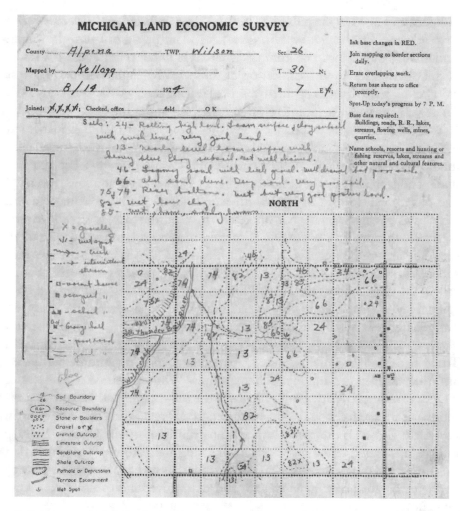

Soil map made by Charles Kellogg in the summer of 1926 in Wilson township of northern Michigan under the direction of Lee Schoenmann. Soil series were being mapped

spent the winter months in the laboratory studying the effects of freezing. He collected hundreds of samples by soil horizon and brought them to the laboratory for engineering tests. The purpose was to investigate the relationship between soil characteristics and the problem of buckling on highways that were paved with concrete. Such pavement expanded during summer and therefore had joints at regular intervals but these joints filled up with gravel and stones which caused the concrete to buckle when it expanded during the summer. Charles applied field and laboratory techniques for designing highways that had less potential for buckling problems.[35] He studied the problem and in 1929, he published his first paper entitled *Soil Type as a Factor in Highway Construction in Michigan* [36]. The paper did not cite much earlier work in Michigan or from elsewhere. He noted that there was a range of soil physical tests and classification of soils used by engineers but: "…most workers have completely ignored the profile character of the soil" [33, 34]. There was the basis for an interest that lasted his entire life.

While Charles was working on his PhD, the Head of the Department of Soils, Merris McCool, clashed with the university about expenses and finances. It was a difficult and tense time, and Charles talked to Lee Schoenmann about the possibility of finishing his PhD research at another university. Lee recommended him to his old advisor Andrew Whitson, who accepted Charles as a PhD student. For his research, Charles Kellogg was to investigate soils in northern Wisconsin, and he accepted the offer and moved to the University of Wisconsin in April 1928. In Madison, he liked the atmosphere of research and scholarship, but there were problems between what he called 'the titans of soil.' Andrew Whitson and Emil Truog had worked on a soil survey in Iowa county and on the benefits of liming soils, but they did not get along. It was a conflict between generations, between a geologist and chemist, and between a Minnesotan and a Wisconsinite. Eventually, Emil Truog became Head of the Soils Department, replacing Andrew Whitson, and that stopped the arguing. When Andrew Whitson retired in 1941 after 41 years at the university, Emil Truog was the toastmaster and had only admiration for his predecessor. At the farewell party, Sergei Wilde played the viola, and Otto Zeasman presented him with a painted portrait. Charles Kellogg remained on good terms with both Emil Truog and Andrew Whitson as that was needed to progress his work. It was a skill that he had developed at home and with the field crew in Michigan, and that proved indispensable for what laid ahead.

Back in Michigan, Merris McCool was asked to resign from Michigan State College: "…as a result of his- failure to comply with rules and regulations as established by the Board and the College relative to the administration of finances of his Department. The irregularities were of such a nature that there was no other way for us to proceed. Some definite and decisive action had to be taken by the Institution in an official way." Emil Truog promptly wrote him a letter: "I was very much surprised, in fact, stunned, to hear that you had resigned your position at the Michigan State College. Under your leadership, the Department of Soils at Michigan State College has developed a well-balanced way into one of the leading Soil Departments of the country. I have no hesitancy in saying that I believe your Department is among the four or five of the leading Departments of its kind. I have had occasion to come into intimate contact with a number of your graduate students and have found them to be extremely well trained along both practical end scientific lines. Your Department has been very much alive in both the scientific and practical phases of work. I believe if one were to go through the scientific literature on soils for the past ten or fifteen years, it would be found that about as many good scientific papers have come from your Department as from any other in the country. These papers have dealt with a wide range of subjects and many new and original ideas have been given to the soil workers of the world."

Emil Truog then suggested that he apply for the open position at the Carnegie Institute or the Experiment Station in Hawaii, but Merris McCool took a position in Plant Physiology at the Boyce Thompson Institute in New York. The institute was established in 1920 and aimed to study "why and how plants grow, why they languish or thrive, how their diseases may be conquered, how their development may be stimulated by the regulation of the elements which contribute to their life." The institute sponsored the *First International Congress of Soil Science*. While working at the Boyce Thompson Institute, Merris McCool sent a letter to Curtis Marbut asking whether he could replace him after his retirement. He had taken geology from Curtis Marbut at the University of Missouri and was an early adopter of his work. They both attended the 1924 meeting in Rome where the *International Society of Soil Science* was founded, and Merris McCool had helped with the 1927 congress. Merris McCool did not get Curtis Marbut his position, the job was given to Merris McCool's former student.

When Charles Kellogg arrived at the Department of Soils in Madison in 1928, it had graduated over 120 students with a soil degree, half of whom with an MS or PhD The PhD program was the largest in the department. In Wisconsin, soils had been surveyed since the 1880s and the first soil map in 1883 showed the topsoil textures across the state [37]. The soil survey in Wisconsin was officially started in 1893 as part of a geological and natural history survey [38]. Andrew Whitson was in charge of the Wisconsin soil survey and the Department of Soils.

After a short stay in Madison, Charles moved to Bayfield County, some 500 km north along the shores of Lake Superior. The soil survey team included Ken Ableiter who was working on his MS degree. Although he was not able to be part of the survey, it was coordinated and directed by Andrew Whitson who had surveyed the soils of Bayfield County with Warren Geib and Lee Schoenmann some 20 years earlier. They had produced a 6 inch to the mile soil map (1–10,600) based on topsoil textures [39]. As director of the soil survey, Andrew Whitson emphasized the need for a greater unification

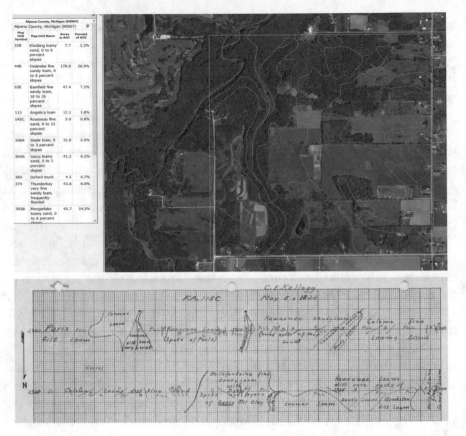

Current web soil survey map of Wilson township in northern Michigan, and cross-section of soil observations made by Charles Kellogg in the summer of 1926

of methods and aims [40]. Trained as a geologist, he resisted the pedological ideas of Curtis Marbut on soil formation and distribution, and when Andrew Whitson visited the field crew in Bayfield toward the end of the fieldwork, he had a critical look at the map legend, and said: "I see you have the Bureau concept of soils." He was a man of a few words.

Charles Kellogg was inspired by Curtis Marbut who had made him aware of the Russian approach: "Since the rise of the Russian school of soil science, soil classification has been recognized as a separate branch of science. Its object of study is the soil, considered as a dynamic body, produced through the operation of natural forces. The action of climate, vegetation, and other agencies upon the surface of the earth has produced soil bodies with certain characteristics. On the basis of these characteristics the classification is made. The soil itself, as the product of soil-building forces acting upon the formations, is the object of classification. The important soils of Wisconsin have been studied in order to determine something of their character and how they are to be classified."

The Bayfield County report was published in 1929. Thirteen soil series with different soil texture phases were distinguished, and the report discussed the geology, climate, and recommendations for agricultural improvement. Andrew Whitson emphasized the economic evaluations of farmland, and the survey was modeled after the Land Economic Survey [27]. Classification of soils was used for land assessment:"The suitability of land for agriculture or other use is determined by a number of factors. Some of these factors, such as chemical composition and water holding capacity, affect the character of crops or tree growth to which they are adapted. Other characteristics such as lay of the land and stoniness affect the expense of clearing and operation as farmland. It is necessary, therefore, to classify the soils in such a way as to express all of these important features. The chief factors which will determine the best use to be made of unimproved land are texture or fineness of grain, topography or lay of the land, and stoniness" [41]. The report of Bayfield County did not have the discussion of the soil-forming factors that characterized later reports.

As director of the soil survey, Andrew Whitson visited all ongoing surveys in Wisconsin over the summer. During a survey in a central county of the state, the field crew, including Warren Geib, suggested that they moved their field operation from the swampy southern part of the county to the northern part, as they were being eaten alive by mosquitoes. Andrew Whitson reprimanded the crew and rejected the move: "The work will continue as planned from south to north in an orderly manner. I can assure you that the things that you complain most about in the present are the things that you will boast about in the future" [42]. And so they did. Andrew Whitson was quiet, steady, and sturdy and saw the soil almost as a holy thing made to serve humans and do God's will [42].

After the fieldwork in Bayfield County, Charles Kellogg returned to the farm in Palo, and to Lansing where his son Robert was born. Back in Madison, he used the data of the soils of Bayfield to investigate relationships between properties and the age of the soil, work which would become his PhD thesis. The soils of Bayfield were compared to those in Europe and Russia, which he knew were not older, but more developed and more podzolized: "...it is not time in the sense of years but the stage of development of the soil in relation to its environment which constitutes 'soil age.' But assuming no change in climate, it can not be doubted that at some future time they will reach the same stage of development, or even a more advanced stage" [43]. He separated the effects of geology, parent material,

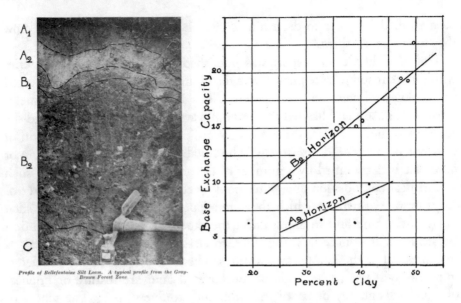

Profile of Bellefontaine Silt Loam. A typical profile from the Grey-Brown Forest Zone

Soil profile of the Bellefontaine silt loam – typical profile from the grey brown forest zone. Relationship between the base exchange capacity and the clay content of two horizons of the grey brown forest soils. From *Preliminary Study of the Profiles of the Principal Soil Types of Wisconsin*, by Charles Kellogg in 1930

and vegetation on soil profile development and soil processes. The soils under prairie were high in bases in the surface horizons but that was easily lost upon cultivation:"How rapid this degradation may be is not known. It may be mentioned that sufficient applications of calcium compounds to render the colloids immobile will prevent the podsolization process" [43]. He studied the relation between the amount of bases and the texture in horizons of a mature soil and postulated that if soils had reached a state of equilibrium, a constant relation between the amount of bases and the percentage of clay in the B2 horizon as compared to the A2 horizon would be expected. He concluded that: "After a soil has developed a mature profile, the horizons have come into equilibrium with the environment and the only change is one of depth unless there is a change of the environment."

Charles Kellogg's thesis was entitled *A study on the profiles of the soils in Bayfield County, Wisconsin*. Andrew Whitson edited it after which it was submitted to the *Michigan Agricultural College*, where it became the fourth PhD thesis from the Department of Soils. He finished the thesis a few months before the country sank into the Great Depression. Unemployment was high, and in Michigan, over one-third of the population had no work, and people moved out of the state [44].

Charles Kellogg in 1929 shortly after finishing his PhD, no glasses, no pipe. Photo from the Kellogg family album

Charles had experience in soil survey and had established some professional connections. In 1929, the American world of pedology was small, and he was determined to get a job, bit in the meantime, Andrew Whitson offered him a short-term position to study and sample the main soils of Wisconsin. Charles accepted the position and traveled with a field crew across the state and sampled tens of soil pits. They made detailed soil profile observations and brought soil samples back to Madison, where they were stored in glass jars. The position did not pay much, and as he was still paying rent for an apartment in East Lansing where Lucille and his son Robert were staying, in Madison Charles slept on a cot in a vacant office in the Department of Soils building.

While working in the laboratory and writing the report, he continued looking for university or research jobs, but few hires were made in academia. Emil Truog alerted him of a job with a fertilizer company that paid $5,000 per year, and a soil vacancy at *North Dakota Agricultural College* that would

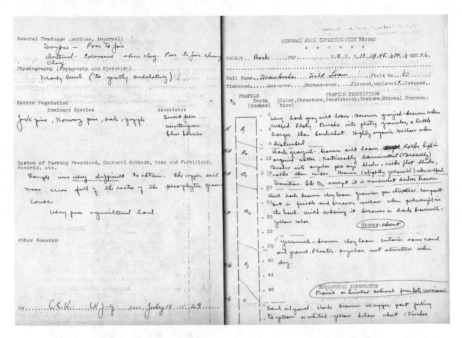

Soil profile description from Charles Kellogg, 19th July 1929, Rock County, Wisconsin. The soil was a Waukesha Silt Loam—a soil series that no longer exist. The nomenclature of the horizons was as follows: A₀₀—Recent forest accumulation, leaf litter, etc. A₀—Forest mold. The compaction portion. A₁—Upper mineral soil containing organic matter. The transition between A₀ and A₂. This horizon is practically absent in the Podsols. A₂—The zone of maximum leaching. A₃—Transitional to B, but with more characteristics of A than of B. Usually absent. B₁—Transitional from A to B, but with more characteristics of B than A. Usually appears as the weathered B₂. Often too thin to be important. B₂—The maximum zone of accumulation. Usually has the most dense structure. It may not always be characterized by a great percentage of clay, but represents the region of "B horizon tendencies." B₃—Transitional to C, but with strong characters of B. C₁—The upper part of C, slightly weathered. (Usually not separated). C—Parent material. D—Lower strata, such as deep clay under sand, which can in no sense be considered as parent material to any portion of the solum

pay $2,500 per year. By the end of the 1920s, Emil Truog had trained tens of MS and PhD students and took interest in their welfare [45]. Emil Truog and Charles Kellogg discussed the two positions, and Charles preferred the academic position in North Dakota. Emil Truog wrote a letter of recommendation to Harlow Walster, Dean of the *North Dakota Agricultural College*, who had studied soils in Wisconsin. In November 1929, Charles Kellogg attended the 2-day meeting of the *American Society of Agronomy* in Chicago, where he met, for the first time, Curtis Marbut. He also met with Dean Harlow Walster, and they talked for a while. They were both Freemasons.

A young Dr. Charles Kellogg with pipe in 1933, and Morrill Hall where his office was at *North Dakota Agricultural College* in the 1930s. The Hall was named after J.S. Morrill who wrote in 1862 the Land-Grant Colleges Act that established public colleges and universities in the USA

It was 2 weeks after the stock market crash which would bring down the American and world economy for a decade, but Charles was invited for an interview in Fargo, which took place toward the end of November 1929.

North Dakota Agricultural College was established in 1890—one year after the state had been admitted to the union. Some 20 years later, when all the land had been taken and was cultivated, the state adopted a flag and the motto 'Strength from the Soil, Reapers of the Deep.' The college started teaching a soil physics course in 1900, and there were fertilizer experiments with wheat. The study of soils gained some reputation and by the mid-1920s, an emphasis on soils was optional for a degree in the Department of Agronomy [46]. The first soil surveys in North Dakota were made by the *Bureau of Soils* and staff from the college in 1902 [47].Macy Lapham mapped soils in North Dakota in 1908, and Thomas Rice in 1910, and by the early 1920s, Harlow Walster had been a member of a soil survey team working in the north-central part of North Dakota [19].Charles Kellogg was hired to further develop the soil survey activities in the state, and to teach soil courses in Fargo. It was a thousand miles from Palo.

He had finished his PhD degree and left Madison on the last day of 1929 arriving in Fargo on 1st of January 1930. In his Model-T, he brought a suitcase with clothes, some books, and a box with the soil samples in glass jars that he had collected in Wisconsin. Lucille and Robert arrived 2 weeks later, and as he wrote in his diary: "We soon got used to the cold." Besides his travel through Michigan, Wisconsin, and Illinois, the land west of the Mississippi was all new to him. Fargo, situated in a small section of the Red River Valley, was flat, windy, mostly treeless, and only the cold winters reminded them of the Wolverine state where trees were abundant and the land was rolling. Being naturally connected to the land, it was years later that Charles Kellogg pondered about moving across America, and how it affected feelings and tempers: "One who has lived on the plains or steppes feels the forest to be a place enclosed, a prison. Yet one from the forest seems exposed, lonely, on the plains. How the people, the women especially, who came from the podzolic soils of the East to the treeless Chernozem and Chestnut soils of the plains suffered from loneliness!" [31].

In Fargo, he was warmly welcomed by Dean Harlow Walster, who had grown up on a farm on the Kettle Moraine in Wisconsin. Harlow Walster had obtained his BS degree at the University of Wisconsin in 1908 under Andrew Whitson and had obtained a MA at Harvard and a PhD from the University of Chicago. With Andrew Whitson, he wrote the books *Notes on Soils,* which aimed to present a brief outline of work in soils for students in agriculture [48], and *Soils and soil fertility,* which presented the idea that the soil was a revolving fund from which nutrients can be removed or lost [49]. The soil was described as a bank with deposits and withdrawals like Justus von Liebig had conceptualized it. They discussed the revolving fund in terms of losses and gains: "This fund suffers some losses and profits by some additions. There are losses by the removal of crops, by leaching of the soil and subsoil, and by erosion of the surface soil, which is richest in soluble matter. There are gains by the weathering of the rock particles of the soil and by fixation of nitrogen from the atmosphere. When the losses exceed the gains, the difference must be made up by the use of fertilizers. The fertility of the soil which is removed in products sold from the farm may be either in vegetable form, as when grain or hay is sold, or in animal form, as in the sale of fat stock or dairy products. The bones of all animals, for example, contain phosphorus" [49]. That concept, based on the theory of Justus von Liebig, has been used in nutrient balance studies across the globe [50], but received criticism from Charles Kellogg, who found that it left much unexplained, like for example why soils were so different in the first place: "…some very poor and others very rich" [31].

A few days after his arrival in January 1930, he wrote to Emil Truog as a 28-year-old assistant professor: "I am here." They corresponded for about 4 years, with letters about magnesium, phosphorus, soil survey, students, teaching, publications, the economic conditions, family matters, and the situation on the job market. Emil Truog wrote to Charles Kellogg on 31st of

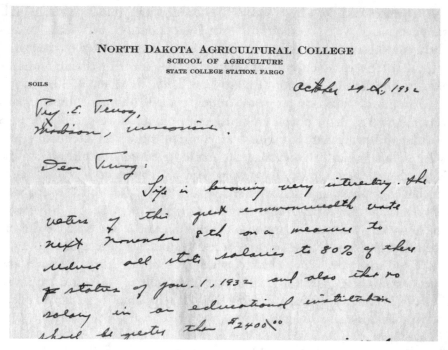

Fragment of a letter from Charles Kellogg to Emil Truog from October 1932. Emil Truog preferred typed letters, and had commented on his handwriting in another letter: "I gave your previous letter to some of the boys to read and they said that they wished you would write a little plainer so that they would not have to spend so much time figuring out your penmanship". The boys were the graduate students in the Department of Soils at the University of Wisconsin

January: "We are having nice spring weather here and I presume you are having mild weather also." The letter came from Madison, Wisconsin. Some winter days were mild.

Charles was given an office in the newly built Morrill Hall where the soil science courses were being taught. Charles found his new colleagues cordial and agreeable but: "…of course, things are on a smaller scale here; however, the organization is housed in a fine new building. The greatest and most apparent need is a laboratory. It's a good thing there is no laboratory the first term for it will take three months to organize the chaos of balances, etc. and millions of soil samples." The laboratory was a concern, and he wrote: "I hope to help the bacteriology people in running H-electrode determinations on soils. They have great difficulty in getting a constant reading in the case of soils above 6.5. That is, the soil sample will vary even when the apparatus is all OK on the buffer solutions. Dr. Walster tells me that the work here shows that alkali soils give no response to phosphate fertilizer!" The alkali soils, characterized by a high soil pH, poor structure, and low water infiltration capacity, were common in North Dakota but not in Wisconsin and Michigan where the rainfall is twice as high, and salts wash out of the soil.

Emil Truog recommended that Charles Kellogg started working on fertilizer experiments, and Charles wrote to Emil Truog in March 1930: "As you know fertilizers have not yet found much place in the agriculture in this section; however, many people are considering phosphates especially for corn in order to hasten maturity. However, the results are very conflicting and as far as this Station is concerned the matter is entirely empirical. On the Station here acid phosphate has always been profitable and hastened maturity, but rock phosphate has never shown any effect. This soil here is a very heavy, alkaline clay with a high content of organic matter (Fargo Series)." The heavy clay soils were common on both sides of the Red River separating North Dakota from Minnesota and occurred from Wahpeton all the way to the Canadian border. They cracked widely and deeply in the summer and were so sticky after rain that it is almost impossible to walk or drive across the field. In Texas, Alabama and along the Mississippi they were named 'black cotton soils' whereas in North Dakota they were named the 'lacustrine soils of Lake Agassiz of the Red River Valley '[51].

As these clayey soils had a high phosphate fixing capacity, Emil Truog suggested hill fertilization because it was more effective than broadcast fertilization. The plans for the experiment were not finalized by the early spring, and Charles Kellogg wrote to Emil Truog: "Dean Walster and I have still another conference on the fertilizer plans (when, I don't know), but I doubt very much if he will do anything this spring. Finances are in terrible shape here. There is hardly money for common office supplies." Although he had just started, his appointment was uncertain, and he was informed that there was some chance he could stay provided he was not on the payroll over the summer. He mentioned that he was looking for a position with some security and that he was much in need of encouragement because of all the uncertainty at the college. Emil Truog replied: "I am hoping that it may be possible for you to find a position which will make it possible for you to utilize and develop the special training which you have along soil physics and soil classification lines."

Emil Truog suggested that Charles write to the Head of the Department of Soils at the University of Missouri, Merritt Miller, and ask him about the position of William Albrecht who would be on a leave of absence for a year. The position would pay $2,000 per year and Emil Truog believed the experience would be good for him: "I think, however, that you would be able to fill the place satisfactorily and perhaps bring some new lines of thought to

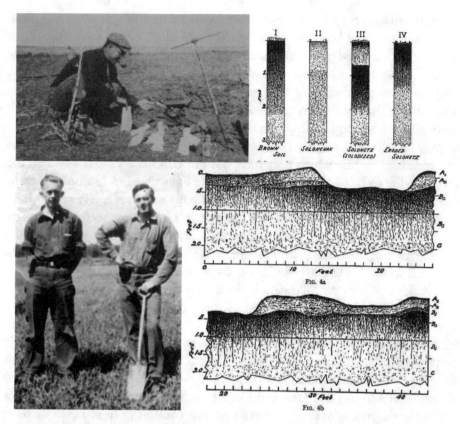

Top photos: Charles Kellogg, with pipe, sampling soils in a pit in McKenzie County in North Dakota – note the surveyor in the soil pit; some of the principal types of soil profiles in western North Dakota. Lower photos: Roy Simonson (left) and Enoch Norum mapping soils of a quarter section as a class exercise for a course at *North Dakota Agricultural College* in the spring of 1932, and a cross section made to scale through an eroded spot on a Solonetz complex. Enoch Norum became the chair of the Soil Department at North Dakota State in the 1960s

the department. I know you are probably just about fed up on 'pinch hitting' and that you would like to become one of the regulars on a team as soon as possible. I feel that there will be some other positions opening before fall. The experience at Missouri, I believe, would be worth while to you and would give you an opportunity to advertise yourself to Professor Miller, and I am sure that would have considerable values as far as further work is concerned." Charles was not excited about the position at the University of Missouri and thought the salary was too low. He declined the position but was delivered a newborn bundle of joy: Mary Alice was born on 28th of February in Fargo and she was named after his paternal grandmother. Sometime later he wrote to Emil Truog: "My little girl is a dandy – but horrors. She has red hair."

In February of 1930, the publisher William and Wilkins sent to Emil Truog an outline of book proposal by Curtis Marbut. Emil Truog and Curtis Marbut had worked together during the preparation and excursion of the *First International Congress of Soil Science* a few years earlier. The book proposed was entitled *Classification of United States Soils* and the publisher asked Emil Truog to review the proposal in strict confidence. He shared it with Charles Kellogg adding the note from the publisher and suggested a change in title to *Soils of the United States—Formation, Classification and Use.* He added: "The title which they have did not cover the subject and also has too little life to it. I believe a book along the lines that Dr. Marbut has indicated would be especially valuable for many purposes and believe would have a considerable sale, not only in this country but in foreign countries. At any rate. It would have the advantage of getting Dr. Marbut's ideas along this line in permanent book form." Emil Truog told the publisher that he was hardly qualified to do this and asked Charles to go over the proposal.

Charles sent him the following reply, only 4 days later: "As to Dr. Marbut's book: Frankly I think his outline calls for too much philosophy. I had hoped that Dr. Marbut would find it expedient to make an exhaustive scientific treatise, including all know worthwhile material on soil morphology. And it certainly is by no means limited to the United States: the big function of Soil Geography is to treat soil morphology on a world basis. Then we know the character of Russian, German and English soils, as compared to our own, we can apply the results of their work to our own conditions in so far as we have similar soil conditions. We can never do this until we have a comprehensive view of the soils of the world. I think this point to be very important and fundamental to the progress of soil science. Marbut is not in a position to discuss agricultural economics. And if he were, the place for it is not in a scientific treatise on the SCIENCE of soils. I think that the last three items, together with portions of the three previous ones, should be cut down. I question the place of the chapter on 'Fundamentals of Soil Productivity.' I think that the chemistry involved in the development of soil morphology should be given the major consideration, together with the discussion of soils as they are. Sir E.J. Russell has a good treatise on Soil Productivity: let this book be as 'meaty' a treatise on soil morphology. I suppose a companion volume could be issued dealing with economics and that phase of agriculture. Suggested title is very poor, besides being improper English. I would suggest: SOILS: THEIR GENESIS AND CLASSIFICATION. I cannot believe that Marbut is in a position to discuss their use in anything like an exhaustive manner."

Much of Charles Kellogg's way of thinking on soils, his somewhat unremitting critical style, and emphasis on soil morphology was embedded in his review of Curtis Marbut's book proposal. He was a few months in his first job, and 28 years old. Curtis Marbut was one of the most respected and knowledgeable soil scientists in the country, but Charles fearlessly dissected the proposal. It must have pleased Emil Truog. The book was never published—perhaps because of the criticism or because time was up: Curtis Marbut died in 1935. Years later, Charles Kellogg reflected on Curtis Marbut's writing: "Somehow his writing lacked the sparkle and eagerness of his speech. His alertness of face, his smile, the sparkling eyes, and the brandishing finger enlivened his talk, made it easy and simple, and still gave it a sense of urgency" [52]. Some of the material that was proposed ended up in the *Atlas of American Agriculture* but it took until 1951 before the book was published and the material was prepared by Guy Smith, Henry Krusekopf and Francis Hole [53, 54]. The book carried the title which was recommended by Charles Kellogg 16 years earlier.

In his first semester, Charles Kellogg wrote up the description and sampling of the soil profiles across Wisconsin that he had conducted in 1929, and that was published as a bulletin *Preliminary study of the profiles of the principal soil types of Wisconsin* [43]. In the spring, he traveled to Madison and gave the manuscript to Andrew Whitson who was disappointed as he had expected that Charles would have questioned the work of Curtis Marbut. Andrew Whitson demanded changes that had to be made if it were to be published. Charles made the changes and told Emil Truog that evening about it. Emil Truog thought for a while and spoke decidedly: "...you have a key to the office, go in and put the text back as it should be. Prof. Whitson will never know the difference." Charles did as suggested, and the bulletin was published as he had originally written it [43]. Here, like perhaps for the rest of his life, he followed his individual will instead of conforming to social expectations as the reading of *Self Reliance* and works of Charles Emerson might have taught him. Importantly, he had the support of Emil Truog who often publicly argued with Andrew Whitson.

As Charles Kellogg was not paid during the summer months, he was offered a position with Lee Schoenmann in Michigan mapping soils in the swamps of the northern part of the state. Lee Schoenmann had been mapping soils in Michigan since the early 1920s, and he had worked with him during several summers when he studied at the *Michigan Agricultural College*. It would pay $175 per month and Charles found "...it seems tough to go to such salary—but the baby must have shoes." His position at *North Dakota Agricultural College* became increasingly uncertain and students worked up a petition to keep him. He thought that would not be wise although it was gratifying that they were willing to write such petition. In April 1930, he ended his letter to Emil Truog: "With a family here, furniture in Madison, no definite position in the offing, and busted—I become philosophical." As a true mentor, Emil Truog ended his letter with: "I shall be pleased to hear further about your work. With best wishes, and regards, I am, very truly yours. E. Truog. Professor of Soils."

Instead of the summer position in Michigan, Emil Truog recommended Charles to a wealthy friend to conduct a soil survey of his estate near Baltimore. Charles was keen to get a well-paid summer job and when it came through Emil Truog wrote to him: "I think you will begin to take seriously what I said several months ago, when I intimated that some jobs would certainly come along before snow flies." Charles responded: "I need no prompting in regard to the possible job of making a soil map of a plantation near Baltimore. I would take it by all means should the chance come. Not only would it replenish the exchequer, but also would give me some much needed experience with the Eastern soils and agriculture."

The search for other positions continued throughout his first semester at *North Dakota Agricultural College*. In May 1930, he wrote to Emil Truog: "We haven't been paid for over two months and no one has any idea when we will be paid. What damn politics this state has. I would not like very much to leave here now as I have the best group of students coming up for the senior year next year I've ever seen anywhere. But how to live? " He applied for a position at the University of Hawaii and Virginia Station. Lucille had some doubts about the position in Hawaii, and Charles wrote: "I'm having a little trouble to convince Mrs. Kellogg that she would like to go to Hawaii if that job came through but I think I can before the time comes." He also applied for a research position in Tucson, Arizona, but it did not offer contacts with students that he enjoyed but he thought that it would be good for a few years: "I have observed, however, that men in universities are chosen more on the basis of their research work than on their teaching ability. This being the case I think that a few years spent in such an organization would put me in line for a better position in a teaching institution. Like many people without the necessary talent, I hope to be able to obtain an administrative position some day in the (not too far) distant future. If I am able to get such a position I hope it will be before I've become 'old and crabby.'".

Dean Harlow Walster tried to obtain a permanent position for Charles Kellogg at *North Dakota Agricultural College*. It was soils, Wisconsin and Emil Truog that connected the two. Emil Truog wrote in May 1930 to Harlow

Walster also encouraging him to make the position of Charles Kellogg permanent, to which Harlow Walster replied: "I have not yet had official action in regard to the appointment of Dr. Kellogg but his name is still on the budget recommendation and as far as I know will be left there by the President. If he is removed from the list it will be through Board action only." Toward the end of May, his appointment was assured, and Charles thanked Emil Truog for all his efforts: "My morale is up and how I thank you!" His salary was set to $2,500 for 9 months with the promise of being appointed to Associate Professor on a 12-month basis within a year, as well as private laboratory. He liked the students at *North Dakota Agricultural College* but added: "The thing which I dislike here very much (Q.T. for heaven's sake) is the extremely low scholarship of the faculty, especially the Agricultural group and those in Education and Literature. Walster is a prime and, in my opinion, easily the best man on the campus." Emil Truog was pleased with that response and wrote: "…I am glad to note that your morale and pep are on the up-grade. Things are never half as bad as they appear."

In the first semester at *North Dakota Agricultural College* Charles wrote philosophical letters to Emil Truog, whose response was always courteous but not of the same theoretical level. His letters were practical and encouraging. Charles wrote to him in February of 1930: "…I never before fully realized or appreciated your remarks: 'Read science' and 'Learn fundamentals.' As you know, I've always been interested in the consideration of knowledge in a broad way and its application to human life. The more advanced men here are very much interested in the USE of knowledge and the putting of two and two together. But alas! There is no two and two? The library has all the advanced and recent work in philosophy but no recent books on science. I'm beginning to come to your way of thinking: if the men get the fundamentals the rest will probably come but if they don't have those nothing can possibly ever come. In my advanced course I've spent more time on fundamentals as applied to soil physics than I have on soils."

Charles Kellogg was not an experienced teacher and expressed his doubts to Emil Truog: "I hope to remember your advice – not too much but thoroughly on a few points. I plan to stress plant physiology all of the time. I will give the sophomores a short quiz nearly every period. My reason is that I feel young students become fatigued in long exams and the shorter ones keep them 'on their toes.' Now if you have any criticisms, I would be pleased to hear them." In the winter semester he wrote to Emil Truog:"I have six seniors for two lectures a week (no lab) and for two terms. I will give them (I think) a good stiff course in soil physics and soil chemistry, particularly plant nutrition and the physiological aspects theory, for the first 8 or 10 weeks and then proceed to the practical applications." There were not many textbooks available to teach soil science: "Dean Walster has made up my mind. Bear's *Soil Management* is the best text to use for the sophomores [55]. Dean Walster didn't like Weir [56], and as I'm in no position to be a very good judge, I made no comments. He wants quite a bit presented on Soil Profiles. I also understand I must give, or help give, a session course in Soil Management." For his teaching Charles Kellogg used the outline of Norman Comber's *An introduction to the scientific study of the soil* [57]. A slim book that presented an introduction to soil science for agricultural students, with an emphasis on English soils and almost exclusively citations to English journals and books.

Since 1928, pedology was taught at the University of Nebraska by J.C. Russel and in 1932 Charles Kellogg started to teach Pedology in North Dakota (Soils 355: Pedology). The course was an effort departing from the prevailing explanation of soil genesis as rock weathering or deposition of weathered rock material [58].Roy Simonson took Soils 355 from Charles Kellogg, who stressed the importance of logic from the beginning to end of the term [58]. Most attention was given to Michigan, Wisconsin, and North Dakota as he knew those soils best. He discussed the soil-forming factors and used the expression 'Soil $= f$ (pm, cl, v, r, t)' to explain soil formation.

Teaching went well and students were interested in his classes. He was worried whether there would jobs for the students after their graduation, and in 1932 he wrote to Emil Truog: "I have now about 12 to 15 major students in Soils. Next June we'll have four or five men to place. It looks bad. There probably isn't any other field of agriculture that is better, however. These men are the cream of the ag school and will be able boys. What would you suggest? Naturally I'm not committed to place them, but nevertheless I feel a certain responsibility." Like Emil Truog, he cared for his soils students and organized parties around the house that were also attended by students majoring in English. The economic depression and the drought continued to make for a difficult financial situation, and the faculty salaries were reduced to 80% of their status at the beginning of the year and no salary in an educational institution was allowed to be greater than $2,400 per year. There were shortages of many goods and one of the land grant colleges provided information on how corn could be used as a fuel in rural schools [10].

In 1930, he had a summer job in the Worthington Valley, north of Baltimore. A soil survey had to be conducted of a 1,200-hectare estate owned by C. Wilbur Miller who was the President of the Davison Chemical Company that produced sulfur and phosphate fertilizers. He was a friend of Emil Truog who had suggested that Charles would conduct the soil survey: "I recall that you expressed a desire to have a soil survey made of your farm and I promised to see if I could find someone who is qualified and who would be available to do a job the coming summer. I have a man, Dr. C.E. Kellogg, who is unusually well qualified and who, I think, could be available for this work the coming summer. This work would consist in making a detailed map showing in colors the different kinds of soil and also, if desired, the topography. Samples of soil would also be taken and analyzed so that you would know exactly what the condition of fertility is on each field and even the different portions of the same field." The job paid $500 per month including all expenses, and in those summer months Charles Kellogg made as much as half of his annual salary at *North Dakota Agricultural College*. It also paid almost three times more than the soil survey summer job that was offered by Lee Schoenmann in Michigan.

A detailed soil map of the estate was made, and he collected samples that were sent to Madison for analyses. The cropping history of each field was recorded and related to the soil data. In Madison, the soil samples had to be analyzed, and he wrote to Emil Truog: "I appreciate your anticipation of my distaste for the labwork; however, I feel that it would be good to do it, especially if you are there to correct errors." But he did not analyze the samples himself and hired a graduate student. Based on the field survey and soil analytical data, he wrote a detailed plan for liming and fertilizer use at the estate. After the survey in Baltimore, he returned to Madison by way of Palo, where he had a severe asthma attack that kept him at the farm for a few days. With the earnings of the soil survey in the Worthington Valley, he bought a two-door 1929 Pontiac and drove it back to Fargo.

Back at the college, he organized courses in soil chemistry and physics, in plant nutrition, and an elementary course in geology and his teaching load was much heavier than what he was told at the interview. The following year, he taught a course in soil management and a seminar for junior and senior majors. Students majoring in soils had their own study room, and on any night, the lights would be on until 10 o'clock. He also talked to farm groups, wrote for farm papers, and continued working on phosphorus fertilization on the black clay soils around Fargo. A phosphorus kit developed by Emil Truog was tested, and Charles found it: "…to be utterly worthless. It is true that it does not check with plot yields from fertilized plots nor does it consistently indicate phosphorus deficient fields." Emil Truog ignored the comments but offered a detailed explanation for the lack of a response as the alkaline and alkali soils lack iron and manganese for good crop growth. He also thought that sulfur supplied with the phosphorus may be responsible for the yield increase and that the calcium in the superphosphate may have given a more favorable ratio of calcium to magnesium or sodium. The soils did not have high sodium carbonate and the alkalinity was due to the high levels of calcium carbonate. Emil Truog suggested conducting some greenhouse tests with the soils and mentioned that Charles could send him 80 kg of the soil for testing in Madison: "…we test it out in the laboratory with our regular laboratory method, from our results up to date with both the laboratory method and the field method, I feel sure that quite reliable results are secured in most all cases. There is, of course, always the possibility of finding a soil which acts in a peculiar way and I am especially interested to find these in order that we may investigate them and find out what the peculiarity is." The 80 kg of soil was promptly sent.

In the first year at North Dakota, Charles Kellogg's research mostly focused on soil fertility, and as he wrote to Emil Truog: "The whole field of soil fertility is wide open in this country and if I can only get the time and a reasonable amount of green house space, there are several important conclusions that only need a little thought and work." He found magnesium deficiency in barley and related it to phosphorus availability and wrote a paper that he sent to Emil Truog: "Acting upon your suggestion I've written the enclosed note for *Science (the journal)* (I am a subscriber). The only data, in addition to what I have now, would be the yields of ripe barley which I can get in a few weeks. These data will show large differences because of the much larger heads of the MgO-treated plants. This would not, however, add much and I question if they should be given in such a note anyway. No more data would be available until next winter. But now early publication is necessary, or someone will beat me to it. I would much like a letter of acceptance before school closes. It would help me here a great deal. A letter to *Science* is enclosed. If you care to, you might make any changes, needful in your opinion, and forward the manuscript directly, thus saving time. It is highly probable that a letter from you would hasten publication and insure acceptance. But this only provided you feel the publication justified. I'm so enthusiastic about it that my judgment may be off, but all the parts fit together so nicely."

Emil Truog wrote him back: "My experience has been that it will take at least six months to have an article of this kind published in *Science*. The *Journal of the American Society of Agronomy* has in recent years made a practice of and has also, I believe, encouraged the publication of short articles which they place under the heading of 'Notes'. I believe these notes are printed rather promptly, by which I mean within a month or two. The regular articles are printed within six months and hence under any circumstances you would at least get the article as quickly in this Journal as in *Science*, and, after all is said and done, I believe it would come to the attention of more agronomists and fertilizer people than if published in *Science*. For these reasons I am now suggesting that you send it to the Journal rather than *Science*. I have never been able to get anything published in *Science* short of six months regardless of how urgent I tried to make the matter appear. I have tried to shorten the article as much as possible because it often works out that the promptness of publication is just about inversely proportional to the length of the article." Charles Kellogg followed Emil Truog's advice and the paper *Magnesium— a possible key to phosphorus problem in certain semi-arid soils* was published in 1931 in the *Journal of the American Society of Agronomy* [59]. It was his second paper. In November, he went to the *American Soil Survey Association* meeting that was held for 2 days in the Stevens Hotel in Chicago, where he presented the study on the relationship between magnesium and phosphorus deficiency. At the meeting he listened to Jethro Veatch talking about soils in relation to the growth of aquatic plants, and papers by Hans Jenny, Richard Bradfield, Sergei Wilde, Jacob Lipman, Macy Lapham, Mark Baldwin, Charles Shaw, Selman Waksman, and Curtis Marbut.

Early 1931, the Board of Supervisors in McKenzie County asked the *North Dakota Agricultural College* if they could survey the soils and classify the land for tax assessment. They asked Charles who instantly responded: "Of course we can do it!" McKenzie County in northwestern North Dakota covered 730,000 hectares of flat and badlands, and thousands of little lakes bordering Montana in the west and the Missouri River in the north. The soil survey offered him work over the summer and in the spring, and he shared the news with Emil Truog, who responded: "I was glad to note that you have finally completed arrangements to be on the Soil Survey. I do not know whether you have heard the news or not but it you have not, the news is that the Wisconsin Soil Survey is entirely wiped out for the present. No appropriation was made for it and there does not seem to be much hope of getting money from any other source at the moment. You can therefore thank your lucky stars that you are up in good old North Dakota." And lucky he was and wrote back: "I finally got an appointment for the summer. The Director put my salary at $25 per month less than he had definitely agreed to do! But as you say, I should be thankful for most any job."

That summer Lucille and the two children went east to Michigan and Charles drove west to Watford City, that was some 600 km west from Fargo. Watford City, the county seat for McKenzie County, became their base for the soil survey. The crew included seven surveyors from the *Division of Soil Survey* (successor of the *Bureau of Soils*) and eight from North Dakota Agricultural Experiment Station. Constantin Nikiforoff, Ken Ableiter, Roy Erickson, and Roy Simonson were part of the survey, and of the 15 surveyors, only three had mapped soils before. McKenzie County was in the drier part of the state with some 350 mm of rain per year. Prior to 1900, most of the land was used for cattle ranching, but rapid settlement occurred with homesteaders from eastern North Dakota, Minnesota, Iowa, and Wisconsin in the early 1900s [60]. Most of the homesteading took place between 1904 and 1910, and the settlers grew wheat but also some oats, barley, rye, and flax. The rolling areas were settled before the plains.

Most of McKenzie County was in the Glaciated Missouri Plateau but the southwestern part missed out on glaciation [61]. Its soils were new to Charles Kellogg, and the land was flat against the backdrop of the eroded badlands and high prairies. He had seen the leached and sandy soils of Michigan and Northern Wisconsin, the soils with clay in the dense drift, and thick black loamy soils covered with grass. Now, there were soils that had ample clay and salt. Traverses were made by a pickup truck, on foot, or on horseback, and they slept at farms and ranches during the week where they were fed as well. The farmers called them 'dirt inspectors' [62]. In order to understand the saline soils, he asked Thomas Rice, the inspector of the *Division of Soil Survey*, to analyze some of the soil samples. They mapped soils using auger observations, a sharpshooter drain spade, field maps, and a plane table on a tripod. In the valley of the Little Missouri River, they found an abandoned homestead cabin that had old furniture and heaps of books including the complete works of William Shakespeare and a novel by Fyodor Dostoevsky.

During the soil survey his salary was further cut; frustrated he wrote to Emil Truog in August 1931: "When I left Fargo, Director Trowbridge agreed to pay me my regular salary $277 plus expenses. After I'd been here and received salary for June, he changed it to $200 + $3 per diem. Now he has changed it to $200 + actual expense (also including July). This means I'll receive $205 less for the summer than he originally agreed or $135 less from what he gave in a letter to Dr. Walster. I was told he was very dishonest but didn't get a letter on the agreement. I have a copy of his letter in regard to the $200 + $3 per diem. I've written to him about it. Unless he changes it back, would you start a legal action to recover the $135.00? There is no hope for the first as I did not have a letter on the matter. I'm too upset to write logically, I guess. This salary deal is the most unethical treatment I've ever received. I'd sure love to quit but I guess one can't do that now. I hope I can find some other place for next summer." Charles Kellogg started looking for other jobs and wrote to Emil Truog: "I wonder if the Carnegie position in Soil Physics will ever open up? I hope so for this looks very good to me." It did not open up.

Curtis Marbut and Thomas Rice visited the survey in McKenzie County and they were accompanied by Dean Harlow Walster. It was an honor for Charles and the crew, but it was also a test to see how they were doing. Thomas Rice had surveyed soils in western North Dakota with George Coffey, and he had mapped soils in 12 states, including North Dakota, where he spent much of 1910. Although Thomas Rice was an experienced soil surveyor, he was quiet and did not say much during the visit. Charles Kellogg showed them the draft maps, and Curtis Marbut had a long look, was silent for a while, and then criticized the map legend which he thought was too lengthy and detailed. Charles convinced him that such detail is required for accurate land appraisal, and a discussion followed. At the end, Curtis Marbut more or less agreed with the legend but found that the mapping progressed much too slowly.

After inspecting the maps and legend they went to the field. Curtis Marbut climbed to the top of a small butte, and looking over the countryside, he remarked that the field crew should be able to generalize more and thereby make faster progress. He was not prepared to demonstrate such generalization [47]. On top of a high conical scoria hill, Curtis Marbut said: "Kellogg what soil difference makes that line down there?" They overlooked plowed fields interspersed with fallow land, and Charles replied: "Dr. Marbut, that line is not due to a soil difference. It marks an old, old field boundary. On one side the gramma grass is the original vegetation. In the old plowed field, the needle grasses have invaded, and these are now yellow." Curtis Marbut was not convinced, and they walked down the hill and dug a small soil pit in both areas. The soils were different, and Charles explained that in this landscape soils could not be mapped from the hilltops but digging soil pits and traversing the land was the only way to map the soils. They returned to Watford City and at dinner that evening Charles said: "Now, Dr. Marbut, that you have seen the soils and you know what is demanded of the survey, which kinds of soil in the legend would you combine?" There was silence, and then he said: "It's all right, all right." The old master conceded.

For Roy Simonson, the first weeks of fieldwork in McKenzie County were an overwhelming experience, and he was unable to remember any of the soil profile observations. He was a young student and Charles Kellogg had instructed him to take field notes and came to check his notes at the end of the week. The mapping continued until snow fell, and as Curtis Marbut had told them the challenge was to develop a map legend and test it. They mapped about 2.5 square kilometers per day and mapping generally went faster in the eroded Badlands. The soils in McKenzie County were new for Charles Kellogg, and many pits and transects were dug to study the soils [65].

As he wrote to Emil Truog: "This walking will do me much good. These soils are also very new and interesting. The native soils belong to the Chestnut group. We have black alkali and saline soils also, but only in small bodies. This particular area is in very bad shape because of prolonged dry weather. We are now having good rain, however."

Charles Kellogg studied the Solonetz and solodized Solonetz in detail. They were striking [47]. He sampled small bodies of soils with natric horizons in a matrix soil without natric horizon which formed in miniature bases of about 15 cm deep, and these were called slick spots. There were up to 12 of these spots per hectare [62]. They were common features in the Great Plains, and some decades earlier, Vasily Dokuchaev had referred to those slick spots as 'smallpox scars on the face of the steppe' [62]. Russian soil scientists had divided the alkali soils into two groups: Solonetz with a definite structure, and the Solonchak without soil structure [8]. A prominent feature of the Solonetz soils in McKenzie County was the abundance of sodium, and Charles Kellogg postulated a pedogenetic pathway for these soils based on the studies of Konstantin Gedroiz which had been given to him by Curtis Marbut. Charles Kellogg distinguished three broad groups; normal soils (Chernozems) had the base exchange was saturated with divalent cations such as calcium and magnesium and these soils had a neutral pH. The second group was the saline version (Solonchak) that had excess soluble salts especially sodium and the smallest particles, named colloids, were flocculated. Thirdly, he distinguished the alkaline soils (Solonetz) that had high sodium but low percentage of soluble salts. These soils had a prismatic structure and a very strong soil structure with unbreakable aggregates [65].

Fieldwork for the soil survey of McKenzie County took two summers and was finished in 1933. The report did not get published until 1942, and it became the most comprehensive soil survey without the use of aerial photography [60]. Such photographs had been used in soil survey in the early 1920s, but they were not available for McKenzie County [12]. The soil map was detailed and four slope classes were recognized and mapped; stoniness and the quality of grass cover had four classes [66]. The soils were classified in 21 series that include 27 types, 20 phases, 8 complexes, and 4 miscellaneous land types. The report was written by Ken Ableiter and M.J. Edwards, who had been with the soil survey since the 1910s, and had conducted surveys in Alabama, Arkansas, Florida, Mississippi, but mostly in Wisconsin. Neither of them had been in the field crew that had conducted the survey in 1931 and 1932. Ken Ableiter later became the principal correlator in the Great Plains based in Lincoln [67].

The 1942 McKenzie County soil survey report included a summary statement of the five factors of soil formation, followed by a summary of the morphology of some soils and placement in one or more great soil groups. Unlike earlier reports, it had a section on soil survey methods and definitions. The general descriptions of soils in map units were longer than those in previous survey reports, and the report had a list of native plants, with both common and Latin names. Much of the credit for their identification belonged to Martin Johnson, a farmer who advised Charles Kellogg on various matters related to grazing and how the land should be rated. Martin Johnson had a son named Bill Johnson who went to study soils with Charles Kellogg at *North Dakota Agricultural College*, and many years later replaced him as chief of soil survey [35].

During the fieldwork of the soil survey in McKenzie County, the crew had collected data on the carrying capacity of the soils, on marketing costs, and on the distance animals could graze from water resources. Back in Fargo, Charles Kellogg started to develop a system for rural land classification based upon the potential productivity of the land. Such systems had been used in Russia and Germany where it drove the need to understand and map soils [63]. He convinced Harlow Walster to hire Ken Ableiter to help with the land classification. For tax assessment, a productivity rating was assigned to the soil map unit during the fieldwork [47]. The best soil rating was 100, and other soils were assigned lower ratings in steps of five, down to zero. Soils that were not suitable for any crops were given a rating of 30 or less and these soils were considered suitable for grazing. After all ratings had been made, the extent of the map unit would be measured, with some correction for the distance from cropland to the market or distance from water for grazing.

Charles Kellogg and Ken Ableiter realized that the land classification method could be adapted elsewhere, but that the details of the method would vary [64]. Besides tax purposes, the land classification system was to raise value on better land and lower them on poorer land but when it was finished and made public, it brought protests from farmers and ranchers. Ken Ableiter was assigned to deal with the protests, and he found some mapping errors, made corrections, and by the end of 1935, protests had ended. The land classification system was used for several decades and considered fair by most taxpayers [62]. The work in McKenzie County advanced the use of soil surveys for taxation purposes and other interpretations, and the approach was adopted by other states [35, 47].

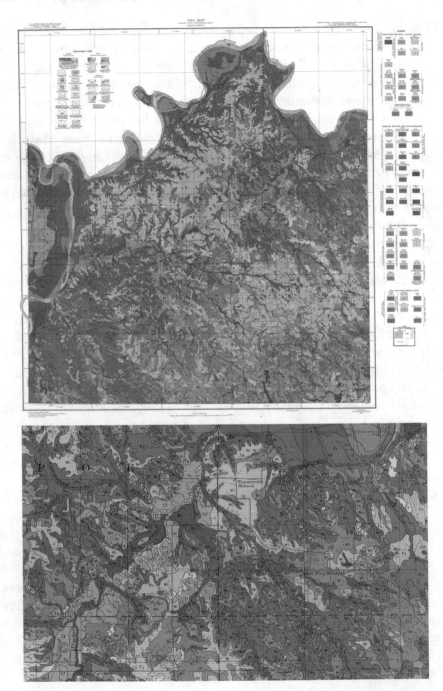

Detailed soil map of McKenzie County in North Dakota published in 1942 (scale one inch to the mile, or 1:64,000). Field work was conducted in the early 1930s by Charles Kellogg, Ken Ableiter, Roy Simonson, Constantin Nikiforoff and several others. Map was made without the use of aerial photographs

The ongoing financial crisis and drought continued to affect his work and his salary. In 1932 he wrote to Emil Truog: "Well, it has happened—our salaries have been cut 10% with the very good possibility of another before next year is over. In addition, there must be other economies, including staff reductions. Chapman is fired after July 1st, and no one knows when he will 'get his'. I do not feel greatly alarmed about my $2,250 job—but one never knows. I certainly feel chagrined to think about this job and the study and work it has taken to get it. But it may be one should be thankful for that and hope that it gets no worse. Have you any suggestions of places for me to write in regard to a better position? Or is that simply out of the question?" Many of his colleagues looked for other jobs, and Charles applied for a position at Clark University in Massachusetts where Curtis Marbut lectured every year. He also considered writing a book on soil classification that could be used by people in geology and economics and was thinking to go into the details of the chemistry of soil genesis since Curtis Marbut was not going to write it.

In the summer of 1933, Thomas Rice from the *Division of Soil Survey* visited and told him that they were creating a position for Charles at the *Division of Soil Survey* as assistant to Curtis Marbut. It was a big surprise, and it would give him the chance to study soils and develop his land classification approach for the entire country. He thought it could be a good position for a few years and assumed that there would be much politics at that level of the soil survey. Excited about this prospect he wrote to Emil Truog: "… of course I would rather have a decent position in a university, but the chances aren't as good. I'm spoiled." Charles Kellogg had also heard that Andrew Whitson from the University of Wisconsin was to retire from the Wisconsin Soil Survey, and he wondered what the chances were for that position. And he had just applied for a soil survey position with the newly established Tennessee Valley Authority.

Later in the summer of 1933, Charles Kellogg and Roy Simonson traveled from North Dakota to visit the World's Fair in Chicago [68]. Roy had become his favorite student, and in a letter to Emil Truog, he characterized Roy as: "…the nearest a genius I've ever seen in soils work." Before traveling to Chicago, Roy Simonson had to see a physician about his ears, as his hearing was becoming impaired. The doctor found that his ears were almost plugged with a combination of wax and dirt, and there was an oat kernel in one of his ears. At the recommendation of Charles Kellogg, he bought a new suit ($12) and leather Gladstone bag ($8) for the trip which was a bit excessive in Roy's view. On their way to Chicago they stopped in Madison, where in order to save money, they slept in the Department of Soils building.

Charles Kellogg and Roy Simonson went to the fair every day; Roy had a hotel on the outskirts of Chicago and traveled on the elevated subway, nicknamed the L, and met with Charles every morning at the front steps of The Field Museum. They strolled through science hall, the futuristic agricultural buildings, and were amazed by the abundance of technology and revolutionary architecture. In the exterior of the Electrical Building, bas-reliefs represented 'the conquest of time and space'– a consciousness that was slowly and deeply embraced in the study of soils. Visitors received the booklet *Giants of the New Age*, that told them: "…beneath the surface of modern life, the gigantic forces of Science move on quietly, steadily." The development of the electric furnace was exhibited, as well as many products created by synthetic chemistry such as anti-freeze, solvents, long-playing phonograph records, dental plates for false teeth, and photographic film. The age of plastics was announced [69].

As opposed to World's Fair in Chicago in 1893, and in St. Louis in 1904, there was no special soils exhibit in 1933. The official tour guide dedicated, however, a full chapter on agriculture and included a sentence that could have been written by Emil Truog: "Chemistry has taught us to banish or to put to good use insect life and fungus growths to analyze the soil and enrich it." The Department of Agriculture had an exhibit and distributed a booklet that illustrated how "The United States Department of Agriculture touches your life in scores of ways." The booklet had some information about soil erosion and soil fertility, and the notion was made that: "…promoting agricultural welfare advances the general welfare. Even in helping the farmer to conserve his soil and to produce better crops at a lower cost, the Department benefits the nonfarmer" [70]. All over the fair, there was the hope that the nation was in transition and that President Roosevelt would end the depression [71]. Several installations from the 1933 exhibition toured the country in subsequent years and promoted optimism in agriculture [71]. Over 39 million people visited the fair, and there was an outbreak of amoebic dysentery with more than a thousand cases and 98 deaths.

Roy Simonson went to a baseball game one afternoon after which he wandered along the shore of Lake Michigan and sat on the masonry wall behind the aquarium watching ships leave and head out into the lake. The joy of watching the ships must be his Viking ancestry, he wrote in his diary. At night colored light illuminated the grand buildings and monuments, and Chicago was the 'City of Lights' and 'Rainbow City.' At the last evening, Charles Kellogg and Roy Simonson reminisced about all they had seen at the fair while they were overlooking the lake and could hear the city roaring behind them. Roy thought that he did not like Chicago: "The individual sounds of the city mingled into one roar and I could no longer sort out the sounds of the trains, cars, the elevated, and the railroads. Together they made large sound as though there were a huge centrifuge sitting behind him sucking in the wheat, the cattle, the ores, and the people and spewing out flour, steaks, steel rails, and broken old men." His ears might have been too clean. Charles Kellogg noted that: "Men as conceited as they both were could not stand Chicago because they were simply too unimportant "[68]. They were both young, had grown up in a rural area, and lived in Fargo that had about 30,000 inhabitants whereas Chicago in 1933 had a population over 3.5 million. It was quite the experience, and some years later Charles Kellogg would write: "To the countryman cities seem to be conscious creations of men, of men without roots, who have grown apart from the soil" [31].

After the fair, Roy visited the farm of Charles Kellogg in Palo as he was interested to see how farming was done in Michigan. He took a bus that traveled through the sand dunes along Lake Michigan, and onward to Grand Rapids and Palo. He was warmly welcomed by the Kellogg family, and the next day they made a soil observation behind the farm in a forest with secondary growth oak and hickory. They distinguished five A horizons within the first 40 cm, and three B horizons up to 1 m depth. Each horizon was described, including color such as yellow or rusty yellow, texture, roots, and gravels. The soil was named a Kent loam, which was described in 1924 in Muskegon County some 100 km to the west. The had a clay-enriched layer but that had partly lost its clay; such layer became much later known as a glossic horizon.

The next day, they traveled thousand miles to Fargo in the two door-Pontiac. Roy sat with the two children in the back seat. The luggage had to be shipped back separately because of a lack of space and Charles Kellogg let Roy pay for that. They drove north through the straits of Mackinac, across them by ferry, and then west along the shores of Lake Superior, through Minnesota and into the plain of the Red River Valley. Back to work, to the calming land blanketed with grass and wheat, with the wind in the willows. And back to their college and its dwindling budgets.

Soil observation at the Kellogg farm in Palo, Michigan, by Roy Simonson, 30th August 1933. The profile sketch showed several A horizons with plant remains, an A2 (now designated E) horizon, followed by several B horizons. Charles Kellogg and Roy Simonson had just returned from the Chicago World's Fair, stayed the night at the farm in Palo, and drove the next day back to Fargo in North Dakota

References

1. Davis CM. Readings in the geography of Michigan. Ann Arbor, Michigan: Ann Arbor Publishers; 1964.
2. Veatch JO. Soils and land of Michigan. Lansing: The Michigan State College Press; 1953.
3. Graff GP. The people of Michigan: a history and selected bibliography of the races and nationalities who settled our state. State library occasional paper no. 1. East Lansing: Michigan Department of Education, Bureau of Library Services; 1970.

4. Branch EE. History of Ionia County Michigan: her people, industries, and institutions: with biographical sketches of representative citizens and genealogical records of many of the old families. Indianapolis: B.F. Bowen & Company, Inc.,; 1916.

5. Tower P. The journey of Ionia's first settlers. Mich Pioneer Hist Scoeity Collect. 1897;28:145–8.

6. USDA-Soil Conservation Service. Soil survey of Ionia county. Washington D.C.: USDA in cooperation with Michigan Agricultural Experiment Station; 1967.

7. Vuisotzkii G. Glei *Pochvovedeniye*. 1905;7:291–327.

8. Glinka KD. The great soil groups of the world and their development (translated from German by C.F. Marbut in 1917). Ann Arbor, Michigan: Mimeographed and Printed by Edward Brothers; 1928.

9. Kellogg RL. Remembering Charles Kellogg. Soil Surv Horiz. 2003:86–2.

10. Kellogg CE, Knapp DC. The college of agriculture: science in the public service. New York: McGraw-Hill Book Company; 1966.

11. Schmaltz NJ. Michigan's land economic survey. Agric Hist. 1978;52:229–46.

12. Gardner DR. The national cooperative soil survey of the United States (Thesis presented to Harvard Univsersity, May 1957). Washington DC: USDA; 1957.

13. Cline MG. Agronomy at Cornell 1868 to 1980: agronomy mimeo no. 82–16. Ithaca, NY: Cornell University; 1982.

14. Robertson LS, Whiteside EP, Lucas RE, Cook RL. The soil science department 1909–1969—a historical narrative. East Lansing: Michigan State University; 1988.

15. Woodruff CM. A history of the department of soils and soil science at the university of Missouri: special report 413 college of agriculture. Columbia, Missouri: University of Missouri; 1990.

16. Shaler NS. The origin and nature of soils. Department of the Interior. U.S. Geological Survey. Washington: Government Printing Office; 1892.

17. McCool MM. Some unusual soils that occur in Oregon. Agron J. 1914;6:159–64.

18. Merkel DM. The curious origin of Podology: the story of a milestone paper. Unpublished manscript, sabattical report; 2013.

19. Holman HP, Pease VA, Smith K, Reid MT, Crebassa A. Index of publications of the Bureau of chemistry and soils. 75 years 1862–1937. Washington DC: U.S. Department of Agriculture; 1939.

20. Joffe JS. Pedology (with a foreword by C.F. Marbut). New Brunswick: Rutgers University Press; 1936.

21. Veatch J. Geography of the soils of Scotland. Michigan academy of science, arts and letters. 1928;X:179–89.

22. Mitchell RL. Sir William Gammie Ogg. Year Book R.S.E. 1980:67–1.

23. Ogg WG. The contributions of Glinka and the Russian school to the study of soils. Scott Geogr Mag. 1928;44:100–6.

24. McCool MM, Veatch JO, Spurway CH. Soil profile studies in Michigan. Soil Sci. 1923;16:95–106.
25. Foss JE. Milestones in soil morphomelogy and pedogenesis. Soil Sci. 2006;171:138–41.
26. Simonson RW. Origin and acceptance of the term pedology. Soil Sci Soc Am J. 1999;63(1):4–10.
27. Schoenmann LR. Description of field methods followed by the Michigan land ecomomic survey. Bull Am Soil Surv Assoc. 1922;IV:44–2.
28. Simonson RW. The United-States soil survey—contributions to soil science and its application. Geoderma. 1991;48(1–2):1–16.
29. Whitney M. Instructions to field parties and descriptions of soil types: field season 1903. Washington DC: U.S. Department of Agriculture. Bureau of Soils; 1903.
30. Truog E. Enlarging the use of soil survey maps and reports. Soil Sci Soc Am Proc. 1950;14:5–7.
31. Kellogg CE. The soils that support us. New York: The Macmillan Company; 1941.
32. Glinka K. Die typen der bodenbildung—ihre klassification und geographsiche verbreitung. Berlin: Verlag von Gebrüder Borntraeger; 1914.
33. Prokhorov NI. Soil science in the construction of highways in USSR. USSR: Publishing Office of the Academy, Leningrad; 1927.
34. Burton VR. Soil science in highway engineering. Vol Commission V. Commission VI. Miscellaneous papers. Washington D.C.: The American organizing committee of the first international congress of soil science; 1928.
35. Helms D, Effland ABW, Durana PJ, editors. Profiles in the history of the U.S. Soil Survey. Ames: Iowa State Press; 2002.
36. Kellogg CE. Soil type as a factor in highway construction in Michigan. Pap Mich Acad Sci, Arts, Lett. 1929;10:169–77.
37. Chamberlin TC. General map of the soils of Wisconsin. Beloit: Wisconsin Geological Survey; 1882.
38. USDA-NRCS. History of Wisconsin soil survey. Madison: Natural Resource Conservation Service; 2007.
39. Whitson AR, Geib WJ, Schoenmann LR, Musback FL, B MG. Soil survey of the Bayfield area, Wisconsin. Vol Bulletin no. XXXI, Soil Series No. 5. Madison: Wisconsin Geological and Natural History Survey; 1914.
40. Truog E. Andrew Robeson Whitson 1870–1945. Soil Sci. 1946;60:272–4.
41. Whitson AR, Geib WJ, Kellogg CE, et al. Soil survey of Bayfield County, Wisconsin. Bulletin no. 72A. Soil series no. 50. Madison, Wisconsin: Wisconsin Geological and Natural History Survey; 1929.
42. Beatty MT. Soil science at the University of Wisconsin-Madison: a history of the department 1889–1989. Madison: University of Wisconsin-Madison; 1991.
43. Kellogg CE. Preliminary study of the profiles of the principal soil types of Wisconsin. Madison: Wisconsin Geological and Natural History Survey; 1930.

44. Freedman R, Hawley AH. Umeployment and migration in the depression (1930–1935). J Am Stat Assoc. 1949;44:260–72.
45. Jackson ML, Attoe OJ. In Memoriam Emil Truog 1884–1969. Soil Sci. 1971;112(379–380).
46. Prunty L. Soil science and pedology disciplines in the North Dakota 1862 land grant college of agriculture: a review Fargo: Department of Soil Science. North Dakota State University; 2004.
47. Thompson KW. A history of soil survey in North Dakota. Dickinson: King Speed Publishing; 1992.
48. Whitson AR, Walster HL. Notes on soils. Madison: Democrat Printing Company 1909.
49. Whitson AR, Walster HL. Soils and soil fertility. St. Paul Minnesota: Web Publishing Co.; 1918.
50. Stoorvogel JJ, Smaling EMA. Assessment of soil nutrient decline in Sub-Saharan Africa, 1983–2000. Wageningen: Winand Staring Centre-DLO;1990. Report no. 28.
51. Willard DE. The soils of North Dakota. Fargo: North Dakota Agricultural College; 1909.
52. Kellogg CE. Introduction. In: Marbut CF, editor. Soils: their genesis and classification. Madison, Wi: Soil Science Society of America; 1951. p. ix–x.
53. Marbut CF. Soils of the United States. Part III. In: Baker OE, editors. Atlas of American agriculture. Washington: United Sates Department of Agriculture. Bureau of Chemistry and Soils; 1935:1–98.
54. Marbut CF. Soils: their genesis and classification (Lecture notes from 1928). Madison, Wi: Soil Science Society of America; 1951.
55. Bear FE. Soil management. 2nd ed. New York: Wiley, Inc.; 1927.
56. Weir WW. Productive soils—the fundamentals of successful soil management and profitable crop production. Philadelphia & London: J.B. Lipincott Company; 1920.
57. Comber NM. An introduction to the scientific study of the soil. New York: Longmans, Green and Co.; 1927.
58. Simonson RW. Early teaching in USA of dokuchaiev factors of soil formation. Soil Sci Soc Am J. 1997;61(1):11–6.
59. Kellogg CE. Magnesium—a possible key to phosphorus problem in certain semi-arid soils. J Am Soc Agron. 1931;23:494–5.
60. Edwards MJ, Ableiter JK. Soil survey. McKenzie County. North Dakota. Washington: United States Department of Agriculture. Bureau of Plant Industry; 1942.
61. NRCS. Soil survey of McKenzie County, North Dakota. Washington: USDA-NRCS; 2006.
62. Simonson RW, Ulmer MG. Initial soil survey and re-survey of McKenzie County North Dakota. Soil Surv Horiz. 1989;30:1–11.
63. Fallou FA. Pedologie oder Allgemeine und Besondere Bodenkunde. Dresden: Schönfeld Buchhandlung; 1862.

64. Kellogg CE, Ableiter JK. A method of rural land classification. Washington D.C. : United States Department of Agriculture; 1935.

65. Kellogg CE. Morphology and genesis of the Solonetz soils of western north Dakota. Soil Sci. 1934;38:483–501.

66. Edwards MJ, Ableiter JK. Soil survey McKenzie County North Dakota. Series 1933, No. 37. Washington D.C.: United State Department of Agriculture. Bureau of Plant Industry; 1942.

67. Grossman R, Lynn W. Remembering Andrew Russell Aandahl. Soil Surv Horiz. 2005;46(2):85–6.

68. Simonson BM, Roy W, Simonson A. Century as a soil scientist. Soil Surv Horiz. 2015;49:63–7.

69. Schonauer JR, Schonauer KG. Chicago in the great depression. Charleston: Arcadia Publishing; 2014.

70. Chew AP. Science serving agriculture. Washington: Government printing office; 1933.

71. White AF. Performing the promise of plenty in the USDA's 1933–34 World's fair exhibits. Text Perform Q. 2009;29:22–43.

5

From a Farm on the Plains—Roy Simonson

"Once a man had thrust his hands into the soil and knew the grit of it between his teeth, he felt something rise within him that was not of his day or generation, but had persisted through birth and death from a time beyond recall."

Martha Ostenso, 1937

"For each of us what we know best tends to become the universe."

Roy Simonson, 1976

The Great Plains of North Dakota were settled several thousand years ago by the Mandan, Hidatsa, Arikara, Sioux, and Chippewa tribes. The Turtle Mountain Band of Chippewa had migrated from the Great Lakes in the late 1400s and used hundreds of different plants for food, medicine, dyes, and rope. They knew the land and all it would give and take. The plains were clothed with grass that was occasionally burned, sometimes on purpose and sometimes by lightning which kept the trees out and maintained feeding grounds for the bison. Not long before that, glacial ice had leveled most of North Dakota, and after the ice melted and retreated, a fine-textured till on ground moraine was left behind. The till was fertile, and over time soils accumulated abundant organic matter that originated from grass roots which died each winter but vigorously regrew in the spring and summer. Fine clay particles and calcium stabilized the organic matter so that the microbes could not further break it down. There was soft powdery calcium in the subsoil that had been brought up from below by rising water in fine pores, and some of calcium had been deposited by the wind and washed down in the subsoil. Western wheatgrass, green needlegrass, and blue grama flourished and supplied organics in return for the inorganic abundance. Scattered across the plains were cottonwood and willow trees that lost their leaves in the fall and greened up in spring. The climate matured to hot and dry summers and very cold winters to let the land remember where it came from. Rain was not more than half a meter per year. The land laid untouched by the European plow until about 1870 [1].

Early European settlers in North Dakota were told that: "…the soil was so fertile that if one planted a spike in the ground in the early spring, by fall the spike would have grown to be a six-foot crow bar" [2]. The rain would follow the plow, according to the Scandinavians that settled in North Dakota in the late 1800s. The American Norwegian novelist Martha Ostenso described the European settlers in her novel, *The Stone Field*: "They were men with a surface hardness baked upon them by the brief, fierce suns of northern summers, frozen upon them by the cold of winters legend-long. But beneath their weathering they were ordinary men, likable or unlikable in part, sensitive or dull to the impact of life upon them – as ordinary as the very rich, the very poor, the unimportant. And the chances great" [3]. Limitations of the land were not well recognized and agricultural experimental stations tried to promote dairying in North Dakota [2]. However, there was an aversion to raising cattle: "…the Scandinavians have worked with fish too much. They do not understand anything with warm blood in it" [4]. The agricultural experimental stations then turned their attention to the promotion of wheat growing.

The European settlers called the state the 'northern edge of the farmers' last frontier.' In the Red River valley, the land was fairly flat, and settlers could break a claim and harvest a crop of wheat in one season. There was a rapid settlement and a speculative type of farming and it attracted settlers who became known as 'Bonanza farmers' [5]. They grew wheat under difficult conditions, prices were fluctuating and there were opportunities for high profits and large losses. Most settlers risked everything they owned or could borrow and some were very successful and by 1890, there were over three hundred farms of more than 400 hectares in the Red River valley [5]. In the 1890s, durum wheat was introduced that was nicknamed macaroni or pasta wheat. High yields were obtained, and as a result, more land went under the plow, and soon no land was left for settlement by the Europeans. By that time, the original settlers who had come thousands of years earlier, had lost all their land.

Compared to the eastern states, settlements in North Dakota were late, as were research and scientific exploration but a state-wide soil map was made in 1909 by the geologist Daniel Willard from *North Dakota Agricultural College* [6]. He knew the state well and recognized the importance of soils: "There are no great mineral deposits in North Dakota; there are no great forests; no extensive water powers on which industries can be built up. North Dakota's resources are, therefore, the resources of the soil. In this she has a rich heritage. North Dakota is probably destined to be one of the leading agricultural states of the great West and that means of the whole United States. One of the greatest questions, therefore, that confronts the practical agriculturist in North Dakota is that of the proper study of the soil as a basis for agriculture." Daniel Willard had the 'as the rock, so the soil' view on the origin of soil: "The soil we have already seen is a form of rock. In other words, the soil has been formed from the rocks of which the earth's crust is composed. In the first place, in what we may call the beginning, probably there was rock but no soil, and no plants. Gradually soil was formed from the surface rock, and somehow plants began to grow. In this primitive condition there probably were no such plants as we now see in the fields, nor was there any such depth of soil overspreading the rocks as we now find covering the landscape" [6]. By the late 1920s, the North Dakota State Legislature authorized counties to use funds for the classification of rural land for tax assessment and an effort was made to make a soil map of the entire state [7]. A soil survey was especially needed as appraisals of the soils ranged from 'great desert' to 'great potential.'

Roy Simonson was born in 1908 on a farm in the unincorporated community of Agate, close to the border with Canada. The settlement was named after a fine crystal silica rock that glaciers had transported and sporadically deposited in the area. Roy's parents had come from Norway, having first settled in Minnesota, and were landless when they moved to North Dakota in 1896. In Agate, there was nothing but open prairie to the west, and they were the first European settlers. They filed homestead claims on two-quarter Sects. (65 hectares each) in the spring of 1897, and those were their first steps toward owning land. To obtain title to a claim, they had to build a house, dig a well, plow and plant several acres, and then they had to prove up the claims that would have sufficient income. They built a shack, plowed the land, but they earned so little that Roy's father took a teachers examination and worked in the frontier schools to secure a small but steady income [8].

The Simonsons had eight children and Roy was their second child. At home, they spoke Norwegian but Roy learned English at school, and also picked up some profanity from the hired workers on the farm. The school was about four kilometers from the farm and had 20 children of all ages and most of them walked to school when the weather was good and came by horse and buggy in autumn and winter. Roy's sister Alice was a few years older and drove the horse, but older boys from nearby farms came on horseback and the school had a barn to stable the horses. Roy learned to read and was fascinated by *Arabian Nights* and the magical events in the book did not seem impossible to him. He also discovered the books of Mark Twain. After 5 years in elementary school and at the age of 11, he quit school, and as far

1. Breaking the Virgin Soil, Velva, North Dakota.

Plowing 16 feet through 7 inch Virgin North Dakota Soil.

Postcards from plowing up the prairies in North Dakota in the early 1900s. Velva was some 160 southwest from Roy Simonson's birthplace. They planted wheat, year after year. Postcards from North Dakota

as he knew, his education was complete. Most farm children did not go to high school.

Roy was small for his age and too small for work in the field, he worked inside the house. His parents decided to send him to the high school in the town of Bisbee, that had 100 students and 6 teachers. There, he boarded during the week in the parsonage of the Lutheran church to which his parents belonged. He went back to Agate for weekends and wandered the fields and hunted ducks or prairie chicken. School did not go well, as he lacked motivation. Roy's shotgun was taken away and his father cautioned that if he failed more courses, the shotgun would be sold. Some of the children in his class did not speak to Roy, because his father was a member of the North Dakota Nonpartisan League—a left-wing populist political party that was pro-farmer and supported state-operated mills and banks. Roy was too small to play baseball or basketball, and too shy for the stage, but he liked skating and thought that anyone who could afford shoe skates was rich. In 1924, at the age of 16, he graduated from high school without any idea of what he might do, and so he worked on the farm [9]. By that time, he was tall enough to operate the grain drill to sow wheat, oats, barley, and flax. The drill was drawn by four horses and he could sow eight hectares per day. Some years later his father bought a tractor that could sow 32 hectares in one long day.

Roy's parents thought that it would be good if he would learn about the outside world and pick up some useful skills for the farm, and he enrolled in a short course in agriculture at *North Dakota Agricultural College* in Fargo. These courses were practical and allowed farm boys to enter college. They were given in the winter so that they could work on the farm during the other seasons. In January 1926, Roy left Agate for Fargo which was established in 1871 and named after a Northern Pacific Railroad director. The town was built on the banks of the Red River valley and the clay loams from former Lake Agassiz sediments that covered glacial drift over Precambrian rocks. Getting off the train in Fargo, Roy felt lost as nothing was familiar and the city was bustling. Fargo had a population of about 25,000. He had signed up for a 10-week course in farm mechanics, where he learned about Model T Ford engines, stationary gas engines, tractors, forge shops, and magnetos. Roy did well, and his instructor encouraged him to enroll in a college-level course rather than the winter course but after 10 weeks, he went back to the farm.

His parents encouraged him for further studies and in 1927, he enrolled at *North Dakota Agricultural College* as a student in civil engineering and dreamt of building bridges. He was motivated less by the prospect of education than by a desire to be away from farm work. Roy was now 19 and restless. He took some engineering courses but often returned to the farm, not to work but to be outdoors. Civil engineering was not what he had hoped for, as they followed handbooks that he found uninspiring and he quit his study and returned to the farm where for 2 years. In the summer when there was less work at the farm, he played baseball with farmers from around Agate. He did much of the fieldwork and plowing of the fields in autumn took up 8 weeks using five–nine horses. In the winter, frost would break up the plowed clods and nature prepared a fine tilth for the sowing of wheat in May. Once the first pair of furrows had been made, there was little for him to do except sit on the plow and think. While the horses pulled the moldboard across the field and the autumn sun rose above the horizon, he stared at the fresh furrows and how the soil differed across the field. He had always thought that soils were much the same everywhere, but in the furrows, he saw differences in color and stoniness, and sometimes the horses had it easier than other times.

Roy was thinking of becoming a farmer, but his father expected that family farms would disappear in the near future, and they would be owned by large corporations with 'resident managers' handling several farms. He thought that Roy's farm experience and a college degree would give him a better chance to become such a manager. So, in January 1929, Roy went back to *North Dakota Agricultural College* and enrolled as freshman in the Division of Agriculture. Each time he returned to Fargo, it seemed less of a frightening place and smaller than before. More than 8% of the students in agriculture had Norwegian names, and most students were born to recent immigrants. Roy took foundation courses, including animal husbandry and livestock judging, and found them a waste of time as he believed that the reason for show animals was the vanity of the owner. For a course in English, he read *John Brown's Body* by Stephen Vincent Benét, which had just been published but he failed to appreciate it. He became disillusioned again with college courses, he had very little money and with spring weather coming, he was restless and ready to get outdoors. Again, he went back to the farm and got a job at the grain elevator in Agate.

Roy returned to college in the autumn of 1930 and picked up his study. Despite the lack of motivation, his grades were good and the Masonic lodge in Fargo awarded him a scholarship of $100. The Dean of the College, Harlow Walster, was a Freemason. Roy remained unimpressed by most of the teaching and had not chosen a major subject for his BS degree. In the spring of 1932, he took the *Soils 355* course taught by Charles Kellogg which he enjoyed, particularly the section on soil classification. Roy decided to major in soils, as did his friends Andy Aandahl and Marlin Cline, who prior to taking *Soils 355*, were majoring in other subjects [10]. Roy thought Charles Kellogg was an excellent teacher but their first encounter did not go well. During a laboratory class, Charles Kellogg noticed that Roy no longer took mathematics, as most students in agriculture dropped that subject. Charles Kellogg was upset by that, and he found that the lack of mathematics was a major deficiency in the education of agricultural students. Later on, Roy learned that Charles Kellogg would go to the Registrar's office and look up a student's grades in chemistry, English, and mathematics. If a student had good grades in all three subjects, he would help them; others, he would scare away from majoring in soils. Charles Kellogg was convinced that curricula put too much emphasis on agricultural techniques, when in fact they should teach fundamental science and language, as those: "…tools are the basic principles of science, logic and language" [11].

To pay for tuition, board, and lodging, Roy enrolled in the Reserve Officer Training Corps (ROTC) and like his fellow students, he resented military training, but they had no choice as it paid for their education. One day, Harlow Walster called him in and offered Roy a summer job as a soil survey assistant in McKenzie County in the northwestern part of the state. Roy had seen the land between Agate and Fargo but had not been west where it was much more arid. The pay was good: $60 per month plus expenses. Roy accepted the offer and with Charles Kellogg as soil survey leader, a field crew from the *Division of Soil Survey*, and a team of students, they made Watford City their base camp. Watford City was founded 20 years earlier by businessmen in anticipation of the arrival of the first Great Northern Railroad train that was to connect Minneapolis with Seattle. The town reminded Roy of what he had read in the 1920 book by the anti-Bolshevik Mikhail Artsybashev, *Breaking Points*: "The little town lay in the Steppes and beyond its outskirts, beyond the vibrating air of the distant countryside, lay the intangible depths of the horizon of towering forests and the remote indifferent sky." He added to that in his diary: "Loneliness should be the normal lot of the inhabitants, and we were about to join them." It was a wide-open landscape, more desolate than Agate, which laid 400 km to the east. The economic crises hit many of the farmers and the price of oats was so low that fields were burned to avoid the expense of harvesting.

In McKenzie country, Roy learned to describe and observe soil profiles, a skill that he developed and mastered over the rest of his life. Charles Kellogg told him that he should describe each profile for what was there, regardless of what he had learned from the books. To study these soils, the field crew dug trenches across slick spots in the Solonetz [2]. Years later, Roy would recall those days in McKenzie County: "…heck it was me that dug all of those holes." They mapped soils until the first snow in early December. Back in Fargo, he helped with developing a land classification system for tax assessment based on the soil maps. Roy also worked 100 h per month at 25 cents per hour grading papers for Charles Kellogg and helping with the laboratory sessions. He sometimes lectured in the *Agricultural Geology* course of Harlow Walster. The fieldwork in McKenzie County and the teaching sparked his academic interests but he was always short of money. He participated in a contest sponsored by the president of General Mills where the winner would get $350—enough to cover a year of his studies. Roy described a proposal for rural zoning where land use would be restricted in some areas and unrestricted for others, whereas land suitable for grazing only and land suitable for cropping was mixed. He thought it was a new idea but learned later that Lee Schoenmann had developed it for Wisconsin 10 years earlier. Still, Roy won the award, and with that, he had enough funds to finish off his BS degree at *North Dakota Agricultural College.*

In the autumn of 1933, Charles Kellogg wrote to Emil Truog: "…next year I will have my best class!!! There will be 3 or 4 absolutely exceptional students and one, who is on our survey, that is the nearest a genius I've ever seen in soils work. This man, Mr. Simonson, has about 94 to 95 on a college average! I hate very much to give this man up. But what is to be done? The two best men in the School of Agriculture, who helped me the most with my boys, have resigned. All of the good men in the college will likely be gone by the new year." Charles Kellogg recommended that Roy wrote to Emil Truog to ask if there was a scholarship open at the University of Wisconsin. Roy wrote the letter in December 1933 and mentioned that he had worked with Charles Kellogg and Ken Ableiter, a former Wisconsin student, and he listed Harlow Walster as a reference who was a graduate from the University of Wisconsin. Emil Truog sent him the application forms but as Roy had a BS degree and wanted to enroll directly in the PhD program, he needed to list manuscripts on the application. But even without manuscripts, he would be admitted in the PhD program because of his grades and the strong recommendation from Charles Kellogg.

A few weeks later, Roy wrote back that he had suffered pneumonia and had incurred a hospital debt, and he asked Emil Truog whether he knew of any summer jobs. Instead, Emil Truog offered him a room in the soils building that he could share with other graduate students: "…there is a good possibility that you might get quarters there without charge. I believe some of the boys will be vacating this fall. In the past the understanding has been that the boys occupying the rooms would compensate for the privilege by looking after the closing of the doors and windows in the greenhouses and Soils Building during the storms and at night when they need attention." Students had been housed in the soil building since it was built in 1914.

Roy Simonson, 25 years old, graduated from *North Dakota Agricultural College* in the spring of 1934 but did not attend the graduation as it was costly; a diploma was $5 and the cap and gown rental were even more. It all seemed frivolous to him and he needed to pay hospital bills from his

A FEW READINGS IN SOIL CLASSIFICATION

1. The great soil groups of the world and their development. (A translation of K. Glinka's Die Typen der Bodenbildung, by C. F. Marbut)

2. Notes regarding the soils along the route of the transcontinental excursion of the First International Congress of Soil Science. C. F. Marbut.

3. The vegetation and soils of Africa. H. L. Shantz and C. F. Marbut. American Geographical Society. Research Series, No. 13, 1923.

4. Soil absorbing complex and the absorbed soil cations as a basis of genetic soil classification. K. K. Gedroix.

5. Proceedings of the First International Congress of Soil Science. Volume IV. 1927. A great many articles of importance. See also Volume I.

6. The classification of soils on the basis of analogous series in soil formation. D. Vilensky. Proc. of the Int. Soc. of Soil Science. Vol. I, No. 4, 1925.

7. Principal features of distribution of soils and vegetation in the United States. D. G. Vilensky. Soil Research, Vol. I, No. 2, 1928.

8. Reports of the American Soil Survey Association. (For technical terms see Bulletin 8, 1927).

9. Several papers prepared for the First International Congress of Soil Science. 1927.

      (a) A brief survey of soil investigations in Laturia. J. Wilyn.

      (b) Contributions to the study of the soils of Ukrania. Several authors. Includes a general article by Vilensky.

      (c) Genesis of soils. S. S. Neustruev

      (d) Dakuchaiev's ideas in the development of pedalogy and cognate sciences.

      (e) The classification problem in Russian soil science. J. N. Afonosiev.

10. Major world soil groups and some of their geographic implications. Louis A. Wolfanger. Geographical Review XIX:1:94:1929

11. Soils. Hilgard. MacMillan. 1914.

12. Soils of Wisconsin. A. R. Whitson. Wis. Geol. and Nat'l History Survey. Bul. 68. 1927.

13. Lectures given by Dr. C. F. Marbut before the graduate school of the United States Department of Agriculture.

14. Reports by C. F. Marbut to the International Society of Soil Science on the classification, nomenclature and mapping of soils.

Readings in soil classification suggested for the soil class at *North Dakota Agricultural College* taught by Charles Kellogg, 1932. University of Wisconsin-Madison, Department of Soil Science Archive

appendectomy and pneumonia. He requested Emil Truog to postpone his scholarship at the University of Wisconsin until the fall of 1935 because of his financial situation and he was hoping for an offer from the *Soil Erosion Service* in Ohio or the *Division of Soil Survey* in Washington, and both positions paid about $2,000 per year. He also applied for a job in Montana with the Agronomy Department where he would be in charge of a detailed soil survey, and that position would pay $2,400 per year.

Emil Truog responded a few days later: "Holders of scholarships or fellowships have, of course, the privilege of resigning them at any time, but cannot defer an appointment for a year with the assurance that they will receive it the following year. I am inclined to believe that you will be better off in the long run to go ahead with your graduate work immediately if that is possible from a financial standpoint. There is always a tendency, when one gets onto a job that pays fairly well, to hold it for several years, with the result that often one does not get back for graduate work until it is rather late." He added that a new type of graduate fellowship had been established in the natural sciences for: "…young men of VERY EXCEPTIONAL TALENT AND ORIGINALITY with a stipend of $1,000 to $1,200 per year." Emil Truog hoped Roy would apply for that fellowship.

Roy Simonson talked with Charles Kellogg, who was about to leave Fargo and take up a position as assistant to Curtis Marbut at the *Division of Soil Survey* in Washington. Charles Kellogg urged Roy to take the scholarship at the University of Wisconsin and said that he would lend Roy the money for the first year. As he had several unpaid hospital bills, Roy was not keen to borrow more money. He thanked Charles Kellogg for the offer, worked all summer, and accepted the fellowship to conduct his PhD studies with Emil Truog. It was not what he wanted but there were very few jobs, he was eager to learn, and followed the advice of Charles Kellogg.

In the fall of 1934, Roy Simonson moved from Fargo to Madison. During the first year at the University of Wisconsin, he took several courses and spent much time in the library. His idea was to work on the formation of Podzols, but as those soils were in northern Wisconsin and no money was available for travel, Emil Truog snubbed that idea. The Podzols of northern Wisconsin had been studied 5 years earlier by Charles Kellogg as part of his PhD. As Roy wrote with some disappointment in his diary: "Emil would probably have accepted a proposal for constructing a synthetic Podzol profile and then leaching it with various solutions. That never crossed my mind. I learned later that leaching of samples had been done elsewhere before 1934 and was done in other places a decade or two afterward." Instead, Emil Truog assigned him to grind feldspars to fine particle sizes, as feldspar minerals were assumed to be responsible for the base exchange capacity of the soil. Roy washed the feldspar samples with a weak acid solution and was instantly reprimanded by Emil Truog as it destroyed the base exchange. He had blundered in Emil Truog's eyes, and Roy had to grind more feldspars. That was all his research in the first year of his PhD study.

Roy Simonson had good interactions with other PhD students and was one of the few that entered the PhD soils program without an MS degree. He learned that his background, the education at North Dakota, and the summer jobs put him ahead of others. He was practical but had some good knowledge of the basic sciences. With Marion Jackson, Eric Winters, and other students, he lived upstairs in the soils building in exchange for looking after the classrooms and greenhouse. After a while, he moved to a student hall where bathrooms were shared, all students were male, and there was no kitchen. Meals were in another building with breakfast served as cafeteria style, while other meals were served by waiters. One evening at dinner, someone asked if soils were truly so important that a Department of Soils was justified. Roy answered that soils were not of great importance apart from the production of the food we ate, the clothes we wore, and the houses we lived in. Otherwise, soils had some but no great importance. They were willing to concede that food, clothing, and shelter were important, but Roy declared that the soil was a physicochemical system which no chemist in his right mind would dream of studying. He listed minerals, particle sizes, organic matter in various stages of decay, micro-organisms, partial pressures of gasses, and everything else that came to mind. Only fools would try to characterize such complex systems. The question came back for the rest of his long life.

In his first semester at the University of Wisconsin, he attended a meeting of the *American Soil Survey Association* in Washington. He stayed at the house of Charles Kellogg in Arlington, Virginia, and together they drove to the meeting, which was held at the Wardman Park Hotel. The hotel was far away from the city center and for that reason nicknamed 'Wardman's Folly.' It had opened in 1918 when all public gatherings were banned because of the flu pandemic. Roy saw for the first time some well-known soil scientists including Curtis Marbut, Richard Bradfield, and Horace Harper. One morning on their way to the meeting, they stopped at the soil survey office and met with Curtis Marbut. He had ignored Roy when visiting the soil survey in McKenzie County in 1932 but now they talked, and Roy wrote in his diary: "Curtis Marbut could be charming in conversation if he wished. He could also be curt."

In Washington, Roy skipped the meeting one afternoon and went to a concert by the National Symphony Orchestra to hear Richard Wagner's

Department of Soils picnic in 1935—sitting left: Otto Zeasman, Emil Truog and Roy Simonson (hat in hand); Standing right Andrew Whitson; Sitting right: Sergei Wilde (hat on). Photo from University of Wisconsin-Madison, Department of Soil Science Archive

Ride of the Valkyries and part of *Die Walküre*. Richard Wagner had a special connection to Norway which inspired him to compose the opera *Der fliegende Holländer*. His cinematic music pleased Roy. Back in Madison, Roy worked in the laboratory on the exchange capacity of the mineral portion of the soil and not much was known on what compounds caused the exchange. He treated samples of a Miami silt loam and Bentonite clay with sodium sulfide and studied the effects upon the exchange capacity. He found that Chernozems had a portion of their base exchange attributable to iron compounds, but that older soils lacked it as they were too weathered; the iron was broken down and the oxides had given the soils their red and yellow colors.

Roy Simonson clashed with Emil Truog, not just once, but on several occasions. Some of that was the nature of the relationship. Clashes often occur between graduate students and their supervisors, as many graduate students tend to go through a period of pubescence with their supervisor. That usually happens halfway through their degree—too far from the beginning and too far from the end, at which obnoxiousness and a degree of separation from their supervisors is natural. Sometimes, perhaps quite often, it leads to independent thinking and some great work, and most students grow through their pubescence period and mature. For Roy, the clashes had to do with the work being restricted to the laboratory, and he was longing for the wide-open fields of North Dakota. The lab-as-a-cage feeling did not apply to all students. For example, Marion Jackson, who grew up on a farm in southern Nebraska and also worked on the soil exchange capacity under Emil Truog, discovered that the laboratory was his natural habitat. When Marion Jackson applied for graduate school in 1937, his thinking was, however, different: "...I am interested in continuing work in land use planning, land classification and soil development or in research in soil physics and soil chemistry. After completing my PhD degree I hope to enter the field of soil classification and land planning or to teach soils with part time to be devoted to research." That did not happen, and he became a scholar in soil mineralogy, silicate crystal chemistry, and mineral transformation by weathering.

Roy Simonson was by nature at home in the field. In the summer of 1935, the *Division of Soil Survey* offered him to become a member of a field crew appraising lands behind the Fort Peck Dam in Montana. He discussed it with Emil Truog, who had no objection; his course work was on schedule, but his thesis research had not progressed much. Roy needed the money as he was paying partly for his sister's education. He boarded the train in Madison and the journey took 2 days and went through much of Minnesota and North Dakota at night. After he woke up, the train followed the Missouri river and he gazed out of the window and contemplated on the landscapes where cut banks marked stream channels, sagebrush was dominant, and stray cottonwoods grew along the river. When he finally arrived in Glasgow in Montana the landscape appeared similar to what he had seen in McKenzie County. It felt like he knew the land.

At Fort Peck, Roy was part of a crew appraising land that soon would be under water when the dam was finished [12]. Missouri was a wild and free river and the purpose of the Fort Peck Dam was flood control and hydroelectric power generation. Meriwether Lewis and William Clarke had passed the site of the dam in 1805. The dam would become the highest of six major dams in the Missouri River, over 6 km long and 76 m high. The lake would cover almost 100,000 hectares and was to become over 60 m deep. The project had started in 1933 and was a part of the New Deal and provided jobs for 11,000 drought-plagued and unemployed people, with most of them earning 50 cents per hour [13]. Sixty of them died during the construction and six were forever entombed in the dam when a large slide occurred in September 1938. President and Eleanor Roosevelt came to visit a few times to inspect progress and to boost morale. That was needed as the work continued during the cold winters. The president was popular and the workers cheered and thanked him for their jobs, and his portrait, under the headline, 'A Gallant Leader,' hung in the bars of the dusty frontier towns [13].

The Fort Peck Dam's spillway was photographed by Margaret Bourke-White for the cover of the first issue of *Life Magazine* in 1936. An article,

10,000 Montana relief workers make whoopie on Saturday night, showed her photographs of the dam and spillway under construction, but mostly, it revealed the towns, townsfolk, the bars, and some of the hardship. The eight-page photo essay was a disappointment to many, the photos were liked, but the story overplayed the wild west angle of the towns and she had sensationalized the project and the living conditions of the workers. Later on, she referred to it as a period devoted to industrial shapes, and people were only being incidental [13]. But the Fort Peck Dam photos made Margaret Bourke-White famous, and in the same year, she and Dorothea Lange took the iconic photos of the Dust Bowl migrants.

Roy Simonson and the surveying crew lived on a houseboat in the Missouri River. The crew members included an experienced soil surveyor, a property appraiser, and an agricultural economist. They spent 1 day to appraise a ranch or other holding. Roy had to map the soils, types of fences, vegetation, and height of the land above the river, whereas the others measured and photographed the farm buildings. They used aerial photographs at a scale of about 1,000 feet per inch (1 to 20,000) as field sheets. Special features on the photos were hollow white squares for section corners and hollow white triangles for quarter corners. These were made of crushed limestone and put down before the photographs were taken. The aerial photographs were of good quality, and Roy had never used them before but by the end of the summer, he was convinced that aerial photographs would be very useful in soil survey [12]. Aerial photographs had been used by the military during the First World War and in the early 1920s, William Cobb, Tom Bushnell, and Mark Baldwin used them in soil surveys in North Carolina and Indiana [14]. In 1931, soil surveyors in Michigan had used stereoscopes and aerial photographs to map slopes.

After Roy Simonson and the crew had surveyed the land, they assigned an appraisal price that ranged from $2.5 to $200 per hectare. The lowest price was for Badlands, sandbars, and any land that they considered worthless. The government was trying to be fair, and it was assumed that even the Badlands should be worth something for providing a land surface and even for holding the world together. They traveled with their houseboat upstream on the river where it was affected by the dredging, and Roy noted: "…the water in the Missouri River in eastern Montana looked like coffee with too much cream. It was easy to see why people living along the river said that it was too thin to plow but too thick to drink. I did learn over the summer, however, that a person could drink the water without ill effects." The survey was completed by early December and the temperatures were below freezing. As the soils were dry, they did not freeze, so they could still use a spade and auger and

The houseboat tied up to a large tree on the Missouri fifty kilometers upstream from the Fort Peck dam. Landscape south of the Badlands with short-grass prairie and a few trees in the intermittent streams, two of the common buttes are in background to the right. Photos from: *Six months along the Missouri,* by Roy Simonson from 2004

make observations. Towards the end of the year, he took the train back to Madison, but stopped in Fargo and met with Dean Harlow Walster, who offered him a position as an assistant professor after he had obtained his PhD degree. It flattered Roy, but after some thinking, he decided that going back to North Dakota would be mistake.

Whilst in Montana, he wrote a long letter to Emil Truog in which he talked about the surveying, and the analytical results on the solubility of feldspars. It was the summer of 1935 and the *Third International Congress of Soil Science* was to be held in Oxford in the UK. Emil Truog was to attend the congress and Roy wished him a good journey: "It must surely be interesting to be there; I would have liked to go but it's quite out of the question." Emil Truog was pleased to see the feldspar solubility results and was: "...convinced that some exceedingly interesting and useful results may be obtained from work along this line." Emil Truog wrote to Roy that he had received a letter from Charles Kellogg who was planning on attending the congress in Oxford: "I certainly hope he will be there and that he will find it possible to go on the tour through the British Isles. Due to the rush of work, it is rather difficult to for him to get away for more than just a short time." Emil Truog then left for Oxford, after which he traced his family roots in Switzerland. Charles Kellogg was busy in Washington, and left at the last minute for the congress, as did his director Curtis Marbut.

Later in the summer of 1935, Roy discussed with Emil Truog the progress in his program, and in particular, the minor, which was a requirement for a PhD degree. Emil Truog wanted him to do a minor in physical chemistry but Roy proposed a minor in geology. He did not want to spend his life in a laboratory, which prompted Emil Truog to burst out: "What's the matter with that?" Emil Truog enjoyed the laboratory and could not see why anyone else would not. It further delayed the approval of Roy's PhD program, and he was already 2 years into it.

In 1936, the publishing firm McGraw Hill sent a book manuscript by Hans Jenny to Emil Truog for review and appraisal. Emil Truog had good connections with the publisher and was often asked to review proposals or to write a textbook himself. He asked Roy to go over the differential equations and partial differentials and apparently, Roy was the only person in the department who could do that [15]. Roy prepared a short statement about the mathematics and their value; he saw little value in it given the uncertainties in the numbers for any equation and as it avoided the interactions among the factors of soil formation. Emil Truog wrote to McGraw Hill and recommended rejection of the manuscript. It was much to the disappointment of Hans Jenny who had just joined the University of California Berkeley. Hans Jenny then contacted Richard Bradfield, whom he had replaced at the University of Missouri, and who had become an influential soil scientist. Richard Bradfield contacted McGraw Hill, and after some discussion, they accepted to publish Hans Jenny's book [16]. It was not the only time Richard Bradfield used his influence, and on several occasions, he helped Hans Jenny to get his papers published in *Soil Science* when they were rejected [16].

The book *Factors of Soil Formation—A System of Quantitative Pedology* finally appeared in 1941 [15]. It was rooted in the work of Eugene Hilgard, and Russian soil science [17]. Hans Jenny formulated the five factors of soil formation, and synthesized the work of Vasily Dokuchaev, Sergei Zakharov, and Charles Shaw [18, 19]. Some 50 years after *Factors of Soil Formation* was published, it was reprinted and Roy Simonson reviewed the reprint [20], in which he restated some of the findings that were in his review for the original proposal in 1936: "The subtitle for the original edition and the reprint is *A system of quantitative pedology*. That is certainly a desirable goal even though it was not reached by the book. A more accurate subtitle would have been *An approach to quantitative pedology* or *An attempt at quantitative pedology*." Roy saw the advantage of bringing together examples and discussions of the factors of soil formation, and he concluded the review with: "Knowledge of the antecedents of current ideas is helpful indeed in appreciating both validity and limitations of those ideas" [20].

In Wisconsin, land was being mapped by the *Soil Conservation Service*, and a meeting was held in Madison. Andrew Whitson, who directed the soil survey in Wisconsin, suggested that Roy attend the meeting led by Etan Norton, who was in charge of the soil surveys and based in Washington. Etan Norton had written the *Soil Conservation Survey Handbook* [21]. At the meeting in Madison, the soil surveyors were told that if they wanted their job to last, they better make the *Soil Conservation Service* look good, and to help farmers was not the prime purpose. That was a bad philosophy according to Roy and the experience quelled his aspirations to work for the *Soil Conservation Service*, and he became aware of the differences and rivalry between the *Soil Conservation Service* and the *Division of Soil Survey*.

The Farm Credit Administration asked Charles Kellogg in 1936 whether he knew a practical-minded soil scientist who could work in land appraisal in North Dakota, Montana, Washington, and Idaho. Charles Kellogg did not have to think long and recommended Roy Simonson. The suggestion was much to the dislike of Andrew Whitson, who wanted Roy to conduct soil surveys in Wisconsin and annoyed by all that, he assigned Roy a B grade for a course in soil classification and survey. It was a hard pill to swallow, as no other graduate student in soils in the preceding 15 years had been assigned a grade below A. Nevertheless, some years, later Andrew Whitson would write a glowing recommendation when Roy Simonson applied for a position at Iowa State College.

The relationship between Roy and Emil Truog worsened and they had not figured out what samples to use, and when Roy proposed to use soils from North Dakota, it further irritated Emil Truog. Roy was lectured on his limited understanding of soil by Emil Truog, who pounded the laboratory desk for emphasis and concluded: "Young man, it's time you learned that what works with one soil, works with another soil." Roy Simonson was not convinced, but said nothing. He continued working on the removal of free iron oxides from soil samples but obtained discouraging results, although they were presented by Emil Truog at the *Soil Science Society of America* meeting in Washington. In his absence, Roy had to lecture on podzolization and laterization, and Emil Truog had offered him some advice: "Keep it as simple as you can," and Roy realized that it was a keystone in his approach and philosophy. Simple concepts were Emil Truog's strong points, although Roy thought that Emil Truog at times carried it too far.

After the *Soil Science Society of America* meeting in Washington, Emil Truog gave Roy two manuscripts on the preparation of sample size for fractionation and removal of iron oxides. Other students were listed as co-authors, but his name was not included. Roy commented on both manuscripts, did not think that the authorship was fair, and felt that Emil Truog assigned authorship based on whom he liked. After some time, Emil Truog brought him the revised manuscripts that took into account Roy's comments, including him as a co-author. They were both published in 1937 [22, 23]. Many years later, he learned that Emil Truog had a great deal of respect for Roy's writing ability, but Emil Truog never told him that.

In the summer of 1937, Roy worked for Charles Kellogg editing manuscripts intended for the yearbook *Soils & Men* [9]. There was pressure to publish this book, and Charles Kellogg knew he could rely on Roy's writing and editing skills. Roy was uncertain about the job because working at the *Division of Soil Survey* did not include field work but he took the position, boarded the train to Washington, and was based in one of the 4,500 offices of the south building of the Department of Agriculture. The recently completed building was the largest office in the world until the completion of the Pentagon in 1943. Roy rented a room in a house a few doors from Charles Kellogg's home in Arlington and had breakfasts at their house, and took the bus to and from work.

Initially, Roy had no formal editing assignment which made him apprehensive. His main task became checking manuscripts and reviewing whether they were technically sound, logically organized, and written in good English. Roy assumed that if the manuscripts were written by established soil scientists, they would be good. The first paper he reviewed was by Selman Waksman of Rutgers University. The paper was on the making of compost, but he could not make head or tail of it, and he thought it was written by someone who thought in Russian and wrote in English. The paper was rejected, and Charles Kellogg had the task of informing Selman Waksman. Roy spent 3 months reviewing 40 manuscripts, eight of which he considered acceptable, eight hopeless, and the rest somewhere in-between. One of the best manuscripts was on soil phosphorus by Bill Pierre who was the Head of the Agronomy Department at Iowa State College and had obtained his degrees under Emil Truog. Bill Pierre had also conducted soil surveys with Andrew Whitson, Warren Geib, and Lee Schoenmann and was one of the first to study the acidifying effects of nitrogen fertilizers. One year later, Roy would meet Bill Pierre when he applied for a position at Iowa State College.

Roy worked on a chapter on the physical nature of soil with Lyle Alexander and Thomas Rice. He also completed a small soil map of the world and helped to write a chapter on the nature and extent of soil losses, as well as two other chapters. For all the work, he received a little footnote in the yearbook and in one of the chapters. Roy refined his editing and writing skills and did get paid, even though Charles Kellogg told him that the summer had given him such a wide-ranging soil science experience that there was no need to pay him. At the *Divisions of Soil Survey*, Roy found most people unfriendly except Charles Kellogg and Ken Ableiter. The others resented the 'youngsters' like Roy Simonson who came in at the coattail of the new chief, Charles Kellogg, whom they also disliked. It did not help that Charles Kellogg referred to Roy as 'the young man.' Roy also met with the appraiser team from the Fort Peck dam. They now all worked in Washington and told him that most of the land that was underwater had been purchased through negotiation and little had to be condemned. It seemed that none of the land along the river was worth $200 per hectare in 1935.

After the summer in Washington, Roy Simonson returned to Madison to continue working on his PhD thesis. Emil Truog was on vacation in Florida for some weeks, but after he came back another difficult period started. Roy was working on data analysis and graphs one afternoon when Emil Truog came in. They had a short conversation and Emil Truog looked at his data and graphs and blasted: "That's the trouble with you, Simonson. You fool around with things that have no significance. If you were in a meadow full of elephants you'd be shooting mice." Roy just wanted to finish his PhD thesis, find a job and leave Madison. He aimed to graduate in the spring of 1938 but still had to do his minor, but it was not possible to minor in Geology and he was not interested in minoring in Physical Chemistry as Emil Truog favored. Roy settled on a minor in Geography and Land Economics and that allowed him to graduate in the spring. Emil Truog did not agree with the choice and felt that a student should not worry about an extra year of graduate studies. In the end, Emil Truog signed off on Roy's minor but told him that he should find his own job after and that no help was to be expected. That did not worry Roy; he counted on the help of Charles Kellogg and had a job offer with the Farm Credit Administration in case everything else failed.

Emil Truog walked into Roy's laboratory one afternoon and told him that his exam, that was named the prelims, was scheduled the following morning, and that there was no time to prepare or even worry. Prelim exams in the PhD program were used to force less capable students into an MS degree or leave the program without obtaining a degree. The next morning there was a committee of five professors, two from Soils and one each from Geography, Chemistry, and Economics. Emil Truog began the examination by asking about his personal history and education which put Roy at ease. He was questioned about the high-analysis phosphate fertilizer at the Tennessee Valley Authority, how to apply calculus and differential equations, about Stokes' law, methods for measuring particle size distribution, weathering of lateritic soils, and a comparison of the soils of the Columbia Plateau of Washington State with those of the Decan Plateau in India. He passed his prelim exams.

Before he had graduated, he had been looking for jobs. Bill Pierre from Iowa State College in Ames had interviewed him in Madison earlier that year. The position at Iowa State College was for a soil surveyor, and it had been created following the death of Percy Brown in the summer of 1937. Roy was invited to Ames for a formal interview that included discussions with the President, Dean of Agriculture, and the Director of the Agricultural Experiment Station. The library and laboratory facilities impressed him, and he thought that more funds were available in Ames than in Madison, although he wondered why their work was so undistinguished. Whilst he was awaiting the decision on the position in Iowa, a job offer came from the Tennessee Valley Authority in Knoxville, based on the recommendation from Charles Kellogg. Roy went to Knoxville for an interview and met with Lee Schoenmann, who had left Michigan and was now a consultant with the Tennessee Valley Authority. Upon his return to Madison, Emil Truog recommended that Roy take the job in Ames, as he considered a job with a state agency better than with a federal agency.

Emil Truog stopped in the laboratory on a Sunday afternoon while Roy was working on his thesis. Roy told him that the position in Iowa was to start in May, which angered Emil Truog as it was two months before his fellowship ended. Emil Truog felt that Roy was not grateful to the university that had supported him so graciously and that his thesis would never be finished before taking up the position in Ames. But Roy finished his thesis, Emil Truog approved it, and Roy left for Iowa and came back one day for the final exam. Roy Simonson graduated on a thesis entitled *Base exchange capacity of soils due to iron compounds* in which soils were treated sodium sulfide to determine their base exchange capacity. He was the 50th PhD graduate from the Department of Soils at the University of Wisconsin. Roy prepared a paper based on his thesis research, but its interpretations differed from the thesis. The paper was intended for the journal *Soil Science,* but Emil Truog never approved it, and the paper was not submitted.

In May 1938, he boarded the train in Madison to start his new position at Iowa State College in Ames. All of his belongings were in a steamer trunk and a Gladstone bag, and he hand-carried a portable typewriter. Roy was 29 and had soil experience in various parts of the country. At Iowa State College, he was given an office in the Agriculture Building that was later called Curtiss Hall. His PhD diploma from the University of Wisconsin arrived in the mail, and though he had expected to feel some exhilaration at seeing the diploma, there was none: "My entire reaction was one of relief—thank goodness that's over." In Iowa, he was paid a salary of $3,360 per year and in the summer he bought a second-hand two-door Chevrolet sedan for $600. The spare tire in the trunk was still wrapped in heavy brown paper. Bill Pierre, the department head loaned him the money for the car, and in turn, Roy lent money to his brother Cliff who now studied at the University of Tennessee after *North Dakota Agricultural College* had lost its accreditation over political meddling and loss of academic freedom.

In Iowa, Roy returned to soil surveys, where his heart and soul were. The surveys were conducted with Andy Aandahl, a former classmate from Fargo and also of Norwegian descent. Andy Aandahl was raised on a farm in North Dakota and had studied under Charles Kellogg [24]. They mapped soils in Tama County and the work was part of Andy's PhD research at Iowa State College. During fieldwork, they were greeted with some hostility by the erosion surveyors from the *Soil Conservation Service* who had wanted Roy's position when it was vacant. Roy learned that the mapping conducted by the *Soil Conservation Service* had to be redone, and he wondered why no aerial photographs were used. During fieldwork, he was struck by the many old men who were sitting around. He had never seen that in North Dakota as the state had been recently settled and there were not many old men yet. They came across a country church with a large cemetery where most gravestones were from the 1800s, and the majority of the gravestones listed young children and women.

In Ames, Roy met the Iowa-born Susan Miller. She had graduated from Grinnell College east of Des Moines and worked as the secretary of the microbiologist and Dean Robert Buchanan at the Experimental Station. Roy visited the station often, and they fell in love. Susan liked birdwatching, classical music, Broadway musicals, and Victor Borge—the Danish comedian, conductor, and pianist who was popular on the radio and television [9]. Her favorite musical was *The Music Man* as it took place in a town just like where she grew up in Iowa. *Ya got Trouble* was often heard in their home.

Roy worked for 5 years at Iowa State College, and 10 years after he had left Iowa, he published the book *Understanding Iowa Soils* that was written with the Guy Smith and the Canadian Frank Riecken [25]. Francis Hole reviewed the book: "...the text is more than accurate; it is interestingly written. They have shared with the reader the refreshing spirit of discovery which soil surveyors experience as they explore anew the soil." It reflected the style of Roy and his ability to write his findings in a grounded and down-to-earth way. Roy Simonson published several papers from his research in Iowa, but did not publish his PhD work, and Emil Truog included his data in two papers in the *Proceedings of the American Society of Soil Science* [22, 23]. Roy's relationship with Emil Truog had been strained, but after he had worked in Iowa for some years, he started corresponding with Emil Truog about a paper in which he and a research assistant Curtis Hutton compared soil carbon analysis by dry combustion and the Walkley–Black method. They sent the manuscript to the *Journal of the American Society of Agronomy*, and Emil Truog was asked to review it, but gave it to Marion Jackson. In his review, Marion Jackson wrote:"Although we usually emphasize the necessity of making papers brief and concise, it seems to me that this paper has probably gone somewhat too far in that direction. In certain places the paper is rather sketchy and the reader is forced to go to the references in order to properly orient himself. I am therefore doing the rather unusual thing of asking that the paper be expanded." Roy revised the paper and it was accepted two weeks later [26]. The paper, seven pages long, has rarely been cited, despite the importance of the comparison made and the number of studies on soil carbon across the globe.

Roy participated in field studies in the upper Mississippi valley that focused on soils derived from loess under a forest cover. Soil researchers from midwestern states and the Department of Agriculture were invited, including Bill Pierre from Iowa, Henry Krusekopf from Missouri, and Charles Kellogg and Mark Baldwin from the *Division of Soil Survey*. Roy was interested in discussing soil profiles with Mark Baldwin, who was quiet and not interested in such conversation. During the field trip Roy Simonson and Charles Kellogg stood on top of a hill overlooking the surveyors, and Roy remarked that the whole trip was a waste of federal money, to which Charles agreed heartily. Roy stayed with friends in Mississippi. In the evening they went squirrel hunting, and the next morning, they prepared their catch for breakfast. Some 12 years after the group meeting and fieldtrip in the upper Mississippi valley, the booklet *Soils of the North Central Region of the United States* was published [27]. Most of the participants of the field trips were listed as authors including Andy Aandahl and Henry Krusekopf. The *Division of Soil Survey*, which had provided funds for the regional soil survey, did not publish the book, and Charles Kellogg and Roy Simonson were not included as authors. The Great Soil Group was used to classify soils, which was a grouping of soils based on zonal, azonal, and intrazonal characteristics. The book was published a few months before the *7th Approximation* was launched, which was to replace all earlier soil classification systems.

The *Soil Science Society of America Meeting* was held in Washington in 1938 and Roy drove there with several of his colleagues. At the meeting, Roy presented his soil survey work in Iowa. Some of his friends spoke to him afterward, noting that his presentation was wooden and that he seemed scared stiff. They told him the talk was poor, and that if he could not do better, he should not be in front of people. Charles Kellogg was also disappointed in his presentation and suggested to Emil Truog that he explain to Roy what was wrong with his presentation. But Emil Truog did not do that and suggested that Roy should first teach some courses before being dressed down. In the beginning of 1939, Roy taught his first course on morphology, genesis, and classification of soils. More than 50 students had registered for the course, and at that time there were about 1,000 students enrolled in agriculture at Iowa State College. Roy had inherited a course that had the reputation of being easy.

In 1938, Walter Kubiëna had lectured for two semesters as guest professor (*research professor Pro Tempore*) at Iowa State College as he replaced Percy Brown who had suddenly died in 1937. Walter Kubiëna liked America. He was of Hans Jenny and Richard Bradfield's generation, and worked most of his life in Austria at the *Hochschule für Bodenkultur* and in Spain. The lecture notes from Iowa were published as his first book entitled *Micropedology* [27]. As a part of pedology, micropedology dealt with the morphology, genesis, general dynamics and biology of soils, and Walter Kubiëna predicted growth in soil micropedology: "The author thinks that the increasing popularity of microscopic pedology with the public as well as with the student is due primarily to the appeal of the wealth of interesting formations and happenings which can be observed in the microscopic world of the soil. To the student who has been accustomed to see and to experience nature only three-dimensionally in his own macroscopic dimensions, the microscopic cavities and the life in them appear as they would were, he actually standing or walking in them. Like the explorer in exotic parts of the earth he roves through the endless systems of ravines and chasms, always eager for new impressions and new discoveries. They always come, because there is much left for everybody to study, describe and explain" [27].

In the last 2 years of the Second World War, more than 100,000 people had died from famine in Austria, and in July 1945, Walter Kubiëna wrote a letter to Emil Truog from liberated but devasted Vienna [28]: "There is no use to stay in Vienna at the present time. The city is destroyed to a great part and has almost no gas, no water, and no electricity. Our laboratories have no furniture and no equipment. People are most horribly suffering from house shortage, hunger, widespread diseases, and almost complete lack of medical supply. I do want to stay in Austria and to become re-established here but I am sure it will take much patience and time. Since Lower Austria and Vienna are still occupied by the Russians I do not want to move out of the section occupied by the Americans and English. I understand that even a temporary stay in America might have some formal difficulties for an Austrian at the present time. I judge it from the fact that we even do not have as yet the possibility to write letters by our own to America. At any rate, I thank you very much for your friendly readiness to help. I am sure that, in some future, it will be possible for me again to pay some more visits to America. I am seeing forward to that occasions with much joy because I love America and because I know that there is a great deal of useful research work left to do for me, particularly in the field of soil erosion (nature, recognition, and control of erodibility)." His love for America was not rewarded by Emil Truog. Instead, he was able to visit the laboratory of Selman Waksman to study the microbiological population using a special soil microscope which he had developed [29]. He then became a frequent guest at Rutgers University and in the Department of Agronomy at Purdue University.

Roy Simonson's first lecture in 1939 was on the nature and scope of pedology, and he did not use any of Walter Kubiëna's lecture material. It was a difficult first semester of teaching for Roy, with half a dozen students dropping the course, and many students receiving low grades in the final exam. In the subsequent years, the course lost its reputation for being easy, and enrollment from students looking for easy credits declined. He ended up with about 12 students each semester. One student showed his notes from the University of Nebraska, where he had taken a course in 1928 that was taught by J.C. Russel. The notes were comparable to Roy's notes from Charles Kellogg's course in Fargo in 1932. Charles Kellogg and J.C. Russel taught the same material but they had never met and had based their course on a few publications that included the bulletins distributed by Russians at the

First International Congress of Soil Science in 1927 [30–41]. These bulletins were influential and according to Roy Simonson: "...it brought forcibly to the attention of Americans the earlier recognition by V.V. Dokuchaev and his students in Russia of five factors of soil formation" [42]. Other sources that were used in the teaching of soil science by Charles Kellogg were *The Great Soil Groups of the World and their Development* by Konstantin Glinka in the translation by Curtis Marbut. For his lectures, Roy also used the 1928 mimeographed lectures by Curtis Marbut [42].

One of the best students in Roy's course was Charles Black and he took the course for five consecutive years. Roy was puzzled by the fact that he repeatedly took his course and asked him why, to which Charles Black responded: "It had the same number but it was not the same course" [9]! Charles Black joined the Department of Agronomy in 1939 and stayed until his retirement in 1985. He founded the Council for Agricultural Science and Technology in 1972, which aimed to get accurate agricultural information to politicians and the general public. Charles Black volunteered with the Iowa Public Radio and compiled a 1,500-page pronunciation guide of music vocabulary, constructed and operated short wave radios, and played the French horn [43].

During his first year at Iowa State College, Roy was invited to the University of Illinois at Urbana-Champaign to present a seminar on estimating crop yields from soil survey data. Charles Kellogg was there too and gave a lecture for the scientific research honor society, Sigma Xi. All the dignitaries wore tuxedos. After the lecture, Roy waited a little before talking to Charles Kellogg, and introduced him to Jerry Meldrum, one of his colleagues. Charles Kellogg and Jerry Meldrum had met before, but Charles denied knowing him which annoyed Roy, and he explained to Charles Kellogg that no one in his own mind was unimportant. Charles Kellogg growled a little and proposed to do better. Jerry Meldrum understood why so many people disliked Charles Kellogg. At the University of Illinois, Roy met with Guy Smith who studied loess and had made two traverses perpendicular to the Mississippi and Mackinaw rivers [44]. It was part of his PhD research supervised by R.S. Smith who was about to retire. Before Roy Simonson returned to Iowa, he met with R.S. Smith who told him to hire young men for the soil survey: "...do not bother with old ones, they have no ideas, hire young men, they will make mistakes but they have new ideas."

At Iowa State College, Roy started to enjoy his teaching, students, and soil survey work. He studied buried soils, and found that not all buried soils had black topsoils as those under forest had thin A horizons [45]. Observations were made at roadcuts, and they made an auger so that they could sample to 10 m depth. It consisted of a worm auger and pipes that were taken apart when they pulled the auger. Years later the geomorphologist Bob Ruhe from Indiana University built a similar drill rig to sample soils to great depth. Roy Simonson also developed methods for estimating the productivity capacity of soils, and soil survey data were used to assess the phosphorus status of soils [46]. Soil profiles were selected that represented the major soil series in each of the important Great Soil Groups. Experimental fields were selected based on soil survey information and a field had to consist of one or, preferably, two of the major soil series [47]. During fieldwork, Roy Simonson became a photographer of soil profiles and his photographs were reproduced in several textbooks [48, 49]. His profile photos of a silty loam soil under prairie and hardwood forest in Iowa were included in *Soil Conditions and Plant Growth*— all the way up to the 11th edition published in 1988 [50].

Roy Simonson organized soil excursions through Iowa, which were attended by researchers, farmers, and people in extension and the private sector. The major soils and their management were shown and discussed during these excursions. Fertilizer company representatives and soil fertility researchers began to wonder if horizon notations were going to replace terms such as topsoil and subsoil. There were tensions during the excursions and that came from the differences in mapping between the *Soil Conservation Service* and the *Division of Soil Survey*. The mapping by the *Division of Soil Survey* had a pedological focus and included a land use capability. The surveys by the *Soil Conservation Service* were practical and focused on farm planning emphasizing slope and erosion classes. Roy Simonson had observed the differences whilst he was in Wisconsin, but now he had to work with both groups. Roy was a follower of the pedologic approach of Charles Kellogg, whereas much of the soil survey work in Iowa had to be done on the terms of the *Soil Conservation Service*. Tensions grew and the *Soil Conservation Service* crew in Iowa reported that Roy was trouble and wrote a letter to the Head of the department, Bill Pierre. Nothing came from it, but it made Roy think about leaving Iowa. He was not the only one and others had left Iowa State College for different reasons. The soil fertility expert R.W. Pearson had studied liming and left for the Department of Agriculture [51], and the soil physicist Lorenzo Richards left for Salinity Laboratory in Riverside, California, where he developed new methods for measuring soil water potential [52]. At Iowa State College, Lorenzo Richards was succeeded by Don Kirkham.

Roy Simonson and Susan Miller married in Albia, Iowa, in November 1942. Shortly after they were married, Roy was offered a position as a soil correlator, also called inspector, for the *Division of Soil Survey* in the southeastern part of the country. He was interested in the position because the area had the poorest soils in the country, and people there more than anywhere else needed to know about their soil resources. There he could study weathered soils, which were different from any soil he had seen in North Dakota, Wisconsin, Montana or Iowa. Roy took the letter to Bill Pierre, but Iowa State College was not able to match the salary of $5,600 per year, and Roy took the position in Tennessee. Roy and Susan Simonson moved to Knoxville in 1943, and Roy was given an office in the New Sprankle Building which was the headquarters of the Tennessee Valley Authority. Here he was reunited with Cliff Orvedal and Marlin Cline, with whom he had studied at *North Dakota State College*.

A year before they moved to Knoxville, the USA had entered the Second World War. It was an uncertain time as Roy could be drafted any time but in order not to lose their employees to the draft, the Department of Agriculture requested deferment every six months which was granted each time, and Roy was never drafted. The war affected their daily lives and work and on one of the long train journeys from Knoxville to North Dakota to visit his parents, Roy and other civilians were not served meals; the dining car was only open to men and women in uniform. He wrote in his diary: "...another day of fasting."

It was August 1944. Paris has just been liberated from the Nazis, Igor Stravinsky presented a public lecture concert in Madison called *Composing, Performing, Listening*, and Emil Truog's older sisters Anna and Lena had died. The Department of Soils at the University of Wisconsin had developed several lines of research that directly aimed at increasing food production, which was requested during the war emergency. The work included fertilizer experiments with corn, small grains, and alfalfa, and a research agenda that was written by Emil Truog. In the middle of the war, Emil Truog wrote to Roy: "Our work here has been going along quite satisfactorily but of course we are short of help in some respects and I doubt that we will have more than two or three graduate students when the fall semester opens. However, I anticipate that when the war closes we will have a flood of graduate students as well as undergraduates." That prediction proved to be correct; the department graduated 58 BS students in soil science in 1950, as many students in one year as in the seven years prior to the war.

As the soil inspector for the southeast, Roy traveled extensively, and he worked well with people in the soil survey, but in some places, he was referred to as the 'Damn Yankee' or the 'God Damn Yankee.' Much of his work consisted of reviewing soil survey plans, legends, field review reports, and checking soil series descriptions. In the field, he started using Truog's soil pH test. It consisted of a porcelain plate on which some soil was placed,

Roy and Susan Simonson in Columbus Ohio visiting Susan's sister Frances on their move from Ames to Knoxville in 1943. Photo from Bruce Simonson

drops of a chemical compound were added, and the color was compared to a standard color chart indicating the pH. After he had used it for a while, he wrote to Emil Truog suggesting replacing the porcelain plate with small plates of waxed cardboards as he lacked facilities for cleaning up the porcelain spot plate whilst in the field. He ended his letter with: "I think that such a spot plate might result in wider use of the test kit, which I think would be desirable."

Emil Truog responded a few days later: "I had thought same of doing this when the test was first designed but did not follow up with the idea at that time. After working in the field with the test, my feeling is the same as yours, namely that it would facilitate field testing greatly if one did not have to contend with the cleaning of the porcelain spot plate." Roy Simonson was pleased that waxed cardboard plates would be considered but had another comment on the porcelain plates: "There is one point that might be worth mentioning. Most of the men who use the test are not aware of the hydrolysis of the glass with its effect upon the pH of the indicator. I find that the indicator gradually becomes more alkaline. This affects the reading somewhat when the test is used with soils that have low base exchange capacities. I think that a little note of caution printed and enclosed in the test kits would take care of this difficulty." That explanatory note, as well as the waxed cardboards, never came.

Despite disagreements during his PhD studies, Roy Simonson and Emil Truog kept a collegial correspondence, and that was mostly at the instigation of Roy. In 1950, Emil Truog organized a symposium on the mineral nutrition of plants that was held at the University of Wisconsin. He had just lost a brother and sister and was thinking of retirement. The international symposium became part of the centennial celebration of the university. The symposium was attended by more than 500 people, and resulted in the book *Mineral Nutrition of Plants*, edited by Emil Truog, and with chapters by Charles Kellogg, Hans Jenny, Edmund Marshall and 14 others [53]. Emil Truog wrote in the Foreword: "Mineral nutrition of plants is a subject of tremendous interest and importance to many people. The plant physiologist must, of course, devote much of his attention to this field; the agronomist, horticulturist, and forester meet problems almost daily which require for their solution specific knowledge dealing with the mineral nutrition of many kinds of plants; and lastly, the fertilizer manufacturer, who is called upon to supply the needed mineral nutrients when they are lacking in soils, must, through his technical expert, keep informed of the latest findings in this field if his enterprise is to attain and maintain a forefront position."

After the book was published, Roy Simonson wrote him a letter in December 1951: "The book will surely be useful to many scientists for a number of years. I am pleased to see that a program and publication of this kind has been sponsored by my *alma mater*. You and the others who made plans for and subsequently handled the symposium are certainly to be congratulated." And Emil Truog responded a few days later: "Thanks indeed, Roy, for your kind words about *Mineral Nutrition of Plants*. Have a number of fine compliments, and I find the book is being adopted partly as a text in advanced and special courses in Plant Physiology and Nutrition in several places. Sales have gone beyond expectations." The book cost $6, a second edition was printed two years later, and an Indian edition was published in 1967. It has been widely used in courses on soil fertility and plant nutrition.

When Emil Truog retired in 1954 a book of letters was compiled. Hundreds of colleagues, friends, and former students wrote him a letter, and so did Roy Simonson: "Word has penetrated all the way to here that you plan to retire at the end of this coming June. I am both sorry and glad to hear that, sorry because of the loss to soil science and glad because you will be able to take things a bit easier. I can hardly believe that the time has come for your retirement. You have always seemed such a permanent part of the University of Wisconsin. I do want to take the occasion of your retirement to express my appreciation for the help given me during my graduate years! My appreciation and gratitude must be shared by scores of others who came under your hands. I am indeed grateful for what I learned at Madison and hope that this gratitude may be a source of some satisfaction to you. Your retirement will be acceptable to soil scientists in this country only if it is to be a change in status. I am sure that everyone would concede you less strenuous activity in the future with the hope that you would not lose interest in soil science nor take your hand out entirely. There seems little danger of that. You will find it easier to quit eating than to quit your interest in soil science. I do hope that you will take care and be able to continue your interest in soil science for many more productive years."

After 6 years as soil inspector in Knoxville and several brief stints of soil survey work in the Pacific, Roy was asked to become the assistant chief of the *Division of Soil Survey* in Beltsville, Maryland. He was reunited with his North Dakota classmate Ken Ableiter, who had replaced Mark Baldwin, and more importantly, with Charles Kellogg. The two had kept close contact, and they worked together until Charles Kellogg retired in 1970. Roy, although criticizing Charles Kellogg at times, was loyal and respected him deeply—he was one of the few who referred to him as 'Chuck.' Only Lee Schoenmann had used that nickname for Charles Kellogg, back in Michigan.

There were large changes in the organization of the soil survey and in 1952, the soil survey activities were brought together in the *Soil Conservation Service*. In the new organization, Roy Simonson supervised the classification, correlation, and nomenclature of soils in different parts of the country, and he had to travel often. In 1950, Charles Kellogg had established a group that would develop a new soil classification system; it included Roy Simonson but was led by Guy Smith. They had written a book about the soils of Iowa [25],

Curtis Marbut in the 1930s and Roy Simonson in the 1950s examining the small reference soil samples. Special Collections, USDA National Agricultural Library

but now they were to develop something more ground-breaking, a new soil classification system, that was launched in 1960 as the *7th Approximation.*

Roy Simonson wrote several influential papers in soil science that were elegantly composed and reflected his diverse experience [54–57]. The 1959 paper, *Outline of a generalized theory of soil genesis*, [54] was amongst his most cited work, and an attempt to offer an alternative to the *Factors of Soil Formation* by Hans Jenny. In 1974, he edited a special issue of *Geoderma* on the applications of soil surveys that were unrelated to agriculture, which was also published as a book [58]. The book was timely, as soils were increasingly used for other purposes in the industrial nations of the world. In the *Preface* he listed the use of soils as construction materials for highways, foundation for houses, vehicles for waste disposal, and that more and more land is needed for home sites, roads, parks, playgrounds for expanding populations. It broadened the soil survey which he thought was needed as: "…the cost of mistakes, both in money and unhappiness, is substantial. Mistakes can be avoided if the kinds, distribution, and usefulness of soils are known."

Roy Simonson was an excellent writer, and it came almost naturally to him. Here is an example from the book on the soils of Iowa: "Soil is a commonplace thing. It is with us from the time that we try to eat it as babies until we are buried in it upon death. It covers the land surface of most of the earth and is absent only from steep and rugged mountain peaks and from the lands of perpetual ice and snow. In spite of its wide distribution, however, the soil does not attract much attention. In most places, it lacks bright colors or other striking features and therefore seems inert and lifeless. Because it is so commonplace, the soil in all its varieties and kinds is often taken for granted, even by the people who make their livings from it. Yet the soil is important to all people. It serves as a foothold and source of nutrients for the many different types of land plants from which people get food, fiber, and shelter. It is the transitional layer between the rock core of the inner earth and the world of living things on the earth's surface.

Soil is sometimes defined as the natural medium for the growth of land plants. This simple statement does not begin to describe soils. The earth has many kinds of soil which differ from one another so greatly that it is hard to see features common to them all. At times, it is hard to tell how soils differ from the various kinds of soft rocks. Nevertheless, soils have one thing which distinguishes them from soft rocks. That is a series of layers roughly parallel to the earth's surface which differ from each other because of the effects of plants, animals, the sun, and the rain. These layers may be as little as a fraction of an inch or as much as several feet thick. There may be as few as two or as many as four or five. It is the presence of these layers that distinguishes a soil from the stuff from which it was made, and so these layers are given a special name. They are called horizons. An example of a horizon familiar to every Iowa farm boy is the so-called 'black topsoil' of many soils in the state. This layer, largely responsible for Iowa's fame as a corn-producing state, is commonly no more than 15–20 inches thick. The different soils of Iowa, and there are many, have individual profiles by which they can be recognized. Each soil profile is made up of several horizons, one below the other. Each soil has its own profile, one that is peculiar to it alone, even as people do. But as there are persons with similar profiles, so there are soils with profiles very much alike though never wholly identical. It is on the basis of similarities in profiles that the relationship of one soil to another can be determined. After the relationships among different soils are known, the knowledge gained from experience with one soil can be applied to other similar soils. Thus, knowing the characteristics of a soil profile becomes a first step in knowing what soil is and how it will respond" [25].

Roy Simonson learned to write from his mentor but had an example in his father, Otto Simonson, who wrote him 265 letters over a 34-year timespan beginning when Roy went to college in 1930. Most of the letters started with *Dear Lad*. Otto Simonson lived most of his life on the farm in Agate, North Dakota that he named 'Only place.' Roy compiled his letters and memoir writings and grouped them by topic such as early life, farming, sports, personal matters [8]. Otto Simonson was born in 1875 on a small farm in Norway and emigrated with his family to America in 1882. Until then, he had always worn wooden shoes and he spoke no English.

The Simonsons lived frugally and in 1931 when the Great Depression was raging, he wrote to Roy: "If conditions do not improve—and there is no indication that will—I will not be able to finance your schooling next fall and winter." Roy had a debt of $1,000 and his income from summer and weekend jobs was much lower than his expenses. Roy was encouraged to find work as the cost for a full year of college was between $300 and $360, but his father discouraged him to borrow money: "That is the quickest route to slavery. By all means, stay away from mortgaging your future. Debt is one of the hells on earth. It takes away your manhood and the freedom of action so necessary for a person to be his best. It creates self-doubt and saps confidence; it makes for worries and robs a person of his peace of mind. Hope is fine when harnessed to common sense but a detriment when running wild. The hope to pay a debt often leads people into greater debt which they never pay and which leads in time to an inferiority complex. That has made farmers the water carriers and wood choppers of the world. In fact, it has had more to do with making them the sons of Martha, the servants, than anything else" [8].

Roy was thinking of all this when he negotiated his PhD position with Emil Truog early 1934.

Otto Simonson commented in 1934 on Roy's move to the University of Wisconsin: "My first remarks are about your work in the university at Madison, Wisconsin. In your letter you wrote that the high ratings given you in the recommendations from Fargo would make it necessary for you to extend yourself to the utmost. That is the right spirit. But enthusiasm alone will not carry you the whole way. There must be a more fundamental motive. A decision to extend yourself is fundamental because it is rooted in both self-interest and a desire to serve. So much for the motive. Now, let us look at the results. You never were physically robust. There is thus the danger that in fulfilling the motive, you will let your health suffer to be able to fulfill the contract, you must see that the needs of your body are met. I have no suggestions to offer on that score because you know your body's requirements better than I do. This is a warning; it is easy to overdo things of great interest. Recognizing that now will give you time to plan your work so that it does not destroy your health. Without health, life provides little joy or pleasure. Last winter you let your vitality get low enough so that pneumonia put you on your back. How to prevent that from happening again is your problem. The best medicines I know are contentment, joy, happiness, good will, fresh air, and sunlight. Destroyers of health are worry, hate jealousy, and greediness. There could be others as destructive of health but I cannot call them to mind right now. You should know them as well as I do" [8].

Otto Simonson touched upon the political climate in 1936: "...that Roosevelt is leading us into a communist dictatorship sounds foolish to me. If any dictator is in the future for this country, he will come from moneyed classes. Moreover it will be a Nazi rather than Communist dictatorship. The reason is self-evident. When in trouble, the moneyed class is more cohesive than other classes. They had the leadership if they are forced to use it. They already control the propaganda. All they need is a popular leader and such a leader could capture and hold the loyalties of the middle classes." About the dismissal of Henry Wallace, the Secretary of Agriculture under President Roosevelt, he wrote in 1944: "Roosevelt ditched the last of the New Deal when he threw out Wallace. Wallace is a dreamer. He went up in the air without adequate footing in reality. But dreamers have first thought of the ideas that practical men later put into use. Radical ideas must first occur as thoughts of one or more individuals. The thought or idea must then be brought into public view. After that, it is on its own. The idea will be criticized, condemned, and ridiculed until most of us wonder what manner of

"...Are you sure that there are clay skin on this ped"—Roy Simonson at his best, in a soil pit near Flint, Michigan in 1971. He was about to retire and edit *Geoderma* for the next 18 years. Photo from Bruce Simonson

crackpot harbored the idea in the first place. But no matter how wild an idea may be, it will have some earnest disciples. They will spread the gospel as best they can, even through several generations. Eventually, some practical men will put the idea into use, possibly with modifications."

At last, an example of Otto Simonson's more poetic writing, here is a letter to Roy from 1961, five years before Otto's death: "It's a beautiful morning with a friendly baby blue vault for a sky and a floor studded with nature's finest diamonds. Mixed with the solid group of sparklers on the ground are the showoffs that catch the eye with intermittent flashes. Every tree and shrub and blade of grass reaching up out of the snow is covered with nature's jewels. As the plants sway in the gentle breeze, bright spots erupt all over to form a milky way of the sheerest jewelry. Mornings like this when the sun enfolds the earth with its shining glamour give my spirits a lift that I cannot get from anything made by man. Love of the wild and free open prairie is what anchored me here. Mornings like this were the charm—and still are. A morning in June rather than in March really captured my heart."

Roy Simonson left his parents farm in the 1920s, lived in Fargo, Madison, Ames, Knoxville, and moved to Washington in 1949. Roy Simonson traveled across the world as he worked for a year for the Military Geology mapping soils on islands in the western Pacific, and to examine soils in Japan and Hawai [9]. His desire to travel and study new soils and landscapes echoed the curiosity and desire that Charles Kellogg and Curtis Marbut had shown. Roy made extensive travels between 1947 and 1978 and kept Foreign Travel Journals from his trips to the Pacific area, Brazil in 1954, Netherlands and Paris in 1956, Sri Lanka, India, and the UK in 1951, Romania in 1964, Rome in 1967, Netherlands 1967, Australia and New Zealand 1968, West Germany in 1971, Honduras and Venezuela in 1974, Brazil in 1977, South Africa in 1978, and Norway in 1984 to accept the degree of *Doktor Agriculturae Honoris Causa* from the Agricultural University. Some of these travels were combined with attending the *International Congress of Soil Science* (France, Romania, Australia). To his regret, he never visited Russia.

At the age of 65 Roy Simonson retired from the *Soil Conservation Service*— 2 years after Charles Kellogg had retired who was only 6 years older. In 1993, Roy and Susan moved to Oberlin, Ohio where their son Bruce taught at Oberlin College. Roy and Susan were music lovers and listened to records of big bands and the Don Cossack Choir—a choir of exiled Cossacks founded in 1921. Roy often brought back records from his overseas trips including a stack of 78 s of samba tunes from Brazil. Roy Simonson did not have the leadership skills and political instinct that Charles Kellogg mastered, nor did he have the laboratory patience for soil investigations or the agronomic drive of Emil Truog. But his work—spread over more than half a century—is among the finest in soil science. His writings are a reflection on what he learned early on from the clarity and style of Emil Truog and Charles Kellogg, but which he exceeded over the course of his long career. We do not have soil scientists like Roy Simonson anymore.

Roy Simonson died in 2008 at the age of 100. Other North Dakota soil scientists, also born on farms lived a long life like Andy Aandahl (1912–2003) and Marlin Cline (1909–2009). Roy's son Bruce and family took his ashes to Ames and placed them in the ground where Susan was buried 12 years earlier. They had met with Charles Black when they buried Susan, some friends from the Conservatory played a string duet from Vivaldi's Four Seasons, and a granddaughter sang. But Charles Black had passed away by the time Roy Simonson was buried, and no music was played.

References

1. Walster HL. The evolution of agriculture in North Dakota. Soil Sci Soc Am Proc. 1955;19:118–24.
2. Thompson KW. A history of soil survey in North Dakota. Dickinson: King Speed Publishing; 1992.
3. Ostenso M. The stone field. New York: Dodd, Mead & Company; 1937.
4. Danbon DB. Our purpose is to serve: the first century of the North Dakota agricultural experimental station. Fargo: North Dakota Institute for Regional Studies; 1990.
5. Geiger LG. University of the Northern Plains: a history of the university of North Dakota 1883–1958. Grand Forks: The University of North Dakota Press; 1958.
6. Willard DE. The soils of North Dakota. Fargo: North Dakota Agricultural College; 1909.
7. Simonson RW, Ulmer MG. Initial soil survey and re-survey of McKenzie County North Dakota. Soil Surv Horiz. 1989;30:1–11.
8. Simonson RW. Letters, letter segments, and memoirs from a farm. Infinity Publishing; 2005.
9. Simonson BM, Roy W, Simonson A. Century as a soil scientist. Soil Surv Horiz. 2015;49:63–7.
10. Helms D, Effland ABW, Durana PJ, editors. Profiles in the history of the US soil survey. Ames: Iowa State Press; 2002.
11. Kellogg CE. The future of the soil survey. Soil Sci Soc Am Proc. 1950;14:8–13.
12. Simonson RW. Six months along the Misssouri. Conshohocken: Infinity Publishing; 2004.
13. Lonnquist L. Fifty cents an hour: the builders and boomtowns of the fort peck dam. Helena, Montana: MSky Press; 2006.
14. Gardner DR. The National cooperative soil survey of the United States (Thesis presented to Harvard Univsersity, May 1957). Washington DC: USDA; 1957.
15. Simonson RW. Historical footnote—factors of soil formation. Soil Horiz. 1992;33:70–1.
16. Maher D, Stuart K, editors. Hans Jenny—soil scientist, teacher, and scholar. Berkeley: University of California; 1989.
17. Hilgard EW. Soils, their formation, properties, composition, and relation to climate, and plant growth. New York: The Macmillan Company; 1906.
18. Florinsky IV. The Dokuchaev hypothesis as a basis for predictive digital soil mapping (on the 125th anniversary of its publication). Eurasian Soil Sci. 2012;45:445–51.

19. Jenny H. E. W. Hilgard and the birth of modern soil science. Pisa: Collana della revista agrochemica; 1961.

20. Simonson RW. Book review of: factors of soil formation. Reprinted by Dover Publications, New York. Foreword by Ron Amundson. Geoderma. 1995;68(334–335).

21. Norton EA. Soil conservation survey handbook: miscellaneous publication no. 352. Washington D.C. : United States Department of Agriculture; 1939.

22. Truog E, Taylor JR, Pearson RW, Weeks ME, Simonson RW. Procedure for special type of mechanical and mineralogical soil analysis. Proc Soil Sci Soc Am. 1937;1(C):101–112.

23. Truog E, Taylor JR, Simonson RW, Weeks ME. Mechanical and mineralogical subdivision of the clay separate of soils. Proc Soil Sci Soc Am. 1937;1(C):175–179.

24. Grossman R, Lynn W. Remembering Andrew Russell Aandahl. Soil Surv Horiz. 2005;46(2):85–6.

25. Simonson RW, Riecken FF, Smith GD. Understanding Iowa soils—an introdiction to the formation, distribution, and classification of Iowa soils. Dubuque: Wm.C. Brown Company; 1952.

26. Hutton CE, Simonson RW. Comparison of dry combustion and Walkley-Black methods for the determination of organic carbon distribution in soil profiles. J Am Soc Agron. 1942;34:586–92.

27. Kubiëna WL. Micropedology. Ames, Iowa: Collegiate Press; 1938.

28. Collingham L. The taste of war: world war two and the battle for food. London: Allan Lane; 2011.

29. Waksman SA. My life with the microbes. New York: Simon and Schuster, Inc.; 1954.

30. Afanasiev JN. The classification problem in Russian soil science. USSR: Publishing Office of the Academy, Leningrad; 1927.

31. Glinka KD. Dokuchaev's ideas in the development of pedology and cognate sciences. USSR: Publishing Office of the Academy, Leningrad; 1927.

32. Gemmerling VV. Russian investigations concerning the dynamics of natural soils. USSR: Publishing Office of the Academy, Leningrad; 1927.

33. Kravkov SP. Achievements of Russian science in the field of agricultural Pedology. USSR: Publishing Office of the Academy, Leningrad; 1927.

34. Neustreuv SS. Genesis of soils. USSR: Publishing Office of the Academy, Leningrad; 1927.

35. Polynov BB. Contributions of Russian scientists to paleopedology. USSR: Publishing Office of the Academy, Leningrad; 1927.

36. Prasolov IL. Cartography of soils. USSR: Publishing Office of the Academy, Leningrad; 1927.
37. Prokhorov NI. Soil science in the construction of highways in USSR. USSR: Publishing Office of the Academy, Leningrad; 1927.
38. Tulaikov NM. Russian Pedology in agricultural experimental work. USSR: Publishing Office of the Academy, Leningrad; 1927.
39. Tiurin IV. Achievements of Russian science in the province of chemistry of soils. USSR: Publishing Office of the Academy, Leningrad; 1927.
40. Yarilov AA. Brief review of the progress of applied soil science in USSR. USSR: Publishing Office of the Academy, Leningrad; 1927.
41. Zakharov SA. Achievements of Russian science in morphology of soils. USSR: Publishing Office of the Academy, Leningrad; 1927.
42. Simonson RW. Early teaching in USA of Dokuchaiev factors of soil formation. Soil Sci Soc Am J. 1997;61(1):11–6.
43. Anon. Charles A. Black 1916–2002. Iowa state university. Unpublished. 2002.
44. Brasfield J. In Memoriam Guy D. Smith. Soil Horizons. 1981;22.
45. Simonson RW. Studies of buried soils formed from till in Iowa. Soil Sci Soc Am Proc. 1942;6:373–81.
46. Pearson RW, Simonson RW. Organic phosphorus in seven iowa soil profiles: distribution and amounts as compared to organic carbon and Nitrogen1. Soil Sci Soc Am J 1940;4(C):162–167.
47. Simonson RW. The United-States soil survey—contributions to soil science and its application. Geoderma. 1991;48(1–2):1–16.
48. Austin ME, Chandler RF, Johnson WM, Powers WL, Simonson RW. Report of committee on exchange of soil pictures and soil profiles. Soil Sci Soc Am J. 1943;7(C):496–498.
49. Krantz BA, Obenshain SS, Simonson RW. Report of committee on exchange of soil pictures and soil profiles. Soil Sci Soc Am J. 1942;6(C):394–395.
50. Russell EW. Soil conditions and plant growth. 8th ed. London: Longmans; 1952.
51. Pearson RW, Adams F, editors. Soil acidity and liming. Madison, WI: SSSA; 1984. Agronomy Monograph.
52. Gardner WR. The impact of L.A. Richards upon the field of soil water physics. Soil Sci. 1972;113(4):232–237.
53. Truog E, editor. Mineral nutrition of plants. Madison: University of Wisconsin Press; 1951.
54. Simonson RW. Outline of a generalized theory of soil genesis. Soil Sci Soc Am Proc. 1959;23:152–6.

55. Simonson RW. Concept of soil. Adv Agron. 1968;20:1–47.
56. Simonson RW. Soil science—goals for the next 75 years. Soil Sci. 1991;151:7–18.
57. Simonson RW. Origin and acceptance of the term pedology. Soil Sci Soc Am J. 1999;63(1):4–10.
58. Simonson RW, editor. Non-agricultural applications of soil surveys. Amsterdam: Elsevier; 1974. Developments in soil science vol. 4.

6

The Mother of the West

"The value of a soil survey is very great.
When the soils of a county have been mapped,
the foundation is laid for many means of betterment.
The cost of a soil map is small as compared
with many other things of a public nature."

Merritt Miller, 1924

"…if you want to be productive in research you have to be immersed in it and
struggling and illumination will come later, if at all."

Hans Jenny, 1978

Agricultural settlement by Europeans progressed from the east to the middle of the USA, and eventually onto the Great Plains and further west. The first settlements were on the Atlantic coastal plain and along rivers where water was available for drinking and transportation [1]. Inland on the extensive foothills of the Appalachian Mountains named the Piedmont, soils were old and rain had leached most of the nutrients [1, 2]. Cotton was grown in the southern Piedmont whilst in the north tobacco was the main crop. Soils eroded from the slopes, their fertility rapidly decreased when cultivated, and the land was not as fabulously fertile as advertised in the brochures sent to prospective settlers [3]. The weather and soils were different from what settlers knew back in the home country where the climate was less extreme and moderated by the sea. Some settlers started experimenting with additions of ash, lime, and marl to improve the soil. Shifting cultivation was practiced in some areas; fields were cropped for some years, then left fallow, cleared after several years, and cultivated again. Such system had been practiced across the low fertility soils of South America and Africa but not in North America. Slaves might have taught the shifting cultivation system to the European settlers [1].

© The Author(s), under exclusive license to Springer Nature
Switzerland AG 2021
A. E. Hartemink, *Soil Science Americana*
https://doi.org/10.1007/978-3-030-71135-1_6

At the time of the Declaration of Independence in 1776, settlers had spread across the Piedmont between the Atlantic Ocean and the Appalachians. Population doubled every 20–25 years with most of the gain from natural increase and not from immigration. Railroads were being built and villages sprouted along the railway and close to rivers where there was wood and water for the household, and peace and protection from American Indians. The fertility of the soil did not determine settlement. At around 1850, settlers crossed the Red River and Missouri River into the Great Plains which were home to many American Indian tribes and 50 million bison that were slaughtered to near extinction. The early settlers were forest dwellers and did not know what to do with the interminable ocean of waving grasses. It was believed that the strong winds in winter would blow away their houses, or that they would be burned in summer. They feared that firewood would be limited and that there would be water shortage as there were few springs and waterways [2]. The settlement of the Great Plains was slow, but once they were plowed and wheat was sown, it was discovered how much fertility these soils harbored underneath the grass. Centuries of accumulated nutrients were released to voracious grain crops. By around 1910, there was no more new land for farm settlement; there were 6.4 million farms with an average size of 55 ha [4]. Knowledge about the land and its soils progressed much more slowly than the settlements.

The first colleges in the USA were established by religious groups in the eastern states. Their main purpose was to train ministers. Harvard University was established in 1636 and was modeled on Oxford and Cambridge Universities in England, and it collected endowments from the beginning. Colleges were formed in Virginia in 1693, in Connecticut in 1701, in Pennsylvania in 1740, in Delaware in 1743, in New Jersey in 1746, in New York City in 1754, and in Rhode Island in 1764. Numerous colleges were started in the nineteenth century. Their character and curriculum came from England, but the concept of the graduate school with emphasis on specialization and a high-level scientific dialogue originated from Germany [4].

The first public university west of the Mississippi was the Missouri State University, founded in 1839. It was established in the center of the state, half-way between St. Louis and Kansas City and a little north of the Missouri River. Like other states, Missouri had a dire need to bring science into agriculture and the university established a College of Agriculture and Mechanical Arts in 1870, followed by an Agricultural Experiment Station in 1888. In the same year, Sanborn Field was started, which was the third experimental field in the world, after the Morrow plots were established in 1876 at the University of Illinois, and the Broadbalk experiment that was established in 1843 at Rothamsted, UK. Sanborn Field had plots with different crops where soil erosion was measured, and crop experiments on poorly drained and clayey soils that crack in the summer [5]. Unexpectedly, two antibiotics were discovered in its soils in the 1940s.

Beginning in 1890, Missouri State University had a progressive President. It was a period of exceptional growth for the university, departments were established, and new hires were made [6]. Merritt Miller and Henry and Krusekopf were its first soil researchers, followed some years later by the soil fertility specialist William Albrecht, the soil chemists Richard Bradfield, Hans Jenny, and Edmund Marshall [7]. The most influential and well-known Missourian who shaped national and international soil developments was Curtis Marbut. Trained as a geologist, he became a physical geographer and pedologist. He developed a field of science, directed national and international institutions that carried those developments, advanced the *Bureau of Soil Science*, the *International Society of Science*, the predecessor of the *Soil Science Society of America*, and he was active in the *Association of American Geographers*. With these institutions, he laid the foundations for the growth and success of soil science as a discipline in the first quarter of the twentieth century, and its flourishing in the subsequent period. He had the intellectual capacity to combine existing ideas and create something novel; he worked untiringly, and his gentle, determined character and charisma stimulated others to follow.

Curtis Fletcher Marbut was born on a farm in the Ozark Mountains of southwestern Missouri in 1863. His father was not home during his birth, as he served with the Union Home Guard, a voluntary military organization that aimed to keep Missouri in the Union. Jane, his mother, named him Curtis, after the Union General Samuel Curtis, and Fletcher, after the Radical Republican Thomas Fletcher, who was elected Governor of Missouri the following year. The Marbuts had come from Hanover in Germany in 1784, and like many Europeans, had their name altered upon arrival in the USA, from Marpord to Marbut [6]. Curtis was a frail baby but became the first of three children to survive to adulthood. He grew up, learned to: "...plow and sow, to reap and mow, and to be a farmer's boy" and to Jane, his mother: "...I owe more to her steady guidance in times of hesitation and doubt than to any other single influence" [6, 8]. She was well-read [6].

Curtis Marbut rode on horseback to the high school in Cassville, and in 1885 he enrolled at the Missouri State University in Columbia. He was: "...still a raw, uncouth country boy" due to his upbringing in the Ozark Mountains. As money was tight and his parents could not pay for his education, he worked at the farm and taught in schools near his home. At the same time, he dated Florence Martin; they had known each other since their school days. Combining his studies with teaching and farm work was hard, and 2 years into his studies at the Missouri State University, he was ready to quit and leave for North Dakota. The area south of Jamestown had opened up for homesteaders, and he wanted to leave the rocky hills and to become a wheat farmer in the Peace Garden State. But Florence was not willing to leave the Ozarks for North Dakota, so they stayed in Columbia, where Curtis finished his BS degree in geology in 1889. They married 2 years later [9].

With his degree from Missouri State University, Curtis continued his studies in geology at Harvard University in 1893. Shortly after they settled in Cambridge, they had their first baby, whom they named Louise. Cambridge was a big town for them, with over 90,000 inhabitants compared to 6,000 in Columbia. Describing his early time at Harvard, Louise noted about her father: "…all his life he had been somehow different. Now he might have mourned in certain reflective moments, I am become a stranger unto my brethren, and an alien unto my mother's children." Louise Marbut knew her Psalms.

In 1894, Curtis completed his Masters of Arts in geology [9]. Most of his courses were in geology but he had learned about soils from Nathaniel Shaler, who had written an extensive soil text in 1892 entitled *The Origin and Nature of Soils* [10]. Nathaniel Shaler had grown up on a farm in Kentucky, and at Harvard, he had been a student of the biologist and geologist Louis Agassiz, whose racist views on polygenism were also held by Nathaniel Shaler. Contrary to Louis Agassiz, Nathaniel Shaler was an adopter of Charles Darwin's evolutionary theory, and liked his book on earthworms from 1881 that was popular in Europe but not in the USA [11, 12]. In the 1880s, Collier Cobb was an assistant to Nathaniel Shaler and he trained students who became the first soil employees of the *Bureau of Soils*.

At Harvard, Curtis took courses from the geologist and physical geographer William Davis, who became his PhD advisor. William Davis had no PhD degree himself, but was a synthesizer of observations and ideas [9]. For his PhD research, Curtis studied the physical geography of Missouri and traversed all counties so he could make map for the entire state. He never took the oral examination of his thesis at Harvard, as he became more interested in field geology than completing the work for his degree [13]. His doctoral dissertation, *Physical Features of Missouri,* was published as a report by the Missouri Geological Survey in 1896 [14]. The report was started at Harvard under the supervision of William Davis, and Curtis acknowledged that: "… enthusiasm for the work and a knowledge of physiographic principles which made the paper possible are due wholly to his teachings" [14]. Curtis was

Topographic sketch with contour intervals of the Flat Creek in Barry County, Missouri, by Curtis Marbut, 1896. The creek is some 20 km from where he was born. It was included in *Physical Features of Missouri* that was published by the Missouri Geological Survey in 1896 and based on his PhD research for Harvard University

awarded an honorary Doctorate by the University of Missouri in 1916, by which he became Dr. Marbut.

In the report *Physical Features of Missouri*, Curtis described landforms, geology, and hydrography. He discussed the history of settlement of Missouri in relation to its geomorphology. Whereas the dissected and mountainous districts in the southeast attracted the mineral-seeking and adventurous French, the alluvial lands along the Mississippi and lower Missouri River attracted farmers from upper Tennessee and Ohio, and the mountains were settled by woodsmen from mountainous parts of Tennessee, Kentucky, Virginia and the Carolinas. The fertile plains bordering the Missouri River attracted tobacco growers of Virginia and the hemp producers of Kentucky [14]. Curtis Marbut made maps for several valleys and escarpments, and cross sections illustrating topographies. Not much was included about soils but he thought that the soil in Missouri produced the greater part of the wealth.

In 1895, Curtis Marbut returned to Missouri and was hired by the Missouri Geological Survey where Charles Keyes was the Director. Geological investigations in Missouri were driven by mining, which had begun in the 1840s with iron mining and smelting, followed by lead in the 1860s, needed for bullets in the Civil War. Coal, zinc, copper, and limestone were also mined. The geologist Charles Lyell had visited Missouri in 1846 to study the effects of the 1811 earthquakes [15]. Charles Lyell had introduced the term 'Pleistocene', and was as important for geology as Darwin was for biology, Carl Linnaeus for botany, and Vasily Dokuchaev for pedology. Charles Darwin had studied under Charles Lyell. During one of his trips, Charles Lyell was shown a mastodon skeleton that a farmer in Missouri had found. The mastodon was dug up with the help of Albert Koch [15], and

Three dimensional soil map of Missouri prepared by Curtis Marbut for the 1904 St. Louis World Fair (Louisiana Purchase Exposition). At the fair, more than 60 countries maintained exhibition spaces and almost 20 million people visited the fair. In the summer of 1904, the Olympic games were also held in St. Louis and they were officially known as the *Games of the III Olympiad*. Today, the map (4 by 6 meters) hangs in Room 117 of Waters Hall, University of Missouri, Columbia

it was purchased by the British Museum and sent to London, never to be returned.

Curtis Marbut was given a sabbatical from the Missouri geological survey and spent the year 1899 in Europe, visiting the Alps, the Hungarian Plain, the Mediterranean Basin, the British Isles, the major river valleys of Europe, and Scandinavia. As his daughter Louise wrote: "These landscapes and landforms were his laboratories, the great clinics, he wanted to see" [6]. He made the journey with the political scientist Isidor Loeb, who worked at the Missouri State University, and in some countries, they traveled by bicycle [9, 16]. While traveling through Europe and meeting geologists, he felt a deep-rooted sense of spiritual kinship with them, and he had a natural way of making friendships [6]. The year in Europe was his first time abroad, and it awakened in him an urge to discover more of the world.

Back in Missouri, Curtis Marbut returned to geological survey as a field geologist, and taught *Physiography and Geology* at the University of Missouri. During the surveys he made his way into the Ozarks by horseback or wagon and whilst observing the geology and the landscape, he developed an interest in soils. Not much was known about soils in Missouri. The first soil surveys were conducted in 1901, but there was no state-wide program. The official soil survey in Missouri was started with the encouragement of Cyril Hopkins from the University of Illinois. He had established a soil survey program independently from the *Bureau of Soils* as Cyril Hopkins did not get along with Milton Whitney, the Chief of the *Bureau of Soils*. In 1905, Cyril Hopkins visited the University of Missouri to talk about the soil survey in Illinois, and he suggested that Missouri should start something similar and that the soil survey be directed by a geologist [16]. The idea was well-received and Curtis Marbut became the director of the Missouri soil survey in 1905 [9].

In 1909, Milton Whitney was a guest speaker at the annual Missouri Farmer's week in Columbia, and downtown Columbia was crowded with people and horses and more than 4,000 visitors. The purpose of the Missouri Farmers Week was to build the reputation of the College of Agriculture, and the agriculture faculty presented their research findings. After his talk, Milton Whitney met with Curtis Marbut and was invited for dinner at his home [17]. The *Bureau of Soils* in Washington was hiring people and Milton Whitney, who had a strong preference for hiring geologists, invited Curtis Marbut to join the bureau. Curtis Marbut accepted the invitation and it was agreed that before he moved to Washington, a detailed soil survey of the Ozarks region would be conducted. Curtis Marbut was excited by this prospect but 2 months later, Florence died of pneumonia; she was only 44 years old. Curtis Marbut was suddenly alone with their five children. Louise was 17, and the youngest was only 3 years old [17].

Curtis Marbut conducted the survey of the Ozarks with the help of Merritt Miller and Henry Krusekopf who were the first soil researchers hired by the University of Missouri. The Ozarks formed a mountain region in between the Appalachians and Rockies and were known as the Interior Highlands, covering most of southern Missouri, and parts of Arkansas, Oklahoma, and Kansas. Curtis Marbut, Merritt Miller and Henry Krusekopf traversed the Ozarks regions and some of the mapping was done from the top of a freight car. It was during the soil survey of the Ozarks, that Curtis Marbut's interest in soils overtook his interest in geology [13].

In 1910, after the fieldwork in the Ozarks was completed, he and his five children left for Washington where Curtis Marbut joined the staff of the *Bureau of Soils*. The children stayed for 2 years in Washington and then returned to Columbia where the family took care of them. Curtis Marbut moved out of their house in Washington, rented a small room, worked 7 days per week, read every soil publication he could get his hands on, and learned to read the language of his ancestors, German [17]. He adopted an ascetic lifestyle and became a somewhat solitary man, crowding his days with work to compensate for the want of his children and deceased wife.

In his first year at the *Bureau of Soils,* Curtis Marbut worked on the soil survey report of the Ozarks that was published in 1910 as *Soils of the Ozark Region. A preliminary report on general character of the soils and the Agriculture of the Missouri Ozarks* [18]. It was his first soil report and included sections on the productivity of the soils and how they were managed. In the Introduction, Curtis Marbut noted: "…a large part of the area has a soil of only moderate fertility and therefore a soil that will not stand abuse indefinitely. It does not smile upon the farmer who refuses to treat it well and supply it with food. It does not lavish great wealth on the soil robber simply because it has not the great wealth to expend upon him. On the other hand it is of such a character that it will not refuse to reward the fullest extent the hand that feeds it. Its responses to good treatment is prompt and full. It returns to the farmer products of many times more value than the value of the food that he gives it. The Ozark farmer, however, has not learned to treat his soil like he treats his horse. He is unable to hear the cry of the starved soil as he hears that of his livestock, though the value of the response of the soil to good treatment is probably greater than that of livestock. The latter can move about from place to place and often gain sustenance regardless of treatment by the owner, the former cannot do so" [18].

Curtis Marbut had collected samples from A and B horizons but not from C horizons which he considered parent material and not part of the soil. In the Ozark report, he discussed soil classification, which at time had barely been developed. He listed various criteria for soil classification, such as the origin of the soil, the kind of rock, the crop to which it seems best suited, or the natural vegetation. For the Ozarks, he developed a classification system which was based on A and B horizons, and his ideas on soil classification in 1910 were "…the ideal classification would be one that takes into consideration as many as possible of the different conditions affecting a soil, or characteristics possessed by it. Until within the last few years there has been in America no well-defined and universally recognized basis of soil classification. The only system of classification used at the present time is based on the subdivision of Physiography—a combination, in a broad sense, of geology and topography. Further sub-classification are in categoric order, Geologic

The summer of 1914, Curtis Fletcher Marbut watching how Merritt Miller digs a soil pit at the edge of a corn field a little east of Columbia in Missouri. Photo Missouri archives

formation (kind of rock) from which the soil is derived, general size of the soil particles and finally the texture and color of the soil" [18].

For Curtis Marbut and many others in 1910, soil classification was largely based on the geologic origin of the soil. Later on, he developed a broader and pedological view on soil classification after he had studied the Russian literature, as he wrote in 1936: "At the same time when western Europe was still engaged in the futile assertion that the soil as soil is dominated in its general features by the materials out of which it has been built, the Russian workers had already shown that the soil is the product of process rather than of material and is, therefore, a developing body rather than a static body. They allied the soil to life rather than to death" [19].

Curtis Marbut succeeded Milton Whitney as Chief of the *Bureau of Soils*. He directed the soil survey for 25 years, during which time about a billion acres or half the agricultural land of the country, was mapped. With the botanist Homer Shantz at the Bureau of Plant Industry and Oliver Baker who was at the Bureau of Agricultural Economics, they formed an influential trio within the Department of Agriculture. Curtis Marbut traveled extensively. With Oliver Baker, he made trips to South America and all over Europe. After attending the *Third International Conference of Pedology* in 1922, he became friends with European soil scientists such as Emil Ramann, Alex. de' Sigmond, David Hissink, George Murgoci, and Konstantin Glinka. In the late 1920s, Curtis Marbut was the best known and well-respected soil scientist from the

USA on account of his leadership in the soil survey, his knowledge and erudition, the organization of the *First International Congress of Soil Science*, and the Transcontinental Excursion that was held after the congress. He received an Honorary Doctorate from the University of Missouri in 1916, and from Rutgers University in 1930, thanks to the efforts of Jacob Lipman who had worked with him during the 1927 congress.

In 1928, Curtis Marbut turned 65, the official government retirement age. He thought of building a house on a wooded hill overlooking Little Flat Creek in Barry County. Building one's own house was a tradition amongst the rural Ozarks. Since 1922, he taught every summer at Clark University in Worcester, where he stayed with the geographer Elmer Ekblaw and the university president Wallace Atwood, who lived in a New England Cape Cod shingle-style house. Curtis liked that house and it inspired him to design his house in the Ozarks. He sent a design to his brother and manager of the apple orchard, who was to supervise the construction [20]. They found a local builder who was from New England, and construction began in 1935. During the early summer of 1935, he spent a week in the Ozarks with Helen, his daughter, examining the final construction. The house was cedar-shingled, had six fire-places, and indoor bathrooms, which was a luxury in the Ozarks in the 1930s [21]. The timber was from trees around the house, and in one of the panels, a lead bullet was encased from a civil war battle [21]. The 11-room house cost $6,000 which was a considerable amount of money in the middle of the Great Depression. Helen Marbut moved to the house in 1948 and farmed the land for a while, despite the reactions of the people in the area—she was unmarried, and a woman from the city.

After Curtis Marbut had spent some time at his new Ozark house, he boarded the boat in Baltimore on 19th July 1935, for his journey to Oxford in the UK to attend the *Third International Congress of Soil Science*. Just before boarding, he was given the proofs of the soil chapter of the *Atlas of American Agriculture* that had been edited by his friend Oliver Baker [22]. It was a monumental chapter that Curtis Marbut had largely completed single handedly. At the congress in Oxford, he presented his ideas on land classification and chaired the Commission on soil classification. Several of his friends suggested that he should become the next President of the *International Society of Soil Science*, but his reply was: "I think you would do better to choose somebody younger." With that notion, characteristic of his unpretentiousness, he left Oxford, took the boat across the Strait of Dover, and boarded a train that took him through Russia to China.

Curtis Marbut had traveled in Russia during the excursion after the *Second International Congress of Soil Science* in 1930, but he had never been to China. He always felt his soil knowledge was incomplete and was looking forward to studying the soils in China [6]. He had been requested to conduct a study by the Soils Division of the National Geological Survey and several American soil scientists who had worked in China. Cooperation was started in the early 1920s and Walter Lowdermilk, John Buck, Charles Shaw, and Robert Pendleton worked in China in the late 1920s and early 1930s [23]. In the summer of 1933, Robert Pendleton was replaced by James Thorp who had worked for the *Bureau of Soils* and went to China on the recommendation from Curtis Marbut [24]. James Thorp was based in Beijing and provided training in soil survey methods for the National Geological Survey with the aim of making a soil map of the entire country [25].

The train journey through Russia was long and Curtis Marbut caught a cold in Siberia. The cold developed into double pneumonia, which he had suffered before, and from which Florence had died. His lungs were weak, and one had been punctured when his Model T Ford went into a ditch in South Dakota some years earlier [6]. He became seriously ill. The other passengers on the train affectionately called him 'the old American professor,' but they grew concerned about his health and gave him all the attention possible. The train stopped in Harbin, some 1,200 km before Beijing, where he was brought to the American Consulate. They took him to the hospital and summoned James Thorp. Curtis Marbut had a high fever but recognized James Thorp, and said: "You've got to get me out of here; I've got ten years of work to do" [26]. But Curtis Marbut's condition worsened that night and he died early in the morning, at the age of 72. It was Sunday, 25th August 1935, some 5 weeks after he had boarded the boat in Baltimore.

Curtis Marbut was cremated in China and the consulate asked some Russian refugees to come in and sing during an improvised ceremony [27]. The ashes were returned by boat to the USA, where they were buried next to Florence in October 1935, some 40 km from where he was born. The Chinese urn blended in color with the brown of the soil and the green of the vegetation [8]. A memorial service was held in the living room of his newly built house where he had hoped to spend his last years. He had left for Oxford and China in July 1935 but died on the way. The journey had ended, but Curtis Marbut had not returned [17].

The unexpected death of Curtis Marbut came as a shock. Obituaries appeared across the world in journals and newspapers. The president of the American Geographical Society noted that he: "…had plowed a deep furrow in the virgin field of soil geography. By his skillful application of the philosophy and principles of modern science to mapping of soils he has given new direction to man's conquest of nature - the better utilization of that precious heritage, the soil, the soil, 'foothold of all things.'" It was through his influence that the fundamental work of Russian pedologists became known in the USA. Russian soil scientists were saddened by his death, and a special issue of *Pochvovedenie* was devoted to Curtis Marbut in 1936. The issue contained 28 articles. The editor, Arseniy Yarilov, described him as a: "Columbus for our time, who had acclimatized the soil science worked out by Dokuchaev and his collaborators to the United States." Charles Kellogg sent copies of the *Atlas of American Agriculture*, which included Curtis Marbut's extensive chapter on the *Soils of the United States*, to Russian scientists [28].

John Russell from Rothamsted in the UK recounted his time on the Transcontinental Excursion: "He took charge of the large and strenuous group that toured the United States and Canada after the formal meetings were over and he was indefatigable in doing everything possible to make the excursion a success. He maintained firmly that the excursion was to be a serious soil study; only reluctantly would he agree to modifications in favour of scenic attractions and not at all if these conflicted with the main purpose of the journey. And although our kindly hosts insisted on giving us numerous banquets and taking us for beautiful drives, and the rather weary and sometimes rather frivolous-minded members willingly accepted these very agreeable diversions, one always felt that Dr. Marbut disapproved and regretted the hours that might have been spent in examining profiles, but instead were used in other ways." John Russell also reflected on some of the personal characteristics of Curtis Marbut: "He believed deeply in his own responsibility for his own fate: he had carefully thought out his position. He will always remain in my memory as a magnificent example of that quiet type of American gentleman one often meets when travelling in the United States: well read, cultured, but modest, never pushing himself to the fore and indeed avoiding the lime-light of publicity." Gilbert Robinson, founder of the Soil Survey of England and Wales, dedicated the second edition of his pedology textbook to Curtis Marbut [29]. Gilbert Robinson noted that Curtis Marbut's influence extended beyond the bounds of his own country and that the soil survey in the USA was a model for the rest of the world.

Charles Kellogg wrote the obituary for *Science* and named him the foremost authority on the soils of the world, and that no single individual had contributed more to soil science [30]. Jacob Lipman noted that he had died with his boots on. Merritt Miller, who had been on soil surveys with Curtis Marbut across Missouri, called him: "...a fine gentleman and a great scientist" [13]. Henry Krusekopf, who succeeded him at the University of Missouri, considered Curtis Marbut to be responsible for the rise of pedology in the USA, and for the establishment of soil science as an independent natural science [16]. Jacob Joffe, whose *Pedology* book featured a Preface by Curtis Marbut, noted that: "American workers, with very few exceptions, have not appreciated what Marbut has done for them and for soil science in the United States. Instead of taking hold and advancing his ideas and extending his methods, his successors have drifted into the business of dramatizing the story of soils as an object of would be national preservation or suicide, or once more a return to the old utilitarian motive" [31]. Jacob Joffe often called Curtis Marbut the 'American Dokuchaev.'

The *Soil Science Society of America* devoted a book to him that was entitled *Life and Work of C. F. Marbut*. Most of the chapters were collected and prepared by Henry Krusekopf and the editorial committee included Merritt Miller, Elmar Ekblaw, Andrew Whitson, and Mark Baldwin. Louise Marbut wrote a biography of her father: "...the hand of destiny had touched his elbow and was urging him towards a future that had terrifying but irresistible promise" [6]. Thomas Rice remembered how Curtis Marbut: "...always had new ideas and was always devising new ways to do things. He could be terribly exasperating to us soil surveyors at times [6]." Curtis Marbut had no respect for public opinion against what he regarded as the scientific truth, and he disliked politics [16].

The botanist Homer Shantz, with whom Curtis Marbut worked on *The Vegetation and Soils of Africa* [32], recalled that Curtis Marbut's early life had inured him to hardship and frugality, that he never grew old, either mentally or physically, and that he had the rare ability of becoming more efficient in his work up to the very end [8]. The 1938 yearbook *Soils & Men* was dedicated to Curtis Marbut, and the book reflected on his legacy: "...although this work by itself is of the utmost significance, perhaps his greatest influence was exerted by more subtle means. No man better exemplified the scientific spirit. His devotion to truth and freedom from prejudice were coupled with a modest, kindly personality that inspired all his associates as well as young men everywhere who were interested in soil science. Although he became a citizen of the world, he retained the simple habits of his early life in the Missouri Ozarks" [33].

Macy Lapham was the first soil survey inspector in the west and reminisced in his 1949 autobiography: "Curtis Marbut came to us as a geologist with an interest in soils and was strongly convinced of the close relationship between geology and soils. Having a strongly critical attitude of mind, he was outspoken in criticism of some aspects of the work and met with some coolness on the part of the personnel. All who knew him later came to appreciate his honesty and sterling character and to value his criticism even though it hurt. As the years went by, he became less of a geologist and more of a soil scientist. He proved himself to be a man who was not afraid to change his mind, even to acknowledging error if convinced he had been in the wrong. As he came under influence of the Russian school of soil science, in which climate and vegetation are stressed in soil development, he completely reversed his attitude as to the dominating influence of parent geological materials on the character of soils" [17]. Macy Lapham recalled that Curtis Marbut was frugal, preferred the horse over the Pullman car, and was a tireless walker.

James Thorp had known Curtis Marbut since 1921 and was in Harbin when Curtis Marbut died. Years later, he recalled that Curtis Marbut had a motto: "…referring to soil genesis: *When in doubt, ask the soil*. He asked the soil, and the soil told him many things. And he inspired the men under his direction to study the soil and let it teach them. He pulled the American soil science of 1910 out of its rut and set it on its way to great advancement. Professor Marbut will always stand out as a brilliant scientist, great teacher, and friendly person. He is rightly honored as a great citizen of Missouri and of the USA as a whole" [26]. James Thorp dedicated his book *Geography of the Soils of China* to Curtis Marbut and Wong When-Hao [34]. James Thorp was greatly influenced by the works of Curtis Marbut, and he had attended his lectures in the Graduate School of the Department of Agriculture in 1928: "…where I attempt to record the details of my indebtedness to him there would be a footnote on nearly every page!" Those 1928 soil lectures were published in 1951, in the book *Soils, their Genesis and Classification* [35]. The book had a pen drawing of Curtis Marbut by Francis Hole, and Charles Kellogg wrote an introduction. He noted that Curtis Marbut was a great teacher with a forceful personality who had inspired young scientists all over the world, including himself. In 1960, Curtis Marbut was elected to the National Agricultural Hall of Fame, which was started by President Eisenhower. The other two soil scientists in this Hall of Fame were Justus von Liebig and Hugh Bennett.

Henry Krusekopf succeeded Curtis Marbut as director of the Missouri Soil Survey. He had obtained his BS in agriculture in 1908, and his MS in 1916 from the University of Missouri in Columbia, and besides directing the soil survey, he had a farm in the southeastern part of the state. He became an expert on the soils of Missouri and knew: "...more about field soils in Missouri than any other man has ever known" [7]. With Merritt Miller, he wrote a statewide soils report and map for Missouri that was published in 1929 [36]. They discussed the factors of soil formation and which ones were responsible for the various soil conditions and caused soil differences. They recognized the following factors: geology, topography, climate, vegetation, and soil age. At any given location, all of the factors were considered operative, but they thought that one factor was dominant over the others [36].

Merritt Miller was born on a farm in 1875 in Grove City, a village south of Columbus in Ohio. He received a Bachelor of Agricultural Science degree from the Ohio State University in 1900, and a Master of Agricultural Science from Cornell University in 1901. Two years later, Cornell University graduated the first PhD in soil science. After a year as a soil surveyor with the *Bureau of Soils* and 2 years as an instructor in agronomy at the Ohio State University, Merritt Miller joined the Department of Agronomy at the University of Missouri. The department was just established, and he served as its Head until it was split into two departments: Field crops and Soils. Merritt Miller was the Head of the Department of Soils until 1938, after which he became the Dean of the college. In 1924, he published the book, *The soils and its management,* which was meant to teach improved soil management for a new generation of farmers [37]. The book became popular, and he stressed that the soil was our greatest resource, that should be cared for: "...we have always been more interested in satisfying present needs and desires than in providing for the future" [37]. He lectured on soil classification based on the division of soil provinces developed in the *Soils of the United States* book from 1913, and he was not in favor of teachings based on the Russian principles of soil science [38]. Merritt Miller believed that west of the Great Plains, there were no soils, but only parent materials [39].

In 1910, Merritt Miller spent two years at the University of Göttingen in Germany. With Andrew Whitson of Wisconsin, he founded the *American Association of Soil Survey Workers* in 1920, and he attended the *First International Congress of Soil Science* in 1927. Merritt Miller became an influential soil scientist and had two sons who became soil physicists. Robert Miller was a member of a military unit near Tokyo, and he raised the first American flag over Japan on 28th of August 1945. After the war, he was hired by Richard Bradfield at Cornell University. Ed Miller was at the MIT Radiation Laboratory and co-developed tactical uses for the radar that were instrumental during the 1944 invasion of Normandy. Ed Miller was hired at the University of Wisconsin by Emil Truog.

Merritt Miller and Frank Duley established soil erosion plots in 1917 [40]. The 1.8 by 27.5 m plots represented 1% of an acre, and these were the first plots for measuring soil erosion. In the late 1920s, the plots and experimental design were replicated at ten research stations nationwide at the recommendation of Hugh Bennett. Results from these erosion plots were used to argue for the establishment of the *Soil Erosion Service* in 1933, and in the development of the Universal Soil Loss Equation in the 1950s. Frank Duley left the University of Missouri, obtained his PhD at the University of Wisconsin in 1923, and joined the Department of Agriculture at the University of Nebraska. In Lincoln, he worked with Jouette Russel, who went by the name of J. C. Russel, and who had one of the first soil science degrees from the University of Minnesota. Frank Duley and J. C. Russel studied how soil management affected soil organic matter levels. They studied the effects of cover crops and tillage and concluded that plowing aerated the soil, stimulated bacterial action, and resulted in the loss of organic matter [41, 42]. Less tillage maintained soil organic matter levels, and the 1940s work of Frank Duley and J. C. Russel was an early promotion of reduced and zero tillage [43].

Less plowing and tillage was also advocated by several others including Edward Faulkner [44, 45], Lionel Picton [46], Jerome Rodale, Louis Bromfield, and Albert Howard [47] and Eva Balfour [48]. The detrimental effects of tillage had been discussed by Nathaniel Shaler in 1892, as he had observed at his parents' farm in Kentucky: "...this slight superficial and inconstant covering of the earth should receive a measure of care which is rarely devoted to it; that even more than the deeper mineral resources it is a precious inheritance which should be guarded by every possible means against the insidious degradation to which the processes of tillage ordinarily lead" [10]. Likely Nathaniel Shaler was referring to the effects of tillage on soil erosion.

In 1929, the University of Missouri had a small but notable staff of soil researchers with Merritt Miller as Head of Department, Henry Krusekopf as pedologist, William Albrecht as soil fertility expert, and Hans Jenny and Richard Bradfield as soil chemists. Richard Bradfield had been hired as an instructor in the Department of Soils in 1920. Like Merritt Miller, he had grown up on a farm in Ohio. When his father died, Richard Bradfield took over the farm to support his mother and sisters, and some of the farm income was used to pay for his graduate studies at the Ohio State University. He studied in the Department of Soils that was established in 1921. Firman Bear was the Head of the department. Richard Bradfield was awarded a PhD degree for the thesis *The Chemical Nature of Colloidal Clay* in 1922. He had found that colloidal clays were complex aluminosilicates and not mixtures of the oxides of silicon, aluminum, and iron. It was the first PhD degree in soils from the Ohio State University, where some of the early teaching in soils was given by Arthur McCall, who in 1904 had succeeded Merritt Miller when he took the position in Missouri [49].

After Richard Bradfield had moved to Missouri, he worked on acid clays, which was researched in the Dutch polders by Jac. van der Spek and David Hissink. In 1927, Richard Bradfield joined the Transcontinental Excursion following the *First International Congress of Soil Science*. His chemistry background had not weakened his appreciation of the soil profile: "The work of the soil surveyor, the classifying and mapping of soils, is one of the corner stones in the foundation of soil science. The history of the soil is written indelibly in the soil profile itself... all we need to know is there if we can learn to read it. Soils are complex but not hopelessly so" [50].

In 1930, when the economic crisis hit, Richard Bradfield left the University of Missouri and returned to Ohio to become a professor of agronomy. In Missouri, he was replaced by his former student Leonard Baver, who stayed until 1937, at which time he also joined Ohio State University. Richard Bradfield had left and was now the head of the agronomy department at Cornell University, teaching courses in soil chemistry as well as soil physics. At Cornell, Richard Bradfield expanded international activities, as he had good connections with the Rockefeller Foundation. In 1941, he served on the team that designed the agricultural program for Mexico, for which Norman Borlaug was hired. That program spread to other countries and was later dubbed 'the Green Revolution.'

Richard Bradfield was an advisor to the United States Department of Agriculture, and together with Bill Pierre of Iowa State College, Bob Salter of the Ohio State University, and Charles Kellogg, he was a leader for agronomic affairs and policies in the USA [51]. Richard Bradfield was the official delegate of the *American Society of Agronomy* for the *Second International Congress of Soil Science* in Russia in 1930, and the third congress in the UK in 1935. During the international soil congress in Madison in 1960, he was the President of the *International Society of Soil Science*. He was the first President of the *Soil Science Society of America* in 1936, and many University of Missouri soil scientists served in that position: William Albrecht, Leonard Baver, Edmund Marshall, Frank Duley, and Hans Jenny.

When Richard Bradfield went on a sabbatical in 1927, his position was filled by Hans Jenny, who was born in Switzerland in 1899 and worked on a farm before starting his studies at ETH in Zürich in 1919. His main interest was animal science, but he was also interested in geology and erosion. He realized that some day all of the Swiss mountains would be flattened and then the Switzerland he knew would not exist anymore. It made him sad [39]. In 1923, he obtained a diploma in agriculture and published a paper on the foundations for animal feeding [52]. Hans Jenny became interested in graduate studies with Georg Wiegner, who was the leading chemist in Europe and an admirer of Justus von Liebig. Georg Wiegner had studied colloid chemistry at the University of Göttingen under Richard Zsigmondy, who invented the ultramicroscope and received the Nobel Prize for chemistry in 1925. Ultramicroscopy demonstrated the crystalline structure of clay before it was confirmed by X-ray diffraction.

Georg Wiegner applied fundamental chemistry to the study of soils, but never held a spade or auger, which would have been below his standing in the German academic tradition [39]. He was an early adopter of the Russian school of soil science, and he taught soils differently from the standard books in Europe. George Wiegner integrated the soil knowledge of the European school with the Russian and American school of Eugene Hilgard [39]. Climate and rainfall in particular were considered the main soil-forming factors. In Russia, rainfall was high in the north whereas the south had little rain, and the soil distribution pattern lined up with this gradient. Georg Wiegner and a graduate student used rainfall and temperature to explain soil geography in western Europe but it yielded no clear pattern because of the large variation in soils across the continent. In 1917, he published a slim but influential book on soil formation from a colloid and chemical perspective that ran through several editions, and in 1926, he published a book on quantitative agricultural chemistry [53, 54]. His books were never translated to English.

Georg Wiegner did not directly admit Hans Jenny as a student and urged him to take some more chemistry courses. After he had taken those courses, Hans Jenny became Georg Wiegner's assistant in soil and colloid chemistry. He cycled around Zürich and collected soil samples from plowed fields and in the laboratory, he measured soil acidity and the amount of calcium carbonate in these samples. George Wiegner often told Hans Jenny that his interests were too broad. There was a rigid scientific climate in the laboratory, and mediocrity was considered worse than a crime. The prevailing thinking of the graduate students at that time was: "facts mean nothing, and facts are boring unless there are theories behind them, or money." That might have been popular in Switzerland at the time, and the Swiss geologist Louis Agassiz had stated a half-century earlier: "Facts are stupid until brought into connection with some general law," or as Justus von Liebig had stated before him: "...facts are like grains of sand which are moved by the wind...principles are these grains cemented into rocks." So, from his earliest research days, Hans Jenny was thinking about data and how they could fit a theory, and that whatever he was to do, it had better rise above mediocrity.

Hans Jenny and another student went hiking most weekends. They were penniless, so they played music in village squares. Hans played guitar and his friend the violin and they needed lodging and slept in hay barns, in exchange for playing music. One day in the high mountains, and much to the excitement of Hans, he stumbled upon a true Podzol that he recognized from Georg Wiegner's lectures. He mentioned this finding to Georg Wiegner and that the climate-soil relationship was applicable in Switzerland and that the Russian soil principles were accurate In 1926, Hans Jenny finished his PhD thesis on the significance of ion hydration on exchange reactions of soil colloids. Shortly afterward, he met a representative of the Rockefeller Foundation, who came to speak about a program by which young European scientists could spend 1 year in the USA. Hans Jenny was keen to enroll in that program although most people he worked with held American science, including soil science, in low esteem—the USA had only one or two Nobel laureates. Hans Jenny requested to work with the plant nutritionist Dennis Hoagland at the University of California Berkeley, but the representative considered the west coast too far away, suggesting instead a university on the Atlantic coast. Fortune would smile upon Hans Jenny: in 1926, Selman Waksman toured Europe and visited Georg Wiegner to see the laboratory where Sergei Vinogradskii had worked in the 1880s. Impressed by Hans Jenny's knowledge and broad thinking, Selman Waksman invited him to New Jersey and wrote a letter of support to the Rockefeller foundation; they accepted Hans Jenny's application.

In 1926, Hans went by boat across the Atlantic, landed in New York, and took the train to New Jersey where he was welcomed by the station director Jacob Lipman. Selman Waksman's laboratory at the New Jersey Agricultural Experiment Station was messy compared to the tidiness at ETH in Switzerland. Selman Waksman described his own laboratory as "My laboratory was rather small and was devoted primarily to the study of the microbial population of the soil and its role in soil processes" [55]. The laboratory might not have been immaculate, but the group of people Hans came to work with was international, and they studied a wide array of topics. He learned about differences in research styles and personalities and observed that Selman Waksman and Jacob Joffe were not friends. Robert Starkey was Selman Waksman's assistant, having completed his PhD in 1924. Robert Starkey worked on nitrogen fixers, microbial sulfur, and nitrogen transformations and microbial degradation of cellulose with the French student René Dubos. Hans Jenny learned about soils from Jacob Joffe, the pedologist at Rutgers University who had come to the USA in 1906. Jacob Joffe was trained by Jacob Lipman and taught the Russian genetic principles of soil science that favored the study of soil profiles. As he wrote in 1929: "…it was the new concept of soils as an independent, natural, historical body which required not only the description of the surface features of soil but also the anatomy of it."

After half a year, Hans Jenny was tired of the conditions in Selman Waksman's laboratory and started to work with John Shive. In 1915, John Shive had published a paper entitled: *A Three-Salt Nutrient Solution For Plants* [56], the findings of which started the soil-less culture and hydroponic movements. That work had the interest of Hand who had initially applied to study with Dennis Hoagland in California. Hans Jenny was not the only one who was unimpressed by the laboratory conditions of Selman Waksman; René Dubos had similar observations and he thought the style was too casual, lacked rigorous intellectual logic, and the presentation of the science was too sloppy. They both looked down on Selman Waksman's laboratory, which, some 25 years later, would receive the only Nobel prize given to a soil scientist. At Rutgers, Hans Jenny and René Dubos obtained insights and training that equipped them do groundbreaking work. René Dubos developed a holistic and ecological concept process of microbiological processes and became a medical scientist and environmentalist [57], whereas Hans Jenny advanced pedology with a theoretical framework of quantitative relations between soils and their properties.

Jacob Lipman, who was director of the Experimental Station and President of the *International Society of Soil Science*, asked Hans Jenny to edit abstracts that had been submitted for the *First International Congress of Soil Science* that was to be held in the summer of 1927. Translations had to be made from German into English and vice versa as well as into some other languages. Hans Jenny attended the congress and Transcontinental Excursion that followed the congress; he was the official delegate from Switzerland [58]. Besides getting to know soil scientists from across the world and meeting again with Georg Wiegner, participation in the congress shaped his future: he was offered a position at the University of Missouri, and he started thinking about the soil–climate relationships in the USA and how they could be explored.

Richard Bradfield also participated in the Transcontinental Excursion and met with Georg Wiegner. He was familiar with Georg Wiegner's work on colloidal chemistry at ETH in Switzerland. After Richard Bradfield had secured a Guggenheim fellowship, it was agreed that he would spend a 1-year sabbatical at ETH with Georg Wiegner. Hans Jenny was offered a position at the University of Missouri when Richard Bradfield went on sabbatical to ETH. Richard Bradfield did not get along with Georg Wiegner, as they had come to soil chemistry from different angles. Richard Bradfield had grown up on a farm in Ohio; he was a practical man and that is how he approached the theoretical aspects of soil chemistry. Georg Wiegner was brought up in the strict European scientific tradition and he advanced the theoretical aspects of chemistry and was less interested in the practical aspects [39]. So Richard Bradfield spent part of his sabbatical in Vienna and in Berlin with the chemist Herbert Freundlich who worked on coagulation and chemical equilibria. Like his colleague Fritz Haber, Herbert Freundlich fled Germany in 1933 to the UK, and came to the University of Minnesota in 1938.

In August 1927, Hans Jenny left the position with Selman Waksman and took the train to St. Louis, where he registered at the Swiss consulate. He felt that he had not achieved much in New Jersey: he still had to learn good English, had shifted his research focus to plant nutrition, and had spent too much time on the translation of abstracts for the first soil congress. He arrived in Missouri with no unfinished work, but had several ideas, or as he would say: "..when I arrived at Columbia in the late 1920s, I was a fresh greenhorn from the Alps" [39]. The University of Missouri allowed him to do what he wanted, but he had to mentor some of Richard Bradfield's graduate students and had to assemble the ultramicroscope that had just been purchased.

Hans Jenny continued to work on the exchange base properties of soils. In 1932, a movie was made about the base exchange with the following purpose: "In order to give students of agriculture a more vivid picture of base exchange, a movie has been produced which demonstrated both technique and mechanism of ionic exchange reactions." The movie was used in the teaching of soil chemistry [59]. The first part of the movie showed the digging of a soil profile (Putnam silt loam), soil sampling, and the preparation of clay suspension. The nature of the colloidal complex, including the adsorption and release of ions, and the nature of base exchange was presented in the second part. The half-hour movie was the first for teaching soil science. No copies remain today.

Hans Jenny became friends with William Albrecht, who studied the relationship between the fertility of the soil and human health. William Albrecht had grown up on a farm in the former prairies of Illinois and had obtained a PhD from the University of Illinois, where he studied symbiotic nitrogen fixation [60]. During his PhD studies, he had found that the nitrogen level of the soil had no effect on the fixation by legumes [61]. In 1916, he was hired my Merritt Miller at the University of Missouri as one of the first professors in soils with a PhD William Albrecht was particularly interested in calcium nutrition, and for him, food was nothing more than fabricated soil fertility [61, 62]. He developed the base saturation ratio as ideal ratios of cations for the soil and taught a no-credit course in glassblowing [39].

After the Second World War, William Albrecht reviewed dental records of 70,000 navy sailors, and linked the health of their teeth to the status of the soil where they had grown up. He concluded that many of our diseases could originate from nutritional deficiencies as a result of low soil fertility [63]. As he summarized the relation between soil and human health in 1948: "If the decay of teeth is linked with the declining fertility of the soil, this concept of tooth troubles may well be a pattern to guide our thinking about other health troubles, not as calls for drugs and medicines, but for conservation in terms of a new motive, namely better health via better nutrition from the ground up" [63]. Over the years, he broadened the idea of plants as biological assays of the soil, to herbivorous animals, and then to humans [61]. He was concerned about the way food was produced and the agricultural practices that were used. William Albrecht discussed 'soil-health' but in relation to human health [61], quite different from the current attempts to define a soil condition. There were others who considered the soil and human health: Lionel Picton had worked for 7 years with Albert Howard in the UK, and in 1949 he reviewed the relations among soil, nutrition, and human health [46].

William Albrecht succeeded Merritt Miller as the Head of the Department of Soils and remained so for over 20 years. In 1950, he invited Charles Kellogg to the University of Missouri, and he gave a talk on: *Our soil resources: their use and limitations.* They had both just returned from the *Fourth International Congress of Soil Science* in Amsterdam and knew each other from the 1938 yearbook *Soils & Men,* for which William Albrecht had contributed a chapter on the loss of soil organic matter and its restoration [33]. In 1933, Charles Kellogg, could have taken a position at the University of Missouri when William Albrecht took a leave of absence for a year, but at the time, he considered the salary too low. Charles Kellogg gave a well-attended talk and after the visit, he noted in his diary: "Albrecht had been furiously writing a great deal of nonsense about the direct effects of liming on food quality. I recall commenting on this in the speech and pointing out that if a diet were made up from crops grown on the best soils in Missouri and a comparable one on the poorest soils, the first would be somewhat more nutritious. But by adding or subtracting a quart of milk a day the small differences would be overwhelmed. Prof. M. F. Miller and the others were highly amused. It is very difficult to understand how M. F. Miller, whom everyone respected, and the others at the University let this quack stay on as Head of the Department of Soils. He wrote bulletins and hundreds of papers, and all sorts of

William Albrecht from the University of Missouri in Columbia. He was one of the first to work on the relation between soil fertility and human health, and for him food was nothing else than fabricated soil fertility. He succeeded Merritt Miller and was the Head of the Department of Soils for over twenty years. Charles Kellogg visited him in 1950, and called him a quack based "...on his bulletins and hundreds of papers and all sorts of curious publications."

curious publications based on 14 animal experiments. He gave no idea of his experimental errors but by giving him the benefit of the doubt the men at Cornell Laboratory analyzed them according to the experimental errors of the best experts with animals. Only one had statistical significance. That one was repeated by the laboratory in cooperation with the North Carolina Experiment Station with opposite results."

William Albrecht did not fit Charles Kellogg's view of the agricultural world which was focused on increased production, reductionist science, and high crop yields, as Charles Kellogg wrote in 1957: "What we seek is not some kind of mythical balance between farmers and the soils they cultivate, but a cultural balance in which we use with understanding and precision all the tools of modern science, engineering and economics" [64]. Charles

Kellogg was opposed to the 'organic matter doctrine' and, some years before his visit to Missouri, he worded this as "…they insist that organic matter is everything, or nearly so; that the usual chemical fertilizers are downright poisonous to soils; that the liberal use of compost gives special qualities to plants-they will be free of insects and disease; and that animal, or even people, will be ever so much more healthy by eating plants grown 'the organic way'" [65]. Despite a rocksteady trust in science and the good that it brought to agriculture, he was sensitive of the changes in agriculture, as he wrote in 1954: "With scientific agriculture, business methods have come into agriculture. Some aspects are bad, such as immediate profit; some are good, such as fertilizers, machinery, electric power, weed control and the like."

William Albrecht had started as conventional soil fertility researcher, but over time he became more interested in the quality of what was produced. Edmund Marshall considered William Albrecht one of the first environmental soil scientists [61], and Hans Jenny thought he had a broad vision; as a new faculty member, Hans Jenny had received much encouragement from William Albrecht [39]. William Albrecht was adopted in the spheres of alternative, sustainable, and organic agriculture. In the late 1940s, he was involved in some controversy with Jerome Rodale, who was a strong advocate of organic farming and re-published Franklin King's *Farmers of Forty Centuries* [66]. Jerome Rodale had created the *Soil and Health Foundation*, which offered funds for researchers who were supportive of organic agriculture. There were only a few applicants for the funding as it was seen as biased, but William Albrecht and a postdoctoral student accepted a grant. The student was enthusiastic about the organic message, but William Albrecht was more cautious, advocating building soil organic matter without condemning all chemicals. But Jerome Rodale legitimized the scientific basis for the organic farming movement by using William Albrecht, and controversies and disagreements were coming to a boil [67].

A public confrontation between Rodale and his scientific opponents took place at the 1950 Congressional hearings on Chemicals in Food [67]. Some of it was related to the use of inorganic fertilizers and their effect on soils and the quality of food. Emil Truog was called in as an expert witness, and they could not have asked for a more outspoken view—Emil Truog had argued against the use of organics only to maintain soil fertility [68, 69]. At the hearing, Emil Truog testified that: "Absolutely no authentic evidence exists to the effect that the application of chemical fertilizers to soils in accordance with approved practices, such as are recommended by the various state agricultural experiment stations and the United States Department of Agriculture, causes injury to these soils with respect to their physical, chemical, or biological condition." Jerome Rodale called that a lie. Richard Bradfield was also called into testify, and the congressional hearing ended with no strong conclusion against the use of fertilizers. In the end, Jerome Rodale no longer sought validation for his ideas from the soil science community [67].

In Missouri, Hans Jenny also became friends with Henry Krusekopf whom he had met during the *First International Congress of Soil Science* in 1927. They went on several field trips in the Ozarks. In the mid-1920s, there were about 3.5 million people in Missouri and about half of them lived on farms. When the Krusekopf family emigrated to the USA and settled in Missouri, they avoided the prairies because they thought that if the soil cannot support a forest it cannot be very good [2]. The prairie soils had thick black topsoils and these magnificently fertile soils became known as Chernozems based on their counterparts that blanketed the Russian steppes. Eventually, they were named Mollisols, belonging to the domain of pedologic royalty [70]. Nathaniel Shaler had noted in the late 1800s that the prairie soils should be cultivated because of the: "…exceeding fertility of the soil and to the unwooded character of their surface." The fertility was caused by the abundant growth of grasses for many centuries and: "…the soil has become richly stored with soluble mineral matter." Nathaniel Shaler thought that leaching was much lower under the prairie grass than under the forests but after a few decades, they should be fertilized, and he concluded that the prairies: "…ought to be reckoned as the largest field of high-grade soils" [71].

Curtis Marbut explained in his 1928 lectures the soil difference between the prairies and those under forest as follows: "…it is known very well that grass does not grow with luxuriance in most regions on perfectly freshly deposited or freshly exposed soil material. Grass is one of those crops which seems to grow best on soils that have reached at least some stage in development beyond immaturity. Trees, on the other hand, will grow on freshly exposed material. Those freshly exposed material of the prairies, therefore, exposed through relatively recent erosion have become occupied by timber because the materials on which they grow are of such character that grass does not grow well on them. Another factor in the case is undoubtedly the influence of grass fire. The grass fires did not or do not 'run' so vigorously over rough country as they do over smooth country, so that much woody vegetation as started out in the prairies would not be as likely to be killed down by prairie fires on the rough lands as on the smooth lands" [35].

Henry Krusekopf made soil maps for 22 counties in Missouri, and in addition, conducted soil surveys in South Carolina in 1918, and in Washington State in 1929. He was an experienced soil surveyor, but for him, soil properties were those that you could see or feel; those that you could not see or feel, he did not consider soil properties [39]. His ideas on classification were in line with Milton Whitney's who viewed that soil analysis had a limited role. As Henry Krusekopf wrote in 1943: "Soil classification units should be established on those features of the dynamic part of the profile, the solum, that are distinct, that are recognizable in the field, and that are fundamental in character....in genetic soil classification, laboratory data can be used only as corroborative evidence. Field classification must precede, not follow, laboratory analysis" [72]. Henry Krusekopf taught Hans Jenny the principles of soil survey, including the use of the plane table, and he became Hans' pedological *guru* and friend. Henry Krusekopf shared his rich knowledge with colleagues, farmers, and everybody else, a service that benefited the university but was not properly rewarded [39]. He worked for 48 years for the University of Missouri.

At the first international soil congress, Hans Jenny had heard J. C. Russel speak about soil organic matter and nitrogen contents along climatic gradients [73]. During the 1927 excursion and the many hours on the train, he thought about soil and climate relationships. Traversing across the Great Plains, he thought about mathematical soil functions that could explain the soil properties as a function of climate: "...the tour experience was a thrill, and it opened a new world. I was impressed when we went from Washington south and saw the southern red soils and, a few weeks later, the black soils of Canada. I searched for a connection between the two. Later, after the tour, I think that the lack of information on the connections pushed me to try to find solutions" [39]. He wondered whether a theory could be developed for the stark differences between the dark colored soils in Canada and the light-colored soils in Missouri. Such thoughts kept him up at night [39].

Henry Krusekopf gave Hans Jenny soil nitrogen and carbon analyses from bulletins and unpublished reports. Hans Jenny used the data to investigate the decline of soil fertility in Missouri. Soil nitrogen levels from cultivated fields, the prairie, and some long-term soil data under continuous corn were compared. He found that that soil fertility had decreased by one-third within 60 years of cultivation. Lower soil nitrogen levels were accompanied by lower corn yields, and that decreased the price of the land. Soil nitrogen decreased more during the earlier periods of cultivation, but then reached a steady state and overall, he thought it would be better to: "...carefully preserve nature's cheap nitrogen than to waste it and then replace it by buying expensive substitutes." The key to maintaining high nitrogen content was a system of crop rotation, especially if manure was applied [74].

Then, Hans Jenny assembled the soil data from Missouri and some other states, arranged them in tables and graphs, and developed equations to describe the trends that he described as 'primitive modeling' [58]. It became clear to him that from Canada to Louisiana there was a temperature gradient with similar moisture conditions, and in the soils across that gradient, the organic matter contents declined exponentially. It took him 2 months to figure that out. Like Curtis Marbut had done in 1927, Hans Jenny called the right-angled intersection of temperature and moisture in the USA, an 'ideal climate checker board,' and he felt that he: "...made the most of Nature's fortunate America Design" [39].

He unveiled the relationships between soil and climatic properties in Europe and the USA [75–77], and 30 years later, for India [78]. With that work, he followed in the footsteps of Eugene Hilgard, who in the 1890s had investigated patterns in the nitrogen and humus content of soils [79]. Hans Jenny might have also taught of the work of Carl Sprengel, who in the 1830s had studied 170 soil samples from several countries, and tried to obtain some geographical order in the distribution of their humus but without success [80]. Besides the relationship between climate and soil nitrogen and organic matter, Hans Jenny investigated the behavior of potassium and sodium during the process of soil formation. There was a need to understand the weathering of minerals as it related to soil development and the fertility of the soil, as "...the problem of soil fertility is as complicated as it is old" [81].

In November 1927, Hans Jenny attended the *American Soil Survey Association* meeting in Chicago and presented the soil organic matter gradient as a function of temperature. Not everyone liked the work, and there were pedologists who thought he was being overly academic. They were further aggravated after he published the results in 1930 in the prestigious *Journal of Physical Chemistry*. He was not the first soil scientist to do so, as Richard Bradfield had published in the same journal in 1924, and Emil Truog in 1916. But Curtis Marbut was impressed by Hans Jenny's work and saw it as proof that the Russian soil climate principles held up in the USA. From then on, Curtis Marbut became friendly and helpful towards Hans, and whenever Curtis Marbut visited Missouri, they went on field trips around Columbia. According to Hans Jenny, there was one unfortunate aspect of their relationship: Curtis Marbut viewed the application of chemistry to geology and soil science to be less useful [39]. In fact, Curtis Marbut often pointed out the need for soil chemistry rather than the chemistry of the soil [16].

In 1933, Hans Jenny admitted Guy Smith as an MS student to the University of Missouri. Guy Smith was born in Atlantic City in Iowa and had graduated with a BS degree from the University of Illinois in 1930. He worked in the Illinois state soil survey, which was separate from surveys of the *Division of Soil Survey* and he liked the work of Hans Jenny Guy Smith conducted a laboratory experiment for claypan formation and he concluded that correlated field and laboratory studies were needed before the relative importance of the various factors in pan formation could be determined. After his MS degree, Guy Smith returned to Illinois where he studied loess across the landscape and was mentored by R.S. Smith [82]. He measured the thickness and properties of loess across two traverses perpendicular to the Mississippi and Mackinaw rivers and found a linear relation in loess thickness to the logarithm of the distance from the rivers. The loess was thinner with increasing distance from its source, and the thinner loess had less carbonate. He had taken a soil functional approach which he had learned from Hans Jenny in Missouri. The study became Guy Smith's PhD thesis: *Illinois loess: variations in its properties and distribution, a pedologic interpretation* that was published in 1940 [83].

Guy Smith joined the army in 1943, and by that time, he was married with three children. He worked on the Ledo road connecting Assam in India to Kunming in China, which was meant to enable Western Allies to supply China with military equipment to be used against Japan. Guy's task was to document progress on the construction of the road. It was a struggle. More than 1,100 Americans and many more local people died during construction. The human cost of the road was described as 'A Man A Mile.' After the war, the road was swallowed up by the jungle, probably at the same rate at which it was built, but without the tragic human loss.

Hans Jenny kept in contact with Guy Smith, and found him a broad and meticulous thinker who wanted to go into depth, not being satisfied with rapid solutions [39]. In 1946, when Hans Jenny was Head of the Department of Soils at the University of California Berkeley, he invited Guy Smith, who had just returned from the war to join his department. Guy Smith accepted the invitation and traveled to Berkeley, where they visited some soil sites and had lengthy discussions. But Guy Smith turned down the job offer and joined the soil survey with Charles Kellogg in Washington. Guy Smith thought that such a position would allow more travel and present more opportunities to see the soils of the world than a position at a university [39]. Guy Smith was an independent scientist, and according to Hans Jenny, he was a positive force and not part of the early mentality of the soil survey group [39]. Guy Smith broadened the concept of soil survey, emphasized more rigorous thinking, and the need to include soil analyses and quantitative criteria. Soil survey and soil classification were in need of these and eventually, Guy Smith became responsible for the development of the *7th Approximation* [84]. He had soil science friends and supporters across the world notably René Tavernier, but

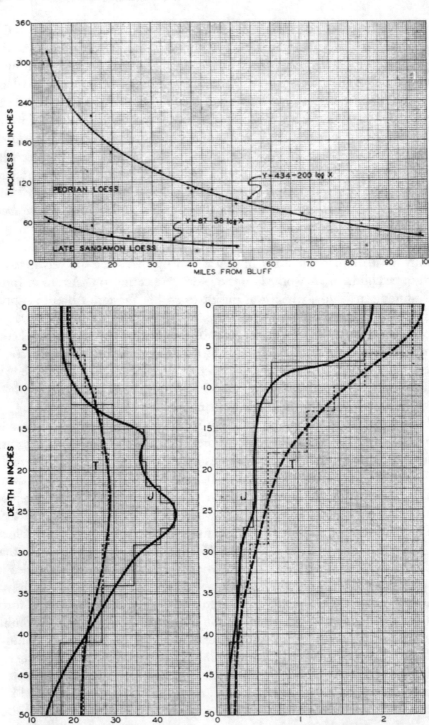

Variation in the thickness of two types of loess with distance from the Mississippi Bluff in Illinois, and depth functions (with a preliminary mass-preserving spline) of the fraction smaller than 1 micron, and soil carbon. From Guy Smith's PhD thesis of 1940 that was published in 1942 as: *Illinois loess: variations in its properties and*

Guy Smith and Hans Jenny on the desert geomorphology-soils project tour in New Mexico in 1961 organized by Bob Ruhe. The project had been started in August 1957 when Bob Ruhe moved to Las Cruces, New Mexico to study soils in the arid and semi-arid regions of the Southwestern USA. The project was coordinated by Guy Smith who was Director of the Soil Survey Investigations Division. Guy Smith was eight years younger than Hans Jenny and had his first MS student at the University of Missouri. At the tour, Stan Buol who worked at the University of Arizona, asked Hans Jenny to classify a soil profile, to which he replied: "I don't classify soils, I study them". Photo by Stan Buol

he was at the receiving end of the criticism of the *7th Approximation*. Gifted as he was, Guy Smith became quite dogmatic [39].

The economic depression that started at the end of 1929 began to affect academic life at the University of Missouri. Some young faculty members were laid off, but Hans Jenny was spared, despite some complaints that the college favored a foreigner over US citizens. Some of the staff who were laid off found work with the *Soil Conservation Service*. As the financial crisis deepened, faculty members were forced to take a leave of absence without pay. Hans Jenny went on leave to Switzerland in the summer of 1933, and wrote to Emil Truog at the University of Wisconsin:

My Dear Prof. Truog:
As you perhaps know we have been working for some time on the problem "exchangeable ions and plant growth" (see Science no 1999, April, 1933). I have always been interested in your viewpoints and your work along this line. Now, I intend to take a years leave of absence, beginning Sept. 1934 and

I should like very much to work in your laboratory. I wonder whether you would be interested in such a cooperative project. Of course some kind of financial aid (perhaps a fellowship) would be appreciated, because my salary will be reduced to one half during the leave. I would appreciate very much to have your opinion in this matter. I shall return to Columbus, Mo., next month.

A month later Emil Truog wrote him back acknowledging the letter, and the reprint from *Science*. He added: "…it is rather late to propose any projects to the administrative offices which require expenditures of money, because the budgets were all made up some months ago." He offered Hans Jenny no position, but an honorary fellowship that came without pay. Hans Jenny stayed in Missouri but had he ended up at the University of Wisconsin, his influence might have been larger, for he always felt that the epicenter of soil science was in the Midwest and not at the coasts [39].

Hans Jenny then contacted a Swiss friend at the University of California, and he was invited to visit Riverside. He met with Walter Kelley who had been at the Riverside since 1914 and had been a chemist at Hawaii agricultural station. Walter Kelley was born and raised on a tobacco farm 60 km north of Nashville in southern Kentucky. He was a soil chemist and well-known for his work on alkali soils that followed earlier studies by Eugene Hilgard and Alex. de'Sigmond. Walter Kelley was the last PhD student of Eugene Hilgard, and thought that Eugene Hilgard was a good scientist and that much of his work would endure permanently: "…in all probability, the work of earlier soil scientists appears no more out of date today than will the researches of the most advanced thinkers of the present time appear to the scientist of the future." Walter Kelley wrote a standard book on cation exchange in soils [85], but he believed that scientific truth was never fully attained; it was ever being sought and the search was eternal [86]. He was concerned about the way soil was treated and about the relationship between fundamental and applied research: "The soil is the very foundation of agriculture and what takes place in the foundation will sooner or later become manifest in the superstructure. Unfortunately, a very high percentage of the American agricultural experiment stations have not yet learned this important lesson. We have been far more concerned about gaining the immediate applause of the farmer than with advancing agriculture on a truly scientific basis" [86]. Walter Kelley was elected a member of the National Academy of Sciences in 1943.

The University of California at Riverside hired Hans Jenny for a year. It was a new environment, and he learned about irrigation practices, alkali soils, soil physics, and the effects of plowing. After six months, Walter

Kelley suggested that he should visit the University of California Berkeley, where more fundamental soil problems were studied. He contacted Dennis Hoagland to see whether there was any possibility to work there. Sometime later, Hans Jenny went to Berkeley, gave two seminars, and had discussions with Dennis Hoagland on colloid chemistry and ion exchange. It became a fruitful visit, and he was excited to visit the university where Eugene Hilgard had worked from 1875 to 1904. After the visit, the university started to create a position for Hans Jenny, which was facilitated by the retirement of the soil chemist Paul Hibbard.

The position at Riverside was only for one year, and in 1935 Hans Jenny returned to the University of Missouri and continued to work on the potassium to sodium ratio, work that he later labeled as geochemistry which he viewed as an extension of soil chemistry. Geochemistry was a field of study founded by the mineralogists Victor Goldschmidt and Vladimir Vernadskii. In 1936, Hans Jenny, 38 years old, became an American citizen, something that he had not envisioned when he moved to the USA in 1926. He had escaped layoffs at the University of Missouri and hoped the citizenship would avoid future dismissals. Selman Waksman had remarked to him once that: "America is an irresistible magnet," and he had predicted that one day Hans Jenny would become a citizen [39]. Shortly after Hans Jenny had become a citizen, he received the offer to become an associate professor in the Department of Plant Nutrition at the University of California Berkeley. He left Columbia in 1936 and headed west, after 8 years in Missouri with a 1-year interlude at Riverside. He considered those eight years his 'Sturm und Drang' period [39].

At Missouri, Hans Jenny was replaced by Edmund Marshall, who studied soil colloids and the chemical environment of plant roots [87]. Born in England, Edmund Marshall obtained his degrees in chemistry at the University of Leeds, which had a strong chemistry department. Edmund Marshall had studied soil organic matter at Rothamsted under John Russell, after which he was a post-doctoral researcher in the laboratory of Georg Wiegner at ETH in Switzerland in 1927 where he shared laboratory space with Richard Bradfield, who was on a sabbatical. Edmund Marshall enjoyed the teaching of Georg Wiegner and learned about Vasily Dokuchaev and Russian soil science, which brought a new perspective to his chemistry and meaning to soil geography; he viewed the exchange complex as the main link between geographical and chemical aspects of soil [88].

After 2 years at ETH, Edmund Marshall returned to the UK and taught agricultural chemistry at the University of Leeds. In the summer of 1935, Geoffrey Milne came to visit before attending the *Third International Congress of Soil Science,* and he was on leave from the East African Agricultural Research Station at Amani in Tanzania. A graduate from Leeds, Geoffrey Milne developed the catena concept in which he related the black and red soils along a topographic gradient [89]. Much of the research in Leeds was focused on the acid soils and podzols in the northern part of the country. Geoffrey Milne brought a different perspective on soil formation, and as a result Edmund Marshall developed an interest in tropical soils.

Edmund Marshall moved to the University of Missouri in 1936 [36, 87], where he used zircon to assess gains and losses in different horizons during soil development. Zircon was suitable for such studies because of its relative immobility [90]. He built on the work of Cees Edelman who in the early 1930s measured heavy mineral for assessing the age and uniformity of soil profiles. Edmund Marshall introduced the idea of isomorphic substitution by which clay minerals become negatively charged and were thus able to hold and exchange cations. He demonstrated the relation between cation exchange capacity, interlayer charge, and ionic substitutions. Edmund Marshall worked for 40 years at the University of Missouri and died in 1982. Posthumously, he was awarded the Distinguished Member Award of the Clay Mineral Society.

At the University of California Berkeley in the 1930s, there were two soil research groups—the plant nutrition department headed by Dennis Hoagland and the soil technology department headed by Charles Shaw. Charles Shaw was born in upstate New York, graduated from Cornell University in 1906, and was hired by the *Bureau of Soils* in 1906. He mapped soils in Louisiana and later with Hugh Bennett in Texas, and between 1908 and 1912, he produced reconnaissance soil maps for most of Pennsylvania. In the First World War, he changed his German last name and joined the University of California as a professor of soil technology. He surveyed extensive areas in California and was a friend of Macy Lapham who had conducted soil surveys in the west since the late 1800s. At Berkeley, Charles Shaw succeeded Eugene Hilgard, who had produced the first agricultural map in 1883, which was in essence a map of major land resource areas [91]. Charles Shaw ignored Eugene Hilgard's work, and often spoke of him somewhat derogatorily. Although he had broader views, Charles Shaw was a 'Whitney man' and did not like Eugene Hilgard because he had overlooked the ion exchange principle essential for the reclamation of alkali soils [39, 92].

Charles Shaw proposed a soil formation formula at the *Second International Congress of Soil Science* in 1930 [93]. In his view, soil evolution could be expressed by the formula: $S = M (C + V) T + D$ whereby soil (S) is formed from the parent material (M), climatic factors (C) and organic life (V) acting over of time (T), and the resultant soil being modified by erosion or deposition (D). The idea of a formula for soil formation had been suggested by Vasily Dokuchaev in the 1880s [94], but erosion as a soil builder or destroyer was not included.

Hans Jenny did not get along with Charles Shaw, although Charles Shaw once proclaimed that Hans Jenny was going to be a new Dokuchaev [39]. Nor did Hans Jenny get along with Earl Storie or Wilbert Weir who were in the same department. He considered them valuable but: "...they were not in the field of what you might say theoretical science" [39]. They were too practical for Hans Jenny. Wilbert Weir was the sixth graduate in soil science from the University of Wisconsin and had worked for the *Bureau of Soils* before joining the University of California as a soil technologist. In the late 1910s, he had edited the *Journal of Soil Improvement*, the official magazine of the Wisconsin Soil Improvement Association. Wilbert Weir and Earl Storie had made a statewide soil map and developed a rating system for the soils of California [95, 96].

Working with Dennis Hoagland was a unique opportunity for Hans Jenny. Dennis Hoagland was professor of Plant Nutrition and had studied chemistry at Stanford University and had obtained an MS degree from the University of Wisconsin. His MS research dealt with the influence of sodium benzoate on the synthesis of urea, but he also published three papers with his supervisor Elmer McCollum on the metabolism of the pig [97]. Dennis Hoagland worked on hydroponics, and aimed to find a solution that provided every nutrient necessary for plant growth while being appropriate for the growth of a large variety of plant species. It was a big topic in the 1920s and 1930s, under different names: growing plants without soil, plant chemiculture, soilless agriculture, water culture, tank farming, hydroponics, or sand culture [98]. At that time, there was no explanation as to why plants were able to absorb elements from dilute solutions and at different rates [99]. Dennis Hoagland established the essentiality of molybdenum for the growth of tomato plants, and he was more of a fundamental soil chemist and less of an agronomist than Emil Truog, whom he knew from his time in Wisconsin. They were of the same generation, and like Emil Truog, Dennis Hoagland had no PhD degree. Dennis Hoagland gave lectures on inorganic nutrition of plants at Harvard University in the 1940s, in which he combined plant physiology, soil science, and inorganic chemistry. In 1945 he became ill but continued to work until retiring in July 1949. He died two months later, 65 years old [97].

Hans Jenny was content to be in Berkeley where he saw himself continuing Eugene Hilgard's effort: "...when I was a student I studied Hilgard's book on

Soil scientists from the west. Left to right: Dennis Hoagland, Charles Shaw and Walter Kelley. Dennis Hoagland (1884–1949) was a plant nutritionist born and raised in a mining-town in Colorado. Charles Shaw (1881–1931) born near Rochester New York, was a pedologist and soil expert for California. Walter Kelley (1878–1965) born on a tobacco farm in Kentucky was an expert in alkali soils, ion exchange, and clay mineralogy. Charles Shaw died young. Dennis Hoagland and Walter Kelley became National Academy Members

soil, but I never dreamed that ten years later (1936) I would be sitting at his old desk in Hilgard Hall teaching his old class. The question is: is this a coincidence or are the forces at work?" [39]. Hans Jenny revered Eugene Hilgard's work, and he found that there was too little appreciation as he had bridged soil research in the USA, Russia, and Europe. Eugene Hilgard was well known in Europe, but not in the USA, where he had battled with Milton Whitney. At the 1927 field excursion, very few of the American participants had heard of Eugene Hilgard, although Georg Wiegner had used his book for his teaching at ETH in Switzerland [100]. In the 1960s, Hans Jenny wrote a biography on Eugene Hilgard entitled *E. W. Hilgard and the Birth of Modern Soil Science,* but he had difficulties finding a publisher [92]. The publisher McGraw Hill said agricultural people do not read historical books [39]. The book was published, though mostly at Hans Jenny's own expense, and did not sell well. Thirty years later, and one year after Hans Jenny had died, the University of California sent around free copies of *E. W. Hilgard and the Birth of Modern Soil Science* to soils departments; the distribution was made possible through a gift from Hans's widow.

In the 1930s, there was animosity between the department of plant nutrition and soil technology at Berkeley which was particularly apparent when advising land users [39]. After the First World War, veterans were given land in California, and Charles Shaw and his group had identified first-class land near Delhi, some 160 km east of San Jose. The plant nutrition people had not been part of the group that had surveyed the land. The veterans planted peach orchards, but after a few years, the trees died. Dennis Hoagland became interested and discovered with Paul Hibbard that the trees suffered from zinc deficiency, which was common on alkali soils. Hans Jenny had knowledge of pedology and plant nutrition and brought the plant nutritionist and pedologists together at Berkeley. Later, he described his early time at Berkeley as "…with the attack infantry, advancing, not thinking what might happen afterwards." When Charles Shaw suddenly died of a heart attack in 1939, the department of plant nutrition and soil technology were merged into the Department of Soils. Walter Kelley became Head of the new department, followed in 1943 by Hans Jenny.

In 1941, Hans Jenny published *Factors of soil formation—a system of quantitative pedology* [101]. It was a step forward in thinking about soils and their formation. Less emphasis was put on Russian and genetic soil science and more attention was given to soil properties and their relation to soil geography. It provided a framework, some unsolvable but sensible equations, and it postulated theory that nourished the discipline of soil science. Soil erosion was absent in the book, for as Hans Jenny said, it was a book about soil formation and not about soil destruction [101]. He mentioned that soil temperature and soil moisture were prime properties and that idea was largely denounced, but years later, both ended up high in the American soil classification system. At the onset of writing the book, Hans Jenny thought of himself: "…since I am an average guy and have ordinary ideas, so a lot of other average people might be interested in these ideas." He had finished the manuscript in 1936 and the publisher, McGraw Hill, sent it to several reviewers, including Emil Truog, who asked Roy Simonson to go over the equations. They considered it not practical. Hans Jenny then talked to Richard Bradfield who convinced McGraw Hill to publish the book in 1941. It was translated and published in Russia in 1949.

Factors of Soil formation—a system of quantitative pedology was more or less ignored by the soil scientists in the soil conservation agencies and the federal soil survey groups, but it was well received by the universities and experimental stations [39]. For many soil scientists, it was too theoretical, and they felt the topic was lost in mathematics [39]. There was also some discussion on the origin of the ideas and whether the works of Vasily Dokuchaev, Nikolai Sibirtsev or Sergei Zakharov had been sufficiently acknowledged. The book was received with mixed feelings in Australia; Robert Crocker liked it, but the Head of the CSIRO Soil Survey and Pedology Section, Charles Stephens, though it was unfruitful. Guy Smith used the book in his teaching at Illinois in the early 1940s, but also thought: "…the whole thing was a personal viewpoint" [39]. The journal *Soil Science* listed the book with the notion: "A unique approach to the study of an old problem."

In the years that followed its publication, several papers were published that offered somewhat different theories or ideas, such as for example *Fundamental formulation of soil formation* by Constantin Nikiforoff [102]; *Soil genesis and the pedogenic factors* by Robert Crocker [103]; *The chronosequence concept and soil formation* by Stevens and Walker [104]; *Soil Formation* by Edward Crompton [105], and *Outline of a generalized theory of soil genesis* [106], which Roy Simonson wrote in 1959. These were well-crafted theories and ideas that became noticeable over the years, but none matched the breadth, depth, and impact of Hans Jenny's book, which had been constructed on a solid foundation composed of the work George Wiegner, the genetic principles of soils, the Transcontinental Excursion in 1927, and Hans Jenny's own work at the University of Missouri. In the late 1980s, Hans Jenny was asked how it was possible that the work and theory in his book had not been superseded, to which he replied: "I simply outlived my enemies."

References

1. Helms D. Soil and southern history. Agric Hist. 2000;74:723–58.
2. Baker OE. The trend of land utilization in the United States and the present situation. Vol Part I. Proceedings. Washington D.C.: The American Organizing Committee of the First International Congress of Soil Science; 1928.
3. Goddard FB. Where to emigrate and why. New York: Frederick B. Goddard; 1869.
4. Kellogg CE, Knapp DC. The college of agriculture. Science in the public service. New York: McGraw-Hill Book Company; 1966.
5. Acikgoz S, Anderson SH, Gantzer CJ, Thompson AL, Miles RJ. 125 years of soil and crop management on Sanborn Field: effects on soil physical properties related to soil erodibility. Soil Sci. 2017;182(5):172–80.
6. Rice TD. C.F. Marbut. Life and works of C.F. Marbut. Madison: Soil Science Society of America; 1936, p. 36–48.
7. Woodruff CM. A history of the Department of Soils and Soil Science at the University of Missouri. Special Report 413 College of Agriculture. Columbia, Missouri: University of Missouri; 1990.
8. Shantz HL. A memoir of Curtis Fletcher Marbut. Ann Assoc Am Geogr. 1936;26:113–23.
9. Tandarich JP. Curriculum Vitae of Dr. Curtis Fletcher Marbut. Soil Horiz. 1985;26(1).

10. Shaler NS. The origin and nature of soils. Department of the Interior. U.S. Geological Survey. Washington: Government Printing Office; 1892.

11. Darwin C. The formation of vegetable mould through the actions of worms with observations on their habits. London: John Murray; 1881.

12. Johnson DL, Schaetzl RJ. Differing views of soil and pedogenesis by two masters: Darwin and Dokuchaev. Geoderma. 2015;237–238:176–89.

13. Miller MF. Progress of the soil survey of the United States Since 1899. Soil Sci Soc Am Proc. 1950;14(C):1–4.

14. Marbut CF. Physical features of Missouri. Mo Geol Surv. 1896;10:14–109.

15. Dott RH. Lyell in America—his lectures, field work, an mutual influences, 1841–1853. Earth Sci Hist. 1996;15:101–40.

16. Krusekopf HH, editor. Life and works of C.F. Marbut. Madison: Soil Science Society of America; 1936.

17. Lapham MC. Crisscross trails: narrative of a soil surveyor. Berkeley: W.E. Berg; 1949.

18. Marbut CF. Soils of the Ozark Region. A preliminary report on general character of the soils and the Agriculture of the Missouri Ozarks. Research Bulletin no. 3. Columbia, Missouri: University of Missouri. College of Agriculture; 1910.

19. Joffe JS. Pedology (with a foreword by C.F. Marbut). New Brunswick: Rutgers University Press; 1936.

20. Morrow L. An Ozarks Landmark—The Curtis Fletcher Marbut House. White River Hist Q. 1981;7:6–9.

21. West D. A visit to the home of a scientific pioneer. Soil Surv Horiz. 1995;36:112–3.

22. Marbut CF. Soils of the United States. Part III. In: Baker OE, editor. Atlas of American Agriculture. Washington: United Sates Department of Agriculture. Bureau of Chemistry and Soils; 1935. p. 1–98.

23. Shaw CF. A preliminary field study of the soils of China (from the "contributions to the knowledge of the soils of Asia, 2" compiled by the Bureau for the Soil Map of Asia. Leningrad: Publishing Office of the Academy of Sciences of the USSR; 1933.

24. Gong ZT, Darilek JL, Wang Z, Huang B, Zhang G. American soil scientists' contributions to Chinese pedology in the 20th century. Soil Surv Horiz. 2010;51:3–9.

25. Wildman WE. Dr. James Thorp, A close Marbut associate—a memorial tribute. Soil Surv Horiz. 1985;26:5–12.

26. Thorp J. Impressions of Dr. Curtis Fletcher Marbut, 1921–1935. Soil Horiz. 1985;26(1).

27. Tandarich JP, Johannsen CJ, Wildman WE. James Thorp talks about soil survey, C. F. Marbut, and China. Soil Horiz. 1985;26(2).

28. The Moon D, Steppes American. The unexpected Russian roots of great plains agriculture, 1870s–1930s. Cambridge: Cambridge University Press; 2020.

29. Robinson GW. Soils—their origin, constitution and classification. An introduction to pedology. London: Thomas Murby & Co; 1932.

30. Kellogg CE. Obituary Curtis Fletcher Marbut. Science. 1935;82:268–70.

31. Joffe JS. Pedology. New Brunswick: Rutgers University Press; 1949.

32. Shantz HL, Marbut CF. The vegetation and soils of Africa. New York: National Research Council and the American Geographical Society; 1923.

33. United States Department of Agriculture. Soils & men. The year book of agriculture. House Document No. 398. Washington United States Department of Agriculture. United States Government Printing Office; 1938.

34. Thorp J. Geography of the soils of China. Nanking: The National Geological Survey of China; 1936.

35. Marbut CF. Soils: their genesis and classification (Lecture notes from 1928). Madison, Wi: Soil Science Society of America; 1951.

36. Miller MF, Krusekopf HH. The soils of Missouri. Bulletin 264. Columbia, Missouri: University of Missouri, College of Agriculture; 1929.

37. Miller MF. The soil and its management. Boston: Ginn and Company; 1924.

38. Marbut CF, Bennet HH, Lapham JE, Lapham MH. Soils of the United States (edition, 1913). Washington: Government Press Office; 1913.

39. Maher D, Stuart K, editors. Hans Jenny—soil scientist, teacher, and scholar. Berkeley: University of California; 1989.

40. Miller MF. Erosion as a factor in soil determination. Science. 1931;73:79–83.

41. Bear FE. Soils and fertilizers. New York: Wiley; 1950.

42. Bennett HH. Soil conservation. New York & London: McGraw-Hill Book Company, Inc; 1939.

43. Peterson GA. Profile of J.C. Russel: pioneer in conservation tillage. J Agron Educ. 1986;15:124–5.

44. Faulkner EH. A second look. Norman: University of Oklahoma Press; 1947.

45. Faulkner EH. Plowman's folly. Norman: University of Oklahoma Press; 1943.

46. Picton L. Nutrition & the soil. Thoughts on feeding. New York: The Devin-Adair Company; 1949.

47. Howard A. An agricultural testament. New York and London: Oxford University Press; 1940.

48. Balfour EB. The living soil—evidence of the importance to human health of soil vitality, with special reference to post-war planning. London: Faber and Faber; 1943.

49. Baver LD, Himes FL. History of the Department of Agronomy 1905–1970. Columbus: Ohio State University; 1970.

50. Bradfield R. What information is necessary for a complete description of a soil? B. Laboratory aspects. Am Soil Surv Assoc Bull. 1927;VIII:104–11.

51. Cline MG. Agronomy at Cornell 1868 to 1980. Agronomy Mimeo No. 82-16. Ithaca, NY: Cornell University; 1982.

52. Jenny H. Foundations of animal feeding. (In German). Schweiz Land Wirtschaft Zeitung. 1923;51:1176–80.

53. Wiegner G. Boden und Bodenbildung in Kolloidchemischer Betrachtung. Sechste ed. Dresden und Leipzig: Verlag von Theordor Steinkopff; 1931.
54. Wiegner G. Anleitung zum quantitativen agrikultuchemischen Praktikum. Berlin: Verlag von Gebrüder Borntraeger; 1926.
55. Waksman SA. The conquest of tuberculosis. London: Robert Hale Limited; 1964.
56. Shive JW. A three-salt nutrition for plants. Am J Bot. 1915;2:157–60.
57. Moberg CL. René Dubos, friend of the good Earth: microbiologist, medical scientist, environmentalist. Washington DC: American Society for Microbiology; 2005.
58. Amundson R. Jenny, Hans. In: Hillel D, editor. Encyclopedia of soils in the environment. Amsterdam: Elsevier; 2005. p. 293–300.
59. Jenny H. The base exchange movie. Am Soil Surv Assoc Bull. 1932;151–6.
60. Albrecht WA. Symbiotic nitrogen fixation as influenced by the nitrogen in the soil. PhD thesis. Urbana-Champaign: Graduate School of the University of Illinois; 1919.
61. Marshall CE. In memoriam William A. Albrecht. Plant Soil. 1977;48(1):1–4.
62. Albrecht WA, Smith CE. Soil acidity as calcium (fertility) deficiency. College of Agriculture: University of Missouri; 1952.
63. Albrecht WA. Our teeth and our soils. Circular 333. Columbia: University of Missouri. College of Agriculture; 1948.
64. Kellogg CE. We seek; we learn. Yearbook of Agriculture. Washington: United States Department of Agriculture; 1957. p. 1–11.
65. Kellogg CE. Conflicting doctrines about soils. Sci Mon. 1948;66:475–87.
66. King FH. Farmers of forty centuries or permanent agriculture in China, Korea and Japan. Emmaus, Pennsylvania: Republished by: Rodale Press, Inc.; 1911.
67. Peters S. Organic farmers celebrate organic research: a sociology of popular science. In: Nowotny HHR, editor. Counter-movements in the sciences. Sociology of the Sciences. Volume III. Dordrecht: D. Reidel Publishing Company; 1979.
68. Truog E. Plowman's folly refuted. Harper's Mag. 1944(July Issue):173–7.
69. Truog E. "Organics only?"—Bunkun! The Land. 1946;5:317–21.
70. Wolfanger LA. The major soil divisions of the United States. A pedologic-geographic survey. New York: Wiley; 1930.
71. Shaler NS, editor. The United States of America. A study of the American Commonwealth, its natural resources, people, industries, manufactures, commerce, and its work in literature, science, education, and self government. (3 volumes). New York: D. Appleton and Company; 1894.
72. Krusekopf HH. Objectives and criteria of soil classification. Soil Sci Soc Am Proc. 1944;8(C):374–6.
73. Russel JC, Engle EB. The organic matter content and color of soils in the central grassland states. Vol Commission V. Commission VI. Miscellaneous papers. Washington D.C.: The American Organizing Committee of the First International Congress of Soil Science; 1928.

74. Jenny H. Soil fertility losses under Missouri conditions. Agricultural Experiment Station Bulletin 324. Columbia, Missouri: University of Missouri; 1933.
75. Jenny H. Relation of climatic factors to the amount of nitrogen in soils. J Am Soc Agron. 1928;20:900–12.
76. Jenny H. The nitrogen content of the soil as related to the precipitation-evaporation ratio. Soil Sci. 1930;29:193–205.
77. Jenny H. Klima und Klimabodentypen in Europa und in den Vereinigten Staaten von Nordamerika. Soil Res. 1929;3:139–89.
78. Jenny H, Raychaudhuri SP. Effect of climate and cultivation on nitrogen and organic matter reserves in Indian soils. New Delhi: Indian Council of Agricultural Research; 1960.
79. Hilgard EW, Jaffa ME. On the nitrogen content of soil humus in the arid and humid regions. Agric Sci. 1894;VIII:165–71.
80. Feller CL, Thuriès LJM, Manlay RJ, Robin P, Frossard E. "The principles of rational agriculture" by Albrecht Daniel Thaer (1752–1828). An approach to the sustainability of cropping systems at the beginning of the 19th century. J Plant Nutr Soil Sci. 2003;166(6):687–98.
81. Jenny H. Behavior of potassium and sodium during the process of soil formation. Agricultural Experiment Station Research Bulletin 162. Columbia, Missouri: University of Missouri; 1931.
82. Brasfield J. In Memoriam Guy D. Smith. Soil Horiz. 1981;22.
83. Smith GD. Illinois loess: variations in its properties and distribution, a pedologic interpretation. University of Illinois; 1942.
84. Soil Survey Staff. Soil classification. A comprehensive system, 7th Approximation. Washington D.C.: Soil Conservation Service. United States Department of Agriculture; 1960.
85. Kelley WP. Cation exchange in soils. New York: Reinhold Publishing Corporation; 1948.
86. Kelley WP. Contributions of Rothamsted to soil chemistry. Proc Soil Sci Soc Am. 1944;8:12–4.
87. Anon. In recognition of C. Edmund Marshall who continues to contribute to basic knowledge of soil science. Soil Sci. 1970;109(1):1–4.
88. Marshall CE. The physical chemistry and mineralogy of soils. Vol II soils in place. London: Wiley; 1977.
89. Milne G. Some suggested units of classification and mapping, particularly for East African soils. Soil Res. 1935;4:183–98.
90. Haseman JF, Marshall CE. The use of heavy minerals in studies of the origin and development of soils. Mo Agric Exp Stn Res Bull. 1945;387:1–75.
91. Smith P. Profiles in history. Soil Horiz. 2015;56(3).
92. Jenny HEW. Hilgard and the birth of modern soil science. Pisa: Collana della revista agrochemica; 1961.
93. Shaw CF. A soil formulation formula. Proceedings Second International Congress of Soil Science. ISSS;1932.

94. Florinsky IV. The Dokuchaev hypothesis as a basis for predictive digital soil mapping (on the 125th anniversary of its publication). Eurasian Soil Sci. 2012;45:445–51.
95. Weir WW, Storie RE. A rating of California soils. Bulletin 599. Berkeley, California: University of California; 1936.
96. Storie RE, Weir WW. Generalized soil map of California. Berkeley: University of California, Division of Agricultural Sciences; 1953.
97. Arnon DI. Dennis Robert Hoagland 1884–1949. Plant Soil. 1950;2(2):129–44.
98. Matlin DR. Growing plants without soil. New York: Chemical Publishing Co., Inc.; 1939.
99. Kelley WP. Dennis Robert Hoagland 1884–1949. A biographical memoir. Washington D.C.: National Academy of Sciences; 1956.
100. Hilgard EW. Soils, their formation, properties, composition, and relation to climate, and plant growth. New York: The Macmillan Company; 1906.
101. Jenny H. Factors of soil formation. A system of quantitative pedology. New York: McGraw-Hill; 1941.
102. Nikiforoff CC. Fundamental formulation of soil formation. Am J Sci. 1942;240:847–66.
103. Crocker RL. Soil genesis and the pedogenic factors. Q Rev Biol. 1952;27(2):139–68.
104. Stevens PR, Walker TW. The chronosequence concept and soil formation. Q Rev Biol. 1970;45(4):333–50.
105. Crompton E. Soil formation. Outlook Agric. 1962;3(5):209–18.
106. Simonson RW. Outline of a generalized theory of soil genesis. Soil Sci Soc Am Proc. 1959;23:152–6.

7

Building an American Soil Survey

"Since soils are the residual product of the action of meteorological agencies upon rocks, it is obvious that there must exist a more or less intimate relation between the soils of a region and the climatic conditions that prevail, or have prevailed therein."

Eugene Hilgard, 1892

A large part of scientific knowledge about soils has come from orderly and refined laboratory analysis and interpretation, the development and testing of sound theory, the adoption and modification of practices and procedures from the other sciences, and countless soil surveys across all parts of the globe. Some knowledge was gained by chance, but most was won through systematic studies. The discipline of soil science grew out of embracing the fundamental and the applied aspects of research, the empirical and theoretical, and the development of ideas, complex schemes, and simple solutions. Unintentionally, soil research celebrated a great deal of trial and error along its winding path and it is hard to think of another sub-discipline within soil science where that is more apparent than in soil survey, in which observations across the landscape led to an organized arrangement of information and deep soil knowledge.

In soil survey, everything had to be learned and discovered—from what was seen to how it was or could be seen. There were no words for many of the observations, and there were only a few people developing methods and theories. It took decades before consensus was created from myriad ideas, and that was accompanied by the rise and fall of some grand concepts. Out there was a seemingly insurmountable amount of variation in soil, and while soil surveyors aimed to recognize patterns, they often lacked the skills and tools to do this properly. Of all the various paths that have been taken in soil science, soil survey has probably more redirections, scientific missteps, and dogmatic millstones than any other. Progress might have been accelerated when less rigidity ruled, but that likely affects all sciences where practices were framed in pre-mature theory. For all the pitfalls, soil survey invented

© The Author(s), under exclusive license to Springer Nature Switzerland AG 2021
A. E. Hartemink, *Soil Science Americana*
https://doi.org/10.1007/978-3-030-71135-1_7

dazzling solutions, unceasing new findings, and stimulated intellectual debate and thinking. From its inception, soil survey formed a dominant foundation for international soil cooperation, and it had a few political enthusiasts who saw it as a vehicle for their agenda or nation.

Soil survey is soil science at its core. It was humanity discovering a material they have lived from for millennia, but that they barely knew; empirically, humans had learned ways to use the soil, but had not managed to comprehend its patterns and complexity. In soil survey, the soil became the object of study, not only as a material for growing plants or harboring life, but in all its facets and dimensions. To make two-dimensional maps of a three-phase system that has four dimensions required ingenuity of the highest sort. No wonder that some think that soil science is on par with research fields such as astrophysics or quantum mechanics. Soil science would not have become the discipline it is were it not for those who attempted to map soils, to bring order to the bewildering, worldwide complexity. In the USA, soil survey was started in the late 1800s, with a pioneering phase that lasted until the 1930s, and a big increase in efforts and knowledge after the Second World War. Three people have played outstanding roles in the USA soil survey as they founded, expanded, and directed a national program: Milton Whitney, Curtis Marbut, and Charles Kellogg. Soil survey also progressed through state and local surveyors and the formation of a professional association that brought together practitioners from all over the country.

Some of the earliest soil mapping was accomplished in Europe, in the first half of the nineteenth century. The Russian Military Department published maps, starting in 1812, that showed items of interest to military operations, including soils, and the Ministry of Government Properties started mapping soils for taxation purposes in 1838 [1]. With the advent of systematic soil investigations in the mid 1800s, soil mapping began in Germany, France, Austria, the Netherlands, and Belgium. Soils were mapped according to geologic material combined with characteristics like color, texture, and drainage; some soil maps were based on the distribution of vegetation. The maps were intended for land use planning or to raise taxes according to the productivity of the soil. In the late 1800s, the need for soil maps became apparent in Russia and the USA as extensive new territories were opened up for agriculture [2].

Early soil maps in the USA were made by the state geological surveys [3]. One of the first soil maps in the USA was made in Wisconsin, in 1882, by the geologist Thomas Chamberlin, who was of the same generation as Vasily Dokuchaev. It showed soil texture and the geological nature of the soil materials and was the result of years of geological exploration while traversing the state on horseback [4, 5]. For Thomas Chamberlin, the character of the soil depended upon the nature of the rock, the degree of weathering, and elements lost by leaching and gained by vegetation or capillary action from beneath [6]. He recognized the difficulties in mapping soils: "There are few natural formations more difficult to map than soils. There is an almost infinite gradation of varieties between which there are no hard-and-fast lines, and it is nearly or quite impossible to represent these gradations on a map" [6]. Between 1873 and 1877, he published, with several co-authors, four voluminous books, over 3,035 pages, under the title *Geology of Wisconsin* [6]. In these volumes, the glacial stages of North America were outlined.

Thomas Chamberlin was a supporter of the geological base for the soil survey as implemented by Milton Whitney of the *Bureau of Soils* [7]. His support for Milton Whitney resonated in an article in *Science* in 1909: "That the origin of the soil body lies chiefly in the granulation of rock; that soils are wasted at the surface by wind and wash; that wind and wash also distribute granules and mix soils and give to nearly all soils some of the essential soil constituents; that progressive granulation of rock adds soils below; that progressive solution removes soil matter from soils and from the rock beneath; that by these composite processes the body of the soil is at once enriched and impoverished." Thomas Chamberlin summed up the main soil physical properties, and some biological properties recognizing that: "...well-known rotation of legumes and cereals that enriches the soil with nitrogen may be supplemented by a long-period rotation of trees and annuals for the enrichment of the soil in potash and phosphorus" [8].

Although he was aware of the problem of soil erosion that he named 'soil wastage', Thomas Chamberlin criticized those who emphasized widespread soil exhaustion: "...there is no substantial grounds for an alarming forecast, applicable to an industrious and intelligent people willing to be guided either by oriental experience or by western scientific research." Likely, he was referring to Franklin King who had traveled through Japan, Korea, and China for about nine months in 1909. The purpose of that trip was to learn how agriculture had been maintained in those countries for several thousand years, which yielded the book *Farmers of Forty Centuries* [9]. Franklin King espoused a broader view of what determined soil productivity; he was interested in how soil exhaustion was absent in places where people cultivated the land for long periods.

There were a few other soil mapping efforts, such as the 1884 soil maps by Eugene Hilgard in Mississippi [10], the soils of Maryland by Milton Whitney from 1893 [11], and the soil survey of Tennessee from 1897 [12]. In the 1880s, Eugene Hilgard worked for the *Bureau of Census* and he attempted to

Monolith sampling for the survey of the soils of Tennessee in the late 1800s by Charles Vanderford (on the right, with hat and moustache) with friends in the late 1800s, and with his wife Florence Anderson. The monoliths were exhibited at the *Exposition Universelle* in Paris in 1899. He died earlier that year

have agricultural surveys, that included soils, mandated within the Geological Survey but that effort did not succeed [13]. Some years later, a soil survey was conducted in Tennessee by Charles Vanderford, who had no college training, fought in the civil war, becoming a Major and after the war worked on his father-in-law's farm. Then he went into the merchant business, became a voluntary meteorological observer, and in 1891, he was appointed Professor of Agriculture at the University of Tennessee. In 1897, Charles Vanderford wrote the bulletin *The Soils of Tennessee* that marked the beginning of the soil survey program in the state [12]. Mechanical analyses of the Tennessee soil samples had been conducted by Franklin King and Milton Whitney [12]. Charles Vanderford strongly believed that soils should be studied and sampled with greater depth: "No man can decide when to plow or to cultivate his fields most effectively unless he knows his land, surface, and under-soil, not to a depth of six or eight inches only, but three or four feet or more. The more accurate such knowing the more intelligently can he adopt the right ways to handle his soil, so as to husband the waters of our abundant rainfall, that the crops may suffer least from our occasional periods of spring and summer droughts." He was the first to take soil monoliths based on a technique exhibited in the Russian pavilion at the World's Columbian Exposition, Chicago, in 1893. Charles Vanderford's Tennessee soil monoliths were exhibited posthumously at the 1899 *Exposition Universelle* in Paris.

In 1893, Milton Whitney made a geological soil map of Maryland and wrote a report with a set of remarkable conclusions. He found Maryland had about 12 or 15 distinct kinds of soil and these had different characteristics and agricultural values. The chemical composition of the soils did not account for their agricultural values, and seasonal weather differences affected crop yield more than fertilizers. He introduced the term 'wheat soils' as well as his favorite: the 'tobacco soils.' In Maryland, Milton Whitney became convinced that moisture and temperature controlled crop production and that soil texture was the best index for the conditions of soil moisture and temperature. He took that conviction into the *Bureau of Soils* and the soil survey program.

Based on these scattered soil survey efforts, the Association of American Agricultural Colleges and the Experiment Stations recommended in 1891 that a soils division be established within the Department of Agriculture [14, 15]. Three years later, a division was created in the Weather Bureau for the study of climatology in its relation to soil. A year later, this division was recognized as an independent office, and in 1897, it became the *Division of Soils*, which was elevated to the *Bureau of Soils* in 1901. In 1899, it began a program that aimed for a nationwide soil survey similar to the geological survey that had started twenty years earlier. In this new survey program, the soils would be mapped, studied, and described in reports that were to be publicly distributed. And so the soil survey in the USA was officially institutionalized [16].

An important purpose of the soil survey was to help settlers. There had been a wave of immigration in the late 1800s, mostly Europeans who settled on the Great Plains. There was not much new land for farm settlement, so the surveys largely focused on serving the existing farmers [17]. From the outset, the soil survey was utilitarian [14]; it not only served to improve land

Soil map of Maryland by Milton Whitney published in 1893. The map showed 12 soils based on geologic formations, and for each soil the clay percentage was given. In the accompanying report the particle size fractions were: clay <5 μm, silt 5–50 μm, and sand 50 μm–1 mm, fine gravel 1–2 mm. Silt had two fractions whereas he distinguished four sand fractions. These fractions were modified in 1928 to: clay <2 μm, silt 2–50 μm, and sand 50 μm–2 mm

management, but also had other purposes so that it broadens the financial support for the *Bureau of Soils*. Charles Williams, an agronomist in North Carolina, listed who benefited from the soil survey: agricultural workers, farmers, county advertising, home seekers, county commissioners, county superintendents of education, other types surveys, teachers of agriculture and sciences, manufacturers and lumbermen, real estate agents, traveling men, canvassers and tourists, the US Post Office Department, and banks lending money to farmers [18]. In practice, the soil survey aimed to help farmers and, in particular, the European settlers.

In the beginning, soil observations were made in the field, and soils were then clustered into soil types, series, and provinces—a bottom-up activity, whereby observations were compared and grouped. After that, some theory was developed as to why certain soils were found in certain locations. It was an iterative process, whereby the kind of field observations was adjusted based on generalizations and soil groupings using those same observations. The repeated cycle of observations was to become convergent, which meant that it aimed to come closer to the desired result as the number of iterations increased. The desired end result was not well-defined, but its main focus was agrarian which made sense as in the 1890s over 40% of the population were farmers. In contrast, soil mapping in Russia focused on processes and factors that explained the soil as a natural body [19]. Broad groups were recognized and established in which soil observations were placed, and these broad groups were split into smaller groups: a top-down approach, although observations and theory were equally essential to defining the broad groups [7]. For the pioneer soil researchers in Russia, soils were to be studied independently of agriculture, but after 1917, the communist party bent the soil survey towards agrarian usage [20].

The history of the soil survey in the USA has been well reviewed [7, 14, 16, 21–24]. The soil survey started with a small budget which steadily grew. Its first Chief, Milton Whitney, was able to convince federal budget committees that soil survey was needed, and he showed, year-by-year, the large increase in acres for which soil maps were made and the number of samples collected. In 1899, Milton Whitney prepared a report entitled *First four thousand samples in the soil collection of the Division of Soils*. The samples were stored in two-liter airtight glass jars and he hoped: "...that this will be a nucleus for a much more extensive and comprehensive soil collection, to be brought together at the national capital, which, if well-arranged and thoroughly classified, with cross references as to the origin, physical properties, and agricultural values, will offer a valuable opportunity for soil investigations where dry samples of soil can be used." He also envisioned collecting samples from other countries [25].

Milton Whitney knew that the emphasis on acreage and the number of samples was a convincing tactic for requesting more funds. The budget increased ten-fold in ten years: the allotment for the soil survey was $16,300 in 1899, $154,000 in 1909, $195,000 in 1914, and $209,220 in 1927. The budget was cut during the two world wars and during the economic depression [14]. The increase in 1933 was part of the New Deal and increased government spending beneficial to the *Division of Soil Survey* that succeeded the *Bureau of Soils*. After the Second World War, the federal soil survey budget was almost one million dollars [26]. Adjusted for inflation, the equivalent budget today would be about $0.5 million when the survey was started in 1899, to $3 million in the early 1900s, and $10 million by 1950.

The start and growth of the soil survey was driven by the intellectual curiosity and ambition of Milton Whitney and Charles Dabney. They met for the first time in the 1870s. Charles Dabney had obtained a PhD in chemistry at the University of Göttingen in Germany, in 1880 [27]. In 1881, Charles Dabney became the director of the North Carolina Experiment

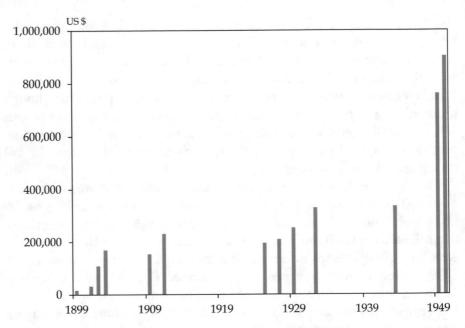

Annual funding for the soil survey in the USA between 1899 and 1950 (13 years of data available). Graph based on information in Merritt Miller's paper *Progress of the soil survey of the United States Since 1899*, published in the *Soil Science Society of America Proceedings* in 1950, and D.R. Gardner's thesis on the history of the National Cooperative Soil Survey of the United States from 1957. The 1950 funding of $902,000 is equivalent to almost $10 million in 2021. By comparison, the *Soil Conservation Service* directed by Hugh Bennett received about $50 million in 1950, equivalent to about $530 million in 2021

Station and, as the state chemist, he was responsible for analyzing the composition of fertilizers. He visited Samuel Johnson at the experimental station in Connecticut and met with Milton Whitney, who had just been hired to work on tobacco. In 1886, Charles Dabney hired Milton Whitney as farm superintendent at the experiment station in North Carolina. Coincidentally, Milton Whitney also had a job offer from Eugene Hilgard to work at the University of California Berkeley, but he preferred to work for the federal government in the Department of Agriculture [15].

Both Charles Dabney and Milton Whitney believed that a federal agency was needed to coordinate the soil survey efforts in the different states, and that soil knowledge would contribute to a healthy agricultural economy. Milton Whitney, or Professor Whitney as he was commonly known [26], became the first chief of the *Division of Soils*. There were others who had interest in the position, like Eugene Hilgard from California though, when it was offered to him, stayed in Berkeley [7]. Had Eugene Hilgard taken the

Wheat stone bridge box. From Milton Whitney, Thomas Means, Frank Gardner and Lyman Briggs U.S. Department of Agriculture Bulletins published in 1897 (Determining the soluble salt content of soils; Determining the moisture content of arable soils)

position, the research direction of the *Division of Soils* would have emphasized soil chemical analyses, and the soil survey would have made a different start [15, 28].

Milton Whitney, born in Baltimore, was trained in chemistry at Johns Hopkins University in Maryland. In the late 1800s, he worked with Lyman Briggs, Frank Gardner, and Thomas Means, and developed instruments for determining moisture and soluble salt content of the soil [29, 30]. The instruments were designed to be portable so that measurements could be made in the field. For soil moisture, they used a Wheatstone bridge box that contained a rheostat for measuring the electrical resistance of the soil. Several electrodes were developed that were buried in the soil and connected to the box bridge. The resistance was converted to conductivities based on the salt and soil moisture content; if the soil moisture content changed by rain or evapotranspiration, so did the conductivity. The box was also used to determine the soluble salt content. With Lyman Briggs, Milton Whitney measured soil temperatures and found that: "…the relative temperature of a soil is certainly one of its important physical properties and we should know the temperatures of the soils adopted to our different agricultural interests" [31].

Milton Whitney studied soils under tobacco and found that there was a connection between the quality of the crop and the texture of the soil. He focused on the soil physical properties because chemical soil analysis often failed to predict productivity, and he viewed soil chemical analysis: "…as a deficiency, not of techniques, but of principle" [14]. He challenged the idea that fertility was mainly a matter of chemical composition of the soil [32], and developed an agricultural classification of soil based on the number of mineral grains per gram of soil: "For example, no crop can be successfully grown, except under highly artificial conditions of manuring with

Alkali soils were an important research topic in the early years of the soil survey. An auger hole, soil on the tarp, and instruments for salt determination including the Wheatstone bridge box (right of the auger) and for sodium carbonate and chloride on the left of auger. From Field Operations of the Division of Soils, Report No. 64 published in 1899

organic matter, or by irrigation, on a soil having so few as one thousand seven hundred million grains per gram" [28]. He ignored most chemical soil properties. Emphasis on soil physical properties was not unique to Milton Whitney. In Russia, Konstantin Glinka considered soil structure and soil color the most important features.

In 1899, the *Division of Soils* had four employees including Milton Whitney; four years later, it had 80 employees. The division was unable to meet even half of the demand for soil investigations [33]; three reports were published in 1899, and by 1903, 72 reports were published [26]. Not all soil surveys in the USA were directed by Milton Whitney; soil surveys also took place outside the *Bureau of Soils*. For example, Andrew Whitson mapped soils in Wisconsin, and the soil chemist Cyril Hopkins had established a soil survey program at the University of Illinois. Still, the national survey was expanding and Milton Whitney was drawing all efforts to the *Bureau of Soils*. The soil survey grew so fast that it had difficulty finding qualified people. Whitney asked recommendations for potential new employees from

Eugene Hilgard in Berkeley, Thomas Chamberlin in Chicago, and Nathaniel Shaler in Harvard [34]. He was convinced that the soil survey was best served by hiring bright, innovative people, and accomplished scientists, and with the increasing federal funds, he hired the soil physicists Lyman Briggs and Franklin King. He also hired Edgar Buckingham who developed the concepts of matric potential and soil–water retention curves [35].

Several new employees of the *Bureau of Soils* had been trained by Collier Cobb at the University of North Carolina; in the late 1880s, Collier Cobb had been an assistant to Nathaniel Shaler at Harvard University [36], and, at North Carolina, he trained Hugh Bennett, Thomas Rice, and George Coffey who later conducted graduate studies with George Merrill in Washington. Milton Whitney also hired graduates from Earlham College, a Quaker college in Richmond, Indiana. They were trained by the geologist Allen Hole, a student of Thomas Chamberlin and father of the pedologist Francis Hole [36, 37]. Mark Baldwin, Earl Fowler, Ralph McCracken, and James Thorp came from Earlham College [34]. While Milton Whitney preferred bright hires, he found it difficult when their views differed from his own or those that were published [34]. As a result, several of these hires did not stay long at the *Bureau of Soils*.

In the beginning, the *Bureau of Soils* worked on several issues other than the development of soil survey. There were studies on the alkali problems of the western lands and how such soils should be reclaimed, the suitability assessment of tobacco growing under shade, and the fixation of atmospheric nitrogen. The bureau also undertook work outside the conterminous USA like reconnaissance soil surveys in Central America, China, the Virgin Islands, Cuba, Alaska, and in the Amazon [14]. China and the USA had soil cooperation since the early 1920s, and Walter Lowdermilk, John Buck, Charles Shaw, James Thorp, and Robert Pendleton worked in China at the invitation of the Chinese government in the 1920s and 1930s [38, 39]. Work in Alaska and Cuba was undertaken by Hugh Bennett [40, 41]. Curtis Marbut worked in the Amazon in 1924. Even so, there were on the whole very few American soil researchers outside the USA when compared to European soil researchers that worked in their colonies in Africa, South America, and Asia.

In 1907, an organizational unit named *Soil Erosion Investigation* was set up within the *Bureau of Soils*. At that time, it was recognized that soil erosion was a serious problem, and some considered it the greatest evil confronting the American farmer. There was concern about the loss of soil, the damage by sedimentation, and: "…soil wash should be considered a public nuisance, and the holder of the land on which it is permitted to occur should be held liable for resulting damages to neighboring lands and streams" [14]. The bureau hired as an expert on soil waters W.J. McGee (William was his first name but he went by W.J.), who had been responsible for part of the exhibit at the World Fair in St. Louis in 1904 [34]. Born on a farm in Iowa, he was largely self-taught and studied the Pleistocene geology of the upper Mississippi Valley [34]. That work was guided by the geologist Thomas Chamberlin [37]. William McGee had worked with Eugene Hilgard in Mississippi and Louisiana in 1891 [42], and for the bureau, he wrote one of the first reports on soil erosion and recommended deep tillage (15 cm), mulching, manuring, retaining crop residues, drainage, contouring, and keeping the soil covered as a measure to reduce soil erosion [43]. William McGee compared soil erosion to a disease: "When the soil is viewed as a suborganic structure exercising normal functions connected with its own circulation, it is easy to see that destructive erosion and other abnormal processes affecting the soil are analogous to the diseases affecting animals and plants, and that, like most diseases, they may be counteracted by treatment tending to retain or restore normal conditions and, as proverbially in other disorders, an ounce of prevention is better than a pound of cure" [43]. One year after the soil erosion report was published, William McGee died in his quarters at the Cosmos Club in Washington [34].

Soil erosion affected soil's physical properties, so it had Whitney's attention, but he ignored the loss of soil fertility resulting from erosion [14]. He did not believe in predictions that the soils of the USA were becoming depleted and eroded, and at the time wrote: "…as a national asset the soil is safe as a means of feeding mankind for untold ages to come. The soil is the one indestructible, immutable asset that the Nation possess. It is the one resource that cannot be exhausted, that cannot be used up" [44]. Later on, he would realize that the soil resource was not infinite [45].

In 1911, a shortage of potash was envisioned as Germany had a monopoly on its supply. Milton Whitney encouraged the fertilizer industry to become a chemical industry, which was even more urgent in 1917 when the USA entered the First World War and potash no longer available [46]. Potash was used as a fertilizer and for industrial purposes, and alternative sources besides potash evaporites were being sought. Milton Whitney summarized the potash situation in the report *Fertilizer Resources of the United States*: "…there are several sources of potash of possible economic importance, one of these overshadows all others. In the kelp groves or beds of giant kelps along the Pacific littoral, the United States possesses an extremely valuable national asset. At their best, and under the most careful and efficient utilization of these groves, they might be made to yield annually an output of potassium chloride approximating the production of potash salts of all kinds from

Old field gully erosion and contouring with meandering balks—the balks were created by ploughing a furrow and vegetating it. Soil tillage followed the contour and so did the planting of the crops. Photos from *Soil Erosion* by W. J. McGee published in 1911 by the *Bureau of Soils*

the world-famous Stassfurt deposits" [47]. The plant nutritionist Dennis Hoagland from the University of California Berkeley started investigating the use of the algae kelp as a source of potassium fertilizer [48].

The soil erosion and potash fertilizer studies were important for the *Bureau of Soils*, but soil survey was the main activity. The initial soil mapping was done on the basis of soil texture and, in 1903, the soil type became the only category of soil classification. Prior to that, soil surveys had a geological base and it was Milton Whitney who based the survey on the recognition of a soil texture, which was seen as innovative [14]. The overall purpose of the survey was to establish series for the different physiographic divisions of the USA. The first field surveyors included Jay Bonsteel, Thomas Rice, Macy Lapham, his brother J.E. Lapham, George Coffey, and Elmer Fippin. Several of them had basic training in geology and geography, as specific soil programs at universities did not exist and were only started in the early 1900s. The field surveyors worked in isolation, without much contact with others in the field and staff at the *Bureau of Soils*. Milton Whitney rarely left Washington and did not visit the field crews [14]. They had no precedents to follow [26], lacked extensive field guides, and different soil mapping approaches were used. Most of all, they lacked a basic understanding of soils and how they were formed and distributed across diverse landscapes.

It was unclear what properties were to be observed in the field except for soil texture, how classes that combined the properties could be established, and how these classes should be put on the map. In 1904, soil color and organic matter became soil properties that were recorded during fieldwork, and in 1906, soil structure, lime content, alkali, drainage, erodibility, physiographic position, nature of subsoil, lithology, and origin and age of parent material were added. A year later, the list was further expanded with aeration, oxidation, and mottling and a rudimentary form of the soil catena was being used [14]. Thus, ten years after the survey was established, a range of soil properties was observed for making soil classes and soil maps. But they were shallow observations: neither soil profile descriptions nor horizons were used to distinguish among soils.

A classification system was developed that used a set of properties observable in the field. During fieldwork, it became clear that there were many soils that did not fit into the soil classification system, or that the existing soil classes were different in other parts of the state or country. With the increased number of surveys and observations came some insight, more classes, and, eventually, a new and more detailed system of classification. Years later, Charles Kellogg reflected on this period and its difficulty: "These early soil scientists were faced with a most difficult dilemma: they could not classify and map soils without knowing their characteristics; and they could not know their characteristics until they examined representative examples in the field and in the laboratory. Not until they had classified and mapped could they understand the significance of combinations of characteristics" [49].

The first field manual was published in 1903; it was entitled *Instructions to field parties and descriptions of soil types*, and Milton Whitney emphasized its purpose: "The descriptions of soil types are given as an aid to the field parties in correlation of soil types and should be carefully studied to this end. Soils of a new area should be correlated with a known type where this is possible" [50]. Soil types were distinguished based on the texture and some morphological features, and if certain features were met when an observation was made, the soil was given a name [16]. Soil texture was particularly important as a criterion or characteristic [14]. Soil types commonly had the location and texture as names, such as the Wabash silty clay loam, or the most well-known: the Cecil sandy loam. There were 260 soil types recognized by the bureau in 1901, with detailed descriptions for each of these types and series that had information on the geographical distribution, surface features and drainage, soil improvements, crop limitations, farm equipment, and the extent of the soil type [50]. The number of soil types increased rapidly; there were 400 by 1904, 715 by 1909, and 1,650 soil types were recognized by 1912 [16].

Descriptions of the soil types were made by the geologist Jay Bonsteel, who, like Milton Whitney, had studied at Johns Hopkins University [26, 34]. He was friends with the soil inspector for the Southern Division, Hugh Bennett, who had joined the survey in 1903. Jay Bonsteel had conducted surveys of Connecticut in 1899, Maryland in 1900, Mississippi in 1901, Illinois and Wisconsin in 1902, Delaware and New York in 1903, Rhode Island and South Carolina in 1904 [51]. In 1903, Jay Bonsteel was professor of Soil Investigations at Cornell University and trained people for the soil survey. He developed a course in soil science and soil surveying, and Charles Shaw was one of the students that took the course. After two years at Cornell, Jay Bonsteel returned to the *Bureau of Soils,* and Elmer Fippin replaced him [52]. At the bureau, Jay Bonsteel was in charge of soil survey interpretation in an organizational unit dealing with *Use of Soils* [14].

The field crew that conducted the soil mapping usually consisted of two people—one in charge and an assistant. The person in charge controlled the fieldwork, prepared the report and maps, and forwarded monthly expense accounts to the office in Washington. There was a maximum of $4 per person per month for laundry. If an accident occurred that resulted in damage to a horse or vehicle because of carelessness, workers were considered personally responsible [50]. The equipment for fieldwork consisted of the following: soil auger, 1 m handle, geologist's hammer, notebooks, compass, plane table, odometer, surveying chain, set of colored pencils, base map, sacks and tags for collecting samples of soil, cards for reporting samples collected, and a copy of the field Instructions. The instructions were that: "...soil samples, as a rule, should not be collected until the party has obtained a very thorough acquaintance with the type conditions" [50].

As more surveys were conducted, it became possible to synthesize and aggregate the information, and in 1909, Milton Whitney published a soil

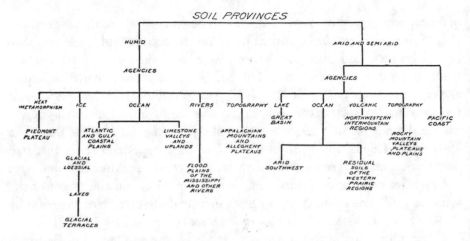

Soil provinces and the dominating factors in their formation. Illustration from *Soils of the United States* by Milton Whitney published in 1909. An accompanying maps showed the distribution of these provinces. At the highest level, soils were separated based on humid or arid and semi-arid conditions; below these were the agencies (ice, ocean, rivers, hear metamorphism, volcanic, topography), and lower level include locations (e.g. Piedmont plateau), parent materials (limestone valleys), and landforms (glacial terraces)

province map of the USA [44]. At that time, less than 5% of the country been surveyed, but 715 soil types had been described [10, 16]. The soil province concept had been introduced in 1906, representing large physiographic units and for the entire country, 14 provinces were distinguished that were aggregations of soil series and soil types [14]. A soil type could only belong to one province. The climate was recognized at the highest level, and there were two broad groups: humid versus arid and semi-arid. There were seven provinces in the arid and semi-arid regions and seven in the humid regions. Below that, 'soil formation agencies' were distinguished such as heat metamorphism, ice, oceans, rivers, topography, lakes, and volcanic, and each of these occurred in one or more soil provinces. The glacial and loess province was considered humid and followed the Missouri River westwards, including most of semi-arid North Dakota and Montana.

Milton Whitney was a tireless advocate for the *Bureau of Soils* and used the number of acres surveyed and the widespread usage of that information as a leverage to obtain more funding. He stressed the importance of the soil and its importance for the country: "The soil is the greatest natural resource of the Nation" and he estimated that the soil produced about $7 billion in products annually and that it formed the base for the industrial wealth and a large proportion of the export [44]. In order to gain interest among a wide audience, the bureau published a booklet entitled *Important Soils of the United States* in 1916. It was meant for teaching agriculture and physical geography in schools and colleges [53]. The booklet was accompanied by 13 boxes with soils and subsoils for each of the provinces and regions and included a map with soil provinces and soil regions based on the work of Curtis Marbut and his colleagues from 1913 [54].

Strongly opinionated, Milton Whitney had disagreements with soil researchers across the country. Much of that disagreement was about his conviction that soil texture was the main factor determining crop production, and that the soil solution was of the same composition in all soils [32]. The activities of the *Bureau of Soils* were critically examined in other countries. For example, John Russell from Rothamsted in the UK disapproved Milton Whitney's papers and the dominant role of soil texture [55]. Criticism came also from within the USA, particularly from Eugene Hilgard from California, Cyril Hopkins from Illinois, and Franklin King from Wisconsin, all influential scientists. There was also critique from Robert Pendleton [7].

Eugene Hilgard was 27 years older than Milton Whitney, and Franklin King 12 years older. Eugene Hilgard called Milton Whitney a "youngster" with "silly propositions" [28]. Some of the arguments with Eugene Hilgard and Franklin King might have been underlain by a difference in seniority, by the fact that Eugene Hilgard had wanted to hire Milton Whitney, and could, himself, have been the first chief of the soil survey. In 1889, Eugene Hilgard did not take the offer to become the Assistant Secretary of Agriculture, and a few years later, Charles Dabney was appointed to that position [7]; Charles Dabney then hired Milton Whitney. Eugene Hilgard was trained in Germany and widely experienced in the west; he emphasized soil chemistry to solve soil problems, whereas Milton Whitney's experience was in the eastern part of the USA and he advocated soil physics to understand how the fertility of the soil could be increased [28]. Eugene Hilgard had: "…no patience with Whitney, and criticized him and his work to anyone whenever the opportunity arose" [7]. In 1907, Eugene Hilgard wrote a letter to his sister, in which he described Whitney as a "humbug" and mentioned that: "…all men of science laugh at his vagaries" [7]. According to Hans Jenny, Eugene Hilgard himself was: "…a fighter, never daunted by opposition, no matter how bitter, he gave no quarter in his striving for what he deemed right and proper…and fighting was his second nature, almost a recreation, and he boasted that he could sniff a battle from afar" [42, 56]. In return, Milton Whitney discouraged the use of Eugene Hilgard's work, and his publications were not made available to the soil survey field crews [57]. As a result, Eugene Hilgard's work was unknown to those working in soil survey.

In 1901, Milton Whitney hired Franklin King from the University of Wisconsin to become the chief of the *Division of Soil Management* within the *Bureau of Soils* [58]. Milton Whitney had a good nose for talent and Franklin King had conducted groundbreaking research in soil physics. He had quoted Milton Whitney's work on soil mechanical analyses in his 1895 book *The Soil* [59], and he had a geological background. At the *Bureau of Soils*, Franklin King studied the level of plant nutrients in soil solution and their relation to crop yields. Chemical analyses were made of soils and plants, and he wrote six manuscripts that were to be published by the *Bureau of Soils:* Milton Whitney rejected three of them as they were contrary to his own ideas that supposed that soil solution concentrations were the same in all soils. They could not agree [58]. Milton Whitney asked Franklin King to resign. In this dispute, Franklin King had the support of Eugene Hilgard who thought Franklin King ought to have lived a century ago, and also felt that Franklin King should replace Milton Whitney as Chief of the Soil Survey [28, 42]. That did not happen. Franklin King died in 1911, at the age of 63.

In 1916, Robert Pendleton at the University of California Berkeley, studied differences in soils mapped by the *Bureau of Soils*. Samples were taken from what were mapped as similar soils and analyzed for physical, chemical, and biological properties. The soils were also used in pot studies. He found that: "…different representatives of a given type are not the same in their ability to produce crops…and all soils mapped under a given name by the *Bureau of Soils* may or may not be closely similar depending on the criteria used" [60]. He concluded that large variation occurred within soils that were considered to be similar; finished his report, and left for a position in India. The report was to be published by the University of California but was delayed for three years. Although the work had been instigated by Charles Lipman, he distanced himself from the report, which he found to be radical and impractical. When it was finally published, the university put a disclaimer upfront, and Robert Pendleton's findings were ignored by the *Bureau of Soils*. Robert Pendleton then worked in the tropics for many years and did not return to the USA until 1946.

In 1903 Milton Whitney and Frank Cameron published the bulletin *The chemistry of the soil as related to crop production*, that caused further disagreements. They dismissed chemical approaches as an indicator of the soil fertility and predictor for crop yield, and yet, they stressed the importance of the soil solution: "Since it has been found that plants grow well and produce normal crops with the water-soluble constituents of the soil, it seems entirely unnecessary in studying the question of the nutrition and yield of crops to introduce artificial digestion media known to attack minerals very slightly soluble in water, while it seems perfectly logical to accept the nutrient solution as it exists in the soil as the basis for the support of plant life and to investigate the question along this line. In other words, it has seemed best to consider the soil as a culture medium containing a nutrient solution—that is, to regard the soil moisture as a proper and sufficient medium for the feeding of plants, and the soil as a reservoir and distributing agent for this solution" [61]. They advocated that rotation and change of cultural methods would maintain the fertility of the soil.

The disagreement between Milton Whitney and Cyril Hopkins from Illinois on the issue of soil fertility was particularly fierce. Cyril Hopkins was six years younger, and a soil researcher with a different perspective. He was born in 1866, on a farm in Minnesota, and grew up in South Dakota where he: "...gathered the buffalo from before the plow that turned the rich virgin soil" [62]. Cyril Hopkins obtained an MS in chemistry from Cornell University in 1884 and a PhD degree in 1889; his doctoral thesis was entitled *The Chemistry of the Corn Kernel*, after which he spent a year in Germany. He was abstemious, would not hire anyone who would use tobacco or alcohol [62], and held an agrarian view on soils. His main interests were soil fertility and permanent agriculture, which he envisioned would be possible through maintaining phosphorus and soil organic matter levels [63]. He viewed soil fertility maintenance as the main problem in the USA and noted that most agriculture practices reduced land productivity.

Cyril Hopkins was an admirer of Justus von Liebig and, like him, strongly believed in agriculture and facts: "Agriculture is, of all industrial pursuits, the richest in facts and the poorest in their comprehension. Facts are like grains of sand which are moved by the wind, but principles are these same grains cemented into rocks" [63]. In 1911, Cyril Hopkins wrote *The Story of the Soil*, which combined a story and scientific soil knowledge: "Truth is better than fiction; and this true story of the soil is in competition with popular fiction" [64]. It chronicles an agricultural student from Illinois in search of a farm where he wished to practice what he has learned. It narrated all he stood for, including the need for phosphorus applications.

Cyril Hopkins challenged Milton Whitney and the bulletin from 1903: "...the injury to American agriculture that may result from the wide dissemination and adoption into agricultural practice of erroneous teaching from one occupying a national position" [62]. Cyril Hopkins bought 125 ha of land that was named 'Poorland Farm' as it would raise nothing besides poverty grass and mortgages [62]. He started to apply large amounts of phosphorus and introduced 'complete fertilizers' that contained nitrogen, phosphorus, and potassium. The land became productive, and he translated the results in terms of practical farming. The work was meant to debunk Milton Whitney's theory on soil fertility, and although the facts upon which he based his permanent agriculture had been known for decades, he promoted them in ways that were easily understood by farmers. He corresponded with soil researchers in Europe regarding the fertility of the soil and what he perceived was: "...the most fundamental problem concerning the future prosperity of America" [65]. After the First World War, he studied soils in Greece under the auspices of the American Red Cross. The main purpose was to increase grain production and, in 1919, he left Greece with a decoration from King Alexander. Cyril Hopkins did not make it back to Illinois; on his way home, he had a hemorrhagic stroke, contracted malaria, and died in Gibraltar at the age of 53 [66].

The controversies between Milton Whitney, Cyril Hopkins, and Eugene Hilgard threatened the existence of the *Bureau of Soils* in the early years [14]. Eugene Hilgard was trying to remove Milton Whitney from his position, and the Secretary of Agriculture was confronted by Cyril Hopkins. Some of the surveyors left the *Bureau of Soils*, and legislators supportive of Cyril Hopkins criticized the work of Milton Whitney in Congress. It resulted in the discontinuance of cooperative soil surveys that had begun in Illinois in 1902, and no federal-state cooperation in California for as long as Eugene Hilgard lived [14]. For many years, the quarrel prevented resumption of friendly relations with the State of Illinois, which instituted a soil survey independent of the *Bureau of Soils* and based on a different system of classification and mapping that, in turn, introduced difficulties in the coordination and interpretation of data [52]. California entered the cooperative soil survey in 1913—a few years before Eugene Hilgard's death; but Illinois did not enter the cooperation until 1942 [14]. In the end, the verdict of history and chemistry was in favor of Eugene Hilgard, Cyril Hopkins, and Franklin King [34].

Milton Whitney had determined that the system of mapping soil provinces and soil series should be used countrywide. Soil classification was based on differences in soil physical properties observed during field examination. Milton Whitney was able to gain support for the soil survey, securing and increasing its funding, much of which was achieved through presenting the usefulness of the survey and the number of acres that were annually surveyed—he had a map in office that depicted the progress of the soil survey by county. Because progress and acres were important for the continuation of funding, there was not much debate for changing the system and the way soils were grouped or classified. The maps were generally well-received but there was also criticism, particularly from Eugene Hilgard: "...it may be seriously questioned whether it would not be better to cover less ground more

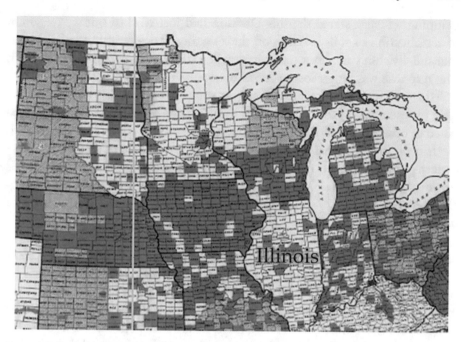

Part of the mid-west US and showing areas covered by the soil survey up to of 1934. Iowa, Nebraska, Wisconsin, Michigan. Indiana and the Dakotas had extensive areas surveyed. In Minnesota and in particular in Illinois little survey was done. In Illinois that was as a result of the feud between Cyril Hopkins and the Chief of the *Bureau of Soils*, Milton Whitney. Map from the soils chapter in the 1935 *Atlas of American Agriculture* by Curtis Marbut.

thoroughly and be content with less extended and less hasty mapping. This superficial method of work naturally excites criticism, not only at home, but also abroad" [14]. The soil surveys slowed down; not because of more thorough investigations as Eugene Hilgard had insisted on, but because of a lack of capacity in cartography and report publishing. Soil surveys had become more detailed, and cartographic costs soared, resulting in a delay of several years before the surveys were published [14].

Milton Whitney opposed the Russian principles of soil science but he was not alone, and according to the historian David Moon: "Russia was not somewhere many agricultural scientists in the United States would look to for such innovations, since many considered Russia to be 'backward'" [7]. After the exhibit at the World's Columbian Exposition in Chicago, in 1893, Milton Whitney wrote to Vasily Dokuchaev for advice on soil classification and mapping, as he had seen: "...the admirable pamphlet you prepared to accompany the collection of the Russian soils at the Columbian exposition," and he requested copies of Vasily Dokuchaev's works on soil classification and mapping. He also stated that his Division had just worked out a: "…system of classification of soils based upon their texture and physical properties, and that the division was about to begin detailed study and mapping of soils" [7]. The response was either lost or never received.

Milton Whitney was a man of strong convictions and had a domineering style of leadership. His insistence on the physical concept of soils and ignorance of other ideas impeded the development of soil survey in the USA [7]. He had no international experience and little knowledge of soils outside the eastern part of the country. Nonetheless, he was dignified and reserved in his manners. Macy Lapham, one of the early soil surveyors, knew him personally, dined at his home, and found that Milton Whitney had a keen sense of humor and was: "…gracious, witty and entertaining" [7, 52].

George Coffey had been with the soil survey since the beginning. In 1913, after he had left the survey, he published *A study on the soils of the United States* in which he showed that soils differ because of variations in the processes by which they have been formed and due to the parent material from which they were formed through weathering [67]. In his view, soil classification should be based upon differences due to factors that he termed 'genetic.' George Coffey questioned what factors should be used for larger groupings and thought that differences in climate were of importance in soil classification only when they produced variations in the nature of the soil. He proposed a system based upon differences in the soil-forming processes, or climatic origin, and his ideas on soil formation and classification followed the Russian school [67, 68]; he had read translations of Nikolai Sibirtsev's work that elaborated the zonal concept of soil distribution [7, 69].

This view was disparate from Milton Whitney's emphasis on soil texture and soil physical criteria that dominated the *Bureau of Soils* [14]. George Coffey stressed the importance of the climatic factor in soil formation: "…while American and other investigators have generally recognized the influence of climatic factors in soil formation, Russian workers, especially Dokuchaev, who is credited with founding a new school of soil investigation, have laid greatest emphasis upon the importance of climate in soil classification. With them it is given first place, because all processes of soil formation or weathering vary more or less with the climate, and therefore soils formed under markedly dissimilar climatic conditions will be very unlike in character. The climatic classification coincides to a certain extent with the chemical and geological classification." It was not to Milton Whitney's liking but, given his earlier problems with Franklin King [58], he decided that George Coffey's work was to be published with a footnote: "…without indorsing the scheme of classification proposed and without accepting all the conclusions drawn from the facts cited" [68]. In the same year that George Coffey published his report, Curtis Marbut published, with the help of others, a massive volume entitled *Soils of the United States* [54]. In that monograph, the *Bureau of Soils* and Milton Whitney's directives were maintained and George Coffey's report was not cited [68].

Curtis Marbut had taken up a position in the, *Bureau of Soils* in Washington, in 1910, to work with Milton Whitney, and he became the Chief of the soil survey in 1913. It is unclear whether the appointment of Curtis Marbut, whose geological background was favored by Milton Whitney, contributed to George Coffey's departure in 1911. George Coffey had a more pedological view than Curtis Marbut and would have been a progressive chief of the soil survey, but he was not selected by Milton Whitney [7]. When George Coffey left, the *Bureau of Soils* lost one of its best minds [14] and it took many years, and the conversion of Curtis Marbut to achieve what George Coffey had laid out in 1912.

By 1913, less than 15% of the soils in the USA had been surveyed and mapped. The surveys showed that up to 90% of the mapped soils were derived from material deposited by the action of water, ice or wind, and those soils were not derived from the weathering products of the underlying hard rocks. Curtis Marbut then wrote that there was: "…a widespread impression from textbooks regarding the origin and formation of soils is that soils are derived directly from rocks through the influence of weathering and the breaking down of the rock in place, leaving a disintegrated mass of material on the surface which constitutes the soil. This is only remotely true" [54]. He did not get the chance to expand on these thoughts for a while. The First World War broke out and, although the USA did not enter the war until 1917, it drastically curtailed soil survey activities [26].

Milton Whitney retired in 1927, by which time the bureau had grown to over 600 people including chemists, bacteriologists, microscopists, pharmacologists, biologists, and engineers. The soil survey personnel had increased from ten in 1895, to 218 in 1927 [24]. Upon his retirement, Milton Whitney was made honorary chairman of the organizing committee for the *First Congress of Soil Science* in 1927, but he was ill and could not attend the congress. In 1925, he had suffered an attack of angina pectoris from which he never recovered. He died at his home in Takoma Park—the Azalea city— on the 11th of November 1927; he was 67 years old. The next day *The New York Times* announced it on page 17: MILTON WHITNEY DEAD. Annie Langdon, his wife, died seven weeks later.

On learning of Whitney's death, John Russel wrote to the bureau: "Your letter grieved me very much. It is sad to think that Professor Whitney is no

The first two chiefs of the soil survey: Milton Whitney with cigar in the top photo, and Curtis Marbut looking at his great group map from 1935. Milton Whitney directed the survey from its inception in 1899 and hired Curtis Marbut in 1910 to help lead the expanding field of soil survey. Milton Whitney was legendary for his ever-present cigar and he could determine the kind of soil in which the tobacco had grown by the aroma of the smoke from these cigars. Photos special Collections, USDA National Agricultural Library

longer with us to help with his advice and encouragement. And yet, knowing how much he had already suffered, we could not have wished that he should continue in his pain, there being no hope of cure. His end was in keeping with his whole life; for he was throughout a man of courage, never fearing to go through what he knew had to be done. So there passes out of our life an interesting figure who made us stop and think what we were doing and why; for this we are grateful and he will long have our respectful remembrance. Please convey to his family my sincere sympathy with them, and that of my colleagues here, in their loss" [70]. John Russell also wrote an obituary that was published in *Nature*, in which he concluded that Milton Whitney had: "…widened the range of subjects and enriched it with ideas and analogies which, if not themselves entirely sound, nevertheless make the investigator stop and think." John Russell had been critical of Milton Whitney's emphasis on soil texture [55].

A few years before his death, Milton Whitney wrote the book *Soil and civilization: a modern concept of the soil and the historical development of agriculture*, which discussed the concept of soil, the soils of the USA, the role of fertilizers, and agriculture throughout the world [45]. The Russian principles of soil formation and distribution were not included, but he noted that soil science was a rapidly growing field: "It is entirely impossible and undesirable in a book of this scope to explain what a soil scientist at present can or cannot do, just as it would be impossible to explain what a physician does. The subject is so complex that books and monographs by the hundreds have been written on every phase of the soil's activities and as with medicine or with surgery a man must keep abreast of the time, watch every advance that has been made, and be prepared to use or to have others qualified in a particular field use the most advanced methods of research for diagnosing the trouble and for prescribing remedial methods" [45].

When Curtis Marbut arrived in Washington, in 1910, he was not exactly welcomed with enthusiasm at the *Bureau of Soils*. He was a geologist, and Thomas Rice, who had been with the bureau since its early days, noted: "We did not think that the soil survey needed a geologist, at least not one of the hard rock or the fossil hunting type. We feared that this was a backward step. Even when Marbut took charge of the soil survey some time later, it seemed to us that he still regarded the soil as an interesting geologic formation that had been overlooked by other geologists" [71]. Macy Lapham thought the new chief would put much more emphasis on geology and started studying his geology textbooks after Curtis Marbut had arrived. Macy Lapham feared for his job [34, 52]. Louise Marbut summarized the early years of her father at the bureau: "…having spent 15 years in a rather exceptionally congenial atmosphere of scholarly thinking at the university of Missouri, he was plunged into what was at first a hostile situation, and one which demanded all his resources of thought and patience and forbearance" [72]. From a progressive university in the dissected till plains north of the Missouri river, he had landed in a bureaucratic government organization that had a rigid way of looking at soils and soil surveys. In addition, there was some competition and antagonism between the United States Geological Survey and the soil survey by the bureau in the Department of Agriculture.

Curtis Marbut was not disturbed by the way he was treated by his new colleagues for he had always felt somewhat different [72]. He trusted that they eventually would recognize and accept him based on his achievements [73]. Like Milton Whitney, he was a man of strong convictions; they were about the same age but Curtis Marbut was open to ideas and had a more creative mind. After he arrived in Washington, most of his time was devoted to the study of soils, and he read the books from Carl Sprengel, Eugene Hilgard, and Emil Ramann. He learned about soil profiles, Chernozems, Podzols, Solonetz, and laterites, and those soils and terms were not used in the *Bureau of Soils* [72]. Shortly after it was published in 1914, Curtis Marbut read the German translation of Konstantin Glinka's *Die Typen der Bodenbildung* [20]. The book was known by very few people in the USA, but it was cited by Hugh Bennett and Thomas Rice in their 1915 reconnaissance soil survey of Alaska [40]. Hugh Bennett and Thomas Rice described the Podzols of Alaska and compared them to soils in northern Russia, and they noted that the *Bureau of Soils* had not yet described such soils. They also discussed the peat soils of Alaska and compared them to those of Finland as described in Konstantin Glinka's book [40].

For two years, Curtis Marbut worked on a translation of the German version and several mimeographed copies were made. It was published in 1928 as *The great soil groups of the world and their development* [73]. In the Preface of the translation, Curtis Marbut wrote: "The great world of Russian soil literature is almost wholly unexplored by scientific men in America. The language is not read, even with difficulty by one percent of the scientific men in this country. Up to the appearance of this book early in 1914, no comprehensive summary of this work had been published in a language available to American readers and the outbreak of the Great War prevented its importation into this country for many years. The soil has been looked upon by scientific men in America and Western Europe largely from one or the other of two points of view. It has been considered either as a medium in which plants grow on the one hand, or merely as a geological formation with slight modification of no great importance on the other. The representatives of the first group have confined themselves in their investigations to questions that concern the getting of immediate results in crop yields. Those of the second group have given the subject very little attention, a mere passing notice. The fact that the soils of our own country are so much like those found in Russia

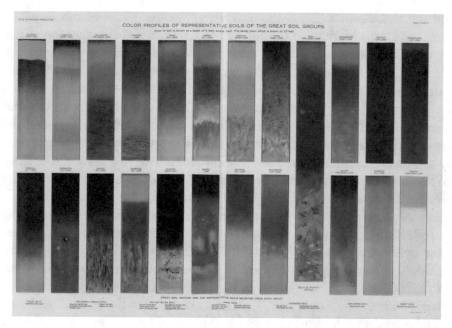

Painted soil profiles representing the Great Soils Groups as defined by Curtis Marbut in the soils chapter of the *Atlas of American Agriculture*, published in 1935. The profiles included Podzols and Chernozems, names that were to disappear from the soil survey in the USA in the mid-1950s. The profiles were painted by Mary D. Arnold and probably inspired Walter Kubiëna as his book *The Soils of Europe* from 1953 included painted soil profiles; those paintings were made by the Austrians Gertrud Kallab-Purtscher and a few by Anton Prazak. Walter Kubiëna preferred paintings over photographs because "...die Zeichnung das Wesentliche besser hervorheben kann"

and Siberia makes the results of Russian soil investigation peculiarly interesting to American investigators. The book however is more than interesting. It is destined to have an important influence on the course of soil investigation in this country, not because all its conclusions are correct or all the principles developed are immediately applicable to conditions here, but partly because of the thought stimulation always effected by the presentation of new points of view and partly also because of its suggestiveness as to methods and the broadening of conceptions by the presentation of the geographic and climatic points of view." Curtis Marbut thought the book would be: "...of great suggestive value for American investigators" [74]. The English translation of Konstantin Glinka's book was influential and shaped pedological thinking in the USA [7].

In his third year in Washington, Curtis Marbut was appointed scientist in charge of the soil survey, and he began working on soil genesis, morphology, and soil classification [26]. Milton Whitney was the Chief of the *Bureau of Soils*, but Curtis Marbut led the Soil Survey division as well as the Committee on the Correlation and Classification of Soils. He paid little attention to administration, but visited all soil surveyors that worked in the northern states during the summer and in the southern part of the country during the winter [52], such as Lee Schoenmann, Thomas Rice, Jethro Veatch, and Warren Geib. They surveyed county by county and helped to unravel the broad soil pattern across the country. The surveyors' work was supervised by a regional correlation and classification committee that consisted of three inspectors: Hugh Bennett for the Southern Division, Macy Lapham for the Western Division, and his brother, J.E. Lapham, for the Northern Division. These inspectors organized and coordinated the soil survey for the provinces or regions. Hugh Bennett was born on a farm in North Carolina; he had been an officer in the First World War, and in his early days as an inspector, he would sometimes go to the field in his army uniform. James Thorp was on the soil survey in New Jersey, in the 1920s, and was often challenged by Hugh Bennet, as he would say "What's over the edge of the woods?" and then would run over [75]. He was inspired by Thomas Chamberlin, who in the late 1800s had coined the term 'soil wastage' for soil erosion [76]. Hugh Bennett contributed to newspapers to make some extra income and discovered that soil erosion stories were publishable. Macy Lapham was born in 1874, grew up in Michigan, saw the change from the 'Wild West' to the 'Mild West,' and was one of the first soil surveyors hired by the *Bureau of Soils*. He was a soil surveyor for over 50 years, mostly in the west, but he also undertook a soil survey of western North Dakota with George Coffey and Thomas Rice [7, 52].

A few years after Curtis Marbut had arrived in Washington, he synthesized all the soil information since the late 1800s in the book the *Soils of the United States* [54]. About one million square kilometer had been mapped, at a range of scales. A somewhat larger area had been mapped in Germany, France, Great Britain, Ireland, and Italy. Soil survey methods differed across the USA although some common methodologies had been developed following the field handbooks. The overall aims of the soil survey were: "A soil survey exists for the purpose of defining, identifying, mapping, classifying, correlating, and describing soils. The results obtained are valuable in many ways and to men of many kinds of occupation and interests. To the farmer it gives an interpretation of the appearance and behavior of his soils and enables him to compare his farm with other farms of the same and of different soils. The soil survey report shows him the meaning of the comparison and a basis for working out a system of management that will be profitable and at the same time conserve the fertility of his soil. To the investor, banker, real estate dealer, or railway official it furnishes a basis for the determination of land values. To the scientific investigator it furnishes a foundation knowledge of the soil on which can be based plans for its improvement and further investigation by experiment. To the colonist [sic] it furnishes a reliable description of the soil" [54].

For *Soils of the United States*, the country was divided into seven soil provinces and six regions [54]. A soil province was considered a large physiographic unit and Curtis Marbut defined it differently than Milton Whitney had done in 1909 [44]. In a province as defined by Curtis Marbut, soils were formed by the same forces or group of forces, and each rock or soil material that had undergone the same forces, yielded the same soils. Forces was a term introduced by Konstantin Glinka but *factors* became the main term, although *agents*, *agencies*, *causes*, *elements* or *conditions* of soil formation were also used [77, 78]. The provinces concept could not explain why certain soils in California and the midwest looked and behaved similarly, and a soil type could only belong to one soil province [19]. A soil region was considered to include several soil provinces [54]. In essence, the soil provinces and regions were geographically delineated, but were not based on soil characteristics. Within a province, soil series occurred that had the same range in color, character of the subsoil such as color and structure, similar relief and drainage conditions, and a common origin. The soil individual was considered to be the soil type, which was defined as a soil with the same texture, color, structure, character of subsoil, general topography, process of derivation, and parent material. By 1912, some 534 soil series and 1,650 soil types were described based on the work of Milton Whitney [54]. The system was in essence a modified geological survey in which parent material and soil texture were the distinguishing principles.

In the *Soils of the United States*, Hugh Bennett. J.E. Lapham and Curtis Marbut described the seven soil provinces covering the central and eastern part, whereas Macy Lapham described in detail the six soil regions in the west. The area and distribution of soil series and soil types was presented, and each of the provinces had a key to the soils that was a geologic and physiological classification. A soil map was included with the *Soils of the United States* book; it was the third nationwide soil map of the USA in four years. Earlier soil maps were produced by Milton Whitney in 1909 and George Coffey in 1912 [44, 67]. It took until 1935 before another nationwide soil map would be published.

Curtis Marbut was in charge of the soil survey from 1913 until 1935. Trained as a geologist, he initially held a geologic view, but that changed following his participation in field tours and after having read works from others, particularly the works of Konstantin Glinka and Eugene Hilgard [14]. Whereas Milton Whitney had never traveled much outside of Washington or the east coast [7], Curtis Marbut traveled to all surveyors across the country and to many countries abroad. He visited almost all of the 3,007 counties in the USA. Within the *Bureau of Soils*, questions were raised about the need for all that travel, but Curtis Marbut had Milton Whitney's support and much of the travel was done cheaply; Marbut traversed by foot, stayed in cheap places, and kept his expenses to a minimum [79]. According to Macy Lapham, Curtis Marbut was: "…frugally minded and had attained his manhood and accomplishments the hard way. He was keenly conscious of his responsibility in the spending of public funds. It really hurt him to feel that he was enjoying the luxury of travel at the expense of others; at times he felt guilty in traveling by Pullman car when the end of the journey might be arrived at by riding in the day coach. He was a tireless walker and disliked paying what he considered an exorbitant price for livery hire. Consequently, a good deal of field inspection was done on foot. Many a long mile have I limped painfully along behind his tireless long legs" [52].

In order to raise the profile of the *Bureau of Soils* and to familiarize himself with soils in other countries, Curtis Marbut made trips to Canada, several countries in Europe, Russia, Brazil, Argentina, and the Caribbean [73]. He never visited Africa but, in 1923, made one of the first soil maps for the continent [80]. Curtis Marbut had developed ideas on soil distribution in the USA, and he assumed that the principles explaining the soil formation and geography in the USA could explain soil distribution in Africa. He used rainfall, vegetation, and parent material to make a 1 to 25 million soil map that contained 16 map units, such as Chernozems, desert soils, laterites, and alluvium. Based on the soil map, a land classification map was made at a scale of 1 to 10 million which showed agricultural productivity potential with ratings from very high to very low.

In 1935, Curtis Marbut compiled a soil map of the USA that was based on hundreds of detailed soil surveys. The map introduced the Pedocal and Pedalfer boundary that was placed partly parallel to the 98 meridian and followed the 750 mm of yearly rainfall boundary. Pedocals and Pedalfers divided the country in a humid and an arid or semi-arid region and was more refined than Milton Whitney's distinction some 25 years earlier [44]. It was Curtis Marbut's most elaborate effort in which he cited the works from Eugene Hilgard, Edward Blanck, Konstantin Glinka, Vasily Dokuchaev but also from young pedologists like Hans Jenny, Victor Kovda, and Charles Kellogg. Terms such as Podzols and Chernozems were introduced and he used the concept of the Normal soil that had been introduced by Vasily

Dokuchaev but had been replaced by Zonal soils by Nikolai Sibirtsev, and by Dynamomorphic soils by Konstantin Glinka. In 1914, Glinka surmised that extensive areas of Chernozems would occur in the Dakotas, Nebraska, and Texas, whereas the soils of Mississippi would be similar to those along the Amu river at the border between Russia and China. Chestnut soils had not been described in the USA, but Konstantin Glinka assumed that such soils would be found west of the Chernozems [74]. Chestnut soils were dry Mollisols that had accumulated calcium carbonate in the subsoil; Curtis Marbut named those soils 'Dark-brown soils'—and indeed, they were found west of the Chernozems.

In 1927, Arthur McCall was selected to head the soils work of the newly formed *Bureau of Chemistry and Soils* that included the *Division of Soil Survey*. Arthur McCall took the place of Whitney, who had headed the bureau since its inception but, now ill, decided to devote his time writing up some research results. Arthur McCall was born in Ohio and taught in public schools until he enrolled at the Ohio State University. After obtaining his BS degree in 1900, he became a scientific assistant in the physical laboratory of the *Bureau of Soils*, and in 1904, returned to the Ohio State University to become an Assistant Professor of Agronomy. In 1916, he obtained his PhD on nutrient solutions in sand cultures from Johns Hopkins University; his entire thesis was published in the journal *Soil Science* [81]. He then became head of the Department of Soils and Geology at the University of Maryland, where he wrote a laboratory guide for determining the physical properties of soils [82]. Arthur McCall was a conservative man, but not too much trouble for Marbut who had subtle but recurrent conflicts with Milton Whitney. In 1927, Arthur McCall became the executive secretary of the *First International Congress of Soil Science*.

In the early 1900s, there was a diverse community of soil surveyors across the country. Many had geological backgrounds, some had some basic soil training and gained experience by mapping soils all year long: in summer in the north, in the winter in the south. Soil surveys were conducted at the national and state level by the *Bureau of Soils*, and by soil researchers from land grant colleges and agricultural experimental stations. Given the size of the country and the different techniques and ideas, the approaches and methods evolved differently, and there was a need for comparison and consensus. Apart from a few meetings that were organized by the bureau, soil surveyors lacked an independent professional organization, unlike the agronomists. In 1907, the *American Society of Agronomy* was established. Some 30 years earlier, the *Society for the Promotion of Agricultural Science* had been established for the promotion of general agricultural science, but the objective of the *American Society of Agronomy* was more specific: "...to increase and disseminate knowledge concerning soils and crops and the conditions affecting them" [83]. From the beginning, the *American Society of Agronomy* had soils members including Merritt Miller, Edward Voorhees, and Cyril Hopkins. George Coffey was President in 1909 and Jacob Lipman was the vice president. Curtis Marbut became a member in 1910, but paid his dues irregularly and became a 'lapsed member' in 1918. The society published a quarterly journal, and it encouraged critical discussions and controversial articles for they: "...will often disclose the need of additional investigations and will stimulate a higher standard of work" [84].

In 1908, the *American Society of Agronomy* established a committee on soil classification and mapping chaired by George Coffey. The committee had 15 members, including Curtis Marbut and Andrew Whitson. The purpose of the committee was to bring uniformity to soil classification and nomenclature [83]. Its first report, in 1914 [85], recognized that the climatic conditions under which the soil was formed determined its properties and its "crop producing power." Precipitation and humidity were the highest level for grouping soils with humid, semi-arid, and arid regions as the main groups. It was envisioned that dynamic agencies such as weathering, biological agents, wind, water, and glaciation should be part of the classification system. Lastly, the committee suggested that the specific characters and conditions (color, drainage, organic matter) and texture should be used as diagnostic criteria. The report was followed by a statement from the chair of the committee, George Coffey, who found that too much attention was given to the factors that produce differences and too little on the soil differences themselves [85]. Other committee members differed in opinion on various aspects and there was little agreement among the members. The committee was terminated.

The next year, George Coffey suggested that a new committee should be formed with only five members who should have experience in classifying and mapping soils in the field. That committee was formed a year later with Curtis Marbut as chair; its sole focus became soil classification. The committee emphasized physical examination of soils in the field and laboratory, as well as chemical analyses, climatic variation, geologic origin and structure, and physiography of the region of occurrence. The committee wrote a report in 1916, which provided detailed instructions for examining each of these, but concluded: "...that an attempt to define soils on the basis of differences in climate and physiography was an attempt to confuse the description of objects with their geography and environment" [86]. The report was the last of the committee on soil classification for the *American Society of Agronomy*.

An impasse had been reached, there was no consensus, and there was no theoretical framework. Too little was known about soils to conceive a system that was detailed enough to be meaningful, yet general enough to be rationally achievable and widely applicable.

Early in 1920, the *American Association of Soil Survey Workers* was formed by representatives from a number of states in the Midwest [14]. The association was formed by Andrew Whitson, who directed the Wisconsin soil survey, and Merritt Miller of Missouri. The association aimed to establish closer cooperation between state and federal soil surveys. Some state soil surveyors opposed Curtis Marbut's ideas that had been spread through the *American Society of Agronomy* committee and the *Bureau of Soils*. The *American Association of Soil Survey Workers* was started as a pressure group in opposition, and they disagreed with the increasing influence of the Russian principles of soil science and the abolishing of geological and agricultural criteria in soil classification [14]. Many state soil surveyors considered those criteria crucial for classifying soils.

The soil survey under Curtis Marbut de-emphasized interpretations that he thought should be done by others. The purpose of the survey was to get the facts about soils, classify, and map them [16]. Curtis Marbut raised the professional and scientific standing of soil survey and in 1920, he had written that the days of geological criteria in soil classification were over and that the soil was to be viewed and classified on its intrinsic properties: "The soil itself must be the object of observation and experiment and the facts obtained must be soil facts before they can be incorporated into soil science. The science of zoology was developed through the study of animals, that of botany through the study of plants, and soil science must be developed through the study of the soil" [14].

The purposes of the *American Association of Soil Survey Workers* were to establish a closer relationship among soil survey workers of the various states, to provide a medium through which there may be a free discussion of problems arising in soil survey work, and to aid in developing cooperation between the various states and the *Bureau of Soils*. All people in the USA or Canada engaged in soil survey work or interested in its utilization were invited to become members. The dues were $1 per year. In addition to an annual meeting, the association planned to hold sectional meetings as the membership increased, so that soil surveyors in all parts of the country had the opportunity to get together for conferences. The association focused on soil survey and mapping, but over time it became a scientific society in which technical papers were presented and published [26].

The *American Association of Soil Survey Workers* held its first two-day meeting in the Geology Building at the University of Chicago, in November 1920, just as the *American Society of Agronomy* had held its foundational meeting in 1907. The association had 55 members, and the meeting was opened by the President of the Association, Andrew Whitson. Warren Geib was elected the secretary and treasurer. He worked at the soil survey in Wisconsin in the summer and in the southern states in the winter. In 1915, he was the first to obtain an MS degree in soils from *Michigan Agricultural College* [87]. The first day of the meeting had several practical reports and experiences. Lee Schoenmann presented an odometer that was attached to a Ford car and that was adapted to the accurate measurement of short as well as long distances. The device became later known as a speedometer and from 1937 onwards: "…every car used in soil surveying must be equipped with this device even in areas where base maps available" [88].

Merritt Miller and Henry Krusekopf, from Missouri, discussed the utilization of soil survey and stressed that the techniques used in soil survey were under development. They made the subtle case that the states could do some of the work more efficiently than the *Bureau of Soils*. The state would be better at obtaining data on the utilization of soils, whereas the classification should be the same across the country and the *Bureau of Soils* should coordinate this effort. They realized that soil was not limited by state boundaries, and differences between survey methods in states caused discrepancies. Merris McCool from Michigan spoke about the need for soil laboratories and what types of analyses were needed in relation to soil moisture and soil solubility. Lastly, Curtis Marbut spoke about the soil reconnaissance survey work in the Great Plains where he had spent several years making soil observations. Two soil characteristics were used to group the soils: soil color and the depth to carbonate accumulation. The grouping was as follows: black soils with carbonates at 63 cm, very dark brown soils with carbonates at 45 cm, chestnut brown soils with carbonates at 38 cm, dark brown soils with carbonates 30 cm, brown soils with carbonates at 23 cm, and lastly, light brown soils with carbonates at about 15 cm. The foundation was laid for the Pedocals that Eugene Hilgard had termed as soils with climatically conditioned lime horizons [42].

At the business meeting, the *American Association of Soil Survey Workers* established committees on Color Standardization, Information and Publicity, and a Committee to prepare a statement to be used for states interested in soil survey. Merritt Miller was elected as new President, and the next meeting was planned for November 1921, in East Lansing, Michigan. The head of the Soils Department Merris McCool was charged with organizing

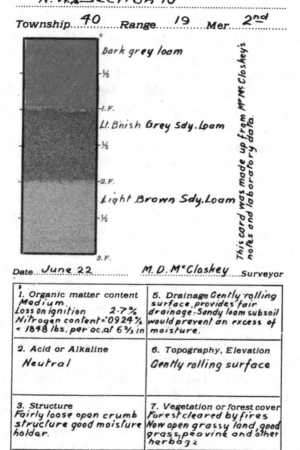

SOIL PROFILE

Cross section of type found in

N. W.¼ Section 10

Township 40 Range 19 Mer 2nd

Dark grey loam

-½

-1.F.

Lt. Bnish Grey Sdy. Loam

-½

-2.F.

Light Brown Sdy. Loam

-½

3.F.

This card was made up from Mr Mc Closkey's notes and laboratory data.

Date June 22   M. D. McCloskey Surveyor

1. Organic matter content Medium. Loss on ignition 2·7% Nitrogen content=·0924% < 1898 lbs. per ac. at 6⅔ in	5. Drainage Gently rolling surface, provides fair drainage. Sandy loam subsoil would prevent an excess of moisture.
2. Acid or Alkaline Neutral	6. Topography, Elevation Gently rolling surface
3. Structure Fairly loose open crumb structure good moisture holder.	7. Vegetation or forest cover Forest cleared by fires Now open grassy land, good grass, pea vine and other herbage
4. Texture Very fine texture (See analysis, over)	8. Remarks Forest fires have seriously depleted organic matter.

Comments: Provisional, Melfort Series
This soil would be Melfort Loam, but the matter has not yet gone far enough to give definite name
Place further comment on other side
We have not sufficient profiles to warrant the naming
H.

Mini soil profile and description published in 1924 by A.H. Hawkins from the University of Saskatchewan in the *American Soil Survey Association Bulletin* (Vol. 5: 160–176). These profiles were used in the field to distinguish soils and their horizons: "These graphic representations enable one much more readily to determine the soil horizons in any locality, than a mass of figures and descriptions, depths and other information, and a soil profile of an area may be drawn giving a very considerable amount of valuable information regarding the locality or the type in which it is taken"

the meeting. The Association published a Bulletin after the first meeting: over 5,000 copies were printed, and these were sent in bulk to 36 soil workers for distribution, and complimentary copies went to numerous others and overseas. Some 2,150 copies were sold. In 1921, there were 136 members in 34 states, 8 Canadian Provinces as well as The Philippines. The Association then held annual meetings in Urbana, Illinois in 1922, where the first Russian soil researcher participated [7], and in Chicago, in 1923, where 85 people attended, including Milton Whitney for the first time. At the meeting in Chicago, the association changed its name to the *American Soil Survey Association.* In 1925, Merris McCool became the President.

The *American Soil Survey Association* was the bridge between the soil survey work at the state level and the *Bureau of Soils.* It created agreements on common terminology and sought harmonization of methods and soil survey approaches. In the beginning, it dealt with soil variation, laboratory methods, soil acidity and its measurement, and the use of airplanes and aerial photography in soil survey. The association contributed to the development of soil science, and in particular, to soil survey in the USA. The 17 volumes of Bulletins published by the association contained a considerable number of original papers. The association was dominated by people from Wisconsin (Andrew Whitson and Warren Geib), Missouri (Henry Krusekopf, Merritt Miller, and later Frank Duley and William Albrecht), Michigan (Jethro Veatch and Merris McCool), as well as J.C. (Jouette) Russel from Nebraska, and Harlow Walster from North Dakota.

Even though the association was formed to counter the dominance of the ideas from Curtis Marbut and the initial resistance came from his former Missouri colleague Merritt Miller, and from Andrew Whitson from Wisconsin, once Curtis Marbut started attending the meetings, their criticism faded. Eventually, Curtis Marbut came to lead the association, such was the nature of his intellect and personality [14]. Over time, Andrew Whitson saw the value of the changes that Curtis Marbut brought to the soil survey and at a special recognition dinner of the *American Soil Survey Association,* he had the honor of presenting a gold watch to Curtis Marbut, who appreciated the gesture but forgot to wear the watch as it seemed unimportant compared to the approval of his fellows for the work he was doing [89].

In 1935, plans were made to unite the *American Soil Survey Association* with the Soils Section of the *American Society of Agronomy*, and this idea was supported by Curtis Marbut [73]. Some members were concerned about duplication of efforts and there was difficulty merging the two organizations. The *American Soil Survey Association* largely consisted of soil surveyors and there was a strong bond between them; not many were interested in being part of an agronomy society [26]. The Soils Section of the *American Society of Agronomy* was chaired by William Albrecht of Missouri, and he suggested that Richard Bradfield and Charles Shaw drafted a constitution for the new organization that was to be named the *American Society of Soil Science*. Emil Truog and Mark Baldwin were tasked with formulating an editorial policy for the journal that this new society would publish. The *Soil Science Society of America* name was proposed by Emil Truog who preferred a name that was like the *American Society of Agronomy*. The *American Society of Soil Science* was considered but found less suitable because of its stroppy acronym. The *Soil Science Society of America* was established in 1936 [90, 91]. Its first president was Richard Bradfield, who had just taken a position at his *alma mater*, the University of Ohio. William Albrecht was secretary *pro tempore*. The *American Soil Survey Association* had 485 members when it transformed to the *Soil Science Society of America*. It had many new members from people hired by the *Soil Conservation Service*.

References

1. Krupenikov IA. History of soil science from its inception to the present. New Delhi: Oxonian Press; 1992.
2. Kellogg CE. Soil genesis, classification, and cartography: 1924–1974. Geoderma. 1974;12:347–62.
3. Brevik EC, Hartemink AE. Early soil knowledge and the birth and development of soil science. CATENA. 2010;83(1):23–33.
4. Chamberlin TC. General map of the soils of Wisconsin. Beloit: Wisconsin Geological Survey; 1882.
5. Hartemink AE, Lowery B, Wacker C. Soil maps of Wisconsin. Geoderma. 2012;189–190:451–61.
6. Chamberlin TC. Geology of Wisconin. Survey of 1873–1877. Volume II. Beloit: Commissioners of Public Printing; 1877.
7. Moon D. The American steppes. The unexpected Russian roots of Great Plains agriculture, 1870s–1930s. Cambridge: Cambridge University Press; 2020.
8. Chamberlin TC. Soil productivity. Science. 1909;33:225–7.
9. King FH. Farmers of forty centuries or permanent agriculture in China, Korea and Japan. Emmaus, Pennsylvania: Republished by: Rodale Press, Inc.; 1911.
10. Brevik EC, Hartemink AE. Soil maps of the United States of America. Soil Sci Soc Am J. 2013;77(4):1117–32.
11. Whitney M. The soils of Maryland. College Park, MD: Mary land Agricultural Experiment Station. Bulletin no. 21; 1893.
12. Vanderford CF. The soils of Tennessee. Knoxville: University of Tennessee. The Agricultural Experiment Station; 1897.

13. Amundson R, Yaalon DH. Hilgard, E.W. and Powell, John, Wesley—efforts for a joint agricultural and geological survey. Soil Sci Soc Am J. 1995;59(1):4–13.
14. Gardner DR. The national cooperative soil survey of the United States (Thesis presented to Harvard University, May 1957). Washington, DC: USDA; 1957.
15. Helms D, Effland ABW, Phillips SE. Founding the USDA's division of agricultural soils: Charles Dabney, Milton Whitney and the state experiment station. In: Helms D, Effland ABW, Durana PJ, editors. Profiles in the history of the U.S. soil survey. Ames: Iowa State Press; 2002. p. 1–18.
16. Helms D, Effland ABW, Durana PJ, editors. Profiles in the history of the U.S. soil survey. Ames: Iowa State Press; 2002.
17. Kellogg CE, Knapp DC. The college of agriculture. Science in the public service. New York: McGraw-Hill Book Company; 1966.
18. Williams CB. How the soil survey is proving most valuable to North Carolina. J Am Soc Agron. 1924;16:447–51.
19. Maher D, Stuart K, editors. Hans Jenny—soil scientist, teacher, and scholar. Berkeley: University of California; 1989.
20. Glinka K. Die Typen der Bodenbildung - Ihre Klassification und Geographsiche Verbreitung. Berlin: Verlag von Gebrüder Borntraeger; 1914.
21. Simonson RW. Historical aspects of soil survey and soil classification. Part III. 1921–1930. Soil Surv Horiz 1986;27(11–14).
22. Simonson RW. Historical aspects of soil survey and soil classification. Part II. 1911–1920. Soil Surv Horiz 1986;27:3–9.
23. Simonson RW. Historical aspects of soil survey and soil classification. Part I. 1899–1910. Soil Surv Horiz. 1986;27:3–11.
24. Weber GA. The Bureau of Chemistry and Soils. Its history, activities and organization. Baltimore, Maryland: The Johns Hopkins Press; 1928.
25. Whitney M. First four thousand samples in the soil collection of the Division of Soils. Washington: U.S. Department of Agriculture. Bulletin no. 16; 1899.
26. Miller MF. Progress of the soil survey of the United States Since 1899. Soil Sci Soc Am Proc. 1950;14(C):1–4.
27. Lawes JB, Morton JC, Morton J, Scott J, Thurber G. The soil of the farm. Washington: Orange Judd; 1883.
28. Amundson R. Philosophical developments in pedology in the United States: Eugene Hilgard and Milton Whitney. In: Warkentin BP, editor. Footprints in the soils. People and ideas in soil history. Amsterdam: Elsevier; 2006. p. 149–166.
29. Whitney M, Gardner FD, Briggs LJ. An electrical method of determining the moisture content of arable soils. Bulletin no. 6. Washington: U.S. Department of Agriculture. Divisions of Soils; 1897.
30. Whitney M, Means TH. An electrical method of determining the soluble salt content of soils. Bulletin no. 8. Washington: U.S. Department of Agriculture. Divisions of Soils; 1897.

31. Whitney M, Briggs LJ. An electrical method of determining the temperature of soils. Bulletin no. 7. Washington: U.S. Department of Agriculture. Divisions of Soils; 1897.

32. Russell EJ. Obituary. Prof. Milton Whitney. Nature. 1928;121(3036):27.

33. Whitney M. The work of the bureau of soils. Bureau of Soils Circular no. 13; 1904.

34. Helms D. Early leaders of the soil survey. In: Helms D, Effland ABW, Durana PJ, editors. Profiles in the history of the U.S. soil survey. Ames: Iowa State Press; 2002. p. 19–64.

35. Nimmo JR, Landa ER. The soil physics contributions of Edgar Buckingham. Soil Sci Soc Am J. 2005;69(2):328–42.

36. Brevik EC. Collier Cobb and Allen D. Hole: geologic mentors to early soil scientists. Phys Chem Earth. 2010;35(15–18):887–894.

37. Tandarich JP. Wisconsin agricultural geologists: ahead of their time. Geosci Wisconsin. 2001;18:21–6.

38. Shaw CF. A preliminary field study of the soils of China (from the "contributions to the knowledge of the soils of Asia, 2" compiled by the Bureau for the Soil Map of Asia. Leningrad: Publishing office of the Academy of Sciences of the USSR; 1933.

39. Thorp J. Geography of the soils of China. Nanking: The National Geological Survey of China; 1936.

40. Bennett HH, Rice TD. Soil reconnaissance in Alaska, with an estimate of agricultural possibilities. Washington: Bureau of Soils; 1915.

41. Bennet HH, Allison RV. The soils of Cuba. Washington, DC: Tropical Plant Research Foundation; 1928.

42. Jenny HEW. Hilgard and the birth of modern soil science. Pisa: Collana della revista agrochemica; 1961.

43. McGee WJ. Soil erosion. Washington: U.W. Department of Agriculture; 1911.

44. Whitney M. Soils of the United States. Bureau of Soils Bulletin no. 55. Washington, DC: Government Printing Office; 1909.

45. Whitney M. Soil and civilization: a modern concept of the soil and the historical development of agriculture. New York: D. van Nostrand Company, Inc.; 1925.

46. Anon. Prof. Milton Whitney, soil scientist, dead. Offi Rec. United States Depart Agric. 1927;VI(46):1 and 5.

47. Whitney M. Letter of submittal. Fertilizer resources of the United States. Message from the President of the United States. Washington; 1912. p. 7–8.

48. Kelley WP. Dennis Robert Hoagland 1884–1949. A biographical memoir. Washington, DC: National Academy of Sciences; 1956.

49. McCracken RJ, Helms D. Soil survey and maps. In: McDonald P, editor. The literature of soil science. Ithaca: Cornell University Press; 1994. p. 275–311.

50. Whitney M. Instructions to field parties and descriptions of soil types. Field season 1903. Washington, DC: U.S. Department of Agriculture. Bureau of Soils; 1903.

51. Holman HP, Pease VA, Smith K, Reid MT, Crebassa A. Index of publications of the Bureau of Chemistry and Soils. 75 years 1862–1937. Washington, DC: U.S. Department of Agriculture; 1939.
52. Lapham MC. Crisscross trails: narrative of a soil surveyor. Berkeley: W.E. Berg; 1949.
53. Anon. Important soils of the United States Washington: U.S. Department of Agriculture. Bureau of Soils. Government Printing Office; 1916.
54. Marbut CF, Bennet HH, Lapham JE, Lapham MH. Soils of the United States (edition, 1913). Washington: Government Press Office; 1913.
55. Russell EJ. The recent work of the American Soil Bureau. J Agric Sci. 1905;1:327–46.
56. Lipman CB. Eugene Woldemar Hilgard. J Am Soc Agron. 1916;8:160–2.
57. Tandarich JP, Sprecher SW. The intellectual background for the factors of soil formation. In: Factors of soil formation: a fiftieth anniversary retrospective. Madison, WI: Soil Science Society of America; 1994. p. 1–13.
58. Tanner CB, Simonson RW. King Franklin Hiram—pioneer scientist. Soil Sci Soc Am J. 1993;57(1):286–92.
59. King FH. The soil: its nature, relations, and fundamental principles of management. New York: The MacMillan Company; 1895.
60. Pendleton RL. Are soils mapped under a given type name by the Bureau of Soils method closely similar to one another? University of California Publications in Agricultural Science. 1919;3:369–498.
61. Whitney M, Cameron FK. The chemistry of the soil as related to crop production. Bureau of Soils—Bulletin no. 22. Washington, DC: U.S. Department of Agriculture; 1903.
62. Moores RG. Field of rich toil. The development of the University of Illinois College of Agriculture. Urbana: University of Illinois Press; 1970.
63. Hopkins CG. Soil fertility and permanent agriculture. Boston: Ginn and Company; 1910.
64. Hopkins CG. The story of the soil—from the basis of absolute science and real life. Boston: Richard G. Badger; 1911.
65. Hopkins CG. European practice and American theory concerning soil fertility. Agricultural Experiment Station. Circular no. 142. Urbana: University of Illinois; 1910.
66. Russell EJ. Dr. Cyril G. Hopkins. Nature. 1920;104(2618):442–443.
67. Coffey GN. A study on the soils of the United States. Bureau of Soils—Bulletin no.85. Washington: Government Printing Office; 1913.
68. Brevik EC. George Nelson Coffey, early American pedologist. Soil Sci Soc Am J. 1999;63(6):1485–93.
69. Sibirtsev NM. Pochvovedenie. St. Petersburg: Y.N. Skorokhodov; 1900.
70. Anon. Sir John Russell pays tribute. Offi Rec. United States Depart Agric. 1927;VI(52):4.
71. Rice TD. C.F. Marbut. Life and works of C.F. Marbut. Madison: Soil Science Society of America; 1936. p. 36–48.

72. Marbut-Moomaw L. Curtis Fletcher Marbut. Life and works of C.F. Marbut. Madison: Soil Science Society of America; 1936. p. 11–35.

73. Krusekopf HH, editor. Life and works of C.F. Marbut. Madison: Soil Science Society of America; 1936.

74. Glinka KD. The great soil groups of the world and their development (translated from German by C.F. Marbut in 1917). Ann Arbor, Michigan: Mimeographed and Printed by Edward Brothers; 1928.

75. Thorp J. Impressions of Dr. Curtis Fletcher Marbut, 1921–1935. Soil Horiz. 1985;26(1).

76. Helms D. Hugh Hammond Bennett and the creation of the soil conservation service. J Soil Water Conserv. 2010;65:37A–47A.

77. Nikiforoff CC. Fundamental formulation of soil formation. Am J Sci. 1942;240:847–66.

78. Neustreuv SS. Genesis of soils. Leningrad: USSR, Publishing Office of the Academy; 1927.

79. Marbut CF. Soils of the United States. Part III. In: Baker OE, editor. Atlas of American agriculture. Washington: United Sates Department of Agriculture. Bureau of Chemistry and Soils; 1935. p. 1–98.

80. Shantz HL, Marbut CF. The vegetation and soils of Africa. New York: National Research Council and the American Geographical Society; 1923.

81. McCall AG. Physiological balance of nutrient solutions for plants in sand cultures (Dissertation submitted to the board of Johns Hopkins University). Soil Sci. 1916;II:207–53.

82. McCall AG. The physical properties of soils—a laboratory guide. New York: Orange Judd Company; 1909.

83. Anon. Business section. Proc Am Soc Agron. 1908;I:6–15.

84. Anon. Controversial articles. J Am Soc Agron. 1913;5:122.

85. Coffey GN. Progress report of the committee on soil classification and mapping. J Am Soc Agron. 1914;6:284–8.

86. Marbut CF. Report of the committee on soil classification. J Am Soc Agron. 1916;6:387–90.

87. Robertson LS, Whiteside EP, Lucas RE, Cook RL. The soil science department 1909–1969—a historical narrative. East Lansing: Michigan State University; 1988.

88. Kellogg CE. Soil survey manual. Miscellaneous Publication no. 274. Washington, DC: Department of Agriculture; 1937.

89. Simonson RW. Historical aspects of soil survey and soil classification. Part IV. 1931–1940. Soil Surv Horiz. 1987;27:3–10.

90. Hartemink AE, Anderson S. 100 years of Soil Science Society in the U.S. CSA News. 2020;65:26–27.

91. Tandarich JP, Darmody RG, Follmer LR, Johnson DL. Historical development of soil and weathering profile concepts from Europe to the United States of America. Soil Sci Soc Am J. 2002;66(4):1407.

8

Of Soils and Men

*"I can say frankly when I came into the Department there was no soil scientist
who had entree to the Secretary's office, and soil data were not considered consistently
in any phase of the Department's program. This condition is now changed."*
*"After all we are working in the interests of the farmers, or should be at least,
and we are spending the money of the American people."*

Charles Kellogg in a letter to Emil Truog, October 1937

Charles Kellogg worked for four years at *North Dakota Agricultural College*
in Fargo. For several years, he had looked for other positions but none were
appealing or had a high enough salary. The country was in an economic crisis
and several of his colleagues had left North Dakota or had been dismissed
because of budget cuts. Emil Truog sent him vacancies when he came across
any—there were not many. Charles Kellogg's work had been noticed by the
Division of Soil Survey where he was seen as progressive, and more importantly, as a supporter of the ideas of the chief, Curtis Marbut. In July 1933,
the *Division of Soil Survey* started creating a position for him, and it was
approved in December.

In February 1934, Charles Kellogg became assistant to Curtis Marbut at
the *Division of Soil Survey* in Washington. Charles Kellogg was pleased with
the position and thought he would stay for a few years and then return
to a university. In 1934, Curtis Marbut was working on the soils chapter
of the *American Atlas of Agriculture* [1]. It was a monumental task summarizing 25 years of soil surveys across the country and including 200 references
that Curtis Marbut named *Pedological Literature* citing everyone who shaped
pedology between the early 1900s and 1932. With the exceptions of Vasily
Dokuchaev and Nikolai Sibirtsev, Curtis Marbut had met them all [2]. The
chapter became 200 pages with descriptions of the soils of over 1,500 counties. It introduced the Great Soil Groups at a map with a scale of 1–8 million,
and the *Pedocals,* in which calcium carbonate accumulated, and *Pedalfers*

© The Author(s), under exclusive license to Springer Nature
Switzerland AG 2021
A. E. Hartemink, *Soil Science Americana*
https://doi.org/10.1007/978-3-030-71135-1_8

A man in a suit—Charles Kellogg in 1935, as Principal soil scientist, Division of Soil Survey, Bureau of Chemistry and Soils. Photo Special collection National Agricultural Library

in which it did not. Curtis Marbut also used Russian soil names such as Chernozems and Podzols.

Pedocals had been coined by Eugene Hilgard to describe soils with climatically conditioned lime horizons; they were also termed as Pedocalic groups and included black earths and chestnut earths [3]. Pedalfers were defined as soils with clayey or sesquioxide-enriched subsoils, also termed Pedalferic Groups, and included gray-brown earths, ferruginous laterites, and prairie earths [3]. In later years, Eugene Hilgard named Pedocals the non-lixiviated soils of the arid regions and Pedalfers the lixiviated soils of the humid regions [4]. The grouping of soils according to climatic factors and subsoil properties became a key of all subsequent systems of soil classification in the USA.

There was not much time for Charles Kellogg to get used to his new role in Washington, as Curtis Marbut died in August 1935. Publicly, there was uncertainty about whom should succeed him. Merris McCool from Michigan, who had been a student of Curtis Marbut at the University of Missouri and an early adopter of his ideas on soil survey, had applied for the position in 1933 when Curtis Marbut was about to retire [5]. But Curtis Marbut did not consider him suitable. Merris McCool had left Michigan in the late 1920s and worked in Plant Physiology at the *Boyce Thompson Institute* and Curtis Marbut thought that he had been out of soils too long.

In October 1935, Curtis Marbut's urn was buried next to Florence and a memorial service was held at his newly built home in the Ozarks. The building had started before he traveled to the UK and China and was hastily finished to prepare for the memorial service [6]. Hans Jenny, who was at the University of Missouri, took part in the service [7]. The living room had ceiling-high book cases that contained Curtis Marbut's lecture notes and books. The walls were covered with soil maps from the 1935 Atlas [1]. Hans Jenny also had the maps framed and hung in his office, and he found it: "…always a pleasure to look at it and appreciate the broad soil patterns." [7]. After the memorial service, there was a discussion about who might be Curtis Marbut's successor. The soil scientists from the University of Missouri thought it would be Merritt Miller, who had been a professor of agronomy and soils since 1904, had worked with Curtis Marbut in Missouri, and had the leadership skills, background, and scientific credentials. He had co-founded the *American Association of Soil Survey Workers* in 1920, but he did not support the Russian principles of soil science that Curtis Marbut had espoused; in his 1924 book *The soil and its management* [8], he reproduced Milton Whitney's outdated soil province map so he was not considered for the position.

At the gathering in Missouri, Arthur McCall announced that Charles Kellogg had been chosen to succeed Curtis Marbut. This surprised quite a few, as Charles Kellogg was not well-known, only 32 years old, and had been assistant to the Chief for just a year. It puzzled Hans Jenny by what "strategy and manipulation" he was selected [7]. Charles Kellogg had participated in soil survey interpretations for highway construction, tax assessment, and land classification [9]; he had worked in the glaciated landscapes of Michigan, Wisconsin, and North Dakota in a range of soils—but that was about it. Importantly, he had embraced Curtis Marbut's ideas that had shifted the soil survey to a pedological approach. Charles Kellogg's studies in North Dakota had attracted the attention of Curtis Marbut. They had met several times,

Charles Kellogg and Curtis Marbut in front of the Great Soil Group map in the Spring of 1935. Curtis Marbut pointing at Springfield: "Now son, you ought to remember who bought a house there in 1844, he was about your age when he bought it. Some 19 years later, I was born 360 miles from there, and 200 miles from the Pedalfers and Pedocals divide. Please remember that too." Photo Special Collection National Agricultural Library

and Charles Kellogg had become his assistant in 1934 [10]. Curtis Marbut had handpicked his successor: it turned out to be quite a superb pick.

In 1935, Charles Kellogg was appointed Principal Soil Scientist and Chief, *Division of Soil Survey*, Bureau of Plant Industry, U.S. Department of Agriculture. He was the third Chief of the Division, following Milton Whitney and Curtis Marbut. Upon his retirement in May 1971, he had 36 years of service and became the longest serving chief of the Division. Milton Whitney established the soil survey, largely expanded the organization, but had a narrow focus. Curtis Marbut, a solitary man, cautiously led the soil survey out of its geological base into an organization that embraced pedology. The change in the organization mirrored his own intellectual development and his influence went beyond the nation, it was truly international.

Charles Kellogg inherited an organization at a crossroads. It could have continued its course in which quantity was somewhat favored over quality but, instead, Charles Kellogg brought science into soil survey, introduced new techniques such as the use of aerial photography, aimed for standardization in soil survey and classification, and worked with soil surveyors across the country and the globe. He delegated authority, decentralized the organization, and held regular staff meetings—none of that happened under Curtis Marbut [10]. Charles Kellogg instituted scientific standards in soil correlation, used funds for basic soil research and as travel funds for soil correlators [10], but his biggest challenge was not the scientific direction of the soil survey but its very survival in the face of rivalry with the *Soil Conservation Service*, established shortly after his arrival in Washington. It was a colossal task for a 33-year-old ambitious soil scientist who was not well-traveled, lacked international experience, and was barely known across the country.

Charles Kellogg arrived in Washington in the middle of the Great Depression, and in the early days of the New Deal. Besides a profound economic crisis, the country experienced some of the driest years on record and crops were failing, unemployment was high, and poverty spreading. President Franklin Roosevelt, who had been elected in late 1932, crafted an ambitious plan to pull the country out of the crisis. The early days of the New Deal comprised a series of programs aimed to restore prosperity, stabilize the economy, and provide jobs and relief to the poor. Charles Kellogg admired the actions of the President and Henry Wallace, the Secretary of Agriculture who was responsible for the start of the Tennessee Valley Authority [11]. This new agency was set up to tackle a slew of problems such as flooding, soil erosion, and a struggling economy. The Tennessee Valley Authority aimed to develop the region's farming, provide electricity to villages, farms, and businesses, and replant forests—the first government agency to address the development of a large region. Henry Wallace transformed the Department of Agriculture from an agrarian institute to an instrument of public policy on agriculture [12].

The New Deal era was a time of big government programs. It appealed to Charles Kellogg, who believed that the principal aim of government programs was to help people and that a big country needs big government programs to advance welfare. Coming from a rural area, and Fargo that had about 30,000 inhabitants, he enjoyed a bustling Washington, which in 1935, had a population of over 1.5 million. Although it was a comfortable and conservative southern American city, it was not a center of art, had no lively theaters, and only a few good restaurants; streets were empty by ten o'clock at night [11]. It was a center of civil servants, Charles Kellogg was one of them—driven to make the soil survey and the New Deal a success.

In his first few years at the *Division of Soil Survey*, he churned out several publications, starting with a review on soil erosion. Curtis Marbut had not been interested in soil erosion, considered it a local problem, and had a geologic view: "...for the world as a whole, erosion is unimportant." [10].

Within the soil survey, soil erosion had received some attention as Milton Whitney had created a Soil Erosion Investigation unit in 1907, but ever since soil erosion was not high on the list of issues to work on. That all changed in March 1935, when a dust storm covered Washington in a yellow haze [13]. Reports of storms had come from various states and were all over the news. The soil scientist Frank Duley send a telegram to the *Soil Erosion Service* announcing that dust storms were an almost daily occurrence in Western Kansas [13].

A human tragedy unfolded and the *Division of Soil Survey* had to do something. Charles Kellogg's first assignment was to review soil erosion by wind, previously reviewed in 1894, by Franklin King [14]. Charles Kellogg wrote a slim report entitled *Soil blowing and dust storms* [15]. Most of the windblown erosion was from wheat growing areas, triggered by a severe drought that plagued the Dakotas, western Minnesota, and extended into Kansas, east to Ohio, and as far west as central Montana and Colorado. The major harm was damage to the newly planted crops, loss of the fertile topsoil, and the accumulation of windblown soil against buildings and on highways. The cause was a combination of a long dry spell with intense soil tillage, and Charles Kellogg stressed that the moldboard plow was not always a desirable implement. He recommended strip farming but noted: "The increased severity of the dust storms, however, is much more nearly correlated with the weather conditions than with the percentage of land under cultivation." Shelter belts were discussed that were common in the Russian steppes and had been introduced in the USA by the forester Raphael Zon, a schoolmate of Vladimir Lenin. Raphael Zon had emigrated to the USA, in 1898, and became friends with Curtis Marbut [16].

Charles Kellogg recommended that appropriate tillage, strip farming, and planned land use were prime measures to control soil blowing; the term 'soil erosion' occurred nowhere in his paper [15]. Charles Kellogg certainly considered soil blowing and dust storms a serious problem, but he had now trodden into the territory of Hugh Bennett, who was to become the doyen of soil erosion control. While working at the soil survey for the *Division of Soil Survey*, Hugh Bennett had published a report in 1928 entitled *Soil Erosion— A National Menace* [17]. Charles Kellogg listed it in his review, but without naming Hugh Bennett, who, in Charles Kellogg's view, was writing: "...dramatically about the growing soil problem." Soil exhaustion, a term commonly used in the 1800s, was replaced in the 1920s by an ominous new threat: soil erosion—the most visible evidence of soil exhaustion.

Shortly after *Soil blowing and dust storms* was published, Charles Kellogg wrote an opinion piece in the New York Times: *Men turn their thoughts to the soil,* with the sub-title: *Uncontrolled exploitation here and in Europe has led to a new attitude towards conservation and production* [18]. He noted that 30 years earlier, people in the USA did not think about failure and everyone was willing and anxious to take a chance. But when individual ruination grew into community ruination because of soil erosion, the nation could not ignore its consequence; a new way of thinking towards rational use was needed. Society, especially the next generation, would have to pay for all the failures. Therefore, the type of soil should be examined, its capabilities should be determined, and Charles Kellogg concluded: "The people of the United States with their tremendous resources of soil, water, forests and minerals, have all the necessary natural requirements for a varied but secure social life on the highest level known to any people." [18]. This was a strong plea for soil survey, land classification, and land-use planning.

The soil survey was Charles Kellogg's main assignment, interest, and passion, but the soil survey had developed jargon and he realized that clarifying information was needed for those that were not directly involved in the soil survey. In 1936, he summarized the knowledge and significance of the Great Soil Groups that had been introduced by Curtis Marbut to group soil series in geographic units smaller than the soil province [19]. He introduced soils and the Great Soil Group distribution to the general reader and provided an overview of some important soil properties, soil processes, and the significance of soil groups and soil types. Soil genesis was introduced by a formula combining the individual factors that had been unraveled by Vasily Dokuchaev some 50 years earlier [20]: Soil = f (climate, vegetation, relief, age, parent rock)

Processes that formed soils were either *destructional* such as physical and chemical weathering, or *constructional* including biological forces that Charles Kellogg considered to be part of pedology. Seven major soil processes were reviewed: calcification, podzolisation, laterization, salinization, solonization, solodization, and gleization. Based on his work in North Dakota, he developed pedogenetic pathways for arid areas that showed soil evolution as affected by soil processes, erosion, drainage, and vegetation. Normal soils could evolve into soils high in salt such as Solonchaks, soils high in sodium such as Solonetz, or into more acid and leached Soloths. The processes would start with normal soils such as Chernozem or brown soil, or the processes began on weathered parent material such as alluvium. Normal soils represented a soil where the rates of soil erosion were in equilibrium with soil formation. The concept of normal soils had been introduced by Vasily Dokuchaev as soils that had not been changed by dynamic processes other than soil-making processes. Charles Kellogg's pathway of soil genesis,

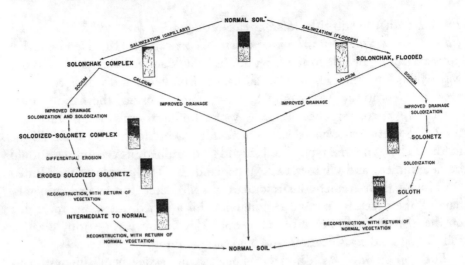

Charles Kellogg's outline of the cycle of soil evolution. Normal soils turn into Solonchaks, Solonetz and Soloths as affected by drainage, erosion or vegetation. The processes may start with normal soils such as Chernozem or brown soil, or the processes may begin on weathered parent material such as alluvium. Normal soil was a term introduced by Vasily Dokuchaev, and used by Curtis Marbut. It represented a soil where the rates of soil erosion was in equilibrium with soil formation. From: *Development and significance of the great soil groups of the United States* published in 1936. Note the use of Russian terms

although focused towards the drier areas, was an innovative way of thinking about soil changes and evolution.

In his early years at the *Division of Soil Survey*, Charles Kellogg focused on tools to improve and advance soil survey and its scientific basis. Since his involvement with soil survey in Michigan in the early 1920s, he considered soil survey more an art than a science. During the Michigan soil surveys, there was no standard vocabulary; different words were used for describing soil horizons and layers were labeled surface soil, subsoil, and substratum [21]. Soil horizon designations were not used, nor did they use the field guides that had been published by the *Bureau of Soils* since 1903 [22, 23]. Three years into his position at *Division of Soil Survey*, Charles Kellogg published the first edition of the *Soil Survey Manual* [24]. The manual was much more detailed than the earlier field guides, but the main difference was the focus on the soil profile as a unit of observation and as a record of soil processes and properties.

Soil profiles were recognized and described before the late 1800s but the scientific study of soil profiles did not occur until the 1870s [25, 26]. The pedologist from Rutgers University, Jacob Joffe, had viewed this in 1929 as: "…the new concept of soils as an independent, natural, historical body which required not only the description of the surface features of soil but also the anatomy of it; for this it is necessary to cut a vertical section and thus obtain a profile view of the exposed vertically dissected body." [26]. Charles Kellogg considered the recognition of a unique soil profile for each kind of landscape the greatest single advance ever made in fundamental soil science. It was analogous to the development of anatomy in medicine, which Curtis Marbut had aptly called 'soil anatomy.' [27]. The 1937 *Soil Survey Manual* provided instructions for a detailed description of soil horizons and the landscape in which it was found. For monolith sampling and preparation, the manual recommended the methods used by Konstantin Glinka [28].

Whereas Curtis Marbut largely avoided soil interpretations, Charles Kellogg promoted the interpretation of soil survey data with the overall purpose to assist people and the management of the land [9]. He had learned this from the soil surveys in Wisconsin, North Dakota, and in particular, in Michigan. Soil interpretation for programs of rural land use was close to his heart, as was the relationships between soils and land use. Soil survey interpretation became an integral part of the survey, and according to Charles Kellogg, only pedologists could take on that task: "…pedologists themselves must devote more attention to summarizing the results of their investigations and making them readily available for public use. The use of soil survey data in land classification and land-use planning is essential, but to be of maximum value, pedologists themselves must make the application." [29]. Charles Kellogg worked with Ken Ableiter, and they established interpretation guidelines named *A method of rural land classification* which were adopted in most states, with the exception of California where a different productivity index for rating soils was developed [10, 30, 31].

The subject of soil interpretation was not new to the *Division of Soil Survey*. Under Milton Whitney, Jay Bonsteel had been in charge of soil survey interpretation in an organizational unit dealing with the use of soils in the early 1910s. They published some reports that dealt with the use and management characteristics of particular soil types. Under Curtis Marbut, soil erosion and soil interpretations had received less attention; perhaps, deep-down, he remained a geologist and physical geographer in his soil explorations. He was born and raised on a farm, but his soil investigations had no direct agrarian usage in mind. But the 1930s was a new time, and the emphasis on the practical application of soil surveys was encouraged by the Secretary of Agriculture, Henry Wallace. The applications went hand-in-hand with increased scientific demands, and Charles Kellogg insisted that the soil survey and its products had to be practical but scientifically sound: "No choice can be made between a utilitarian soil survey and a scientific soil survey. Of course, soil

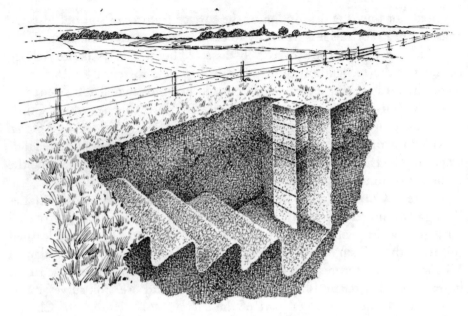

How to prepare a soil profile for sampling according to the 1937 *Soil Survey Manual* by Charles Kellogg. The explanatory text was: "Sketch showing a convenient method of preparing an excavation for collecting soil samples. The samples should be collected, so far as possible, one directly above the other."

surveys are made for predictions about land use and management—and most of them are—must be practical. But they will not be practical unless they are also scientifically sound." [32].

The 1937 *Soil Survey Manual* included a bibliography: "The following books and papers are suggested for the reference shelf," and listed the proceedings of the *First International Congress of Soil Science* and books by Eugene Hilgard, Jacob Joffe, Curtis Marbut, George Merrill, John Russell, Konstantin Gedroiz, and Konstantin Glinka. It was the beginning of a reading list for soil scientists that Charles Kellogg compiled in 1940 and updated several times [33–35]. The 1937 *Soil Survey Manual* was the first and last *Soil Survey Manual* that carried Charles Kellogg's name as sole author [24], although he acknowledged the help of Ken Ableiter, Mark Baldwin, Macy Lapham, Constantin Nikiforoff, and Thomas Rice. The authorship in subsequent manuals was referred to as *Soil Survey Staff* or *Soil Survey Division Staff* [36, 37].

It is impossible to overstate the contribution of the *Soil Survey Manual* for the subsequent evolution and application of soil surveys worldwide. The Manual has been used, adopted or translated by soil survey centers across the world, many of them commencing soil surveys decades after it was started in the USA. The manual was freely distributed and had more lasting influence than the soil classification system that eventually came with it—*Soil Taxonomy*—although that was also widely and freely distributed. American soil survey methods have proven more widely acceptable than its soil classification. Over time, the soil survey became highly detailed and somewhat complex and under Charles Kellogg's leadership, soil cartography started to use units known as associations which were not the same as the Russian soil complexes [38]. Soil associations were geographical groupings of soil series used to create small-scale maps. The soil associations put less emphasis on the genetic relationships among the soil series [39].

Towards the end of 1936, Henry Wallace asked Charles Kellogg for a comprehensive statement about the soils of the USA and their uses, to be prepared as the 1938 Yearbook. The first yearbook of agriculture had been published in 1849 [40], several years before the Department of Agriculture was formed under Abraham Lincoln. *Soils & Men* was the third yearbook dealing with a special subject, and that idea had come from Henry Wallace. *Better Plants and Animals* was published in two volumes in 1936 and 1937, and this was to be followed by the 1938 *Soils & Men* yearbook of agriculture. The rationale for the 1938 yearbook was: "…because our knowledge of soil has expanded greatly and emphasis and needs have changed, this Yearbook of agriculture is limited to the management of soil, itself a big and burgeoning subject."

Charles Kellogg expostulated that the knowledge was hardly adequate for the task, but the Secretary felt that one of the values of such a book would be that it would detail what the research needs were [41]. It became a kind of 1930s Apollo 11 program for American soil science. Charles Kellogg decided that it had to become a collaborative effort from the best soil scientists across the country. He developed a structure for the book, invited authors, compiled data, and contributed to several chapters. He asked Roy Simonson, his former student from North Dakota, to come and help edit manuscripts. Roy was a gifted writer, a PhD student under Emil Truog at the University of Wisconsin, and was happy to be working with Charles Kellogg and to be away from the rigor and attention of Emil Truog's laboratory.

Chapters in the 1938 yearbook were grouped in five sections. Charles Kellogg oversaw the section that dealt with the fundamentals of soil science and the section that described the soils of the USA. The *Soils & Men* yearbook was published two years after Henry Wallace had made the request; it contained 58 chapters, 1,232 pages, and a 44-page summary. There were hundreds of authors. It was viewed: "…by citizens and commentators to be one of the most valuable works ever issued by the Department." *Soils & Men* was dedicated to Curtis Marbut.

The Foreword was written by Henry Wallace: "…Nature treats the earth kindly. Man treats her harshly. He overplows the cropland, overgrazes the pastureland, and overcuts the timberland. He destroys millions of acres completely. He pours fertility year after year into the cities, which in turn pour what they do not use down the sewers into the rivers and the ocean. The flood problem insofar as it is man-made is chiefly the result of overplowing, overgrazing, and overcutting of timber. This terribly destructive process is excusable in a young civilization. It is not excusable in the United States in the year 1938. We know what can be done and we are beginning to do it. As individuals we are beginning to do the necessary things. As a nation, we are beginning to do them. The public is waking up, and just in time. In another 30 years it might have been too late. The social lesson of soil waste is that no man has the right to destroy soil even if he does own it in fee simple. The soil requires a duty of man which we have been slow to recognize. In this book the effort is made to discover man's debt and duty to the soil. The scientists examine the soil problem from every possible angle. This book must be reckoned with by all who would build a firm foundation for the future of the United States. For my own part I do not feel that this book is the last word. But it is a start and a mighty good start in helping all those who truly love the soil to fight the good fight." The 1938 yearbook summarized much of what the New Deal stood for: there is a problem, we have the duty to fix it, we can do it.

The 1938 yearbook covered a diversity of topics, including: the soil and the law, cover and green manure crops, the loss of soil fertility, the nitrogen cycle, farm tenancy, and crop yield on eroded soils. It provided a broad overview of soil knowledge, including economic and human aspects. For example, *Soils & Men* showed that the birth rate was usually high in areas of low agricultural productivity where soil erosion was also serious. Economic problems drove migration and, in 1927, almost four million people moved from urban areas to farms or *vice versa*, and between 1920 and 1936, more than 50 million moves were made. The population of the USA was about 106 million in 1920 and 129 million in 1937. To cope with the increase in population and migration of people, an emphasis was put on the need for more permanent and rational use of the soil by applying scientific principles and studies. It aimed to break a pattern of overexploitation: "Habits of thought toward land and its resources were formed in an environment where the consequences of waste and destruction could always be avoided by moving to fresh lands." [42].

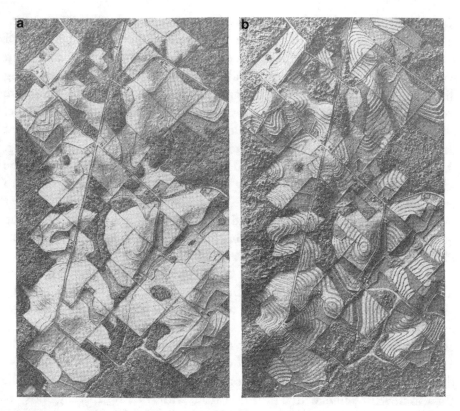

Model for proposed land management change of a 1,300 farm in Coushatta, Louisiana—before and after treatment. From: *Soils & Men*, Yearbook of Agriculture published in 1938

Soils & Men elevated soils and soil science in the USA; it was somewhat comparable to the massive soil science volumes by Edwin Blanck published two decades earlier [43]. Above all, it was a grand piece of cooperation by American soil scientists. *Soils & Men* was widely distributed and over 230,000 copies were printed and sold. In 1950, Roy Simonson found *Soils & Men* in a bookstore in Cincinnati at a time when the book was expensive and hard to find.

The world was at war in 1941. Geoffrey Milne died in Nairobi, James Joyce died in Zürich during surgery for a perforated ulcer, Hans Jenny published *Factors of soil formation*, and Charles Kellogg, 39 years old, became the Fifth President of the *Soil Science Society of America*. He also published his first book: *The soils that support us* [44]. The book highlighted that soils were the product of different combinations of rock, climate, slope, vegetation and time, and it was not very quantitative, but an eloquently written and broad essay on the connection between soils and humans. In the preface, Charles

Kellogg wrote: "The bond of relationships between man and soil is so deep and so fundamental that there is great danger of superficial recognition of a few symptoms and the application of superficial remedies that may even hasten the process of degradation. To all people whether they gain their living directly from the soil or not, it is important to know that these things are so and to think deeply upon what our own course, as a people should be. Nature has endowed the earth with glorious wonders and vast resources that man may use for his own ends. Regardless of our tastes or our way of living, there are none that present more variation to tax our imagination than the soil, and certainly none so important to our ancestors, to ourselves, and to our children." [44]. At the end of the book there was a section called *Where to Look* that listed some 15 books and publications in soil science, including those by Curtis Marbut, Gilbert Robinson, John Russell, Jacob Joffe, Eugene Hilgard, Konstantin Glinka, and *Soils & Men* [42].

Charles Kellogg revealed some of his philosophical thinking when discussing that humans live for much besides food: "The landscape of any place has a deep influence upon the artistic expression of people. The slow, monotonous music of the plains or steppes stands in contrast to the sharp variations of that from the mountains, or the rhythm of the tropical jungle. The careful detail of the English novelist somewhat suggests the detail of the landscape of forested regions clothed with small fields and farms. The Russian novelist paints with a broader brush, with wide sweeps, suggestive of the steppes themselves." [44].

The soils that support us ran through several editions, including one for the military in 1944. It was republished in 1956. Marion Jackson received it from his wife as a Christmas present in 1944. Edward Faulkner acknowledged the authoritativeness of the book in his *Plowman's Folly*. While Emil Truog had criticized *Plowman's Folly* and Charles Kellogg agreed with that critique, Edward Faulkner nevertheless wrote: "No attempt will be made to clarify the highly technical matter of soil classification. For such information the reader can now be directed to an extremely readable book on the subject, written by a man whose acquaintance with the subject is probably unmatched in this country. Charles E. Kellogg, Chief of the Soil Survey, United States Department of Agriculture, published late in 1941 *The soils that support us*. In my opinion there is no easier source from which the layman can obtain correct information on the subject at hand. After reading Mr. Kellogg's book, the reader who wishes more detailed information about the characteristic soil types of a given area of the country will find much helpful data in *Soils & Men*." [45]. Those kind words were not reciprocated by Charles Kellogg, for whom science was sacred and alternative approaches to agriculture were to be regarded with disdain.

Charles Kellogg wrote a second book that grew out of his hobby and professional passion: *Our Garden Soils* [46]. He was an avid gardener and saw the need for a book about the soils of gardens, as opposed to horticulture, landscape gardening, insect and disease control, or engineering. The garden was seen as personal to many people and part of their home, and it was defined as all the combination of cared-for soils and plants such as the kitchen garden, the flower beds, the lawn, and the trees and shrubs. The book was meant for people who do not have good soil in their garden, which included soils that were too sandy, wet, sloping, thin, clayey, dry or too infertile for gardeners. Charles Kellogg explained that soils were complex and different depending on locations, and that: "…continually the gardener must be on his guard against recommendations that apply to a different type of soil than the one he has. Before he attempts to follow his successful neighbor's example, he should be sure the soils are similar." A good garden has plenty of organic matter ("the gardener's elixir"), and plenty of well-behaved water ("water thoroughly but gently"). The book ended with an overview of soil and climatic requirements of vegetables, herbs, small fruits, fruit trees, herbaceous perennials, annual flowers, shrubs, vines, ground covers, bulbs, and ferns. It was dedicated to Lucille and their two children: "… for helping hands in summer and quiet evenings in winter." [46].

Charles Kellogg was 33 years old and had not been outside the USA. His first overseas trip was to the UK, in 1935, to attend the *Third International Congress of Soil Science*. Like his predecessor, he became an avid traveler and visited Puerto Rico, Hawaii, and Canada, as well as many states to inspect the progress of the soil survey. The Second World War reduced soil survey activities and his overseas travel, but soil survey interpretations were expanded to include military and engineering purposes [10]. No reports and maps were being printed, and a backlog accumulated, which was published in the late 1940s and early 1950s [47]. At the same time, there were inquiries about the costs of the soil survey. The cost to map a county had more than doubled in 20 years and: "…in the minds of some rather conservative people and some not so conservative, the question arises as to the amount of time and expense that can be justified in constructing and publishing standard soil survey maps for the country." [47].

In 1942, Charles Kellogg traveled to Mexico for the Rockefeller Foundation with a team that designed an agricultural program for which Norman Borlaug was hired in 1944. In June 1945, Charles Kellogg traveled to Russia with the aim to resurrect the *International Society of Soil Science* or to form a new association under the United Nations. It was his first visit to Russia. He represented the USA and met with several Russian, British, and French soil scientists. In September 1945, he visited Canada for the foundational meeting of the Food and Agricultural Organization of the United Nations.

In the summer of 1946, he visited Alaska and together with Iver Nygard, he studied the soils of what was considered the last frontier, or the frozen north. Iver Nygard was born in Minnesota and had obtained his MS in 1926 and PhD in 1933, from the University of Minnesota [48]. In 1927, he attended the *First International Congress of Soil Science* and he joined the soil survey in 1943. In Alaska, the purpose of the study was to suggest definitions and names for the broad soil groups and to explore their relationship to the environment. Charles Kellogg and Ivar Nygard collected samples and used published material including the Alaska study by Hugh Bennett and Thomas Rice from 1915 [49]. There was excitement in doing fieldwork in Alaska, as Charles Kellogg wrote: "Here in Alaska the soil scientist can see the first-hand effects of glaciers. Landscapes are reshaped, rocks are pulverized, rocks and rock flours are mixed together, old soils are buried, along with the plants growing on them, and new soil material is left by retreating glaciers and by the streams fed from the melting ice." [50]. They distinguished physiographic provinces, discussed the five soil forming factors, and produced a soil map that showed the Great Soil Groups distribution across Alaska.

Charles Kellogg worked well with Iver Nygard, but in his diary, he noted: "…I loved Iver Nygard but not his procrastination about getting his material to me. Iver was brought up on the Swedish syntax and backed into every paragraph like a mule into a thorn bush." The Alaska report appeared five years after Charles Kellogg and Iver Nygard had visited America's icebox [50]. The landscapes and soils of Alaska made Charles Kellogg reflect on glaciated Michigan, where he had grown up but on soils that were ten thousand years older. The soils in Michigan had weathered and leached somewhat, had formed a darker layer with organic matter, and were browned up by the release of iron that formed an oxide in the soil. Soils in Alaska were still in the making, some of the parent materials were fresh from underneath the ice, and millions of hectares of wet and partly frozen volcanic soils occurred along its southern rim.

After the survey in Alaska, Charles Kellogg requested funds from the US International Cooperative Administration to work in Iceland where it was perceived that similar conditions existed—cold and volcanic soils. In the summer of 1950, Charles Kellogg traveled to Iceland to discuss the soil survey, and a year later, Iver Nygard was sent to Iceland to start a soil survey program [51]. Iver Nygard hired the Icelandic soil scientist Björn Johannesson, who had obtained his PhD in 1945, at Cornell University under the supervision of Richard Bradfield. Iver Nygard and Björn Johannesson were to write a report on the soils together but Iver Nygard died suddenly and the report was to be finished by Björn Johannesson, who spent four months at the soil survey office in Washington, working with Cliff Orvedal [52]. Björn Johannesson had brought with him soil samples that were analyzed for mineralogical properties. Charles Kellogg oversaw the production of the soil map of Iceland while his son Robert, a professor of English literature, edited the report. The similarities between the soils of Alaska and Iceland were described, and it was noted that some of the soils in Iceland were similar to the volcanic soils of Japan. In Iceland, the properties of the soils were close to the parent materials, and three major soil groups were established: soils with vegetation, soils with little or no vegetation, and land with neither soil nor vegetation [52]. The land was cold and young, ideal ground for studying soil formation and the interaction between parent material and climate.

Some 30 years before Charles Kellogg and Iver Nygard studied the soils of Alaska, Hugh Bennett, and Thomas Rice had conducted a soil reconnaissance study of Alaska [49]. Charles Kellogg and Iver Nygard referred to that study in their 1951 report which was a touch of courtesy, even generosity, as Hugh Bennett and Charles Kellogg were not friends [50]. The chilly relationship can be traced to the establishment in 1933 of a parallel organization by Hugh Bennett, the *Soil Erosion Service* that conducted land surveys. Hugh Bennett was able to obtain generous federal funding and subtly, and sometimes less subtly, discredited the work of Charles Kellogg's *Division of Soil Survey*. The establishment of the parallel organization had its basis in the ecological and human tragedy of the Dust Bowl across the southern Great Plains. Severe drought and strong winds blew fine soil particles from the Great Plains all the way to the east coast. The dust reached Washington and stimulated the political will to create the *Soil Erosion Service* [15, 53]. A call for such a national soil erosion control program had been made several times before, last in 1928: "The time is at hand when state and federal agencies should join forces in a cooperative nation-wide effort to establish fundamental principles and promulgate practices that will check the enormous losses occasioned by unrestricted soil erosion." [54]. But they did not and it took the Dust Bowl, a new President, and the persuasiveness of Hugh Bennett to establish the *Soil Erosion Service* as a nationwide agency.

Hugh Bennett started in 1903 as a soil surveyor under Milton Whitney. He mapped soils in 11 states in the east and south but, gradually, became more interested in land and, in particular, in soil erosion. He replicated the erosion plots that were established by Frank Duley and Merritt Miller in Missouri at ten research stations nationwide in the late 1920s [55]. The results from those erosion plots showed how serious soil erosion could be under different conditions across the country. When the *Soil Erosion Service* was established in 1933, Hugh Bennett hatched the idea of a Soil Conservation District for the implementation of soil and water conservation. Such districts were meant as pilot studies, and a national model was developed for establishing conservation districts, called the Standard State Soil Conservation District. The 'complete farm plan' was the essence of the conservation district and the plan included terracing, contour farming, and strip cropping [56].

President Roosevelt wrote personally to all state governors, encouraging them to adopt the 'complete farm plan' for the Soil Conservation Districts and he made the widely quoted statement, likely written by Hugh Bennett "The nation that destroys its soil destroys itself." One of the first soil conservation districts was the Coon Creek watershed in the Driftless Area of Wisconsin, some 100 km south of Independence [57]. The district was proposed by Otto Zeasman, born in Kiev but schooled in engineering and soils at the University of Wisconsin. Otto Zeasman worked on soil drainage and was an early advocate for soil erosion control measures in the Driftless Area of Wisconsin. He was one of the first to recognize that soil erosion from melting snow can be severe on frozen ground [58].

The *Soil Erosion Service* became the *Soil Conservation Service* in 1935 with Hugh Bennett as Director and Robert Allison as a senior conservationist. Together, they had published the first soils book on Cuba [59]. Hugh Bennett then hired Jay Bonsteel, who had retired from the soil survey in 1920 and was now appointed to help with the promotion of soil conservation. Constantin Nikiforoff, born in Russia and deeply loyal to Charles Kellogg, called the people in Hugh Bennett's group the 'soil erosionists', others called Hugh Bennett a 'soil evangelist.' [60]. Hugh Bennett put a different emphasis on soil, believing that society was charged to maintain, protect and preserve the soil. Charles Kellogg, on the other hand, focused more on what was present and people had to live with [56].

Soil erosion was not only a prominent topic in the USA, but also in other parts of the world. It had been recognized as a serious problem in, for example, India, in the early 1900s, in South Africa, and in Australia which suffered a long dry spell in the 1930s [61]. In 1939, Graham Jacks and Robert Whyte published a global overview of soil erosion, entitled *The Rape of the Earth—A World Survey of Soil Erosion* [62], published in the USA as *Vanishing Lands: A World Survey of Soil Erosion*. In 1948, one of the first reports from

FAO reviewed soil conservation across the world [63]. The main contributors were Mark Baldwin, Charles Kellogg, and Robert Pendleton. Hugh Bennett provided information on the USA but did not contribute to this report. In 1950, the approaches in the FAO report were adapted to Australia where, in parts of New South Wales, almost half the land suffered erosion which was attributed to intensive tillage in the wheat growing areas and overgrazing of the rangeland, not least by rabbits [64].

Hugh Bennett focused on the impact of soil erosion within the USA, though he had some interest in the international aspects of the erosion problem. He traveled to Central America in 1926 [65], Cuba in 1926 [59], South Africa in 1944 [66], and several countries in Europe. He attended only one international soil congress, in Amsterdam in 1950. In South Africa "he talked pretty rough" and made his hosts uncomfortable, telling them "…you do everything to impoverish your land, nothing to build it up. There's no sense or economy ruining farms in your eagerness for gold and diamonds. Don't talk about your difficulties. For a change, let me hear just one thing that you might think might possibly be practicable " [67]. The war on soil erosion became headline news in South Africa and Hugh Bennett was cartooned, interviewed, and on the radio. In his home country, he became famous, and his books were used across the world [53, 68]. Hugh Bennett preferred that visitors came to America where he saw the *Soil Conservation Service* as: "…the beginnings of the world attack on land misuse." [67]. He did not aspire to international cooperation: he was convinced that the American model and approach he had developed was universal. In contrast, Charles Kellogg needed partnerships across the world to advance soil survey and develop a comprehensive soil classification system that could be used globally. Charles Kellogg relished international partnerships, having learned from his predecessor how much was to be gained from collaboration and friendships across the globe. Soils were more diverse than soil erosion and its control.

After 1935, there were two soil mapping programs, that in the *Division of Soil Survey* in the Bureau of Chemistry and Soils headed by Charles Kellogg, and in the *Soil Conservation Service* led by Hugh Bennett. This did not go down well. There was rivalry, distrust, and problems of coordination arising from the different approaches in mapping [10]. The mapping by Charles Kellogg's *Division of Soil Survey* had a pedological focus but included land-use capability. The maps were sometimes criticized as needing to be more practical—or needing to be more scientific and not concerned with crop yields, water control, or farm planning. The surveys by the *Soil Conservation Service* were of a practical nature and focused more on mapping the land than on the soil and its characteristics, emphasizing slope, and erosion classes.

Hugh Bennett and Charles Kellogg who rivaled for over 15 years within the Department of Agriculture for funds, recognition and the way soils should be mapped. Hugh Bennett, 20 years older, was one of the first soil surveyors, and then led a highly visible and successful *Soil Conservation Service* from the mid 1930s onwards. They both valued the land, its farmers and saw the need rationale for land use. That is where their parallels ended

The *Soil Conservation Service* questioned the suitability of the soil survey maps for farm planning and Hugh Bennett used some of the early work to discredit the soil survey, even though he had been hired by Milton Whitney in 1903 [10, 47]. However, the soil erosion surveys had a strong appeal and the *Soil Conservation Service* received increased funding from the emergency public works, whereas funds for the soil survey were cut back in the early 1930s [10]. Charles Kellogg, however, had the land-grant colleges on his side, as they favored developing soil surveys throughout the country, and they preferred a pedological approach that could be used beyond soil and water conservation [10].

For Charles Kellogg, the mapping by the *Soil Conservation Service* was too narrow, lacked a scientific base, and provided no answers to questions that might arise in the future. It was practical, fast, and lacked detail, whereas the soil survey was detailed, slow, and required many observations and laboratory data. In the late 1940s, Charles Kellogg hoped that: "…the period of extreme statement on soil erosion is nearly over, else it might stimulate some counteraction just as extreme." [69]. While there were more scientific developments in the soil survey than in the soil erosion mapping, Hugh Bennett was much better at putting his message across, which was easier than explaining the need for pedology and soil survey.

July 1946 saw the launch of the *Journal of Soil and Water Conservation* which became an important vehicle for the *Soil Conservation Service*. Hugh Bennett questioned in the first issue: "… is there an alternative to the kind of soil conservation program I have described?" and he thought other answers were absolute government regulation, or abandonment of hope. The *Soil Conservation Society of America* was also established, and its President wrote in the journal: "…the development of soil conservation thought and knowledge in recent years has been nothing short of phenomenal. What we are witnessing today, however, is only a beginning." Within a few years, the *Journal of Soil and Water Conservation* had over 3,000 subscribers [70].

The *Soil Conservation Society* published a large number of semi-glossy brochures aimed at land users, farmers, the general public, and politicians in particular. They had alarming titles such as *Our American Land—The Story of its Abuse and its Conversation, Erosion of topsoil reduced productivity, Winning the Battle for the Land, Soil Depleting, Soil Conserving and Soil building Crops, We can all help to Save our Soil,* and *From the Dust of the Earth* that included pictures of deeply eroded fields, gullies as well as contour planted and community planned activities. The message was: soil erosion will bring us down, it is everywhere, but there is hope, it is named: soil conservation. Overall, that terrifying yet forward-looking message resonated more than the need for pedological investigations to produce detailed soil maps at a scale that, although not of direct use to farmers, were extremely useful as a regional planning tool, and for understanding soils across the landscape. The soil survey lacked a journal or magazine solely devoted to its findings, although it produced some brochures meant for land users and the general public. The *Soil Science Society of America Journal* and its predecessors were not read by land users and, indeed, was not widely read by soil surveyors across the country. In 1960, the periodical *Soil Survey Horizons* was started by Francis Hole to share experiences and news among soil surveyors and pedologists [71]. It never attained the impact or professionalism of what the soil erosionists published.

After the Second World War, the *Soil Conservation Service* experienced large growth and a doubling in its budget compared to the budget before the war. In 1950, the *Soil Conservation Service* surveyed over 12 million hectares and, by that time, there were about 700 field workers working in 48 states. By comparison, the soil survey had about 100 soil surveyors, and they were active in less than half of the states. The budget for the soil survey was about $900,000 by 1950, whereas the *Soil Conservation Service* received nearly

$50 million. A reorganization of the soil survey and the various groups that conducted soil research with the Department of Agriculture was considered. It had been brewing for a while.

In September 1937, the Secretary of Agriculture, Henry Wallace, wrote to six agencies within the Department of Agriculture about the soil and erosion surveys: "I have today approved in a separate memorandum a co-ordinated procedure for making soil and erosion surveys. I hope that under the new plan the Department and cooperating State agencies can move forward more rapidly in obtaining accurately and efficiently the information on soils, slope, degree of erosion, and other physical factors which we need for land-use planning, research, and other purposes. This improvement in survey procedure emphasizes another need—that of coordinating the basic soils work of the Department. Various phases of soils research are now conducted in at least four bureaus. This in itself is not to be condemned; many research programs cross bureau lines. But apparently the present disorganization in soils research is causing that research to lag just at a time when basic soils information is most needed. The difficulty may be in the way the work is organized, it may be in our failure to have sufficiently close cooperation between the agencies which handle different aspects of soils research, or it may be that the principal difficulty is inadequate funds."

Following this memorandum, Charles Kellogg wrote a six-page confidential letter to his friend, Emil Truog in Wisconsin. He outlined his frustration with the situation and desire to move ahead with the soil survey and the direction it had taken: "A cooperative arrangement has just been developed between the *Soil Conservation Service* and this Bureau in regard to soil and erosion surveys. Naturally, I cannot help but feel that a higher quality of work could be done at somewhat less cost if it were all done by this Division. From the point of view of the Secretary's office, however, the dilemma is this: The red flag for soil conservation is out in Congress and it is easy to get money under that flag. The scientific work and the trained specialists are in the Bureau of Chemistry and Soils and without their cooperation and assistance it is impossible for the *Soil Conservation Service* to make dependable maps. Since the *Soil Conservation Service* is not a scientific agency the Secretary hesitates to throw the work in that direction. For financial and political reasons, he hesitates to throw it in the other direction. It has been a very serious problem all the way along the line and I think that our present arrangement, the details of which, of course, do not interest you probably, is the best that could be made in view of all of the circumstances." He emphasized that the *Soil Conservation Service* was not a scientific organization, their methods of mapping inadequate, but that their message resonated better with the Congress which allocated the budget. He ended his letter with: "You must appreciate that all of this is highly confidential. I do not need to emphasize my respect for your judgment in this whole field. You are well aware of it. I have explained this matter to you so that you may see what my general thinking is and be in a position to advise me and to assist, if necessary. I don't have my mind completely made up and I am trying the best I can to enter these deliberations with an open mind as to what should be done. After all we are working in the interests of the farmers, or should be at least, and we are spending the money of the American people."

Emil Truog responded a few days later doubting that he could give anything worthwhile by way of letter and that he held comments until they met in person. A month later, he traveled to Washington before attending the annual meeting of *Soil Science Society of America*. Charles Kellogg advised a stay at the Cosmos Club: "This is a club composed of scientific men and I think you would make a good many interesting contacts. Rates are similar to those in the hotels, possibly a little cheaper." Charles Kellogg, like his predecessors Milton Whitney and Curtis Marbut, was a member of the Cosmos Club, an elite private club for men distinguished in science, literature, the arts, or public service; it remained men-only until 1988. Charles Kellogg and Emil Truog discussed the differences between the *Division of Soil Survey* and the *Soil Conservation Service*. After he had returned to Madison, Emil Truog wrote a supporting letter to Charles Kellogg: "There certainly never existed a more favorable time for setting up a real comprehensive soil program for the whole country than the present. The whole country is 'Soil Conscious' to an extent that that one would never have believed possible only a few years ago. That is at least something that the *Soil Conservation Service* has put over and I believe the whole soil program should now make the most of this accomplishment. It has cost a lot of people's money to do this and the most should be made of it. There is of course, no danger that the soil will be over-emphasized."

In 1938, the *Division of the Soil Survey* was transferred from the *Bureau of Chemistry and Soils* to the *Bureau of Plant Industry,* directed by Bob Salter from Ohio. The transfer brought little change to the soil survey. By the late 1940s, the *Soil Conservation Service* vastly outproduced the area mapped by the *Division of Soil Survey*. The surveyors in both organizations and those working in the land-grant colleges and experimental stations observed the bitter fight with discontent. In 1950, Charles Kellogg reflected: "...the Soil Survey was severely criticized during the 1935–1945 period by those who maintained that it was being ruined by becoming dominantly utilitarian and by those who maintained that it was being stifled by the academic and theoretical. Although both voices were strong, fortunately they were nearly equal." [32]. The work done by Charles Kellogg and his office was also supported by Chief who was a defender on the need and usefulness of soil survey [10].

Hugh Bennett addressed a Congressional committee and named the work and maps by the *Soil Conservation Service*: "...the most comprehensive chart ever made of land conditions in this country." He ignored the earlier countrywide soil maps such as the map published by Curtis Marbut in the *Atlas of American Agriculture* [10]. For Charles Kellogg the sole emphasis on soil erosion was too narrow, as he summarized it in 1948: "It would be unfortunate for soil erosion to be neglected as it was in former years. It would be unfortunate for it to be exaggerated or overemphasized at the expense of other problems of soil use. Above all, concentration on the erosion itself to the neglect of its causes could be wasteful or even harmful." [69]

Hugh Bennett and Charles Kellogg were rivals—both scientifically and in the pursuit of funding for their organizations that fostered willful mutual ignorance. In 1939, Hugh Bennett published *Soil Conservation*, a 958-page overview of soil erosion and conservation in the USA [53]. The soil survey was not mentioned, and Curtis Marbut was only cited as the translator of the Konstantin Glinka book. Two years later, Charles Kellogg published *The soils that support us*, and discussed soil conservation districts, and the special attention these districts were given by the newly created *Soil Conservation Service* [44]. No further mention was found in the 358-page book. In 1947, Hugh Bennett's book *Elements of Soil Conservation* ignored the soil survey altogether [10, 68]; and there was no mention of the soil survey and Charles Kellogg in the biography of Hugh Bennett: *Big Hugh—The Father of Soil Conservation*, authored by Wellington Brink—the founding editor of the monthly *Soil Conservation* magazine. He called Hugh Bennett "the man with the answers." [72]. The Preface of the biography was by Louis Bromfield, a Pulitzer Prize winner, who founded an experimental farm in Ohio and became an early advocate of organic agriculture. When Louis Bromfield met Hugh Bennett: "... I knew that I was in the presence of a great man and a man to whom this nation and the whole world owed a very great debt. Like all great men, he was simple and direct, with no time for shiftiness or pomposity." [72].

The *Division of Soil Survey* published a detailed *Soil Survey Manual* in 1937, and two years later the *Soil Conservation Service* published its *Soil Conservation Handbook* [24, 73]. The handbook was written by Ethan Norton and he was hired from the Illinois Agricultural Experiment Station and the Illinois soil survey. He had studied the effects of topography on soils and had chaired a soil horizon criteria committee in the late 1920s

[74, 75]. Ethan Norton wrote with practical applications in mind, and the *Soil Conservation Handbook*: "…sets forth a procedure for making a soil conservation survey and includes instructions for mapping the major physical land features essential to the development of a coordinated soil conservation program and for interpreting those features in terms of land use capability." [73]. By contrast, the purpose of the soil survey as defined in the 1937 *Soil Survey manual* was: "…to determine the morphology of soils, to classify them according to their characteristics, to show their distribution on maps, and to describe their characteristics, particularly in reference to the growth of various crops, grasses, and trees. The ultimate purpose is to provide accurate soil maps, necessary for the classification, interpretation, and extension of data regarding agricultural production, the classification of rural lands, and for the factual basis in the development of sound programs of rural land use, whether planned by public or private agencies, or by individuals." [24]. For the non-expert, the ultimate goals of both organizations—maps for land-use capability, and maps for rural land-use planning—were difficult to distinguish.

The controversy between the soil survey and the *Soil Conservation Service* ended on 14th of October 1952 with a top-down political decision to merge the two agencies. It was heralded as a great merger, as both agencies had complementary strengths and experiences. The merger was based on a carefully crafted recommendation of a national committee appointed by the Secretary of Agriculture. The committee composed of Richard Bradfield from Cornell University, Bob Salter from the Ohio State University, and Bill Pierre of Iowa State College [76]. In the 1950s, they were leaders for agronomic affairs and policies in the USA [76]. The soil surveys of the two federal agencies were combined into the *Soil Conservation Service*, and Charles Kellogg was made Assistant Administrator for Soil. Bob Salter who had been Chief of the Bureau of Plant Industry, became the Administrator, succeeding Hugh Bennett, who had directed the organization since its beginning in 1933 [76]. Hugh Bennett, 71 years old, retired and was made Special Assistant to the Secretary of Agriculture, in charge of conservation and resource matters. The pooled resources included 900 people, and a budget of six million dollars [10].

Charles Kellogg was demoted to an Assistant Administrator and his soil survey organization was swallowed by the *Soil Conservation Service*. He did not take this as a demotion: "You cannot be humiliated or put down unless you allow it." [60]. Most importantly, he was put in charge of soil surveys and that was all that mattered for him. Hugh Bennett, who was still influential, was convinced that Charles Kellogg and the new *Soil Conservation Service* were: "…wrecking soil conservation." [10]. Hugh Bennett died shortly before the *Seventh International Congress of Soil Science* in 1960, still recognized as an evangelist who made soil erosion and conservation one of the most prominent programs in the USA. *Soil conservation* was a term that appealed to the general public, administrators, and politicians alike. For that reason, Charles Kellogg, shrewdly aware of his dependence on favorable political winds, accepted the *Soil Conservation Service* as the name for the agency that he now directed.

Internationally, there was some misperception on the use of the term. At the *Fifth International Congress of Soil Science* in Congo, in 1954, a heated discussion was started by Vladimir Ignatieff from the FAO who asked for a more precise definition of soil conservation. He had noted that Charles Kellogg was a promotor of the term, but he found that it was too pessimistic and he suggested *soil improvement*, because: "Man causes either deterioration or improvement but the soil never remains in the same state as the word conservation implies." Charles Kellogg, at his most diplomatic, replied that soil conservation had become a symbol to many people and was already well-known and used, so it would be better to change the meaning of the term than use another. Soil and water conservation should be closely related as it was the case in the USA, and soil conservation had a broader meaning that included securing crop harvests and farm profitability. The term *soil conservation* anchored more deeply in American soil than the practices it stood for.

Like Curtis Marbut who traveled extensively, Charles Kellogg developed an interest in the tropics. Some of Charles Kellogg's travels had a political and administrative motivation. In 1945, Clinton Anderson became the Secretary of Agriculture, after he had chaired the Special Committee to Investigate Food Shortages after the Second World War. Clinton Anderson believed that: "The United States has a unique responsibility in the world program for peace, and for the building of an expanding world economy. One of the obligations that goes along with being the richest, most productive nation in the world is that we set the pace. In this country we have only about six percent of the world's people, but we produce somewhat more than half of the world's goods. We have more than half of the power, we turn out about sixty percent of the manufactures, and ours is the mightiest, the most productive agriculture. Fortunate as we are in having a highly developed, integrated economy, we cannot expect to prosper indefinitely in a depressed world. United States agriculture has certain definite broad assignments, against which we should measure all of our policies, plans, and action programs." [77]. Such was the thinking of the Secretary of Agriculture after the Second World War. The

Charles Kellogg was the third Chief of the soil survey from 1935 to 1971. As a child on the farm in Michigan, he longed for fancy clothes. In his early years as chief, he dressed very well, and was always accompanied by his pipe. Photo Kellogg family photo album

assertiveness trickled down through the organization and came naturally to Charles Kellogg.

In 1947, Charles Kellogg traveled to the Congo, which had been declared a private colony of the Belgium King Leopold II in 1908. It was on the allied side in the Second World War and provided raw materials, notably gold and uranium, to the UK and USA. The purpose of Charles Kellogg's visit was to suggest descriptions, nomenclature, and capabilities for the principal soils of the Congo Basin based on the American experience in soil survey [78]. It was his first visit to Africa, and he had only seen tropical soils in Puerto Rico and Hawaii in 1936. Tropical soils were absent in the contiguous USA,

Charles Kellogg lightning a pipe in an elephant grass field on an alluvial soil with glei in Luki. In 1976, Luki became an Unesco Biosphere Reserve representing the humid tropical rainforest ecosystem. Photo from the Special collection National agricultural Library

and Charles Kellogg had the desire to understand those soils better for the development of a soil classification system that could be used across the globe. He was also interested in the tropics for humanitarian reasons, and he knew how difficult it was for poor people to improve their soils and make a living. In 1948, he described it as: "...when a rural population becomes poverty-stricken, it fails to maintain its soil. An exploited people pass on their suffering to the land." [69]. Or, as Hugh Bennett had put it: "Poor soils make poor people, and poor people make the soil worse." [53].

In Congo, Charles Kellogg traveled with the Director General of the national agricultural research institute by plane, car, and railroad. They studied soils on experimental fields, farms, and along miles of country roads. In his diary, he wrote: "I met chiefs who had known Stanley." Towards the end of his visit, he urged the Congolese and Belgians to set a date for independence. Independence came 13 years later, after which the country became a battleground of cold war politics, and some 100,000 people lost their lives. While traversing the country, 50 soil profiles were sampled; the samples were analyzed in the Beltsville laboratory that was part of the *Division of Soil Survey*. Charles Kellogg named and defined the soils, emphasizing that all of it was tentative, but: "...to attempt none at all would call for the reading and remembering of an enormous mass of material." Soils were grouped as zonal (Latosols, Chernozems, Podzolic soils), Intrazonal soils (Hydromorphic, Halomorphic, Latosolic), and Azonal soils (Alluvial, Lithosols). The report offered suggestions for research and stressed that some areas had high potentials and that highly productive agriculture under scientific management existed in other parts of the tropics. The universality of principles was also highlighted: "...much time has been wasted teaching techniques used in temperate regions to agricultural technicians and advisors in the tropics that could have been used to very good advantage in teaching principles from which appropriate techniques can be developed." [78]. The Soil Map of Africa published in 1964 with an explanatory monograph made no mention of Charles Kellogg's report [79].

In 1950, Charles Kellogg summarized research questions for tropical soils at the *Fourth International Congress of Soil Science* in Amsterdam, where he had organized a special session on tropical soils. In his view, there was much work to be done: "Hundreds of problems with tropical soils could be mentioned: Where does the nitrogen in the soil come from? The neatly drawn nitrogen cycle of our textbooks won't do. If we knew how the nitrogen came into the soil, perhaps we could influence the amounts greatly through management. Many of the 'self-mulching black clays' have less than one-half the organic matter of the associated Red Latosols. Why? How much of the benefit from composts on many tropical soils is due to effects on structure and moisture and how much is due to fertility, because, in our ignorance, composting is the only way to add a properly balanced nutrient supply? Under the corridor system of shifting cultivation with, say, 12 years of forest and 6 years of crops in an 18-year cycle, how much can the forest fallow be reduced by mineral fertilizers? Or are the trees essential for shade and soil structure anyway? We know generally that the laterite clay of the Ground-Water Laterite hardens irreversibly with dehydration. How much dehydration? How can it be controlled?" [32]. The list of questions revealed some of his own interests such as shifting cultivation, organic matter, structure, and the formation of laterite, but it also highlighted other research areas. The list also provided answers: "Answers to questions like these depend first upon the establishment of the principles of soil genesis and behavior. With luck, we may answer a few by mass plot work along purely empirical lines, but the results will be very costly and slow in coming; and even then, they will be applicable only to soils like those in the plots. Thus, in the expansion of the Soil Survey we must provide for fundamental researches, using both the experimental method and the method of scientific correlation, else we may find ourselves promising more than we can deliver." [32].

Following his visits to the tropics, Charles Kellogg became interested in the question of feeding the world. He developed the idea that the capacity to feed the world was more a problem of human relationships than of scientific riddles: "The social, economic, and political problems are many and difficult. The technical problems of soils, plants, and animals, great as they are, are small by comparison." [79]. In the 1940s, he found that: "…people must make a go of things of where they are, or else move into areas where a great deal of planning is needed for successful agriculture or industry. There is no more fertile soil waiting only for the plow. But there is a great deal of unused land in the world that can be made productive through the application of modern science, land that is made up of thousands of unique types of soil." [27]. He believed that education was needed in many parts of the world; education on soils, their origin, use, and management.

Charles Kellogg held no formal teaching position since he had left *North Dakota Agricultural College* in 1934. But he had ideas on teaching that he shared widely. Often, graduates told him that they found soil science boring and their teachers uninspiring. He thought that soil science should never be boring or devoid of the human element: "The general soil scientist has more human-interest material to draw on than any other teacher in a college of agriculture". [80]. Many courses were focused on temporary agricultural issues and systems, which would be of little use to a professional several years later when circumstances had changed. He believed in a firm grounding in science that would lay the foundation for learning and adapting [9]. Coming from the farm himself, he emphasized that most government leaders were from towns and knew little about agriculture. They assumed agriculture was a simple enterprise, and: "…they find it extraordinarily difficult to work with cultivators as fellow citizens" [81]. Charles Kellogg respected Henry Wallace and Emil Truog who were both raised on a farm.

After the Second World War, universities in the USA increased instruction in international agriculture and allowed students from economically developing nations to enroll. It made the universities more international, although it was realized that professors in the USA lacked experience, as well as field sites comparable to those in tropical countries. With the help of the federal government, there were considerable investments in overseas programs. For example, in the 1950s and 1960s, the Department of Soils at the University of Wisconsin established soil science departments in Brazil, Indonesia, and Nigeria. Students from economically developing nations studied in the USA, and as Charles Kellogg summarized it: "…the student may learn readily, what he learns and its application here may help little in his own country. Yet he will have the prestige of an American college degree." [12].

Charles Kellogg built and expanded the soil survey and led the program for 37 years, longer than anyone before or after him. He recruited several former students whom he had inspired to follow a career in soil science, or as Hans

Jenny wrote: "...he brought with him most of his North Dakota graduate students. They occupied, for decades, leading positions in soils work." [7, 9]. Charles Kellogg hired Roy Simonson in 1949 to become the assistant chief of the *Division of Soil Survey*. When in October 1952, all survey activities were consolidated into the *Soil Conservation Service*, Roy Simonson was in charge of classification, correlation, and nomenclature of soils. Charles Kellogg hired Bill Johnson, another North Dakota student of his, and who succeeded Charles Kellogg after he retired. He hired his former student Cliff Orvedal who headed the World Soil Geography Unit [82]. Cliff Orvedal studied the potentially arable soils of the world [79], and compiled a five-volume bibliography on the soils of the tropics. Charles Kellogg also hired Ken Ableiter whom he had worked with at the survey in Wisconsin and in North Dakota, and he kept close contact with Marlin Cline at Cornell University. He hired Guy Smith as a soil coordinator, but he sensed he could not do much until there was a better system of soil classification. Thus Guy Smith convinced Charles Kellogg that the soil classification system was illogical, and Charles Kellogg instructed him to make a better one [7].

Charles Kellogg was in close contact with most of his North Dakota students, and people that were loyal and whom he trusted. In that sense, he was no different than Hugh Bennett, who hired his friends and colleagues including Walter Lowdermilk, Jay Bonsteel, and Robert Allison, or Milton Whitney, who tried to hire the best people, and in particular, those that were sympathetic to his ideas. But Charles Kellogg strongly believed in teamwork, which he deemed essential, as even the simplest survey required cooperation among soil scientists and cartographers, geologists, agronomists, botanists, foresters, horticulturists, and economists [32]. Curtis Marbut was more of a solitary but charismatic man, and a bit of an outsider in the soil survey [5]. Milton Whitney had developed a reputation for fiercely opposing change and ruled in an autocratic way. Charles Kellogg directed the phenomenal expansion of the soil survey between the mid-1930s and late 1960s. In 1934, in the middle of an economic and ecological crisis, he stepped into an organization that was small and on the brink of being swallowed up by Hugh Bennett and the 'soil erosionists.' He provided the scientific and political leadership that eventually made the soil survey successful; it grew and, in the early 1950s, there were about 130 soil surveyors, and by the 1970s, the soil survey employed 1,500 researchers [9, 83].

It was Milton Whitney who brought soil properties into the soil survey starting with the texture of the soil [16]. That was a first step away from soils as a geological material. The list of properties that were observed and sometimes measured increased in the first ten years of the soil survey, and the

Some American and Russian soil scientists in 1962 at the house of Charles Kellogg on 4100 Nicholson Street, Hyattsville, Maryland. Left to Right: Roy Simonson, Charles Kellogg, Innokenti Gerasimov, and Guy Smith. Roy Simonson had been a student of Charles Kellogg at *North Dakota State Agricultural College* early 1930s; Guy Smith was hired in 1949 and became the main architect of *Soil Taxonomy*; Charles Kellogg met Innokenti Gerasimov in June 1945 in Moscow, at the meeting that aimed to resurrect the *International Society of Soil Science*. Innokenti Gerasimov was interested in the field soil diagnostics and considered them poorly developed. He called them 'the Achilles heel' of pedology

geologic origin of soil became less important. Classification was largely based on the nature of the parent material and once the concepts of soil provinces and soil types had been established, rigidity halted progress. Curtis Marbut brought the soil profile and soil morphology into the survey, and he based soil classification on properties and soil morphology, because the theories of soil formation were incomplete and evolving. In 1928, the *American Soil Survey Association* advocated that soil characteristics should always be kept differentiated from the factors responsible for their development. Curtis Marbut transformed the soil survey into a pedological survey and introduced soil anatomy as a branch of soil science.

In 1950, Charles Kellogg realized the risk of an emphasis on the morphological approach in soil survey and he found it problematic that "…the soil scientist feels no responsibility for discovering the genetic laws and relationships." [32]. Soil understanding came from knowledge about their formation and why soils differ from one location to another. For him, it was not only about the mapping and classification, it was also about soil understanding and the use of soil survey for better management. Under Charles Kellogg, 'field men' were to become 'scientists.' [5]. Charles Kellogg had a keen interest in the use of the soil survey, as he summarized it in 1950: "As scientists and technicians, we must never forget that our technology will be understood and applied effectively only by alert people—people with opportunities for health, for education, for the full use of their labor and genius, and for participation in the affairs that affect them." [32]. For that to happen, the work had to be practical and be guided by sound scientific theory. The scientific mission of the soil survey was not well defined, and it was unclear who determined the priorities, methods of working, and overall emphasis. The soil survey had no external or internal advisory board, and its vision and courses were set by only a few. Alternative suggestions for grouping soil series such as Hans Jenny's based on functions of the soil forming factors from 1946 [84], were ignored. Throughout the history of the soil survey, it seemed that when a course was set, it proved arduous to change.

Some six years after Charles Kellogg had arrived in Washington, the Second World War broke out. He was told not to serve despite his cavalry background in Michigan. The Kelloggs put a Victory Garden in his back-yard in Hyattsville, that was planted with flowers after the war was over [11, 46]. At the end of the war, his father Herbert died at the age of 65 years. He had lived most of his life on the farm in Palo and had hoped for a long time that Charles would take over the farm. But Charles was different, went to college, and broadened his horizons beyond the glacial drift of Michigan. After they buried his father, his mother Eunice moved off the farm and into the village of Palo, and sometime later they brought her to Hyattsville to live with Lucille, him and their two children – just like his parents had done with his grandparents. Eunice died in 1953, at the age of 77 years. She was buried next to Herbert at the cemetery at the south side of Palo.

References

1. Marbut CF. Soils of the United States. Part III. In: Baker OE, editor. Atlas of American agriculture. Washington: United Sates Department of Agriculture. Bureau of Chemistry and Soils; 1935. p. 1–98.
2. Rice TD, Marbut CF. Life and works of C.F. Marbut. Madison: Soil Science Society of America; 1936. p. 36–48.
3. Wolfanger LA. The major soil divisions of the United States. A pedologic-geographic survey. New York: Wiley; 1930.

4. Jenny HEW. Hilgard and the birth of modern soil science. Pisa: Collana della revista agrochemica; 1961.

5. Merkel DM. The curious origin of Podology: the story of a milestone paper. Unpublished manuscript, sabbatical report; 2013.

6. West D. A visit to the home of a scientific pioneer. Soil Surv Horiz. 1995;36:112–3.

7. Maher D, Stuart K, editors. Hans Jenny—soil scientist, teacher, and scholar. Berkeley: University of California; 1989.

8. Miller MF. The soil and its management. Boston: Ginn and Company; 1924.

9. Helms D, Effland ABW, Durana PJ, editors. Profiles in the history of the U.S. soil survey. Ames: Iowa State Press; 2002.

10. Gardner DR. The National Cooperative Soil Survey of the United States (Thesis presented to Harvard University, May 1957). Washington DC: USDA; 1957.

11. Kellogg RL. Remembering Charles Kellogg. Soil Surv Horiz. 2003; 86–92.

12. Kellogg CE, Knapp DC. The college of agriculture. Science in the public service. New York: McGraw-Hill Book Company; 1966.

13. Helms D. Hugh Hammond Bennett and the creation of the Soil Conservation Service. J Soil Water Conserv. 2010;65:37A–47A.

14. King FH. Destrcutive effects of winds on sandy soils and light sandy loams: with methods of protection. Wisconsin Agricultural Experiment Station; 1894. Bulletin no. 42.

15. Kellogg CE. Soil blowing and dust storms. Miscellaneous publication no. 221. Washington DC: United States Department of Agriculture; 1935.

16. Moon D. The American Steppes. The unexpected Russian roots of great plains agriculture, 1870s–1930s. Cambridge: Cambridge University Press; 2020.

17. Bennett HH, Chapine WR. Soil erosion a national menace. Vol Government printing office. Washington; 1928.

18. Kellogg CE. Men turn their thoughts to the soil. New York Times. 1935; 6, 19.

19. Kellogg CE. Development and significance of the great soil groups of the United States. Miscellaneous publication no. 229. Washington DC: United States Department of Agriculture; 1936.

20. Dokuchaev VV. Russian Chernozem (Ruskii Chenozem). Selected works of V.V. Dokuchaev. Volume I (translated in 1967). Washington DC: Israel Program for Scientific Translations. U.S. Department of Agriculture; 1883.

21. Simonson RW. The United-States soil survey—contributions to soil science and its application. Geoderma. 1991;48(1–2):1–16.

22. Whitney M. Instructions to field parties and descriptions of soil types. Field season 1903. Washington DC: U.S. Department of Agriculture. Bureau of Soils; 1903.

23. Whitney M. Instructions to field parties and descriptions of soil types. Washington DC: U.S. Department of Agriculture. Bureau of Soils; 1914.

24. Kellogg CE. Soil survey manual. Miscellaneous Publication No. 274. Washington D.C.: Department of Agriculture; 1937.

25. Tandarich JP, Darmody RG, Follmer LR, Johnson DL. Historical development of soil and weathering profile concepts from Europe to the United States of America. Soil Sci Soc Am J. 2002;66(4):1407.

26. Joffe JS. Soil profile studies: I. Soil as an independent body and soil morphology. Soil Sci. 1929;28(1):39–54.

27. Kellogg CE. Modern soil science. Am Sci. 1948;36:517–35.

28. Glinka K. Die Typen der Bodenbildung - Ihre Klassification und Geographsiche Verbreitung. Berlin: Verlag von Gebrüder Borntraeger; 1914.

29. Kellogg CE. Report of the land-use committee. Am Soil Surv Assoc Bull. 1935;XVI:147.

30. Kellogg CE, Ableiter JK. A method of rural land classification. Technical Bulletin no. 469. Washington: United States Department of Agriculture; 1935.

31. Weir WW, Storie RE. A rating of California soils. Bulletin 599. Berkeley, California: University of California; 1936.

32. Kellogg CE. The future of the soil survey. Soil Sci Soc Am Proc. 1950;14:8–13.

33. Kellogg CE. Reading for soil scientists, together with a library. J Am Soc Agron. 1940;32:868–76.

34. Kellogg CE. Ours soils in books. Part one. Am Book Collect. 1967;10:17–21.

35. Kellogg CE. Ours soils in books. Part two. Am Book Collect. 1967;10:13–9.

36. Soil Survey Division Staff. Soil survey manual. Washington DC: USDA; 1993.

37. Soil Survey Staff. Soil survey manual. Washington DC: USDA; 1951.

38. Gennadiev AN, Bockheim JG. Development of the soil cover pattern and soil catena concepts. In: Warkentin BP, editor. Footprints in the soils. People and ideas in soil history. Amsterdam: Elsevier; 2006. p. 167–186.

39. Simonson RW. Evolution of soil series and type concepts in the United States. In: Yaalon DH, Berkowicz S, editors. History of soil science international perspectives. Armelgasse 11/35447 Reiskirchen/Germany: Catena Verlag; 1997. p. 79–108.

40. Food and Agriculture Organization of the United Nations and Intergovernmental Technical Panel on Soils. Status of the World's Soil Resources—Main Report (FAO & ITPS, 2015).

41. Soil Survey Staff. Soil classification. A comprehensive system, 7th Approximation. Washington D.C.: Soil Conservation Service. United States Department of Agriculture; 1960.

42. United States Department of Agriculture. (1938). Soils & men. The year book of agriculture. House Document No. 398. Washington United States Department of Agriculture. United States Government Printing Office.

43. Blanck E. Handbuch der Bodenlehre (vol. I-X). Berlin: Verlag von Julius Springer; 1929–1932.

44. Kellogg CE. The soils that support us. New York: The Macmillan Company; 1941.

45. Faulkner EH. Plowman's folly. Norman: University of Oklahoma Press; 1943.

46. Kellogg CE. Our garden soils. New York: The MacMillan Company; 1952.

47. Miller MF. Progress of the soil survey of the United States Since 1899. Soil Sci Soc Am Proc. 1950;14(C):1–4.
48. Anon. The Department of Soil, Water, and Climate: 100 years at the University of Minnesota. Minneapolis: University of Minnesota; 2013.
49. Bennett HH, Rice TD. Soil reconnaisance in Alaska, with an estimate of agricultural possibilities. Washington: Bureau of Soils; 1915.
50. Kellogg CE, Nygard IJ. Exploratory study of the principal soil groups of Alaska. Agriculture Monograph no. 7. Washington DC: United States Department of Agriculture; 1951.
51. Arnalds O. The soils of Iceland. World Soils Book Series. Dordrecht: Springer; 2015.
52. Johannesson B. The soils of Iceland. Department of Agriculture Reports Series B - no. 13. Reykjavik: University Research Institute; 1960.
53. Bennett HH. Soil conservation. New York & London: McGraw-Hill Book Company, Inc; 1939.
54. McCall AG. A national program for soil research. J Am Soc Agron. 1928;20:1241.
55. Miller MF. Erosion as a factor in soil determination. Science. 1931;73:79–83.
56. Hardin CM. The politics of agriculture. Soil conservation and the struggle for power in rural America. Glencoe, Illinois: The free press; 1952.
57. Helms JD, Pavelis GA, Argabright S, Cronshey RG, Sinclair HR. National soil conservation policies. Agric Hist. 1996;7:377–94.
58. Bates CH, Zeasman O. Soil erosion—a local and national problem. Research Bulletin no. 99. University of Wisconsin, Madison: Agricultural Experiment Station; 1930.
59. Bennet HH, Allison RV. The soils of Cuba. Washington DC: Tropical Plant Research Foundation; 1928.
60. McCracken RJ. Evagelists, scholars, historians, lab types, computer buffs, map makers and auger pullers in the soil survey. Soil Surv Horiz. 1993;34:61–71.
61. Sauter S. Lessons from the US: Australia's response to wind erosion (1935–1945). Glob Environ. 2015;8(2):293–319.
62. Jacks GV, Whyte RO. The rape of the earth—a world survey of soil erosion. London: Faber and Faber Ltd; 1939.
63. FAO. Soil conservation. An international study. Washington DC; 1948.
64. FAO. Soil conservation. Adaptation for Australian conditions of "An international study on soil conservation" by the Food and Agricultural Organisation of the United Nations. Melbourne: Department of Commerce and Agriculture; 1950.
65. Bennett HH. Agriculture in Central America. Assoc Am Geogr. 1926;XVI:63–84.
66. Dodson B. A soil conservation Safari: Hugh Bennett's 1944 visit to South Africa. Environ Hist. 2005;11(1):35–53.
67. Brink W. Big Hugh. The father of soil conservation (with a Foreword by Louis Bromfield). New York: The MacMillan Company; 1951.

68. Bennett HH. Elements of soil conservation. New York: McGraw-Hill Book Company, Inc.; 1947.
69. Kellogg CE. Conflicting doctrines about soils. Sci Monthly. 1948;66:475–87.
70. Pritchard HW. Where have we been? Where are we going? J Soil Water Conserv. 1956;11:21–7.
71. Hartemink AE, Watson HD, Brevik EC. On the soil in soil survey horizons (1960–2009). Soil Horiz. 2012;53(4):30–8.
72. Brink W. Big Hugh. The father of soil conservation. With a preface by Louis Bromfield. New York: The MacMillan Company; 1951.
73. Norton EA. Soil conservation survey handbook. Miscellaneous Publication No. 352. Washington D.C.: United States Department of Agriculture; 1939.
74. Norton EA, Bodman GB, Conrey GW, et al. Report of the horizon criteria committee. Am Soil Surv Assoc Bull. 1934;XV(2001):119.
75. Norton EA, Smith RS. The influence of topography on soil profile character. J Am Soc Agron. 1930;22:251–62.
76. Cline MG. Agronomy at Cornell 1868 to 1980. Agronomy Mimeo No. 82-16. Ithaca, NY: Cornell University; 1982.
77. Kellogg CE. Russian contributions to soil science. Land Policy Rev. 1946;9:9–14.
78. Kellogg C, Davol F. An exploratory study of the soil groups in the Belgian Congo. Congo: Institut National pour l'etude agronomique du Congo Belge; 1949.
79. Kellogg CE, Orvedal AC. Potentially arable soils of the world and critical measures for their use. Adv Agron. 1969;21:109–70.
80. Kellogg CE. A challenge to American soil scientists on the occasion of the 25th anniversary of the soil science society of America. Soil Sci Soc Am Proc. 1961;25:419–23.
81. Kellogg CE. Soil use for abundance in the newly developing countries. Food for peace. ASA Special Publication Number 1. p. 1963:28–35.
82. Helms D. Early leaders of the soil survey. In: Helms D, Effland ABW, Durana PJ, editors. Profiles in the history of the U.S. soil survey. Ames: Iowa State Press; 2002:19–64.
83. Contactgroep Opvoering Productiviteit (COP). De bodem- en landclassificatie in de Ver. Staten van Amerika. 's-Gravenhage: Studierapport Landbouw; 1951.
84. Jenny H. Arrangement of soil series and types according to functions of soil forming factors. Soil Sci. 1946;61:375–92.

9

Pochva Americana II

"There is a more or less definite relation between man in his physical and mental makeup and his soil environment. Systematic studies are still lacking on the part that soils have played in the past and are now playing in determining the peculiarities of races and of individuals."

Jacob Lipman, 1927

In the 1870s, migrants from the Russian steppes settled in the Great Plains of the USA [1]. Some 9,000 of them were Mennonites known for their pacifism, and they were descendants of people from the Netherlands who had settled in West Prussia in the early 1500s, where they established farms and businesses. The second group of migrants from Russia, Ukraine, and Latvia came to the USA in the late 1800s and early 1900s, and after the Russian revolution of 1917, when nearly two million Russians emigrated including Vladimir Zworykin, the inventor of television, Igor Sikorsky who created the helicopter, and the composers Igor Stravinsky and Sergei Rachmaninov [2]. Some emigrated for education, which in Russia was the privilege of the land-owning nobility, government servants, and city folks [3]. Many Jewish families fled because of pogroms, plundering, and persecution [4], but in the USA, anti-Semitism was also proliferating and they were not particularly welcomed [2, 5, 6]. Several immigrants became prominent scientists who individually and collectively shaped American and global soil science, including Jacob Lipman, Jacob Joffe, Constantin Nikiforoff, Selman Waksman, Otto Zeasman, Sergei Wilde, and Vladimir Ignatieff.

Jacob Lipman was born in 1874 in a family of nine children in Friedrichstadt, a small town in a Baltic Province that became Latvia [4]. In 1888, the Lipman family emigrated to the USA, where Jacob's father tried to establish a business in New York City. The business failed and they moved to New Jersey and joined a settlement of Jewish farmers. When Jacob Lipman was 14, he worked as an aide in a law office, and some years later he attended an agricultural school, and then Rutgers College in New Brunswick in New Jersey.

A. E. Hartemink, *Soil Science Americana*
https://doi.org/10.1007/978-3-030-71135-1_9

The director of the New Jersey Agricultural Experiment Station, Edward Voorhees, took an interest in Jacob and encouraged him to study soil chemistry, plant nutrition, and: "...especially the role of microbes in soil processes and plant growth" [7]. Jacob's undergraduate years were characterized by poverty [8]. After his BS degree, he obtained an MS in 1900, and a PhD in 1903 at Cornell University. An important subject was the incorporation of atmospheric nitrogen by legumes, and its biochemical pathway was not well-understood. For his MS research, Jacob studied nitrification, and he expanded on the subject in his PhD studies that focused on nitrogen fixing bacteria. Both degrees were supervised by George Caldwell, who was the first professor of agriculture and analytical chemistry at Cornell University. In 1903, Jacob Lipman and James Bizzell received the first PhD degrees in soil science in the USA [9]. James Bizzell became a professor of soil technology at Cornell University, whereas Jacob Lipman returned to New Jersey and became an instructor in agricultural chemistry and bacteriology at Rutgers University.

Jacob Lipman married Cecilia Rosenthal, and they had three sons, Leonard Hertzel, named after the father of political Zionism, Theodor Herzl, and the twins Edward Voorhees and Daniel Hilgard, named after the experimental station director and Eugene Hilgard [10]. The Lipmans lived in a large, rambling twenty-one room house on the experimental farm, where they offered rooms for female students [4]. Jacob Lipman's initial work was on soil bacteriology and he wrote an extensive review with his director Edward Voorhees, who studied soil fertility [11, 12]. In 1908, Jacob Lipman published one of the first books on bacteriology, entitled *Bacteria in relation to country life* [13], and it was dedicated to Antonie van Leeuwenhoek, who had discovered bacteria in the 1670s. In the book, Jacob Lipman reviewed the structure of bacteria, their chemistry, and food requirements, followed by bacteria in air and water, and bacteria in relation to soil fertility, and all other forms of bacteria on the farm. It was a comprehensive review with a deep appreciation for the bacterial world: "It must be remembered that there is reserved for them a certain task upon the proper performance of which depends the well-being of more highly organized creatures. They are the connecting link between the world of the living and the world of the dead. They are the great scavengers intrusted with restoring to circulation the carbon, nitrogen, hydrogen, sulfur, and other elements held fast in the dead bodies of plants and animals. Without them, dead bodies would accumulate, and the kingdom of the living would be replaced by the kingdom of the dead. And yet the soil bacteria are not mere destroyers, for there are among them species that do constructive work, also indispensable" [4].

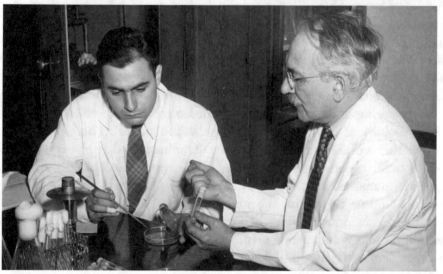

Jacob Lipman (1874–1939) was Director of the New Jersey Agricultural Experiment Station, founder of the journal *Soil Science*, and first President of the *International Society of Soil Science*. Jacob Joffe (1886–1963) came to the USA in 1906 and wrote a pedology textbook that was influenced by Russian soil science. Lower photo: Selman Waksman (1888–1973) (right) and graduate student Albert Schatz (1920–2005) (left) discovered micro-organisms excreting a substance that stopped the growth of tuber-culosis bacteria. That discovery led to the streptomycin anti-bioticum, and the Nobel Prize in Physiology or Medicine in 1952. Photos from Smithsonian Institution Archives, and Rutgers University

In 1911, Jacob Lipman, 37 years old, succeeded Edward Voorhees as the Director of the New Jersey Agricultural Experiment Station. He became interested in the chemistry of inorganic fertilizers and developed the concept of ammonification: the transformation of organic nitrogen to ammonia that was then available for plants [8]. In the process he discovered that the soil harbors free-living bacteria that fix nitrogen without root nodules. Most of his attention went to the direction and expansion of the Agricultural Experiment Station, and he hired and trained some excellent people and built an international network through attending and organizing conferences and meetings. Jacob Lipman had a brother, Charles Lipman, who became a professor of soil chemistry and bacteriology at the University of California Berkeley. They grew up in poverty and believed in the power of education: "...to liberate the human intellect from the shackles of ignorance, provincialism and fanaticism" [14].

At the outbreak of the First World War, Jacob Lipman learned that scientific laboratories were too dependent on Germany for chemicals, glassware, and instruments. He also lamented the dependence of American scientists on literature from Germany and he thought the USA should have its own journals [4]. After he had attended the 1916 meeting of the *Society of American Bacteriologists* where the newly founded *Journal of Bacteriology* was presented, he hatched the idea for the journal *Soil Science* that was to become a monthly journal devoted to problems in soil physics, soil chemistry, and soil biology. It was a daring undertaking, and the question was if soil science was sufficiently important that it warranted a specialized journal. There were doubts on whether sufficient papers would be submitted, and if the new journal would drain papers from the *Journal of the American Society of Agronomy* that was published since 1908 [12]. Jacob Lipman was able to convince Rutgers College to publish the journal: "...for the Benefit of Agriculture and the Mechanic Arts." He then invited acquaintances and friends to be consulting editors and among them were Martinus Beijerinck, Thomas Lyon, Alex. de'Sigmond, John Russell, Oswald Schreiner, Charles Lipman, and two Germans: Eilhard Mitscherlich and Theo Remy.

The first issue of *Soil Science* appeared in 1916, and in the introduction, Jacob Lipman wrote that papers dealing with problems in plant physiology, agronomy, bacteriology, or geology would be included if the paper contributed directly to the knowledge of soil fertility which was broadly defined as the ability of the soil to sustain grow plants. Studies in soil biology convinced Jacob Lipman that it was to become a large research field. The first issue was published shortly after the death of Eugene Hilgard, and it was dedicated to him, with his name and dates in *Fraktur* followed by an obituary

[5]. The *Scientific American* was positive when the first issue of *Soil Science* appeared and wrote that the journal conveniently presented research results which they considered beneficial for all who studied soils [15]. During the first year of *Soil Science*, 59 articles were published, and Selman Waksman accounted for more than 10% of the published pages [12]. The subscription price was $3 per year, and 50 cents was additionally charged for subscriptions in other countries.

From the beginning, the journal became a vehicle for disseminating Russian studies to an American and international readership, which was facilitated by the fact that Jacob Lipman spoke Russian [5]. There was an extensive report on the *First International Congress of Soil Science* and the Transcontinental Excursion in 1927 [16, 17]. In the first few years, there were many papers from Jacob Lipman, Jacob Joffe, and Selman Waksman, and over his lifetime, Robert Starkey published more than 30 papers in the journal. In the 1930s, the journal had about 30 soil scientists as consulting editors from most parts of the world, including Merris McCool, David Hissink, Selman Waksman, and Sergei Vinogradskii. *Soil Science* was a leading journal for many decades, and it was the first American journal devoted to the scientific study of the nature and properties of soil. The impact stagnated in the late 1990s, and the journal ceased after publishing its 9,308th paper in 2019. The 184 volumes of *Soil Science* contained numerous seminal papers, including the classical methodologies for measuring carbon, phosphorus, hydraulic conductivity, moisture flow, and the turnover of soil organic matter [18–22].

Some 17 years before *Soil Science* was launched, the first journal solely dedicated to soil research was published in Russia. The journal *Pochvovedenie* (Почвоведение—the scientific study of soil) was established in 1899 under the auspices of the Free Economic Society. It published Russian soil science which had developed numerous theories and methodologies, and a rich terminology [23]. *Pochvovedenie* published papers in Russian, with summaries in other languages, and the occasional paper in French and English. The journal started to accept papers in English after the *First International Congress of Soil Science* in 1927. Arseniy Yarilov was the editor of *Pochvovedenie*, and he had attended the first congress where he invited: "…American colleagues to make use of this journal and in that way establish closer contact between soil science workers of the United States and Russia" [24]. Both *Pochvovedenie* and *Soil Science* were important for connecting American and Russian soil science.

Jacob Lipman was the editor of *Soil Science* from 1916 until his death in April 1939. He died of heart failure and was 64 years old; Cecilia had died in 1928. Jacob Lipman died a few months before the Commission on

Soil Microbiology meeting at Rutgers in New Brunswick, and the German invasion of Poland. Jacob Lipman was much missed during the meeting, and Selman Waksman, who chaired the Commission on Soil Microbiology, also reflected on the German invasion: "It was tragic that even the few who came to the New Brunswick meeting did not believe, or did not want to believe, or merely said they did not believe, that the war was coming and that Germany had been preparing intensely for it since Hitler's rise to power, until the German hordes actually began to destroy, kill, and loot the Polish republic" [25].

The death of Jacob Lipman was mourned across the world. John Russell from Rothamsted wrote an obituary for *Nature*: "He was a man of unbounded kindness and geniality, wise in counsel, and a delightful host; he attracted and kept together a brilliant band of scientific workers and he made the New Jersey Experiment Station one of the leading agricultural institutions in the world" [26]. Henry Wallace, the Secretary of Agriculture, wrote: "…agricultural science and education throughout the world suffered a heavy loss in the death of Dean Lipman" [4]. Arthur McCall from the *Bureau of Soils* who worked with him on the *First International Congress of Soil Science* wrote in *Science*: "…a young scientist who has not had the friendship of Dr. Lipman, is spared the grief that comes with his loss, but his life is lacking one of its greatest joys and the satisfaction that comes out of such associations" [27].

Jacob Lipman was a humanitarian and an international soil scientist—by birth, and by his way of working [4]. Selman Waksman, like many others, found him to be a gentle man. They had traveled to Rome together in 1924, and in Jacob Lipman's biography, Selman Waksman summarized his character: "…he did not appreciate great music, nor did he become enthusiastic over fine paintings or sculptures or even great natural beauties. What he did observe was the life of the common man, and he was most anxious to do what he could to help improve his lot. He utilized his travels not to visit museums and cathedrals, as many tourists do, but to study methods of farming, problems of food distribution and marketing, and organization of co-operatives" [4]. From the early 1900s until his death in 1939, Jacob Lipman was seminal in the shaping of the soil science discipline. He also named it.

Jacob Lipman and Curtis Marbut attended the soil conference in Prague, in 1922, which was the first American participation in an international soil meeting. The meeting in Prague led to the meeting in Rome where the *International Society of Soil Science* was founded in 1924 after which the journal *Soil Science* received paper submissions from Europe. In 1924, Jacob Lipman was elected the President of the *International Society of Soil Science* and charged with organizing the *First International Congress of Soil Science*. The world of soil science came to the USA and Jacob Lipman brought the Americans into the world of soil science. He had an effective way of working that David Hissink termed 'simple' [28]. In 1926, Jacob Lipman attended a meeting in Groningen where he presided over the chemistry committee. David Hissink had expected that Jacob Lipman would take the opportunity to make an opening speech and talk about the newly formed *International Society of Soil Science*, but Jacob Lipman did not, and started by saying: "The meeting is opened; please Dr. Hissink, what is the first point on the agenda" [28]. An intellectual of great breadth, and a charismatic leader who was well-liked. He built soil science and promoted it globally.

Jacob Joffe was born in 1886, in Kupiškis, Lithuania and emigrated to the USA in 1906—the same year *The Jungle* by Upton Sinclair was published about the Lithuania immigrant Jurgis Rudkus who landed in the Chicago meat industry [29]. Jacob Joffe landed in New Jersey and initially worked in a clothing and shirt factory. He received his education at Rutgers University, took a soil class from Jacob Lipman in 1915, and started to work as a soil microbiologist and soil chemist. He was interested in pedology, had studied the Russian works, and wrote a pedology book that was published in 1936, with a second edition in 1949 [30, 31]. Jacob Joffe positioned pedology in relation to the other sciences based on the work of Sergey Zakharov, who had been a student of Vasily Dokuchaev. He defined Pedology as "…the science dealing with the laws or origin, formation, and geographic distribution of the soil as a body in nature" [32] Constantin Nikiforoff reviewed Jacob Joffe's book and was disappointed by the chapter on soil organic matter and the halftone reproduction of the world soil map that was partly in Russian, and he concluded: "A textbook of this caliber, certainly, deserves a better map."

Jacob Joffe admired the Russian origin of pedology, but he was very critical of Soviet Russia [3]. He translated Russian books into English, like the book by Rode [33], and his view on soils was influenced by Vasily Dokuchaev as well as Curtis Marbut [34]. Jacob Joffe was from the same generation as Selman Waksman, and they came to the USA in the early 1900s, and both ended up at Rutgers University, but they did not get along. In 1957, 5 years after Selman Waksman had received the Nobel Prize, Jacob Joffe retired and moved to Haifa, Israel, where he was active in agricultural research. He died in 1963.

Constantin Constantinovich Nikiforoff was born in 1886 in Pskov, Russia near the border with Estonia—in the same year Jacob Joffe was born some 400 km to the southwest in Lithuania. Constantin Nikiforoff completed his doctoral studies in 1912 under Konstantin Glinka at the Novoalexandrovsk

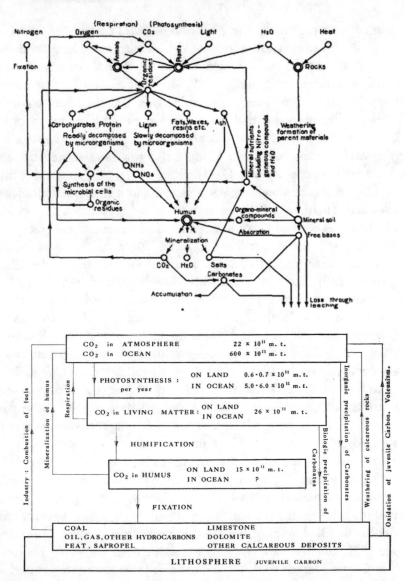

A general scheme of the biopedogenic process of Chernozem formation (top diagram), by Constantin Nikiforoff published in *Soil Science Society of America Proceedings* in 1936. He distinguished three interdependent pedogenic processes: humification, carbonization, and calcification that were responsible for the formation of the Chernozems. Lower diagram: synthesis of global geochemical cycle of carbon and their different pools, numbers in metric tonnes ($\times 10^{11}$)

Institute built on the banks of the Rasshevatka River. He then joined the staff of the Russian Department of Agriculture and Land Improvement and studied thermokarst formation in Siberia and how it affected the building of railroads. He was one of the first to describe the hollows produced by the selective thawing of permafrost that caused the periglacial landscape appearing like karst. He named the karst hollows 'blisters' [35].

Constantin Nikiforoff served as an officer in the Imperial Russian Army and was based in Turkey when the First World War broke out [35]. During the Russian Revolution, he fought with the anti-communist White Army which was defeated and captured by a Red Army detachment, he was court-martialed and sentenced to be shot at sunrise. Shortly before dawn, a detachment of the White cavalry raided the town and released him [35]. In 1917, he escaped from Russia through Turkey and came to the USA in 1921 on a League of Nations pass. The League of Nations was an early attempt towards the establishment of the United Nations. Although President Woodrow Wilson was a proponent of the League, the USA did not join the League of Nations due to opposition from isolationists.

After arriving in the USA, Constantin Nikiforoff drove a taxi in New York City, but he missed the countryside and moved to northern Minnesota, where he became a farm labourer [36]. He was a refugee with no academic or professional documentation, but had a PhD in soil science. Early in 1927, his expertise was discovered by a visiting soil surveyor. Constantin Nikiforoff was then hired by Frederick Alway at the University of Minnesota where he became an instructor and assistant professor of soil science and a colleague of Iver Nygard [37]. At the University of Minnesota, there was little information on Russian soil science except for the translation of Konstantin Glinka's book [38]. Constantin Nikiforoff started an exchange of publications with Sergei Neustreuv of the Dokuchaev Soil Institute in St. Petersburg [1], but the exchange did not last long, as Sergei Neustreuv died in May 1928.

In 1927, Constantin Nikiforoff attended the *First International Congress of Soil Science* where he met again with Konstantin Glinka. Constantin Nikiforoff had been in the USA for six years, whereas it was the first, and last, visit for Konstantin Glinka. Constantin Nikiforoff's background and talents were noted by Curtis Marbut, who asked him, in 1931, to join the soil survey. Constantin Nikiforoff became a member of the field crew that conducted the reconnaissance soil survey of the Red River valley in Minnesota [39]. In the valley, wheat and flax had been grown on the cracking and swelling clay since the late 1800s. His presence attracted attention and soil surveyors from several states visited the soil survey in Minnesota [1].

Constantin Nikiforoff, or Niki as he was commonly called, influenced not only the American soil survey, but also the soil survey in Canada. He was contacted by Joseph Ellis, who conducted soil surveys in the Canadian part of the Red River valley [40]. Joseph Ellis was born in the UK and emigrated to Canada when he was 14, where he worked on a farm. He obtained a BS degree in agriculture in 1918 from Manitoba Agricultural College and then worked for several years as a soil surveyor. In 1927, he obtained his MS in the Division of Soils at the University of Minnesota. Inspired by Russian pedology and working alongside Constantin Nikiforoff, it was Joseph Ellis who brought the Russian way of soil mapping to Canada, where he became the founder and first chief of the soil survey [41, 42]. Joseph Ellis also established the soils department at the University of Manitoba [43].

In 1932, The *Division of Soil Survey* sent Constantin Nikiforoff to North Dakota, where he joined the soil survey crew of McKenzie County with Charles Kellogg and Roy Simonson. Having been trained by Konstantin Glinka, one of the best disciples of Vasily Dokuchaev, Constantin Nikiforoff brought Russian ideas into the soil mapping of North Dakota. It was well-received by Charles Kellogg, who like Curtis Marbut, had admiration for Russian soil science. After Constantin Nikiforoff had worked in North Dakota, he laid out methods for the description of the soil profile, which he called 'recording soil data' and he had done that a few years earlier for soil horizon designation [44, 45]. For recording soil data, he listed most of the field observations that were used for observing a soil profile, such as horizons, their boundaries, texture, consistency, and the physical features of the landscape.

Constantin Nikiforoff was knowledgeable but was at times seen as difficult and critical of his colleagues. In the 1930s, he spent some time in Puerto Rico and disagreed with James Thorp about how the soils were formed on the island. Some of the arguments were the result of English being his second language, and some of it was his view on soils and the Russian schooling in pedology that did not match the traditional and anti-Russian views held by some soil surveyors. In the disagreements with James Thorp, Curtis Marbut had to intervene, and Constantin Nikiforoff knew he was regarded as "criticizingly minded" but viewed himself as: "...not one of these self-conceited Europeans." He also had friends and Roy Simonson was one of them.

Like other Russian pedologists, Constantin Nikiforoff studied hardpans and he knew the works by Eugene Hilgard. He developed a morphological classification system for soil structure, and paleopedology—a field of pedology that had its origin in Russia [46, 47]. Paleopedology was the study of soils of past geologic eras, by which insights were gained about long-term

Russians who shaped American and international soil science: Constantin Nikiforoff (1886–1979); Sergei Wilde (1889–1979); Vladimir Ignatieff (1919–1994). Photos by the *International Society of Soil Science* and Department of Soil Science, University of Wisconsin-Madison

changes in climate, land use, and civilizations [46, 48]. Constantin Nikiforoff was also interested in soil organic matter and contributed a chapter, *Soil Organic Matter and Soil Humus*, to the 1938 *Soils & Men* yearbook [49]. In 1936, he published a paper on the formation of Chernozems and compiled one of the first carbon geochemical pathways [50, 51]. It included estimates of the carbon pools in the atmosphere, living matter, soil and fossil fuel, and rocks. The amounts in the atmosphere and living matter were about the same and considered to be three-times more than in the soil. That was an underestimation.

In 1959, Constantin Nikiforoff sought reappraisal of the soil, as he noted that for: "…agronomists soil was just the 'dirt' supporting their crops. This simple utilitarian concept of soil is so deeply entrenched in people's minds that one may wonder whether it would not be less confusing to leave the term soil entirely to agronomy and coin some other name for the geochemical surface formation which is referred to in agronomy as 'the soil'" [51]. Constantin Nikiforoff stressed the need for basic research and for a shift of attention in soil study from a descriptive morphology to the study of genesis and dynamics. For him, soil science belonged to the family of earth sciences and not to the biological sciences [51].

Constantin Nikiforoff retired in 1959 and became a consultant for research agencies in the federal government, and translated the legend of the soil map of Soviet Russia. In 1960, he attended the *Seventh International Congress of Soil Science* and served as interpreter for the Russian delegation and the organizers. Except for Victor Kovda, who had traveled widely, none of the Russian participants of the congress in 1960 spoke English [52]. Some members of the Russian delegation regarded Constantin Nikiforoff as a betrayer of Russia. This did not bother him. After the *Seventh International Congress of Soil Science*, he had a heart attack and started walking 20 km per day to keep healthy [53]. That helped for many years but in 1979, Constantin Nikiforoff had a fatal heart attack in his house in Hyattsville and died, he was 93 years old. With him, the American soil science community had, in their midst for over 50 years, a scientific grandson of Vasily Dokuchaev and a student of Konstantin Glinka. Constantin Nikiforoff gained some prominence, but his contributions were not broadly accredited. They might have liked his ideas, but perhaps not the person.

Selman Waksman was born in 1888, in Odessa, the Kiev region in Ukraine. The land was draped by Chernozems and the growing of wheat, rye, barley, and oats. He was brought up in a matriarchal home and lost his young sister to diphtheria. His mother died in 1909. A year later, he emigrated to America, where he had relatives. A cousin met him in New York, and Selman worked for some years on their chicken farm in New Jersey [54]. He studied agriculture instead of medicine and that decision was guided by Jacob Lipman, who was 14 years older. Selman Waksman obtained his degrees from Rutgers College, and his MS thesis, supervised by Jacob Lipman, focused on soil fungi and their activities.

After his MS graduation, he wanted to study at a larger university and applied for a fellowship in the Soils Department of the University of Illinois. The Head of the Department, Cyril Hopkins, refused to grant such fellowship to him on the grounds that he: "...could not waste public funds by supporting a student who did not come from a farm and, therefore, could hardly be expected to make an important contribution to practical or even scientific farming" [25]. Selman showed that letter to Jacob Lipman, who smiled and said: "How little he appreciates that help to the farmers must come from the fundamental sciences and from those trained in them." Jacob Lipman added: "You had better stay another year here. It will give you a chance to find out something about the soil microbes; they will have to listen to you then." Selman Waksman followed that advice and then left for the University of California Berkeley, where he studied for a PhD in biochemistry. Again, it was at the recommendation of Jacob Lipman, whose brother, Charles Lipman, was a professor at Berkeley. Charles Lipman had worked for some years with Eugene Hilgard [55], and he studied remnants of spores of bacteria or fungi or resting bodies of other micro-organisms in ancient rocks. He claimed to have found such spores in Precambrian rocks of Canada and the Grand Canyon [25, 56, 57]. Selman Waksman's PhD thesis was entitled *Studies on proteolytic activities of soil micro-organisms, with special reference to fungi*. After obtaining his PhD degree, he was hired by Jacob Lipman and returned to New Jersey. On the banks of the Old Raritan, soil scientists inspired by Sergei Vinogradskii established the strongest soil microbiology program in the country.

In 1924, Selman Waksman traveled to Rome to attend the *Fourth International Conference of Pedology*. After the conference, he visited Martinus Beijerinck, and microbiological and soil laboratories in several countries. He returned to Russia for the first time since he had emigrated to the USA in 1909. In Paris, he visited the *Institut Pasteur*, where Waldemar Haffkine had developed the vaccines against cholera and the bubonic plague some 30 years earlier [58]. Selman Waksman was familiar with the work of Waldemar Haffkine, who was also from Odessa and had fled the Ukraine in 1888, 22 years before Selman Waksman. In 1924, Waldemar Haffkine had retired and moved to Switzerland, became deeply religious and returned to Orthodox Jewish practice [58]. At the *Institut Pasteur*, Selman Waksman spent some time with Sergei Vinogradskii, and they had met one month earlier at the *Fourth International Conference of Pedology*. Sergei Vinogradskii was 32 years older than Selman and had identified bacteria that cause nitrification.

Sergei Vinogradskii and Martinus Beijerinck were born in the 1850s. They studied the biological aspects of soils, and their findings increased a broader ecological understanding and impacted the management of soil fertility [59]. They met once in Paris, in 1896, at the 70th birthday celebration of Louis Pasteur [60]. With 150 years in between, Martinus Beijerinck was viewed as the successor of Antonie van Leeuwenhoek, and he built a school of bacteriology and virology in Delft in the Netherlands. He was a tempered and relentless researcher and known for being difficult with students. The main aim of his work was to create order in the chaos of the microbiological world. Experiments were his passion, and like many scientists at that time, Martinus Beijerinck had broad interests and knowledge. Besides Antonie van Leeuwenhoek, he admired Louis Pasteur, Charles Darwin, and the physicists Isaac Newton and Michael Faraday [61]. When he turned 70, he was forced to retire, or as he bitterly expressed it: "forcibly removed from his laboratory" and he never visited Delft again. He kept an interest in soil microbiology and his greeting to the next generation became: 'Fortunate are those who now start.'

Martinus Beijerinck was a socially eccentric figure with an ascetic lifestyle. According to Selman Waksman, he was:"…austere in appearance, strict in his attitude toward others, apparently lacking in artistic or literary interests, and his whole life was devoted to science. Every word he uttered was along scientific lines" [25]. Like Jacob Lipman, Martinus Beijerinck had no appreciation for music, which he found fatiguing and bad for scientific achievements. Similarly, he maintained an aversion to history and theology [61]. Martinus Beijerinck, like Waldemar Haffkine, had never married. A shy man, Martinus Beijerinck moved with his two sisters Johanna and Henriëtte to the village of Gorssel in the east of the Netherlands. His sister Johanna, who had been a teacher most of her life, died shortly after their move. They had a greenhouse beside the house, a small laboratory, and a garden which was full of botanical wonders that he showed to his visitors with great enthusiasm [61].

Gorssel was a wooded village on the Pleistocene sands surrounded by small dairy farms that charmed and attracted artists, retired politicians, and others from the busy western part of the Netherlands. Martinus Beijerinck became well-known in Gorssel, and letters to him were simply addressed 'Professor, Gorssel.' They were all delivered and none went missing. Martinus Beijerinck stuck out, with a cape and a slouch-hat, and his temper also did not go unnoticed and in the village, he was nicknamed 'the grouch.' In his final years, he lived with his sister Henriëtte, who was deaf. They could scarcely exchange thoughts, and hungry for debate and intellectual stimulus, Martinus Beijerinck founded a local society for scientific lectures, where he spoke on subjects like 'Life and Death' and 'Imagination and Science.' Henriëtte was an artist and made drawings and pictures of plants and microbes, which were used for teaching purposes in the laboratories of Microbiology and Botany in Delft [61].

When Selman Waksman toured Europe in 1924, he was able to visit Martinus Beijerinck in Gorssel, where not everyone was welcome, but Selman came with the regards of Jacob Lipman. The visit was an influential event for the 36-year-old Selman Waksman, who considered Sergei Vinogradskii and Martinus Beijerinck the founders of non-medical microbiology. He and his wife Bobili took the bus from Deventer to Gorssel, which stopped in front of Martinus Beijerinck's house, where they were greeted by Henriëtte. Martinus Beijerinck came to the door and greeted him: "You are the actinomycetes man!" [25]. Martinus Beijerinck was dressed in a black suit with a large black tie. They had long discussions about a range of scientific topics and according to Selman Waksman, his English was good, but he had to search for a word sometimes, and often mixed in Dutch words [25]. Martinus Beijerinck was critical of Sergei Vinogradskii because he never had to work for a living, and he was glad to learn that Selman Waksman was not a communist. They had a frugal Dutch lunch and Selman Waksman took several photographs. In his biography, Selman Waksman noted: "Although I spent only one day with him, this day left a profound impression upon my subsequent work" [25].

Martinus Beijerinck died from an intestinal hemorrhage on the first day of 1931, he was 79 years old. He was cremated a few days later and buried at sea. A burial was avoided because he had been worried that there would arguments, and he insisted that there should be no speeches. His scientific friends held a service in Gorssel and in his birth town Amsterdam where tributes were delivered. At his request, Henriëtte burned all his correspondence. Henriëtte died six years later and left the house in Gorssel to an organization that used it as a holiday home for female teachers. Martinus Beijerinck's successor in Delft, Albert Jan Kluyver, placed the framed photo of Selman Waksman above his desk.

Sergei Vinogradskii was a cultured European, broadly trained and educated and equally at home in the art and sciences. He had only one student, Vasily Omelianskii, and unlike Martinus Beijerinck, he did not leave behind a school or a student pedigree [60]. Sergei Vinogradskii had studied in St. Petersburg, first two years of piano, and then chemistry. He grew up in a banking family and became a man of the land and a philosopher. While in Paris, he kept an interest in scientific life in Soviet Russia, which meant maintaining relations with the communist bosses. Others who had left Russia did not do that, but they remained Russian loyalists and helped their friends who had stayed in Soviet Russia [2]. Sergei Vinogradskii became a world-renowned scientist, and the Russian government found him to be a useful propaganda tool [7]. In 1924, Selman Waksman visited Sergei Vinogradskii, who was also from Ukraine. They became friends and corresponded for over thirty years. Selman Waksman visited him several times in Brie in France and in Russia. Despite frequent invitations, Sergei Vinogradskii did not come to visit the USA. Selman Waksman wrote his biography in 1953, the year Sergei Vinogradskii died, at the age of 96 [60].

Sergei Vinogradskii was in his mid-seventies when he reflected on his own work: "...to judge oneself, especially at the end of one's days, is something totally different. One can see especially clearly how much more could have been done than has been done" [60]. That notion was not that different from Martinus Beijerinck who, neither as student, nor teacher, nor professor attained what he should have attained, into his own opinion: "If I had been ambitious, I might have gained some glory" [61]. So spoke the scientist who, among others, discovered plant viruses and first described biological nitrogen fixation.

While he toured soil laboratories in Europe, Selman Waksman noted that many microbiologists were not living up to the field's possibilities [54]. He found that the microbiological properties of the soil were understudied, and was convinced that: "...there is no chemical reaction in the soil that is not initiated and catalyzed by bacteria or fungi" [62]. By the end of his 1924 tour, he had developed a network of friends including Sergei Vinogradskii, Martinus Beijerinck, and John Russell of the Rothamsted Experiment Station. He received his first exchange student from Marjory Stephenson who discovered in 1933 that soil bacteria produced methane as a metabolic byproduct in the absence of oxygen [63]. After Selman Waksman had returned from his tour across Europe, he reviewed the status and progress in soil microbiology in a paper that offered numerous research ideas [64].

When Jacob Lipman became the Dean of the Agricultural College at Rutgers University, Selman Waksman and Robert Starkey took over the teaching and research in soil microbiology. They studied the microbial transformation of sulfur and wrote a book together [65, 66]. Selman Waksman read in German, Russian, French, and English, and in the 1920s and 1930s, he wrote books on soil humus, enzymes, soil microbes, and various soil microbiology laboratory manuals [66–68]. He did not forget his roots and dedicated the Russian edition of the humus book to the pedologist Konstantin Glinka, the chemist Konstantin Gedroiz, and microbiologist Vasily Omelianskii. His research interest focused on antagonisms among micro-organisms and in particular the actinomycetes, which he had studied since 1914 [69]. Through extensive screening of various groups and species of soil microbes he discovered streptomycin that inhibited the growth of bacteria causing infectious diseases such as tuberculosis. Actinomycetes produced substances that topped the growth of tubercle bacillus and several other penicillin-resistant bacteria on a petri dish [70]. It was a significant finding, as tuberculosis killed over 50,000 people in the USA each year.

Selman Waksman trained over 70 graduate students, including René Dubos from France, who obtained his MS in 1926 and PhD in 1927 at Rutgers University [71]. Born in 1901, in a little village north of Paris, René Dubos attended the Agronomical Institute and was interested in working in the French colony of Vietnam. That desire was dampened by his poor health, and in 1923, he joined the *International Institute of Agriculture* in Rome, where he became an editor of an agricultural journal. In 1924, he attended the *Fourth International Conference of Pedology* and was made responsible for guiding visiting scientists around Rome [59]. At the conference, he listened to papers presented by Sergei Vinogradskii, Selman Waksman, and Jacob Lipman. René Dubos was particularly fascinated by the work of Sergei Vinogradskii, and after a session, he spoke to Jacob Lipman, who mentioned a possible position at the New Jersey Agricultural Experiment Station. Eager to see the USA and without a real plan, René Dubos bought a boat ticket to the USA, and in September 1924, he boarded in Le Havre the *Rochambeau*, named after a French Count who fought in the American Revolutionary War. Coincidentally, Selman Waksman was also on the *Rochambeau*, returning from his microbiology trip through Europe and Russia. After several conversations, Selman Waksman offered René Dubos a position in his laboratory at Rutgers University [59]. For three years, René Dubos worked on the problem of cellulose decomposition by bacteria. After he finished his PhD in 1927, René Dubos moved to the Rockefeller Institute for Medical Research, where he studied microbial diseases and the analysis of the environmental and social factors that affect the welfare of humans [59]. Later on, he coined the slogan 'Think globally, act locally' signifying that ecological consciousness should begin at home.

A large group of international and aspiring students and scientists spent time in the laboratory of Selman Waksman. During René Dubos, his time at the laboratory, there were graduate students from Denmark, Russia, Brazil, the UK, and a few Americans [59]. More international visitors came after the *First International Congress of Soil Science* in 1927. David Hissink arranged for Frederik Gerretsen to spend six months at the laboratory. Frederik Gerretsen had studied under Martinus Beijerinck and was one of the first to study nitrification and denitrification in tropical soils. At the laboratory of Selman Waksman, Frederik Gerretsen studied the influence of temperature and moisture on the plant residue decomposition by micro-organisms. The Austrian Walter Kubiëna visited the laboratory to study soil microbes using a special microscope which he had developed [25]. Hans Jenny came on a Rockefeller scholarship in 1926. Fritz Scheffer visited Selman Waksman in the early 1930s and trained a generation of soil scientists in Germany after the Second World War. In 1933, Fritz Scheffer became a member of the Nazi party NSDAP and that ended their association. Selman Waksman was well aware of the anti-Semitic activities of the Nazis and resigned from two German soil microbiology journals in 1935. He also voted against holding the *Fourth International Congress of Soil Science* in Germany in 1940 [25].

Selman Waksman bought a car in the 1930s. This was seen as exceptional, if not decadent, for an academic, and in particular for someone working in soil microbiology. As he wrote in his biography: "…the point of view prevalent in many European countries and also in America is that this field of research must be considered a luxury science, which at best could only explain certain soil practices, and not a fundamental science which could contribute new interpretations of natural processes and which might yield new practical applications" [25]. In the 1940s, some state senators in New Jersey requested the dismissal of Selman Waksman because of his presumed failure to contribute to the agriculture of the state [72]. That must have reminded him of the refusal of Cyril Hopkins some 30 years earlier to admit him as a graduate student as he did not come: "...from a farm and, therefore, could hardly be expected to make an important contribution to practical or even scientific farming" [25]. A few years after the senators' request for dismissal, he and several graduate students including Albert Schatz discovered micro-organisms excreting a substance which they named streptomycin. The discovery led to the streptomycin antibiotic that soon went into production and ended widespread tuberculosis. Over the years, the university received millions of dollars in royalties. There were no more requests for dismissal.

Selman Waksman was on the cover of *Time* magazine in 1949, and a stamp with his photo was given out in Ukraine. He received the Nobel Prize in Physiology or Medicine, in 1952. In his speech at the Nobel Prize dinner, he said: "With the removal of the danger lurking in infectious diseases and epidemics, society can face a better future, can prepare for a time when other diseases not now subject to therapy will be brought under control. Let us hope that in contributing the antibiotics, the microbes will have done their part to make the world a better place to live in." The soil was not mentioned. In the same year, he published his acclaimed book *Soil Microbiology*, in which he considered soil microbiology a 'borderline science' since it involved problems in ecology, physiology, and biochemistry [73]. For Selman Waksman, the soil was a living system, and he was convinced that soil microbiology had made only a start [73]. Soil as a living system had been promoted by Eve Balfour in the 1940s [74], and Selman Waksman made 'the soil as a living system' a popular perspective.

Selman Waksman's autobiography, published in 1955, described his journey from the Russian steppe to the work on antibiotics, and he gave much credit to Jacob Lipman. His search for a solution to tuberculosis may have been motivated by the work of Waldemar Haffkine—the zoologist turned bacteriologist who injected himself to test a cholera vaccine [58]. Selman Waksman was a broad scholar, and Hans Jenny, who worked in his laboratory for a few months, considered him an intellectual [62]. Like Jacob Lipman, Selman Waksman was interested in ancestry, wrote several biographies and many obituaries. He dedicated his own autobiography to his grandchildren whom he considered 'Native Americans' and wrote the following words: "Your grandparents came to this country as pioneers to help build a new world. Just as earlier pioneers who came to clear the forests, cultivate the virgin land, fight the undesirable animals and transplant the desirable ones, so your grandparents come to avoid persecution, to find greater freedom, and to contribute their share in making this country a better place in which to live. They came from an old race, one that has given the world its highest

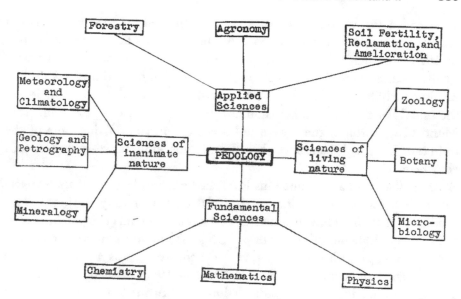

Pedology at the center of other scientific disciplines, from *Pedology*, 2nd edition, by Jacob Joffe published in 1949. Jacob Joffe wrote: "Scrutinizing the achievements of the chemists, geologists, and agronomists in the study of soils, one cannot but recognize that our knowledge has now advanced in proportion to the efforts made"

code of ethics and morals, they, in turn, have tried to create more knowledge, to help alleviate human suffering, and to make the life of man a happier one. They have labored so that you will find the world perhaps somewhat freer from prejudices, freer from suffering, than they themselves have found it" [25]. Selman Waksman became a celebrated scientist. His contributions to soil science include some 60 papers in the field of soil microbiology [71]; and they have been overshadowed by his paper describing the discovery of streptomycin.

Vladimir Ignatieff was born to an aristocratic family in 1905, in Kiev, Ukraine. His father, Count Paul Ignatieff, was Minister of Education in the Czarist Government, but was imprisoned after the 1917 revolution. In 1919, the Ignatieff family fled to England where they bought a 80 hectare farm near Hastings along the south coast [75]. Vladimir Ignatieff was the second of five sons, and attended the Agriculture College of London University, and spent the summer of 1926 on a dairy farm near Montreal in Canada. The family emigrated to Canada, in 1928, which was quite the change, as Vladimir Ignatieff would say later: "Canadians cured my family of any illusions of grandeur they might have had" [76]. They settled two hours west of Montreal. Vladimir wanted to become a homesteader and grow wheat, but wheat prices were low and declining, so instead of becoming a farmer, he attended the University of Alberta. He obtained an MS degree in soil science and worked on the Breton experimental plots that were established in 1929. Research at the Breton plots focused on yield maintenance with small amounts of fertilizers and crop rotation. Vladimir continued his studies, earning a PhD degree in biochemistry at the University of Toronto. It was the middle of the Great Depression and while he was riding the train from Montreal to Toronto, there were hundreds of farmers on the roofs of the train cars. They had abandoned their homesteads and went searching for jobs elsewhere.

After Vladimir had obtained his PhD, he became a lecturer at the University of Alberta but when France capitulated to Germany in June 1940, Vladimir Ignatieff and his brothers joined a reserve infantry regiment named the Calgary Highlanders [77]. He was 36 years old and sent to England, in September 1941. With his degree in biochemistry, it was assumed that he was an expert in chemical warfare. In October 1943, he went to Sicily and among the Canadian forces, Vladimir Ignatieff became known as 'The Mad Russian.' In December 1944, he was sent back to England and to the front again to investigate chemical and petroleum industries. He moved behind the allied forces and was shocked by the destruction in the Ruhr. Near Münster in Germany, he examined a factory where tabun, a nerve gas, had likely been produced. He had difficulty identifying tabun as he was only trained in the use of nitrogen mustard [75], a cytotoxic organic that was the first chemotherapeutic agent for treating cancer.

In September 1945, Major Vladimir Ignatieff boarded a ship in England and returned to Canada where he was decorated for his war efforts. He went directly to Toronto to see his wife and children for the first time in four years, but his mother had died in 1944, and his father, a few weeks before his return [77]. Two days after he had arrived in Toronto, he was seconded to the Department of External Affairs and sent to Washington and then to Chateau Frontenac in Quebec to participate in the meeting that led to the establishment of the Food and Agricultural Organization. This organization had been in the making since the International Institute of Agriculture was formed in 1905, and the League of Nations in 1920, and the basis for this new international agricultural organization was laid in the midst of the war.

In 1943, the US Government hosted the first United Nations Conference on Food and Agriculture in Hot Springs, Virginia. President Franklin Roosevelt spoke at the conference, that was attended by 274 diplomats from 44 countries. The theme of the conference was *Freedom from Want* and the focus was primarily on food production after the war [78]. It was concluded that freedom from want was difficult to achieve without concerted action among likeminded nations to expand and improve food production, to increase employment, to raise levels of consumption, and to establish greater freedom in international commerce. The main resolution of the conference was that no nation can do it alone. The Department of Agriculture sent out a brochure after the conference reviewing the role of the American farmers, and the department feared that feeding the world meant: "...the United States is going to play Santa Claus to the rest of the world" [79]. Before the war, the USA had over 6% of the global population and 5–10% of the food was exported, while considerable food imports were made. It was in the national interest: "...to give a hand to the people of the rest of the world when they are down and out," but the department emphasized that farm productivity had to increase, and that there was a new farm generation for which there was no place in farming. In addition, unemployed people from cities moved to rural areas making the rural population even larger [79].

At the 1943 United Nations Conference on Food and Agriculture, a commission drafted a constitution and assembled ideas for activities of the new organization [80]. Charles Kellogg was asked by the American representative Gove Hambidge to develop ideas for a 'Department of soil science and management' within the World Organization on Food and Agriculture. He wrote a proposal which he sent for comments to Emil Truog, Richard Bradfield, and Bill Pierre. Charles Kellogg envisioned that the new soil science and management department had advisory commissions for regional soil problems, including soil fertility problems, soil erosion control, fertilizers and distribution, irrigation, drainage, and tillage. He recommended that the department would undertake a world survey of soil conditions, and the survey should begin with assembling and compiling the existing soil information which he thought was considerable. At last, the department should have an educational program that would standardize nomenclature and 'transliterate' soil terms [81]. An annual budget of about $685,000 was needed.

The FAO was officially founded in October 1945 and established its first headquarters in Washington by early 1946. Some of the first activities of the FAO were advisory and focused on the collection and exchange of agricultural information [80]. There was a fear by some members of the *International Society of Soil Science* that the FAO as a governmental organization was to dominate the society and its congress [82]. But the FAO never developed a focus on research, and mostly concentrated on international extension programs for rural development. Over time, the FAO established a 'Department of soil science and management' as Charles Kellogg had perceived it in 1943, but it was named 'Division of soil survey and fertility' and later the 'Land and Water' division [80].

Vladimir Ignatieff was hired as the first FAO employee [76]. He moved from Toronto to Washington, where he became friends with Charles Kellogg, and they had met at the meeting in Quebec. Vladimir Ignatieff's task was to provide an overview of fertilizer use in the world and how it should be optimized. A questionnaire, in the form of a circular letter, was sent to member countries requesting information and it became the first worldwide inventory on the use of manures and fertilizers aiming to increase crop production. Charles Kellogg oversaw the work and Emil Truog was a member of the advisory committee. The report was to be called *Efficient Use of Fertilizers*, but Emil Truog wondered whether it should be entitled *Monographs on Efficient Use of Fertilizers and Soil Amendments*. That did not happen, and in 1949, the first edition of *Efficient Use of Fertilizers* was published [83]. Two earlier FAO soil reports had focused on saline soils and soil conservation [84].

Efficient Use of Fertilizers reviewed the role of fertilizers, plant nutrients, the need for organic matter, use of fertilizers, crop rotations, and plant nutrient relationships in different soil regions. The report was timely as there was a dire need to increase food production after the Second World War [85]. Over 50 people contributed to the report, including Charles Black, Edward Crowther, Cees Edelman, Frederick Hardy, Floribert Jurion, Charles Kellogg, Bill Pierre, and Emil Truog. The monograph contained photographs that Charles Kellogg had taken in the USA, Canada, Congo, New Zealand, and the UK. It also included the soil pH—nutrient availability diagram of Emil Truog [86], and one of the first liquid manure drills from Denmark that aimed to reduce gaseous losses of nitrogen when applying manure.

In 1958, the second edition of *Efficient Use of Fertilizers* was published, with 500 references, and highlighting several strategies to encourage fertilizer use by farmers. For the second edition, Vladimir Ignatieff had the help of Harold Page, who had retired from Imperial College of Tropical Agriculture in Trinidad and Tobago. Both had experienced the Second World War. Harold Page had studied at the University of London, in Berlin, and at the *Institut Pasteur* in Paris, and in 1920, he was hired by John Russell at Rothamsted. He became the chief chemist and head of the chemical department and worked on nitrogen fixation by green algae, lignin in humus, and carbon and nitrogen cycling in soils [87, 88]. In the mid-1920s, Harold Page summarized the work of Konstantin Gedroiz on base exchange and absorption which had been translated from Russian into English by Selman Waksman [89]. In 1936, he became the director of the Rubber Research Institute in Malaysia, and during the Second World War, he was interned in a Japanese camp in Sumatra for three and a half years. After the war, he became the principal of the Imperial College of Tropical Agriculture in Trinidad and Tobago, and then joined FAO as a consultant on soil fertility and fertilizers [90].

The second edition of *Efficient Use of Fertilizers* was a marvelous piece of international cooperation and included authors working in Africa, but also Luis Bramão, Guy Smith, and Charles Stephens. Charles Kellogg co-authored three of the eleven chapters, and Emil Truog had retired, but his pH-nutrient diagram was included. *Efficient Use of Fertilizers* contained sections on plant nutrients, the necessity of organic matter and organic manures, commercial fertilizers, factors affecting its use, time and method of application, crops and fertilizers, different soils and the plant nutrient relationships, economics and the farmer, and agricultural services. The book was well-received and much needed, as the preface stated: "Much of the world is still hungry and ill clad. If the soils, the people, and the skills of modern science and economic resources can be brought into proper relationship, an efficient agriculture can supply the food and clothing. One can see how enormous the task is and still assert that is manageable." Likely, those words were written by Charles Kellogg.

Two years after the second edition of the *Efficient Use of Fertilizers* was published, the FAO created the Fertilizer Program, as part of the Freedom From Hunger Campaign and distributed and subsidized fertilizers in low-income countries [80]. It was a standard work on soil fertility, and was some years later overshadowed by the fertilizer guides of Helmut von Uexküll and Jan de Geus, but those guides, published by the fertilizer industry, focused on tropical crops, and lacked the recommendations for different soils [91, 92].

The second task that Vladimir Ignatieff undertook at FAO was the *Multilingual vocabulary of soil science* [93]. The need for such vocabulary had been made by Charles Kellogg, in 1943, as to facilitate exchange and create a common understanding of basic soil terms. The collection of multilingual terms was started by Herbert Greene in 1949, and they were presented at the international soil congresses in the Netherlands, in 1950, and in Congo, in 1954. The languages were English, French, German, Spanish, Portuguese, Italian, Dutch, and Swedish and for the 1960 revised edition, Vladimir Ignatieff added Russian to the vocabulary. Like the 1954 edition, the title of the book was in English, French, and Spanish, but not in German [94].

The FAO had North American roots, with the meeting in 1943, in Virginia, the foundational meeting in Chateau Frontenac in Quebec in 1945, and the headquarters in Washington. In 1951, the FAO member states decided that the new headquarters was to be based in Europe, and Rome was preferred, as it had the International Institute of Agriculture. The institute was founded in 1905 by King Victor Emmanuel III at the instigation of David Lubin, a Californian businessman and farmer of Polish origin.

The mission of the institute was to help farmers with knowledge, establish a system of rural credit, and assistance in trading their produce. The International Institute of Agriculture ceased operations when the FAO took over the mandate of international coordination in agriculture. For the FAO headquarters, the Palazzo at Via delle Terme di Caracalla was chosen that was built under the Fascist government of Benito Mussolini to house the Ministry of Italian Africa. Vladimir Ignatieff was the first FAO employee in February 1946, and in 1951, he and his family moved to the Eternal City. He eventually became the Deputy Director of the 'Land and Water Development Division.' Vladimir Ignatieff retired at the age of 61 in 1966, and they returned to Canada and settled in a rural community near Montreal close to where his parents had settled in 1928. At the *Eleventh International Congress of Soil Science* in Alberta in 1978, he was made an Honorary Member of the *International Society of Soil Science*. Count Vladimir Ignatieff died in 1994, he was 89 years old.

Emil Truog was of the same generation as Jacob Joffe, Constantin Nikiforoff, and Selman Waksman. He occasionally corresponded with them in relation to papers for *Soil Science* and for the organization of the *First International Congress of Soil Science*. Emil Truog was an anti-communist and worked well with Russians who had similar views. There were two Russians and a Ukrainian in the Department of Soils at the University of Wisconsin: Sergei Wilde, Dmitri Pronin, and Otto Zeasman.

Otto Zeasman obtained his BS degree in 1914 and was hired in 1919 by the Department of Soils at the University of Wisconsin. He worked on soil drainage and technologies for controlling soil erosion and recognized that melting snow water on frozen ground caused runoff and soil erosion [95]. Otto Zeasman was born in Kiev, in 1886, the same year as Jacob Joffe and Constantin Nikiforoff. The family emigrated to Wisconsin when he was two years old. At that time, the passing of a physical and mental test was needed to enter the country. When Otto Zeasman was 12, he worked in lumber mills and farms, and he finished his school when he was 13 years old. He joined a debate team at a high school although he was not enrolled, and his love for argumentation was trained and reaffirmed [96]. After working different jobs, he received a scholarship to the University of Chicago, but after a year, he enrolled at the University of Wisconsin, where he received his BS degree.

Many of the soils in Wisconsin were not well drained. The glacial ice had flattened the land, and the till-subsoils were dense and hard. Ponding was common during spring snowmelt and rains, and there was a need for designing drainage systems. Otto Zeasman studied the marshes and wetlands of the state, and he was appointed as the State Drainage Engineer. He worked on the gully erosion problem and noted in 1930 that soil erosion was a local and national problem [95]. When the New Deal programs were rolled out in 1933, young college graduates who led the program asked Otto Zeasman for advice [96]. He disagreed with the approach of the *Soil Conservation Service*, as well as Aldo Leopold's method of soil erosion control that relied on vegetation and land use. Otto Zeasman emphasized the engineering and structural approach and found that planting trees on steep and eroding slopes was useless and a waste of taxpayers' money. His method to curb soil erosion included dams and permanent structures that aimed at reducing runoff. Both Aldo Leopold and Otto Zeasman became members of the Wisconsin Conservation Hall of Fame; Aldo Leopold became the patron saint of the ecological and environmental movement [96]. Otto Zeasman vanished from the scientific horizon.

Sergei Wilde was born in Moscow, in 1898 to parents of Tartar and Dutch ancestry. At an early age, he was introduced to the boreal forest which he came to love [97]. He was schooled in Moscow and Prague, where he worked on his habilitation for a professorship. In 1929, he received a notice from the American Consulate that preferential visas would be granted to agricultural and forestry workers that wished to emigrate to the USA [72]. He applied, obtained a visa, took the boat and arrived on the 7th of May 1929 at Ellis Island, where an officer asked him the standard question: "Whom do you know in America?" to which he answered: "Mr Franklin and Mr Lincoln." That was sufficient. Sergei Wilde worked in Minnesota for a while and in 1934, he was asked by Andrew Whitson to join the Department of Soils at the University of Wisconsin.

By the early 1900s most of the natural forest in Wisconsin had been logged [98]. It was Sergei Wilde's task was to provide much needed soil management information to timber managers, and to work with forestry camps where the labor force consisted of prisoners who had served part of their sentences [72, 97]. Not much was known about soils under forests in Wisconsin and research on soils under forests had been conducted by Peter Müller in Denmark, Ferdinand Senft and Emil Ramann in Germany, and Ivan Tiurin, who was the chair of soil science in the Forest Faculty of the Kazan University [99]. Those studies brought the importance of soils into forestry that led to studies on tree–soil interactions and the role of soils in increasing wood production. In the USA, students in forestry took classes in soils, but these were usually given in the agricultural departments but that changed in the 1940s when several universities including the University of Wisconsin emphasized the role of soils in forestry [100]. It was Sergei Wilde, who brought soil research in forests to the forefront in Wisconsin.

His research aimed to interpret forest soils as carriers of definite floristic associations, as media for the growth of nursery stock or forest plantations, and as dynamic systems that react to different forms of silvicultural practices [97]. The primary aim was to enhance the production of wood without depleting soil fertility or contaminating the environment. He studied tree nutrition, tree–mycorrhiza relationships, and reforestation, and wrote a widely used reference book on forest soils [101]. Semantics was important to him [102], and he found 'parent material' an unacceptable term as lifeless geological strata could not be the material from which a living soil is derived. The O-horizon of soils under forest was barely recognized in pedology and Sergei Wilde emphasized its importance for soil formation [101].

When Sergei Wilde joined the Department of Soils, there were six professors and ten graduate students who taught the introductory courses in soil science. He found that the professors were non-drinking, non-swearing fragments of the Victorian era dedicated primarily to the production of crops [96]. Sergei Wilde mentored Jaya Iyer, who, in 1969, was the first female to obtain a PhD from the Department of Soil Science. He became friends with Aldo Leopold, who had an office in the basement of the soils building for some years. Sergei Wilde played viola together with Francis Hole during lunch hour and at Emil Truog's retirement dinner at the Memorial Union in June 1954, under the name Fine Earth String Ensemble. Sergei Wilde was well-read and loved to drink and swear. He was a pre-revolution Russian aristocrat, had a distaste for the communist revolution, and found in Emil Truog a similar-minded colleague. In 1947, he sent Emil Truog a letter with a verse by the humanist poet Walt Whitman who was a supporter of the Free Soil Party that in the 1850s opposed the expansion of slavery into the western USA.

In the early 1950s, Emil Truog hosted Dimitri Pronin, who had come to the USA in 1947 as a Second World War refugee. He was born in Gorki near Moscow, in 1900 and like Constantin Nikiforoff, fled to Turkey at the time of the revolution in 1917. Dimitri Pronin studied at the College of Agriculture and Forestry in Prague and worked at the Ministry of Agriculture and Land Reform in Poland. During the war, he was caught between Nazi Germany and Communist Russia. In 1947, he joined the Department of Agricultural Engineering at the University of Wisconsin and his two daughters started working in the university library. He joined the Forest Products Laboratory and like Otto Zeasman, he was elected to the Wisconsin Academy of Sciences, Arts, and Letters.

Dimitri Pronin found that Polish chauvinism, German Nazism, and Soviet Communism were equally monstrous forces that aimed at stamping out all

American-Russian harmony in 1954: Francis Hole (1913–2002), and Sergei Wilde (1898–1981) played at Emil Truog's retirement dinner on 3rd June in the Memorial Union in Madison under the name *Fine Earth String Ensemble*. Right picture: Dmitri Pronin (1900–?) who found that Polish chauvinism, German Nazism, and Soviet Communism were the same monstrous totalitarian force, aimed on stamping out all individual freedom. Photos Department of Soil Science archives

individual freedoms [103]. He was a fervent opponent of communism and wrote a Letter to *Life Magazine* in 1956: "It is time the U.S. stopped waiting on Communist action and pursued a resolute strategy of its own in an open ideological drive on all fronts" [104]. At the University of Wisconsin, he found a comrade in Emil Truog. Dmitri Pronin wrote him a letter in 1949: "…to build a new world, where the poor and material weak will be the master of life – that was the slogan. The workers will receive factories, peasantry will receive more land, which was partly in that time in possession of estate owners. And consciously they gave promises about which they have known that they will never fulfil them and that the future development will bring quite opposite things. That the owner of factories and land will be the state, that the situation for a worker and farmer will be worse than during the Zar regime, that for all working people will come the worst kind of slavery."

Dimitri Pronin reviewed a 1949 special issue of the journal *Land Economics* that focused on *Soil Conservation in the USSR*. The papers were from the Russian Research Centre and focused on the drier part of Russia where there was a need to raise agricultural productivity: "Our leaders in government, industry, agriculture and in other phases of our national life must be in possession of true facts and of unemotional appraisals and evaluations of facts" [105]. Dimitri Pronin disagreed with that and saw communism exploiting the majority of the population through collectivization that he termed the 'militarization of agriculture' [106]. Dimitri Pronin's review had a

I swear there is no greatness or power that does not

  emulate those of the earth,

There can be no theory of any account unless it corroborate

  the theory of the earth,

No politics, song, religion, behavior, or what not, is of

  account, unless it compare with the amplitude of the earth,

Unless it face the exactness, vitality, impartiality, rectitude

  of the earth.

          A Song of the Rolling Earth (1855)

    It seems to me this is one of the statements that should be
used by soils people.

              Very truly yours,

              S. A. Wilde

              S. A. Wilde
              Professor of Soils

From a letter by Sergei Wilde (1898–1981) to Emil Truog (1884–1969), 7th July 1947, part of the poem *A Song of the Rolling Earth* by Walt Whitman (1819–1892). From Department of Soil Science archives

headline borrowed from Charles Darwin: *Natura Non Facit Saltos*. But nature did not jump in Communist Russia.

Emil Truog and Dimitri Pronin wrote a lengthy paper entitled *A great myth: The Russian granary*, which aimed to debunk the idea that Russia was the great granary of food grains. In the first paragraph, they wrote: "Russia has never been, is not now, and probably never can become a really great and dependable producer of food grains, such as the United States whose corn production alone equals in food value all of the grains produced in the land of the Soviets." They listed a series of arguments to prove the point. Firstly, they noted that the leaders since the advent of communism have had little or no knowledge of agriculture: "Lenin, Trotzki, and Stalin were never close to the soil." They analyzed the rainfall pattern and stated that the climate was less favorable in much of Russia compared to the Chernozems and the black soils of the Dakotas. They noted that Russia did not have the favorable climate of Iowa or Western Europe where crop yields were much higher.

They wrote that yields in Russia were drastically reduced because of collectivization—the process whereby farm families were uprooted from their old homes and were re-established in hastily built quarters named 'Kolkhoz'. Soil fertility had declined, and limited farm mechanizations were listed as important causes for the dwindling grain production. Some wheat exports were being made but that was done: "…to gain much needed foreign exchange and prestige." The push westwards by the Communist regime was not meant to spread communism, but to gain territory that was more favorable for agriculture like the wheat and corn lands of Romania and Hungary, and the rye and potato lands of Poland and East Germany.

Emil Truog and Dimitri Pronin offered the paper to several magazines including *Harper's Magazine* in 1951. Emil Truog had published in *Harper's Magazine* before with a refute to the *Plowman's Folly* by Edward Faulkner [107], but this time *Harper's Magazine* did not accept their paper. They then submitted the paper to *Successful Farming*, *Collier's*, and *Country Gentleman* but all three magazines rejected it, or as *Collier's* put it: "Right now we are well stocked, scheduled and assigned with articles on Russia and communism." They were unable to get it published in a popular magazine despite the anti-communist wind blowing across the country. Finally, they sent *A great myth: The Russian granary* to the journal *Land Economics* which had been published by the University of Wisconsin since 1925.

Emil Truog and Dimitri Pronin were not the only soil scientists critical of Russian agriculture. In 1952, Jacob Joffe noted: "…that there is a great discrepancy between the boisterous claims of the Soviet press, for popular consumption within and outside of Russia, and the actual state of affairs reflected in the scientific journals and other technical publications. The general picture given by the Soviet officials is that the advanced Russian agronomic science and practice have nothing to learn from the 'deteriorating West,' and that their products are far superior" [3].

When *A great myth: The Russian granary* was published in 1953 [108], Emil Truog sent a reprint to Charles Kellogg, and it yielded a heated exchange of letters. Emil Truog and Charles Kellogg had agreed on refuting to the *Plowman's Folly*, but they did not agree on *A great myth: The Russian granary*. The letters were headed by: *Dear Truog* and *Dear Kellogg*. Charles Kellogg acknowledged the article and noted that it was very hard to arrive at a satisfactory judgment about the potential of agriculture in Russia: "…hard to get facts, harder to interpret many of them." He agreed that Russian agriculture was backward under the Czars. During his visit to Russia in June 1945, he had seen that no agricultural machinery was produced during the war, that there was little education in agriculture, and that fertilizers were unavailable. He expected little food export because of the rapidly growing population and industrialization, and wrote to Emil Truog: "…if the Soviets could get over their ridiculous militarism and throw the same proportion of industrial output and education into agriculture as we have done in this country, I don't really believe they would have too much trouble with their food problem for a relatively good or even high standard of living."

Emil Truog wrote back two weeks later. He found that pedologists, and Charles Kellogg and Curtis Marbut in particular, dealt too much with soil science and too little with crop production. He noted that both of them were: "…unduly awed by the vast areas in Russia, particularly of the Black Soils, but did not appreciate the limitations in crop production due to lack of precipitation." Emil Truog was worried about education in Russia and wrote that it took 75 years in the USA to advance agricultural education and that Russia was far behind in training a new generation of soil scientists. Emil Truog concluded that under the present system agriculture in Russia was bound to get worse.

Charles Kellogg instantly wrote back and was annoyed by the comments over the black soils: "I shall have to assert, Professor, that I have had a bit of experience in soil geography in many parts of the world. Although I have made errors, I am not easily mislead." He listed some of the soil properties that he observed around Moscow where Konstantin Gedroiz had conducted his research, and Charles Kellogg stated that those soils were quite similar to the soils in Ohio. But most soils in the Russian steppe were in a much drier climate, as Eugene Hilgard had already noted: "The climate of the black-earth country of Russia is, though not properly arid, yet one of rather deficient and uncertain rainfall" [109]. For Charles Kellogg the comparison of the soils between the USA and Russia was similar to the debate about the difference of the soils of the temperate and tropical regions: all soils could be highly productive if they were properly managed. Emil Truog answered him with an example from Congo, where the soils needed phosphorus, but transportation costs and market value of the products were so unfavorable that phosphorus fertilization was very difficult if not impossible. They could not agree on the status of the soil resources in Russia and whether enough food could be produced. Emil Truog finalized the letter exchange by saying: "Well, I am afraid we will not get far in the discussion unless we go to Bermuda or some place like that where the climate is very agreeable and then probably that will have a good influence on our thinking." That never happened and their exchange of letters dried up for a while.

Emil Truog (1884–1969) and Charles Kellogg (1902–1980) at their desks in Madison and Washington, thinking, writing. Note the communist hammer-and-sickle stand at Charles Kellogg's desk on the left. His first visit to Russia was in 1945, and in 1958 he became the chairman of a United States mission on soil and water use to Russia. In the 1950s, he was investigated by the FBI for his Russian connections. Photos from Department of Soil Science archives and National Agricultural Library

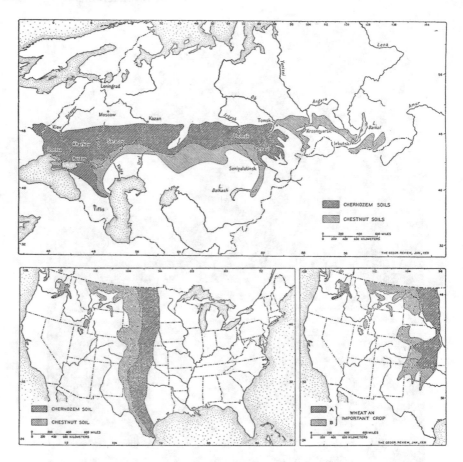

The belts of Chernozem and Chestnut soils in Russia and the USA. Chernozem (Russian for black earth) are mostly Mollisols whereas Chestnut soils are Mollisols formed under grassland in arid areas. The areas within the Russian belt unsuitable for growing grain were not distinguished. It was estimated to be less than 20% of the area in European Russia and not much more in Siberia. For the USA, limitations by the climate in the south and unsuitable soils in Nebraska and South Dakota were shown. Scale of the maps is approximately 1:36 million. From Curtis Marbut his article in *The Geographical Review* from 1931

Emil Truog became increasingly anti-Russia, whereas Charles Kellogg did not depart from his respect for Russian soil science. He thought that the soil studies conducted by Konstantin Gedroiz were superior to all other work done elsewhere, and he considered most soil chemistry studies prior to the works of Konstantin Gedroiz useless [110]. Charles Kellogg's friend Edward Crowther, with whom he had been in Russia, in June 1945, had a similar esteem for Russian science, and wrote in August 1945: "An outstanding feature of Russian agricultural science is the elaborate organization of team-work in a large number of institutes responsible to independent authorities" [111].

Anti-communist sentiments were high in the USA after the Second World War. In August 1949, Soviet Russia exploded its first atomic bomb, and later that year, Communist forces declared victory in the Chinese Civil War and established the People's Republic of China. Then North Korea's army, backed by Russia, invaded South Korea that had the USA on its side. The anti-communist Wisconsin senator, Joseph McCarthy, declared that he had a list of members of the Communist Party who were 'working and shaping policy' in the State Department. In the 1950s, when the anti-communist sentiments were intensifying, Charles Kellogg was visited by the Federal Bureau of Investigation (FBI). He had spoken favorably about Russian soil science and had visited Moscow, in 1945. On his desk, he had a hammer-and-sickle stand that symbolized solidarity between the peasantry and working-class. The FBI questioned him about his Russia connections and his views on collective farming. He answered that in some countries the system seemed to work like in Israel and in farm communities around the world, including Soviet Russia, and he thought it might even work in some places in the USA. They left, and nothing came from that FBI visit [112].

References

1. Moon D. The American steppes. The unexpected Russian roots of Great Plains agriculture, 1870s–1930s. Cambridge: Cambridge University Press; 2020.
2. Soyfer VN. Setting the record straight. Nature. 2002;419(6910):880–1.
3. Joffe JS, editor. Russian contribution to soil science. Washington: American Association for the Advancement of Science; 1952. Christman RC, editor. Soviet science; a symposium presented on December 27, 1951, at the Philadelphia meeting of the American Association for the Advancement of Science.
4. Waksman SA. Jacob G. Lipman. Agricultural scientist and humanitarian. New Brunswick, New Jersey: Rutgers University Press; 1966.
5. Moon D, Landa ER. The centenary of the journal soil science: reflections on the discipline in the United States and Russia around a hundred years ago. Soil Sci. 2017;182(6):203–15.
6. Dinnerstein L. Antisemitism in America. New York: Oxford University Press; 1994.
7. Ackert L. Sergei Vinogradskii and the cycle of life. From the thermodynamics of life to ecological microbiology, 1850–1950. Dordrecht: Springer; 2013.

8. Sackmann W. In memoriam: Jacob Goodale Lipman. Soil Sci. 1980;129(3):134–7.

9. Cline MG. Agronomy at Cornell 1868 to 1980. Agronomy Mimeo No. 82-16. Ithaca, NY: Cornell University; 1982.

10. Jenny HEW. Hilgard and the birth of modern soil science. Pisa: Collana della revista agrochemica; 1961.

11. Voorhees EB, Lipman JG. A review of investigations in soil bacteriology. Bulletin 194. Washington: Government Printing Office; 1907.

12. Tate RL. Soil science: the beginning years. Soil Sci. 2006;171(6):S3–8.

13. Lipman JG. Bacteria in relation to country life. New York: The MacMillan Company; 1908.

14. Reed HS, Charles B. Lipman 1883–1944. Science. 1944;100(2604):464.

15. Anon. Periodical devoted to the study of soil. Sci Am. 1916;114:468.

16. Joffe JS, Antipov-Karataev I. American soils as seen by Russian investigators. Soil Sci. 1929;27:159–66.

17. Truog E. General exhibits. Soil Sci. 1928;25:89–95.

18. Bray RH, Kurtz LT. Determination of total, organic, and available forms of phosphorus in soils. Soil Sci. 1945;59(1):39–45.

19. Campbell GS. A simple method for determining unsaturated conductivity from moisture retention data. Soil Sci. 1974;117(6):311–4.

20. Gardner WR. Some steady-state solutions of the unsaturated moisture flow equation with application to evaporation from a water table. Soil Sci. 1958;85(4):228–32.

21. Jenkinson DS, Rayner JH. The turnover of soil organic matter in some of the Rothamsted classical experiments. Soil Sci. 1977;123(5):298–305.

22. Walkley A, Black IA. An examination of the degtjareff method for determining soil organic matter, and a proposed modification of the chromic acid titration method. Soil Sci. 1934;37(1):29–38.

23. Arend J. Russian science in translation. How Pochvovedenie ws brought to the west c. 1875–1945. Kritika: Explorat Russian and Eurasian Hist. 2017;18:683–708.

24. Joffe JS. Soil science publications in Russia. Science. 1928;67:105.

25. Waksman SA. My life with the microbes. New York: Simon and Schuster, Inc.; 1954.

26. Russell EJ. Dr. J. G. Lipman. Nature. 1939;143(3633):1012.

27. McCall AG. Jacob Goodale Lipman. Science. 1939;89(2313):378–9.

28. Hissink DJ. Jacob Goodale Lipman and the International Society of Soil Science. Trans Third Comm Int Soc Soil Sci. 1939;A:1–3.

29. Sinclair U. The jungle. New York: Doubleday, Jabber & Company; 1906.

30. Joffe JS. Pedology (with a foreword by C.F. Marbut). New Brunswick: Rutgers University Press; 1936.

31. Joffe JS. Pedology. New Brunswick: Rutgers University Press; 1949.

32. Joffe JS. The ABC of soils. Somerville, NJ: Somerset Press Inc.; 1949.

33. Rode AA. Soil science (Translated from Russian). Jerusalem: Israel Program for Scientific Translations; 1962.
34. Joffe JS. Russian studies on soil profiles. J Am Soc Agron. 1931;24:12–39.
35. Simonson RW. Adventures of C.C. Nikiforoff. Pedologue. Mid-Atlantic Assoc Prof Soil Sci. 2017;Winter:13–15.
36. Stelly M. In Memoriam: C.C. Nikiforoff, retired soil scientist with USDA (1867–1979). Int Soc Soil Sci Bull. 1979;56:13.
37. Anon. The Department of Soil, Water, and Climate: 100 years at the University of Minnesota. Minneapolis: University of Minnesota; 2013.
38. Glinka KD. The great soil groups of the world and their development (translated from German by C.F. Marbut in 1917). Ann Arbor, Michigan: Mimeographed and Printed by Edward Brothers; 1928.
39. Nikiforoff CC, Hasty AH, Swenson GA, et al. Soil survey (reconnaissance) the Red River Valley area Minnesota. Washington: United States Department of Agriculture. Bureau of Chemistry and Soils; 1939.
40. Elwell JA, Baldwin M, Strike WW, et al. Soil survey of Wadena County, Minnesota. Washington, DC: United States Department of Agriculture. Bureau of Chemistry and Soils; 1930.
41. Rutherford GK. The role of J.H. Ellis in the early development of soil research in Manitoba, Canada. In: Yaalon DH, Berkowicz S, editors. History of soil science international perspectives. Armelgasse 11/35447 Reiskirchen/Germany: Catena Verlag; 1997. p. 407–414.
42. Anderson DW, Smith CAS. A history of soil classification and soil survey in Canada: Personal perspectives. Can J Soil Sci. 2011;91.
43. Ellis JH. History and development pf the soils department in the Faculty of Agriculture at M.A.C. and the University of Manitoba. Unpublished Report; 1955.
44. Nikiforoff CC. Method for recording soil data. Soil Sci Soc Am Proc. 1937;1:307–17.
45. Nikiforoff CC. History of A, B and C. Bull Am Soil Surv Assoc. 1931;12:67–70.
46. Polynov BB. Contributions of Russian scientists to paleopedology. Leningrad: USSR: Publishing Office of the Academy; 1927.
47. Retallack GJ. A short history and long future for paleopedology. New Front Paleopedol Terr Paleoclimatol. 2013. https://doi.org/10.2110/sepmsp.104.06.
48. Yaalon DH. Soil-forming processes in time and space. In: Yaalon DH, editor. Paleopedology—origin, nature and dating of paleosols. Jerusalem: Israel Universities Press; 1971. p. 29–39.
49. United States Department of Agriculture. Soils & men. The year book of agriculture. House Document No. 398. Washington United States Department of Agriculture. United States Government Printing Office; 1938.
50. Nikiforoff CC. Some general aspects of the Chernozem formation. Soil Sci Soc Am Proc. 1936;1:333–42.
51. Nikiforoff CC. Reappraisal of the soil. Science. 1959;129:186–96.

52. Yaalon DHVA. Kovda—meeting with a great and unique man. Newslett IUSS Comm Hist Philos Sociol Soil Sci. 2004;11:4–9.

53. Fanning DS. Niki stories by Roy W. Simonson. Pedologue. Mid-Atlantic Assoc Prof Soil Sci. 2017;Winter:11–12.

54. Hotchkiss RD. Selman Abraham Waksman. Biograph Memo. 2003;83(321–343).

55. Lipman CB. Eugene Woldemar Hilgard. J Am Soc Agron. 1916;8:160–2.

56. Lipman CB. The discovery of living microorganisms in ancient rocks. Science. 1928;68(1760):272.

57. Lipman CB. Living microorganisms in ancient rocks. J Bacteriol. 1931;22:183–98.

58. Waksman SA. The brilliant and tragic life of W.M.W. Haffkine, bacteriologist. New Brunswick, NJ: Rutgers University Press; 1964.

59. Moberg CL. René Dubos, friend of the good earth: microbiologist, medical scientist, environmentalist. Washington, DC: American Society for Microbiology; 2005.

60. Waksman SA. Sergei N. Winogradsky. His life and work. The story of a great bacteriologist. New Brunswick, NJ: Rutgers University Press; 1953.

61. van Iterson G, den Dooren de Jong LE, Kluyver AJ. Martinus Willem Beijerinck. His life and works. Delft: Alles Komt Teregt; 1940.

62. Maher D, Stuart K, editors. Hans Jenny—soil scientist, teacher, and scholar. Berkeley: University of California; 1989.

63. Štrbáňová S. Holding hands with bacteria. The life and work of Marjory Stephenson. Heidelberg: SpringerBriefs in Molecular Science; 2016.

64. Waksman SA. Soil microbiology in 1924: an attempt at an analysis and synthesis Soil Sci. 1925;19(3):201–249.

65. Pramer D, Lechevalier HA. A salute to Dr. Robert L. Starkey on his seventy-fifth birthday. Soil Sci. 1974;118(3):139–140.

66. Waksman SA, Starkey RL. The soil and the microbe. New York: Wiley; 1931.

67. Waksman SA. Humus. Origin, chemical composition, and importance in nature. Baltimore: The Williams & Wilkins Company; 1936.

68. Waksman SA, Davison WC. Enzymes. Properties, distribution, methods and applications. Baltimore: The Williams & Wilkins Company; 1926.

69. Starkey RL. In recognition of Selman A. Waksman on his Eightieth Birthday. Soil Sci. 1968;106(1):1–5.

70. Waksman SA. The conquest of tuberculosis. London: Robert Hale Limited; 1964.

71. Woodruff HB, editor. Scientific contributions of Selman A. Waksman. New Brunswick, New Jersey: Rutgers University Press; 1968.

72. Wilde SA. Dr. Werner's facts of life. New Delhi: Oxford & IBH Publishing Co.; 1973.

73. Waksman SA. Soil microbiology. New York: Wiley; 1952.

74. Balfour EB. The living soil—evidence of the importance to human health of soil vitality, with special reference to post-war planning. London: Faber and Faber; 1943.

75. Hibbert J. Fragments of war. Stories from survivors of World War II. Toronto: Dundurn Press Limited; 1985.

76. Bentley CF. In memoriam Vladimir Ignatieff. Bull Int Soc Soil Sci. 1994;86:62.

77. Ignatieff M. The Russian album. New York; 1987.

78. Anon. A start toward freedom from want. The story of the United Nations Conference on Food and Agriculture. New York: United Nations Information Service; 1943.

79. Bureau of Agricultural Economics. American farmers and the United Nations Conference on Food and Agriculture. Washington, DC: U.S. Department of Agriculture; 1943.

80. Phillips RW. FAO: its origin, formation and evolution. Rome: Food and Agricultural Organization of the United Nations; 1981.

81. Kellogg CE. Proposal for Department of Soil Science and Management of a World Organization on Food and Agriculture. Unpublished Confidential Report 9/25/43; 1943.

82. Ignatieff V. Professor F.A. van Baren. In: van Baren H, Bal L, editors. This special number of horizon marks the 70th anniversary of F.A. van Baren. Utrecht: University of Utrecht; 1975. p. 36.

83. Ignatieff V, editor. Efficient use of fertilizers. Washington, DC: Food and Agricultural Organization of the United Nations; 1949.

84. FAO. Soil conservation. An international study. Washington DC; 1948.

85. Bouma J, Hartemink AE. Soil science and society in the Dutch context. Neth J Agric Sci. 2002;50(2):133–40.

86. Truog E. Soil reaction influence on availability of plant nutrients. Soil Sci Soc Am Proc. 1946;11:305–8.

87. Bristol BM, Page HJ. A critical enquiry into the alleged fixation of nitrogen by green algae. Ann Appl Biol. 1923;10(3–4):378–408.

88. Arnold C, Page H. Studies on the carbon and nitrogen cycles in the soil. II. The extraction of the organic matter of the soil with alkali. J Agric Sci. 1930;20(460–477).

89. Page HJ. The investigations of K.K. Gedroiz on base exchange and absorption. Vol A. Groningen, Holland: Transactions of the Second Commission of the International Society of Soil Science; 1926.

90. Anon. Imperial College of Agriculture: Mr. H.J. Page, M.B.E. Nature. 1946;158:410.

91. de Geus JG. Fertilizer guide for the tropics and subtropics. 1st ed. Zurich: Centre d'Etude de l'Azote; 1967.

92. de Geus JG. Fertilizer guide for the tropics and subtropics. Zurich: Centre d'Etude de l'Azote; 1973.

93. FAO. Multilingual vocabulary of soil science. Vocabulaire multilingue de la science du sol. Vocabulario multilingue de la ciencia del suelo. Edited by G.V. Jacks, R. Tavernier, D.H. Boalch. Rome: Land & Water Development Division; 1960.

94. FAO. Multilingual vocabulary of soil science. Vocabulaire multilingue de la science du sol. Vocabulario multilingue de la ciencia del suelo. Edited by G.V. Jacks. Rome: Agriculture Division; 1954.

95. Bates CH, Zeasman O. Soil erosion—a local and national problem. Research Bulletin no. 99. University of Wisconsin, Madison: Agricultural Experiment Station; 1930.

96. Beatty MT. Soil science at the University of Wisconsin-Madison. A history of the department 1889–1989. Madison: University of Wisconsin-Madison; 1991.

97. Hartemink AE. Some noteworthy soil science in Wisconsin. Soil Horiz. 2012. https://doi.org/10.2136/ssh2012-53-1-3hartemink.

98. Whitson AR. Soils of Wisconsin. Bulletin no. 68, Soil Series no. 49. Madison: Wisconsin Geological and Natural History Survey; 1927.

99. Tiurin IV. Genesis and classification of forest-steppe and forest soils. Paper presented at: Second International Congress of Soil Science. Leningrad; 1930.

100. Gessel SP, Harrison RB. A short history of forest soils research and development in North America. In: Steen HK, editor. Forest and wildlife science in America: a history Durham. NC: Forest History Society; 1999.

101. Wilde SA. Forest soils and forest growth. MA: Chronica Botanica Company; 1946.

102. Wilde SA. Soil science and semantics. J Soil Sci. 1953;4:1–4.

103. Pronin D. Europe in flames. The horrible years 1939–1945. New York: William-Frederick Press; 1978.

104. Zawacki E, Pronin D. Letters to the editors. Life. 1956;30 January 1956:4.

105. Krimgold DB. Conservation plan for the steppe and timber-steppe regions. Land Econ. 1949;25:336–46.

106. Pronin D. "Soil conservation in the USSR" A reply by Dmitri Pronin. Land Econ. 1950;26(97–99).

107. Truog E. Plowman's folly refuted. Harper's Mag. 1944(July Issue):173–177.

108. Truog E, Pronin DT. A great myth: the Russian granary. Land Econ. 1953;29:200–8.

109. Hilgard EW. Soils, their formation, properties, composition, and relation to climate, and plant growth. New York: The Macmillan Company; 1906.

110. Kellogg CE. Russian contributions to soil science. Land Policy Rev. 1946;9:9–14.

111. Crowther EM. Agrochemistry. Nature. 1945;156(3956):227–8.

112. Kellogg RL. Remembering Charles Kellogg. Soil Surv Horiz. 2003:86–92.

10

Building an International Soil Science

"Our future histories of the Twentieth Century will be certain to. recognize the contributions of soil science to human welfare."

Jacob Lipman, 1936.

"In volume and quantity the pedosphere is relatively insignificant. when compared with other spheres of the earth,. but it is the source of all living existence."

Alex. de'Sigmond, 1938.

The sciences have progressed through the curiosity, intellect, and devotion of researchers and the debate, acceptance, refusal, and spreading of their findings. Many of the early soil studies consisted of observations and descriptions, and there was little to no theory to build on. Soil researchers worked in isolation, scientific communications were limited, and there was some degree of copying without critical analysis of methods and results. Some of the scientific ways, such as building on a body of existing work, acknowledging the source, institutional and international cooperation, were emergent. Findings were exchanged by letters between individual researchers, and in 1665, the first scientific journals were started: *Le Journal des Sçavans* and *The Philosophical Transactions of the Royal Society of England* [1]. The first journal solely dedicated to soil studies was published in Russia in 1899. Some ten years after that, plans were made for an international soil conference that aimed to bring together soil researchers from Europe, USA, Russia, and any country where soil research was conducted. The first international soil meeting was held in 1909 in Budapest, and ideas were made for a series of meetings, but the First World War interrupted further progress. The world slowly recovered from the Great War, and it was not until 1924 that the *International Society of Soil Science* was formed and soil science became international.

In the first half of the nineteenth century, there were people who studied soils in the field, and they often had a geologic background. Others studied soils in the laboratory and were trained as chemists, and some were bacteriologists or microbiologists. These groups operated in different continents and they had little interaction. In the old, long-settled areas of Europe, farmers had learned much about their soils by experimentation and experiences [2]. Population was relatively dense and there was not much new land to extend the farm area so research focused on how to improve the soil conditions of existing fields. As a result, agricultural chemistry developed in Europe. In the USA and Russia, there were large areas that could be used for agricultural expansion, and there was a need to study soils in the field and assess their potential for agricultural development, and so soil mapping was developed in Russia and the USA, although along different lines [2, 3]. The blending of the groups that studied soil conditions and those that mapped soils helped to establish soil science as a unique scientific discipline and fostered international exchange and cooperation.

It took about 50 years before 'soil science' became the overarching term for soil studies, and then to be recognized as a scientific discipline. In the 1800s, several terms described the study of soil, including agrology, agrogeology, agricultural chemistry, and pedology. Most books on soil were compilations of geology, geography, agricultural chemistry, and plant physiology. In 1862, Friedrich Fallou published *Pedologie oder allgemeine und besondere Bodenkunde*, which focused on the relation of soils to geography, geology, and chemistry, and he recognized the importance of the soil profile and distinguished layers and their thickness. Friedrich Fallou introduced the terms 'pedology,' 'solum,' and 'soil quality' [4]. Pedology was defined as natural soil science or the description of the nature of soils no matter what their relation to the vegetation might be [5]. The term pedology was ignored in Germany, possibly as it was seen as jargon and not well-defined, but more likely it was disregarded because Friedrich Fallou was not recognized as a scientist, and had no student followers [6]. The term pedology fell into disuse until Russian soil researchers started to use it in the 1870s [7].

In Germany, 'Bodenkunde' was first used by Carl Sprengel in 1837 and by Friedrich Fallou in 1862 [8, 9]. 'Bodenkunde' or 'Bodenlehre' (soil knowledge) became commonly used terms and were preferred over 'Boden Wissenschaft' (soil science) [10]. In the Netherlands, the term 'bodemkunde' was preferred similar to German usage. In France, 'science du sol' was used as well as 'pedologie' which had a broader meaning than pedology in English. In Russia, 'pochvovedenie' (почвоведение) was the term which meant the scientific study of soil or soil knowledge, and 'pochvovedenie' has also been translated as 'soil science' and 'pedology' [7, 9].

In 1899, Vasily Dokuchaev found pedology: "…to lie at the very core of all the most important divisions of modern natural science such as geology, orohydrography, climatology, botany, zoology, and the study of mankind in the most extensive sense of the word. This branch of science, while still very young, is nevertheless full of extraordinarily high scientific interest and significance; it advances anew every year and daily attracts new energetic, active, and enthusiastic workers and adepts" [11]. He was convinced that pedology was to develop into a new discipline: "The time is nearing when this science, rightfully and owing to its great importance for the destiny of humanity, will be independent and honored, with strictly defined tasks and methods, and no longer confused with the existing branches of natural sciences and even less so with the sprawling science of geography." Vasily Dokuchaev called that a 'happy future' [11].

In the late 1800s, one of the founders of German soil science, Emil Ramann, used the term 'agrology' that had been coined by Friedrich Fallou [7]. Emil Ramann studied the relationship between soils and trees, but his view on soils was geological, and he considered soil studies to be a branch of geology and called it: "geology of the upper crust of the earth" [12]. In 1909, European soil researchers proposed the formation of an International Agrogeological Committee and in 1921, the geologist Jan van Baren advocated for the term 'agrogeology' [13]. He was the first professor of agrogeology in Wageningen, the Netherlands, and had coined a credo that many of his generations embraced: "In the beginning was rock, and the rock was the mother of all soils" [13]. Agrogeology studied: "..the upper part of the land surface which is the seat of all transmutations of energy brought about by sun-heat, atmospheric moisture, and the organic world. If geology in a limited sense investigates the dead rock, Agrogeology makes researches into the outer earth-rind that bears and sustains Life; if Geology is limited in some sense to the Past, Agrogeology is the Geology of the Present" [13]. In 1922, 'agropedology' was used for soil studies, but in Europe only [13].

Although it may have appeared in print earlier, 'soil science' was first printed as a term in the 1901 edition of the *Chambers's Encyclopaedia*, which was a widely read English language encyclopedia. For the lemma 'soils,' the encyclopedia provided a one-page description on the geologic origin of the soil, stating: "…the origin of soils may indeed have been the origin of life itself, and until we can clearly define the one there must of necessity be indefiniteness about the other" [14]. The author of the lemma, John Hunter of Edinburgh, was convinced that the biology of the soil was critical: "The chemical composition and physical conditions of soils have until quite recently been about the only features which received consideration, but it is now beyond doubt that the biological condition is of at least equal importance." John Hunter was certain that: "…many of the heretofore established certainties of *soil science* and of agriculture are destined to be overthrown" [14].

In 1908, John Hunter gave a lecture that was later published in the *Transactions of the Scottish Agricultural Association* in which he reported on the functions and contents of nodules on plant roots; he called them warts, which he thought was as good a description of their appearance as nodules. In his view, soil studies should be on a biological basis, and if that were to happen: "…from that period may be said to date the birth of the New Soil Science" [15]. He saw soil science as a science that was purely biological and needed to understand plant nutrition: "We have to ascertain what essential constituents our soils are deficient in, so that we can feed the advantageous organisms and make them multiply and do their work. We cannot feed a man successfully on, say, whisky alone" [15].

In 1911, *Scientific American* published the article *The new science of the soil*, in which Cyril Hopkins and Eugene Hilgard elaborated on the applications of bacteriology and chemistry in soil studies. They did not write the term 'soil science' but the 'science of the soil' which was also used in the UK, France, and Russia [16, 17]. Edward Free who worked at the *Bureau of Soils* used 'soil science' in a paper on soil physics in 1911: "Soil science has begun its introduction to the quantitative, and its votaries are newly but rapidly learning to 'observe' a little less and to measure a little more and [to be] a little more accurate. It is a much needed change, one of great promise and already of some fruition, but perhaps its greatest service has been to call attention to the weaknesses of present methods and the inadequacy of present tools. There are so few measurements that can be applied to problems of the soil" [18]. The *Journal of Ecology* published an editorial in 1913 in which the journal encouraged submissions from: "…more special branches of science which are now making such great strides and are of immense importance to ecologists, such as the science of soils (*Bodenkunde*) with the investigation of soil-floras and faunas." The *Journal of Ecology* equated 'soil science' to the study of soil bacteria, fungi, algae, protozoa, etc. [19].

In 1913, Thomas Wood reviewed the last twenty years of agricultural research in *Science* [20]. He was a professor of agriculture at the University of Cambridge in the UK and a friend of John Russell of Rothamsted. Thomas Wood thought that the most striking result of the last 20 years was that phosphorus increased the yield of hay and caused an even greater increase in the feeding value of pastures. He discussed contributions of the various sciences to agricultural research and noted: "The first of these is the development of what I may call *soil science*." He equated soil science more or less to bacteriology and thought it had all started with the discovery of atmospheric nitrogen fixation by Sergei Vinogradskii [20]. None of the soil textbooks from the late 1800s and early 1900s used the term 'soil science' and the term is absent in books by Franklin King [21], Eugene Hilgard [22], George Merrill [23], Andrew Whitson [24], Cyril Hopkins [25], and John Russell [26].

The term 'soil science' gained usage from the early 1900s onwards, and initially, it was equivalent to soil biological studies. It became common and more broadly used after the journal *Soil Science* was started in 1916 by Jacob Lipman and when the *International Society of Soil Science* was established at the *Fourth International Conference of Pedology* in Rome in 1924. For many years, the *Bureau of Soils* preferred 'physical geography' instead of soil science or pedology, and it was Curtis Marbut who favored that term as he had learned from William Davis at Harvard University [27]. In 1928, Curtis Marbut noted that 'soil science' was the new term whereas 'the use of the soil' was old [28]. Jacob Joffe used the term 'the science of the soil' as was common in the UK, France, and Russia [16]. In 1928, Emil Ramann used 'science of soil' [29].

Although 'soil science' was more widely used after 1916 and in particular after 1924, the *American Soil Survey Association* aimed to equate 'soil science' to 'pedology' in 1929. At its business meeting at the Rienzi Hotel in Chicago on the 12th of November, the association passed the following resolution [30]:

Whereas
  The term Pedology is derived from the Greek root pedov meaning ground
Whereas
  Its use by Fallou clearly establishes its priority of use and
Whereas
  The use of the term by child specialists is inaccurate, inasmuch as a word derived from boy or child should properly be spelled paedology.
  Be it resolved
  That it is the opinion of the American Soil Survey Association that soil science
  may be properly termed Pedology.

The resolution was prepared by a committee chaired by Charles Shaw from the University of California Berkeley. Percy Brown from Iowa State College who had obtained his PhD under Jacob Lipman at Rutgers University noted at the meeting that: "Soil science or Pedology is now coming to be recognized as a true applied science. Even the 'diehards' among our friends, the pure scientists, are being forced to an appreciation that of the fact that the Science of Soils is not Chemistry, nor Physics, nor any other of the so-called pure sciences but a scientific discipline, of and in itself." He thought that soil survey laid the foundation for 'pedology' [31]. The Hungarian soil scientist Alex. de'Sigmond and other leading soil researchers had already stressed in the early 1900s that 'soil science' was equivalent to 'pedology' [32]. In 1936, Jacob Joffe from Rutgers University wrote a textbook named *Pedology*, and for him and the generation of students that grew up with it, the first sentence of the book made it clear: "Pedology is the science of soils" [33, 34]. By the 1930s, pedology had been narrowed to soil genesis, morphology, classification, and cartography.

When the *Soil Science Society of America* was established in 1936, it was not named the *Pedological Society of America*. The society name was proposed by Emil Truog who had limited appreciation for pedology, and the name was in line with the recognition of 'soil science' internationally. For unclear reasons, the 1936 constitution and byelaws of the *Soil Science Society of America* established that Section or Division 5 (S5) was named Soil Genesis, Morphology, and Classification, and nót pedology. According to the soil physicist Lorenzo Richards, the custody of Division 5 was the only justification for the *Soil Science Society of America*—the body of principles and theory about pedology were unique and provided a claim for the independent status of soil science. He foresaw that the soil physicists would meet with physicists, the soil chemists with chemists, and so forth. Division 5 was distinctive for soil science and the core of the society, but the utilitarian and agrarian view of soils that dominated the leadership of the society failed to recognize that.

There was fear that pedology would be misunderstood and confused with paedology or medial pediatrics or child studies [5, 7]. Wilbert Weir wrote in 1930 a letter to *Science* highlighting the fact that the child study definition of pedology originated in 1896, whereas pedology in the soils denotation had been coined in 1862 [5]. In the same year that the *Soil Science Society of America* and the S5 Pedology Division was established, Wilbert Weir declared 'pure soil science' to be synonymous with 'pedology' [35]. There were others that considered 'soil science' an incorrect term as we then could also have 'rock science' or 'plant science' for which the terms petrography and botany were given. Gilbert Robinson, a pedologist from Wales, found 'soil science' a barbaric term [36]. It did not help. The world ended up with soil science and not pedology for the study of soil, and future attempts to equal pedology to soil science failed.

Another term that was used in the 1920s was 'podology' and Merris McCool and Jethro Veatch had coined it in a paper on soil profile studies [37]. Their paper was one of the first in the USA that used soil horizon designations and Curtis Marbut thought it was among the best work in soil science [38]. Merris McCool and Jethro Veatch had combined field and laboratory investigations for soil profile studies and concluded that: "...profile study of soils constitute a step forward in soil science...it compels a new evaluation of soil in ecologic studies...it places soil study on a natural basis and in fact lays the foundation of a new science which we might name *podology*" [37]. Roy Simonson thought that 'podology' was a misprint, but they had also used 'podology' in an earlier publication, and it was used by others [39].

Thomas Lyon from Cornell University with his former student Harry Buckman reviewed in 1924 the need for a new word that described soil studies [40]. They had just published the textbook *Nature and Properties of Soils*, which had a title that had already been used for a soil book in 1843 [41]. They noted that 'podology' and 'agrogeology' were in use and that neither connoted the plant relationship, and they thought a term was needed for the science of soils in its relations to plant growth and suggested 'edaphology' as it conveyed an agronomic implication [40]. It was derived from *Edaphos*, the Greek word for ground, foundation, or base, and edaphic was used by botanists to denote the relation of the plant to its soil environment [40]. Their interpretation was similar to what Friedrich Fallou had termed 'agrology' [4, 7].

Edaphology became concerned with the influence of soils on living things, particularly the growing of plants, and *Soil Conditions and Plant Growth* by John Russell first published in 1912 and the *Nature and Properties of Soils* by Thomas Lyon and Harry Buckman were edaphology books [26]. They were in contrast with textbooks such as the 10 volumes *Handbuch der Bodenlehre* by Ernst Blanck from the early 1930s [10], *Pedology* by Jacob Joffe [33], and the books by Alex. de'Sigmond [32] and Gilbert Robinson [42]—all of which focused on soils but not so much on its function to support plant growth. Over time, many other words were introduced for various fields of soil studies. In the 1940s, Russian researchers used the term 'agrochemistry' which encompassed crop nutrition and soil fertility. Edward Crowther supported the use of the word [43], but it never took root with that meaning, and was later used for agricultural inputs such as pesticides and insecticides.

The soil science discipline gained prominence when journals solely devoted to soil studies were established exemplifying unceasing studies and a growing body of knowledge. The first soil journal was *Pochvovedenie* in 1899, followed by *Internationale Mitteilungen für Bodenkunde* in 1911, and *Soil Science* in 1916. The journal *Internationale Mitteilungen für Bodenkunde* published mostly in German. Its advisory board included Jan van Baren, Konstantin Glinka, David Hissink, Alex. de'Sigmond, and Eugene Hilgard. It aimed to publish essays and original work from the whole area of theoretical and practical sciences, papers and excerpts, scientific news and communications, and an overview of the new publications. Some years later, it carried the subtitle *Revue Internationale de Pedologie* and *International Reports on Pedology,* and Friedrich Schucht became the editor.

Pochvovedenie published papers in Russian with summaries in other languages and few papers in French and English. After the *First International Congress of Soil Science* in 1927, the journal included papers in English. Konstantin Glinka contributed to the first issue, and the fourth issue of 1899 included a paper by the American Lyman Briggs on electrical methods for determining soil moisture, temperature, and salinity. A series of papers on the contributions by Nikolai Sibirtsev were published after his untimely death in 1900 [12].

At universities across the USA, soil courses were included in agricultural curricula by the early 1900s. There was a need for people trained in soils, particularly in the field of soil survey and soil fertility, and departments were established that were often named Departments of Soils. Cornell University established the Department of Soils in 1908 [44] the University of Wisconsin renamed its Department of Agricultural Physics from 1889 to the Department of Soils in 1909 [45] Minnesota had a Divisions of Soils formed in 1913 [46] Missouri in 1914 [47] Ohio in 1928, and the University of California Berkeley had a Department of Plant Nutrition and a Department of Soil Technology, whereas at Purdue, Iowa, and North Carolina, the study and teaching of soil resided within agronomy departments. In the 1950s, most soil departments were renamed to the Department of Soil Science, and several new soil science departments were established across the USA.

Soil science had a late start in the USA, but it grew steadily from the early 1900s onwards. In a period of 20 years, it established the *Bureau of Soils,* soil departments at universities, and it founded the *American Association of Soil Survey Workers.* By the 1930s, soil science in the USA was better organized than in most European countries, and these developments did not go unnoticed. John Russell from Rothamsted in the UK followed all publications from the *Bureau of Soils* and made numerous visits to the USA. The Nestor of German soil science, Emil Ramann, noted that there were no professorial chairs in soil science in Germany, and concluded: "…Germany is behind other countries; as she is unfortunately also, in the lack of possession of such a Research Institute as, for instance, the United States of America have the

Bureau of Soils, which is chiefly concerned in the scientific investigations of soils" [29]. It took another ten years before Germany formed a soil science society, which was the second society formed after the *American Association of Soil Survey Workers*. Denmark, France, India, the Netherlands, Poland, and Russia established national societies before the Second World War [48]. In most other countries, national societies, soil departments, and soil research centers were established after the war.

Although soil science as a discipline gained distinction in the first few decades of the twentieth century, soil science as a science was not well-defined. There were many definitions for soil, and for decades, the soil was seen as the top 10 to 20 cm, the stratum below was the subsoil and not really part of the soil [42, 49]. The soil was regarded as a production factor or medium for crop production that needed to be understood before it could be improved, or the soil was defined as disintegrated rocks mixed with organic matter [50]. The definition of the soil was relevant for soil survey and in soil classification because it affected how soils were viewed in the field and represented on a soil map. Most of the early definitions stressed the organic and inorganic parts of the soil as well the origin, complexity, and some of its functions [50]. The view of soil as a material and an independent object of study was gaining ground, as was soil science itself.

Wilbert Weir defined soil science in his textbook from 1936: "Soil science, like other natural sciences, is systematized knowledge, embracing theories, facts, generalizations, and principles. Thus, soil science is systematized knowledge of soils and soil fertility, with a view to an understanding of the nature of soils themselves, their relation to each other, and their role in plant nutrition and crop production" [51]. Wilbert Weir's definition from 1936 had some edaphic underlining, but it defined the science of soil studies. For the soil physicist Walter Russell, soil science was: "...the basement of the plant sciences, the attic of the earth sciences, and a pillar of the environmental sciences" [52]. Herman Stremme had a geological view on soils, and for him: "...soil science was the concept of weathering on a climate base" [12]. The pedologist Cees Edelman defined the discipline in the 1950s as: "Soil science is everything a soil scientist is interested in."

Soil science became an organized international science with the formation of the *International Society of Soil Science*, but its beginnings were indistinct, and in 1924, Selman Waksman stated: "...soil science was in its infancy and one might rightfully ask whether such a science exists at all at the present time" [53]. The formation and organization of soil science as a discipline and as an international society was a slow process, but it was a marvelous piece of collaboration among European, American, Russia, and some Asian soil researchers. There were several international learned societies, and International Geological Congresses had been held every four years since 1875. Soil science had to establish itself as an independent discipline separate from agricultural chemistry, bacteriology, or geology. The basis of the formation of the international society was laid in a series of meetings across Europe and started in Budapest in 1909.

In 1908, Peter Treitz from Hungary and Gheorghe Murgoci from Romania traveled to southern Russia to participate in an excursion led by Konstantin Glinka [12]. They knew about the Russian soil studies and how it differed from other work. During the excursion, they hatched the idea of bringing together researchers from various countries to present methods and approaches [54]. With the help of the Royal Geological Institute of Hungary, the *First International Conference of Agrogeology* was organized in Budapest in April 1909 [55]. Participants included soil researchers, geographers, geologists, mineralogists, botanists, bacteriologists, agricultural chemists, agronomists, and foresters. Some 100 delegates attended from Germany, Italy, Norway, Austria, Romania, Russia, Netherlands, and Hungary, including Konstantin Glinka, Peter Treitz, Gheorghe Murgoci, Alex. de'Sigmond, Emil Ramann, and Friedrich Schucht. The organizers had invited the Swedish chemist Albert Atterberg, but he could not attend. There were no participants from the USA, but Eugene Hilgard, who was at the University of California Berkeley, had submitted a paper on the unification of chemical analyses [56]. Konstantin Glinka had brought Russian soil maps and monoliths and gave the opening address: *Soil zones and soil types of European and Asiatic Russia* [12]. Gheorghe Murgoci presented a soil map of Romania based on Russian methods and nomenclature. Papers were presented on agrogeology, soil types in different countries, and research methods for the field and the laboratory. Toward the end of the conference, suggestions were made to develop a uniform system of soil classification.

The conference was followed by an excursion, and during the excursion, the question was asked: "what is soil?" A variety of answers was obtained that reflected the nationality and background of the authors [57]. The way soils were defined followed the views of Albrecht Thaer, Justus von Liebig, Friedrich Fallou, Vasily Dokuchaev, or Eugene Hilgard. Russian and German participants had different views based on how soils were studied. In Russia, Vasily Dokuchaev had a plain extending for thousands of kilometers occupied by a relatively uniform, partly uncultivated soil and varying conditions of climate, vegetation, parent rock, and relief. Friedrich Fallou, from a small area in Saxony, Germany, on the other hand, had soils from mountains to sea level developed in different parent materials [4, 57]. Arseny Yarilov observed that there were some national and geographic aspects: "...did not these teachers of the agrogeologist, soil scientists, agricultural chemists, agronomists, in their own turn reflect in their teachings the conditions of their own country, the peculiar characters of the soil cover of the latter?" [57].

At the 1909 conference, it was apparent that methodologies in soil studies were diverse. For example, the depth and boundaries of the soil were undecided and for some, the soil was in the top 10–15 cm, the stratum below was seen as the subsoil and not part of the soil [42, 49]. Soil was seen as that portion of the ground that was tilled or the portion that was black or dark in color. The subsoil was the ground below the tilled or dark soil [58]. Soil studies were slowly liberated from the other sciences and practices, and the conference was an important point in the development of soil science [32]. At the end of the conference, it was decided to form an International Agrogeological Committee that would organize regular conferences.

The next international soil conference was held in Stockholm in 1910 and was jointly convened with the Eleventh Geological Congress. There were sessions dealing with the mechanical analysis of soils, soil colloids, preparation of soil extracts for chemical analysis, soil cartography, soil classification, nomenclature of soil types, and soil conditions in various countries [59]. The conference was attended by 170 participants from 20 countries; there was no participation from the USA and there was only one Russian delegate [60]. A committee was established that would report on the status of soil investigations in different countries. The next conference was to take place in St. Petersburg in 1914 [61] but in December 1913, the chair of the Russian Academy of Sciences, Alexander Karpinsky, and Konstantin Glinka issued a statement that: "...owing to a number of unfavorable circumstances which it has been impossible to foresee, and to the amount and complexity of the preparatory work, it had become evident that the difficulties were insurmountable" [62]. The organizing committee abandoned the idea of holding the conference in Russia in 1914, but the Italian Government then offered to hold the conference in Rome. A planning meeting was held in Zürich in 1913 [63], and on the 28th of July 1914, the First World War broke out. The sprout of international soil collaboration stopped growing and withered for years.

In 1920, David Hissink from the Netherlands and the Czech Josef Kopecký invited soil scientists from across Europe, USA, and Russia to meet in Prague for an agropedology conference. Both Josef Kopecký and David Hissink were well-known soil scientists, had attended the 1909 conference in Budapest, shared an international vision, and spoke several languages. They resurrected international soil science after the First World War.

David Jacobus Hissink was born in a large family in the Dutch reformed city of Kampen where his father was the municipal secretary. He studied chemistry in Amsterdam and obtained his PhD in 1899. After teaching chemistry for one year, he moved to Bogor on the island of Java in Indonesia where Jules Mohr directed the soil research. David Hissink studied the soils under tobacco near Medan, but his family remained in Bogor—some 2,000 km away from Medan. Unhappy with the separation, they returned to the Netherlands in 1903. David Hissink was hired by the agricultural experimental stations that were established in 1877, and he was appointed as a chemist in Goes and some years later became the station director [64]. When in 1906 several low-lying lands that had been reclaimed from the sea were inundated, he started studying the effect of seawater intrusion and the successive drying up of the land, that was later termed 'ripening.' In these reclaimed lands, named polders, he found that sodium from the seawater reduced water percolation, but that liming maintained the structure and improved water movement through the soil.

In 1916, David Hissink became director of the newly established Soil Science Institute at Groningen. Soil structural degradation by sodium, cation exchange capacity, and soil aging or ripening needed research in the new polders of the Netherlands. With Jac. van der Spek, they devised quantitative methods to determine exchangeable bases in soils, and they calculated the rate of calcium loss in soils that had been reclaimed from the sea. David Hissink demonstrated the role of sodium in soil structure, defined the term base saturation, and studied the connection between base saturation and the pH value of soil. He admired the work of the soil chemist Jakob van Bemmelen, who in the 1870s had established the theory of absorption from the soil solution, and had studied cat clays or acid sulfate soils. David Hissink's studies on base saturation and cation exchange aligned with those of Konstatin Gedroiz in Russia and George Wiegner in Switzerland. His work was fundamental yet practical, and for him: "…the primary duty of man was to conquer the earth solely by improved cultivation of the soil for the benefit of mankind" [64].

David Hissink traveled widely through Europe, and he made the Soil Science Institute in Groningen an international center of soil science. It hosted for many years the secretariat of the *International Society of Soil Science*. Two mornings per week, he worked for the society and he had French and English assistants who helped him with the correspondence, but the German correspondence he did by himself [65]. David Hissink had a somewhat brusque demeanor and was considered by some to be an 'intellectual aristocrat.' Eugene Hilgard, Hans Jenny, Jacob Lipman, and Selman Waksman mailed him reprints of their papers, and quite likely, he returned reprints of his own papers. In the first half of the twentieth century, he became the most renowned soil scientist from the Netherlands [66].

Josef Kopecký was born on a farm in Austria-Hungary and studied *Kulturtechnik* or landscape engineering in Vienna, and his studies focused on the physical properties of soils such as porosity, water and air capacity, permeability, and infiltration [67]. He developed an instrument for determining soil grains via the floating-off method that was used for water flow in land drainage, and he invented the stainless steel cylinder to determine the volumetric mass and bulk density [67]. In 1913, Josef Kopecký drafted a textural triangle using clay and silt, and limits for each fraction that had been established in 1900 by Albert Atterberg [68]. A textural triangle had been attempted in France in the early 1900s, and in the USA where Milton Whitney had plotted soil class boundaries using the silt and clay fractions [69]. In 1927, Hugh Bennett and R. David developed an equilateral textural triangle with sand, silt, and clay fractions on equal axes [70].

The purpose of the conference in Prague was to prepare the organization for the *Third International Conference of Pedology*. The conferences in Budapest and Stockholm had used the term 'agropedology' but the conference in Prague used 'agropedology' which was more widely used in Europe in the early 1920s [13]. Some 50 soil scientists attended the conference in Prague including Alex. de'Sigmond and Friedrich Schucht, and it was the first post-war meeting that was truly international [71]. Representatives from different countries spoke in their own languages [72]. The USA had delegated Jacob Lipman from Rutgers University and Curtis Marbut from the *Bureau of Soils*. According to David Hissink, they both put their stamp on the discussions, on the resolutions that were passed, and on the international cooperation in the field of soil science [71]. There were no meetings on Saturday, and Jacob Lipman and Curtis Marbut visited the old Jewish cemetery in Prague that had been in use from the early 1400s until the late 1700s. They peeped through the gates and hedges, and it was closed because of the Sabbath [72].

Curtis Marbut had translated Konstantin Glinka's book and it was in Prague that they met for the first time. They were about the same age and had numerous discussions during the conference [56]. While in Prague, Curtis Marbut wrote a letter to his chief, Milton Whitney, but did not mention his contacts with Konstantin Glinka, as Milton Whitney was opposed to Russian soil science [56, 73]. Curtis Marbut also became acquainted with Emil Ramann, Peter Treitz, Alex. de'Sigmond, David Hissink, Gheorghe Murgoci, and Herman Stremme. They were leading soil scientists in Europe. At the 1922 conference, five committees were formed: physical soil analyses, chemical soil analyses, soil bacteriology, soil nomenclature, and a committee on cartography [63]. A subsection on soil biology was also established, and the emphasis on soil biology and bacteriology showed the influence of Jacob Lipman. It was decided that the *Fourth International Conference of Pedology* would be held in 1924 at the invitation of the Italian Society of Agronomy. During the 1922 conference, 'agropedology' was shortened to 'pedology' probably under the influence of Konstantin Glinka and Curtis Marbut who both saw pedology independent from agricultural applications.

The *Fourth International Conference of Pedology* was held in Rome in May 1924. It took place under the patronage of King Victor Emmanuel III, who was present at the opening of the conference [74]. His friend Benito Mussolini chaired the *Comité d'Honneur*. Soil scientists from Russia, most European countries, and the USA settled for a week in Villa Lubin near the landscape park of Villa Borghese. Villa Lubin was built between 1906 and 1908 to house the International Institute of Agriculture, and for the construction of the villa, several centuries-old pines had to be felled. The villa was named after the Polish American David Lubin who had convinced King Victor Emmanuel III to establish the International Institute of Agriculture.

The *Fourth International Conference of Pedology* was attended by 463 people from 39 countries. It was a diverse group that included official government delegations, some 300 soil researchers, 120 members of scientific institutions, and 16 industry delegates. Italy, Germany, Czech Republic, and Hungary had the largest representation. Curtis Marbut, Merris McCool, Selman Waksman, and Jacob Lipman attended from the USA, Japan was represented by Arao Itano, and Konstantin Glinka led the delegation from Russia. Sergei Vinogradskii from the *Institut Pasteur* participated; it was one of the very few international conferences he ever attended [75]. From the UK, Bernard Keen, Harold Page, Gilbert Robinson, and John Russel participated. Herman Stremme from Germany was the honorary president [63], and most sessions at the conference were chaired by David Hissink.

The conference in Rome included a scientific program with a keynote presentation on nitrification and its consequences for agriculture and a presentation of recent works in soil physics. Benjamin Frosterus from Finland and Konstantin Glinka provided an overview of soil nomenclature and the classification of soils in 13 European countries, as well as Russia and Egypt. Gheorghe Murgoci presented the state of mapping in several European countries [76]. The work resulted in the soil map of Europe project, which was led for many years by Herman Stremme, after Gheorghe Murgoci had died in 1925 [77, 78]. Sergei Vinogradskii presented a paper on the development of soil microbiology, and Georg Wiegner presented studies on base exchange.

Jacob and Cecilia Lipman, and Deborah (Bobili) and Selman Waksman with their son Byron, on their way to the *Fourth International Congress of Pedology* in Rome in 1924. Photo from the biography of Jacob Lipman

The use of lime and soil acidity were important subjects at the conference. Jacob Lipman gave an overview of the fertilizer industry in the USA. Prior to the conference, two soil samples had been sent to soil researchers across Europe to test and compare the various methods for dispersing soil prior to mechanical analysis. The results were presented at the conference, and considerable variation was found between the different methods and also for the same method, but used by different workers [74]. It was concluded that much work remained, and the study was to be continued and extended. All in all, 250 papers were presented in German, Italian, French, or English.

A small exhibition was held with soil maps and apparatuses for soil analyses. According to Bernard Keen, it were mostly Germans researchers who developed new instruments such as an *in situ* soil moisture measurement device based on electrical resistance and an *in situ* measurement apparatus for pH (written as P_H). There was daily entertainment organized by the Municipality of Rome and the Italian Minister of Agriculture. A field excursion was held in the quiescent volcanic Alban Hills and the Pontine marshes which were partly drained, but where malaria was prevalent. There was another excursion around Naples at the foot of Mt. Vesuvius, which had been rumbling since 1913.

Group picture at the *Fourth International Conference of Pedology* on the steps of Villa Lubin in Rome in 1924. There were 463 participants from 39 countries, including Konstantin Glinka, David Hissink, John Russell, Jacob Lipman, Curtis Marbut, Merris McCool, Selman Waksman, Alex. de'Sigmond, Friedrich Schucht, Herman Stremme, Téodore Saidel and René Dubos

The *Fourth International Conference of Pedology* of 1924 was the largest gathering of soil researchers to date, and for many, it was the first time that they met and learned about different methods and approaches. There were language barriers, but several participants spoke multiple languages and helped with translations. Bernard Keen found that the chief value of the conference was the opportunities for informal discussions with workers from other countries [74]. An example of a new contact was the young French microbiologist René Dubos who was on the same boat as Selman Waksman returning to the USA after the conference [79]. They had lengthy conversations, and Selman Waksman offered him a position at Rutgers University. The conference had all the elements of a contemporary scientific congress, including presentations, discussions, exhibits, and excursions. There was general appreciation for the Italian organizers, although Selman Waksman found that: "…the Italians have done a very poor job in arranging the meetings. There were mix-ups everywhere, from hotel reservations to scientific programs. But the city is beautiful and the people are charming" [80].

On the last day of the conference, 19th of May 1924, the *International Society of Soil Science* was established [59]. There had been several names for the society that included terms such as 'agrogeology,' 'agropedology,' and 'pedology' and early 1924, the *International Committee of Soil Science* had been suggested [76]. In the 1920s, the term 'soil science' was favored over 'pedology' and the influence of Jacob Lipman contributed to that decision. The participants at the conference agreed that soil science needed more than a committee, and after some discussion, the name for the new organization became the *International Society of Soil Science*, which was translated in German as the *Internationalen Bodenkundlichen Gesellschaft* and in French as *l'Association de la Science du Sol*. English, German, and French became the official languages, and the new society had seven goals [61]:

1. Standardization of methods of soil analysis.
2. Standardization of soil microbial research.
3. Program for the development of nomenclature and classification of soils.
4. Preparation of agrogeological map of Europe at the scale of 1:0.5 to 1:2.5 million.
5. The organization of soil investigations in countries where these had not yet started.
6. An introduction to the study of soils into the curricula of intermediary and higher schools.
7. Formation of Plant physiology in relation to pedology.

The objective of the *International Society of Soil Science* was the study and promotion of soil science in general by means of the organization of Congresses and Conferences; the formation of Sections and Committees; the publication of a review; and the institution of a Central Office for Soil Science bibliography (documentation). The headquarters of the society was to be the International Institute of Agriculture in Rome, and the administration of the society was organized by national members or country representatives, but the scientific activities were carried out by commissions and sections [81]. In 1924, there were no national societies of soil science except the *American Soil Survey Association* that had been established in 1920. National societies of soil science were formed in 1926 in Germany, followed by Denmark in 1928, France and India in 1934, the Netherlands in 1935, Poland in 1937, and Soviet Russia in 1939 [48].

The *International Society of Soil Science* had six commissions for which the basis was laid during the conferences in Stockholm and Prague. The commissions grouped all activities and somewhat mimicked the basic sciences: Soil physics, Soil chemistry, Soil biology, Classification and nomenclature and mapping of soils, and Plant physiology in relation to pedology. Except for the last commission, all others remained in use for 75 years and were only changed in 1999 [59]. These six commissions formed the core structure of the new scientific society which was officially founded during the morning session of the last conference day.

Curtis Marbut became the chair of the committee on classification and mapping, and he called for a collective effort from: "...all persons in all countries [for] the study of soil profiles and the accumulation of as much as data as possible on the subject to the end that a scheme of soil classification, based on these features, may be devised." He requested that carefully drawn descriptions of soil profiles, sketches, and photographs supplemented by small samples from each soil horizon were sent to him in Washington. Soil classification was not well-developed, but he stressed that: "...any soil classification that will meet what may be called worldwide demand must be fundamentally scientific. By this statement, it is meant that the classification must be based on studies of the soil, made for the single purpose of finding out the truth about its characteristics with no reference whatever to other considerations. The scheme must be based on soil characteristics, rather on a series of causes, assumed, with or without reason for such assumption, to have produced those characteristics" [82]. Curtis Marbut had become convinced that soils had to be classified based on morphology and properties, not on their mode of formation.

One of the issues that had to be resolved at the conference was where to hold the first congress. After some discussion, and with the diplomatic talent of Curtis Marbut, the USA was selected as the first meeting place of the *International Society of Soil Science*. The Romanian soil scientist Gheorghe Murgoci had also advocated that the first congress was to be held in the USA. He was an influential Romanian soil researcher and had participated in meetings leading up to the 1924 conference [54]. At the last meeting day in Rome, Gheorghe Murgoci was made an honorary member of the newly established *International Society of Soil Science,* as were Lucien Cayeux, Konstantin Glinka, Emil Ramann, John Russell, and Sergei Vinogradskii.

It was resolved to apply the name 'Congress' to the meetings of the *International Society of Soil Science* and that the President was to be elected from the country that organized the congress. The first President of the society was not appointed yet, and the choice was between Jacob Lipman and Curtis Marbut. According to David Hissink: "...both were eminent soil scientists, both of

Vierde Internationale Bodemkundige Conferentie, Rome - Mei 1924. Zitting 2de Commissie

Meeting at the *Fourth International Conference of Pedology* in Villa Lubin in Rome in May 1924, presided by Alex. De'Sigmond, whereas David Hissink (hands on his ear) sits behind the table with papers. Photo and handwriting by David Hissink

an absolute honorable and upright character, both enjoying the regard and affection of their colleagues" [71]. Jacob Lipman from the Rutgers University was appointed as the first President because of his great talent for organization, and because: "...the young society had to cut a good figure at its first congress in 1927, both from the scientific and from the social point of view" [71]. David Hissink was elected as the first Secretary General. The presidency changed with every new congress, but David Hissink was the Secretary General until the Second World War [83]. Now there was a structure for the *International Society of Soil Science*, a group of officers, and a location for the first international soil congress. It was in the roaring 1920s, economies were booming, countries had recovered from the Great War, and soil science excitement filled the hearts and minds of men and a few women, across Europe, Russia, Asia, and the USA.

Vierde Internationale Bodemkundige Conferentie, Rome - Mei 1924.
Bodemkundige Tentoonstelling, Geol. Instituut.

Group picture from the excursion to the Geological Institute of the *Fourth International Conference of Pedology* in Rome in May 1924, Curtis Marbut in center (unfolded papers in hand). Photo and handwriting by David Hissink

A few months after the 1924 congress in Rome, David Hissink wrote to Jacob Lipman about the planning of the international soil congress in Washington. Jacob Lipman passed on the letter to Andrew Whitson, who had founded the *American Soil Survey Association*. As Jacob Lipman wrote to Andrew Whitson (replacing 'pedology' with 'soil science'): "In a letter which has just come to hand from Dr. D. J. Hissink I am reminded that the *Fourth International Conference of Soil Science* [sic], held in Rome last May, approved the organization of an *International Society of Soil Science*. It is now expected that persons interested in soil science will be given the opportunity to become members of the organization. You may have heard that the *Fifth International Congress of Soil Science* will be held in the United States probably in 1927. The American delegates to the Fourth Conference, myself included, are naturally anxious to secure a large membership in the Society from our country. Hence, I am taking the liberty to write to you and to ask that you help me secure members at your institution and elsewhere in your state. Persons who are teachers and investigators or other persons interested in soils and crops are eligible for membership. The annual dues for 1924 have been fixed at $2." Andrew Whitson discussed the letter with Emil Truog, who was interested to become involved in the organization of the congress and meet soil scientists from Europe. Emil Truog had not traveled much and saw it as an opportunity to promote his studies and American soil science. Much work remained for the organization of the congress, but the foundation for a thriving world community of soil scientists was laid—they just had to meet, and that happened in the summer of 1927.

References

1. Hartemink AE. Publishing in soil science—historical developments and current trends. Vienna: International Union of Soil Sciences; 2002.
2. Kellogg CE. Soil genesis, classification, and cartography: 1924–1974. Geoderma. 1974; 12:347–62.
3. Simonson RW. Early teaching in USA of Dokuchaiev factors of soil formation. Soil Sci Soc Am J. 1997;61(1):11–6.
4. Fallou FA. Pedologie oder Allgemeine und Besondere Bodenkunde. Dresden: Schönfeld Buchhandlung; 1862.
5. Weir WW. Soil science. Science. 1930; LXXI:218.
6. Asio VB. Comments on "Historical development of soil and weathering profile concepts from Europe to the United States of America". Soil Sci Soc Am J. 2005;69(2):571–2.
7. Shaw CF. Is pedology soil science? Am Soil Surv Assoc Bull. 1930; XI:30–3.
8. Sprengel C. Die Bodenkunde oder die Lehre vom Boden. Leipzig: Muller; 1844.
9. Tandarich JP, Sprecher SW. The intellectual background for the factors of soil formation. Factors of soil formation: a fiftieth anniversary retrospective. Madison, WI: Soil Science Society of America; 1994. p. 1–13.
10. Blanck E. Handbuch der Bodenlehre (vol. I–X). Berlin: Verlag von Julius Springer; 1929–1932.
11. Dokuchaev VV. The place and significance of modern pedology in science and life. (Translated from Russia by the Israel Program for Scientific Translations). Izbrannye Sochineniya. 1949; III:330–38.
12. Krupenikov IA. History of soil science from its inception to the present. New Delhi: Oxonian Press; 1992.
13. van Baren J. Agrogeology as a science. Mededeelingen van de Landbouwhogeschool en van de daaraan verbonden instituten. In: Wulff A, editords. Bibliographia Agrogeologica, vol XX. Wageningen; 1921. p. 5–9.
14. Hunter J. Soils. Chambers's encyclopaedia: a dictionary of universal knowledge, vol. IX. William & Robert Chambers, Edinburgh, J.B. Lippincot Company, Philadelphia; 1901. p. 556–57.
15. Hunter J. Soil science. Trans Scottish Agric Assoc. 1908;1:66–70.
16. Joffe JS. Notes on the international conference of pedology in the mediterranean region. Soil Sci. 1948; 65(5).
17. Beal WH. The new science of the soil. Sci Am. 1911;104(7):168–87.

18. Free EE. Studies in soil physics, IV: the physical constants of soils. Plant World. 1911;14(7):164–76.
19. Anon. Editorial notice. J Ecol. 1913; 1(1):3–4.
20. Wood TB. The result of the last twenty years of agricultural research. Science. 1913;38:529–40.
21. King FH. The soil: its nature, relations, and fundamental principles of management. New York: The MacMillan Company; 1895.
22. Hilgard EW. Soils, their formation, properties, composition, and relation to climate, and plant growth. New York: The Macmillan Company; 1906.
23. Merrill GP. A treatise on rocks, rock-weathering and soils. New York: The MacMillan Company; 1906.
24. Whitson AR, Walster HL. Notes on soils. Madison: Democrat Printing Company; 1909.
25. Hopkins CG. Soil fertility and permanent agriculture. Boston: Ginn and Company; 1910.
26. Russell EJ. Soil conditions and plant growth. London: Longmans; 1912.
27. Anon. Important soils of the United States Washington: U.S. Department of Agriculture. Bureau of Soils. Government Printing Office; 1916.
28. Marbut CF. Soils: their genesis and classification (Lecture notes from 1928). Madison, WI: Soil Science Society of America; 1951.
29. Ramann E. The evolution and classification of soils. Translated by C.L. Whittles. Cambridge: W. Heffer & Sons Ltd; 1928.
30. Hayes FA, Hearn WA, Shaw CF. Report of committee on resolutions American soil survey association bulletin. 1929; 11.
31. Brown PE. The value of research in connection with the soil survey. Am Soil Surv Assoc Bull 1930; XI:15–9.
32. de'Sigmond AAJ. The principles of soil science. London: Thomas Murby & Co.; 1938.
33. Joffe JS. Pedology (with a foreword by C.F. Marbut). New Brunswick: Rutgers University Press; 1936.
34. Joffe JS. Pedology. New Brunswick: Rutgers University Press; 1949.
35. Weir WW. Soil science. Its principles and practice. Chicago: J.B. Lipincott Company; 1936.
36. Robinson GW. Mother earth. Being letters on soil addressed to Sir R. George Stapledon C.B.E., M.A., F.R.S. London: Thomas Murby & Co.; 1937.
37. McCool MM, Veatch JO, Spurway CH. Soil profile studies in Michigan. Soil Sci. 1923;16:95–106.
38. Marbut CF. Soils of the United States. Part III. In: Baker OE, editors. Atlas of American Agriculture. Washington: United Sates Department of Agriculture. Bureau of Chemistry and Soils; 1935. p. 1–98.
39. Merkel DM. The curious origin of Podology: The story of a milestone paper. Unpublished manuscript, sabbatical report; 2013.
40. Lyon TL, Buckman HO. Edaphology. J Am Soc Agron. 1924;16(1):24–5.

41. Morton J. The nature and property of soils; their connexion with the geological formation on which they rest. London: James Ridgway; 1843.

42. Robinson GW. Soils—Their origin, constitution and classification. An introduction to pedology. London: Thomas Murby & Co; 1932.

43. Crowther EM. Agrochemistry. Nature. 1945;156(3956):227–8.

44. Cline MG. Agronomy at Cornell 1868 to 1980. Agronomy Mimeo No. 82-16. Ithaca, NY: Cornell University; 1982.

45. Beatty MT. Soil science at the University of Wisconsin-Madison. A history of the department 1889–1989. Madison: University of Wisconsin-Madison; 1991.

46. Anon. The Department of soil, water, and climate: 100 years at the University of Minnesota. Minneapolis: University of Minnesota; 2013.

47. Woodruff CM. A history of the department of soils and soil science at the University of Missouri. Special Report 413 College of Agriculture. Columbia, Missouri: University of Missouri; 1990.

48. Larson WE. Soil science societies and their publishing influence. In: McDonald P, editor. The literature of soil science. Ithaca: Cornell University Press; 1994. p. 123–42.

49. Coffey GN, Rice TD. Reconaissance soil survey of Ohio. Washington: U.S. Department of Agriculture; 1915.

50. Hartemink AE. The definition of soil since the early 1800s. Adv Agron. 2016; 137:73–126.

51. Weir WW. Soil science. Its principles and practice. Revised edition. Chicago: J.B. Lipincott Company; 1949.

52. Greenland D. In memoriam Professor Walter E. Russell 1904–1994. Bull Int Soc Soil Sci 1995; 86:63.

53. Waksman SA. Soil microbiology in 1924: an attempt at an analysis and synthesis. Soil Sci. 1925; 19(3):201–49.

54. Florea N. The contribution of Gheorghe Munteanu-Murgoci (1872–1925) and his Romanian colleagues to soil science. In: Yaalon DH, Berkowicz S, editors. History of soil science—international perspectives. Reiskirchen: Catena Verlag; 1997. p. 365–75.

55. Szabolcs I. The 1st international conference of agrogeology, April 14–24, 1909, Budapest, Hungary. In: Yaalon DH, Berkowicz S, editors. History of soil science international perspectives. Armelgasse 11/35447 Reiskirchen/Germany: Catena Verlag; 1997. p. 67–78.

56. The Moon D, Steppes American. The unexpected russian roots of great plains agriculture, 1870s–1930s. Cambridge: Cambridge University Press; 2020.

57. Yarilov AA. A quarter of a century in the service of soil science and of propaganda amongst soil scientists. Soil Res. 1936; V:184–86.

58. Weir WW. Productive soils—the fundamentals of successful soil management and profitable crop production. Philadelphia & London: J.B. Lipincott Company; 1920.

59. van Baren H, Hartemink AE, Tinker PB. 75 years the international society of soil science. Geoderma. 2000;96(1–2):1–18.

60. Moon D. The Russian academy of sciences expeditions to the steppes in the late Eighteenth Century. Slavon E Eur Rev. 2010; 88(1–2):204– + .
61. Anon. History of the organization of the international society of soil science. Soil Sci. 1928; 25:3–4.
62. Anon. The second international congress of soil science held in the Soviet Union, 1930. In: Proceedings and papers of the second international congress of soil science, vol. VII. Moscow: State Publishing House of Agricultural, Cooperative and Collective Farm Literature (Selkolkhozgis); 1932. p. 11–8.
63. Lipman JG. Report of the advisory committee from the American Society of Agronomy to the National Research Council. J Am Soc Agron. 1924;16:806–9.
64. Tovborg Jensen S. Dr. D.J. Hissink, in memoriam. Plant Soil. 1965; 8:6.
65. Hartemink AE, editors. D.J Hissink (1874–1956). Wageningen: NBV; 2010. J. Bouma, A.E. Hartemink, H.W.F. Jellema, Grinsven JJMv, Verbauwen EC, editors. Profiel van de Nederlandse bodemkunde. 75 jaar Nederlandse Bodemkundige Vereniging (1935–2010).Hartemink AE, editors D.J Hissink (1874–1956). Wageningen: NBV; 2010. J. Bouma, A.E. Hartemink, H.W.F. Jellema, Grinsven JJMv, Verbauwen EC, editors. Profiel van de Nederlandse bodemkunde. 75 jaar Nederlandse Bodemkundige Vereniging (1935–2010).
66. Edelman CH. II. Inaugural session. Address by Prof. Dr. C.H. Edelman. Paper presented at: Fourth International Congress of Soil Science, Amsterdam; 1950.
67. Šarapatka B. The contribution of Czech soil science at the turn of the 19th and 20th centuries to knowledge of soils: in memory of professor Josef Kopecký. Soil Water Res. 2015;10:207–9.
68. Kopecký J. Die klassifikationder bodenarten. Prag: Druck von Anton Purkrabek; 1913.
69. Whitney M. The use of soils east of the great plains region. Bureau of Soils Bulletin No. 78. Washington: United States Department of Agriculture; 1911.
70. Davis ROE, H.H. Bennett grouping of soils on the basis of mechanical analysis. Department Circular 419. Washington: United States Department of Agriculture; 1927.
71. Hissink DJ. Jacob Goodale Lipman and the international society of soil science. Transactions of the third commission of international society of soil science, vol. A; 1939. p. 1–3.
72. Waksman SA. Jacob G. Lipman. Agricultural scientist and humanitarian. New Brunswick, New Jersey: Rutgers University Press; 1966.
73. Moon D. The international dissemination of Russian genetic soil science (pochvovedenie), 1870s–1914 J Reg History. 2018; 2:75–91.
74. Keen BA. International conference on soil science. Nature. 1924;114:25–6.
75. Waksman SA. Sergei N. Winogradsky. His life and work. The story of a great bacteriologist. New Brunswick, N. J.: Rutgers University Press; 1953.
76. Frosterus B, Glinka K. Memoires sur la nomenclature et la classification des sols. Helsinki: Comite International de Pedologie; 1924.
77. Stremme H. International soil map of Europe, 1:2,500,000. Berlin: Gea Verlag; 1937.

78. Stremme HE. Preparation of the collaborative soil maps of Europe, 1927 and 1937. In: Yaalon DH, Berkowicz S, editors. History of soil science international perspectives. Armelgasse 11/35447 Reiskirchen/Germany: Catena Verlag; 1997. p. 145–58.

79. Landa ER. Hooked! Seductive details in soil science. Soil Horizons. 2015. https://doi.org/10.2136/sh2015-56-2-rc4.

80. Waksman SA. My life with the microbes. New York: Simon and Schuster, Inc.; 1954.

81. Keen BA. An international congress of soil science. Nature. 1927;3019:385–6.

82. Marbut CF. Outline of a scheme for the study of soil profiles. Washington: International Society of Soil Science; 1924.

83. Hartemink AE. 90 years IUSS and global soil science. Soil Science and Plant Nutrition. 2015;61:579–86.

11

First International Congress of Soil Science
1927

*"The most fundamental soil fact that has been discovered
by the study of soils in the field is that they are not static
unchangeable bodies with no response whatever to forces."*

Curtis Marbut, 1927.

The year 1927 was extraordinary. The economy was booming, the first
Transatlantic phone call was made between London and New York, the
Holland tunnel opened up underneath the Hudson River, African Americans continued to migrate to northern cities escaping rampant racism in the
south, and, in the middle of the country, rains for months on end caused
havoc and deaths. Southern Illinois received over 600 mm of rain in three
months, and parts of Arkansas had over 900 mm. There was the flooding
of the Mississippi River and its tributaries, and an area of the size of South
Carolina was under water for almost half a year. The floods affected 700,000
people and killed an unknown number.

President Calvin Coolidge, elected in 1924, showed no interest in the
disaster. He had grown up on a maple syrup and cheese farm in New
England. As the economy soared, Calvin Coolidge had an easy job and
worked only four hours per day, and he left the relief efforts to Herbert
Hoover, who was born to a Quaker blacksmith farm family in Iowa. Herbert
Hoover had acquired the name of 'food czar' with the assistance to Belgium
after the First World War. For the victims of the Mississippi floods, tent
camps were established, and in Herbert Hoover's view, it: "…was the first
real holiday they had ever known." Herbert Hoover became the President
in 1929, and by lowering the interest rates, he steered the country into an
economic calamity and was voted out in 1933.

Calvin Coolidge, indifferent to the flood tragedy, was however promi-
nently present to host a 25-year-old Minnesotan who had flown across the
Atlantic. It had taken Charles Lindbergh 33.5 h in a flimsy airplane, made

© The Author(s), under exclusive license to Springer Nature
Switzerland AG 2021
A. E. Hartemink, *Soil Science Americana*
https://doi.org/10.1007/978-3-030-71135-1_11

largely of canvas, that was named *The Spirit of St. Louis*. The streets of Washington were packed and it was the largest gathering the capital had ever seen [1]. Charles Lindbergh became an instant celebrity and made a tour across Europe, where hundreds of thousands awaited wherever he showed up. His flight across the Atlantic became celebrated in the USA, where aviation was behind compared to Europe; by the mid-1920s, one could fly almost everywhere in Europe, but no scheduled air service existed in the USA [1]. Just a few days before Charles Lindbergh's stardom, he was an unknown shy, non-drinking Minnesotan who had not finished his studies at the University of Wisconsin. But he knew how to fly planes.

May 1927 will be also remembered for the deadliest school killing in the USA. Andrew Kehoe, a graduate from Michigan State University in East Lansing, had become a disgruntled farmer, and he blew up a school in Bath, some 80 km from Palo where Charles Kellogg was born. He had a large debt and the bank foreclosed on his farm, and in the early morning, he placed boxes with dynamite and military explosives in the basement of the school. A few hours later, the school exploded and in total 44 people died including 37 children [1]. Three days after the school bombing, Charles Lindbergh crossed the Atlantic, and the newspapers had no more attention for the school killing.

The year 1927 was also distinctive because the *First International Congress of Soil Science* was held in Washington. It brought together soil scientists from 31 countries and sparked a worldwide surge in soil science, shaping the interest in soils and its significance across the planet. Those that studied the soil met not as practitioners in geology, chemistry, microbiology, or agronomy, but as an independent and diverse group of soil scientists. The vision for the congress was articulated by President Jacob Lipman and Secretary David Hissink of the *International Society of Soil Science*: "This Congress will bring together in America, for the first time in the history, all those that are interested in the different problems of soil classification, soil analysis, fertilization, and treatment, as well as the relations of the soil to plant growth. Extensive exhibits of various soil types (monolithic columns, in respective horizons), from Europe and America, apparatus used in soil analysis of the soil microflora and microfauna, etc. will be held during the Congress" [2].

The congress was planned as a series of presentations, and David Hissink had recommended that the congress was structured by the six commissions: Soil physics, Soil chemistry, Soil biology, Nomenclature and classification of soils, Soil cartography, and Plant physiology in relation to pedology. The structure proved valuable for the congress, and as it turned out, for many successive congresses. At the *Fourth International Conference of Pedology* in 1924, there had been no time for questions after presentations, and David

Hissink thought that discussions should be stimulated. In the program for the 1927 congress, more time was allotted for discussions and for bringing people together. The congress organizers made a list of delegates from each country with which the USA maintained diplomatic relationships. Jacob Lipman asked David Hissink how many delegates they could expect from Europe and: "…to the extent which the governments concerned will be able and willing to provide traveling expenses." There came no direct answer from the Dutchman, and instead, David Hissink asked Jacob Lipman for funds: "It is of course essential that the chairmen of the various commissions as well as the Editor, Prof. Schucht, should attend the congress. As most of them will not be in a position to defray the attendant expenses, some arrangements will have to be made to get over this difficulty. The same remark applies to my case also." Jacob Lipman had no idea yet how to finance the congress and what funds might be available for travel expenses.

The dates for the congress were not set until the summer of 1925. Jacob Lipman proposed late May or early June and thought that there would be a preference for June as the congress was to be followed by a field excursion. Together with Curtis Marbut and David Hissink, it was determined that the *First International Congress of Soil Science* would be held from Monday 13th of June until Wednesday 22nd of June 1927. The congress would be jointly organized by the *International Society of Soil Science*, the *American Society of Agronomy*, and the *United States Department of Agriculture*. Milton Whitney, who had just retired from the *Bureau of Soils*, was appointed as honorary chairman, and Jacob Lipman and Curtis Marbut were in charge of the organization. Oswald Schreiner served as the chair of the organizing committee and Arthur McCall as its Executive Secretary. They were both employed at the *Bureau of Soils*, and for the congress, all worked well under the leadership of Jacob Lipman; David Hissink named the organizers (Lipman–Marbut–McCall–Schreiner) the 'Big Four' [3]. Jacob Lipman was an effective chair, averse to ostentation, and he had a good relationship with Milton Whitney, as opposed to many of his contemporaries [3, 4].

In order to organize the congress and invite foreign delegates, the committee had to obtain permission from the government, and in December 1925, President Calvin Coolidge sent a report from the Secretary of State to the Congress requesting special legislation to hold the international congress. It read:

To the Congress of the United States:

I transmit herewith a report by the Secretary of State, concerning a request made by the Secretary of Agriculture that legislation be enacted that will give congressional sanction to the holding of an International conference on soil science in the United States in 1927 for which I request the favorable consideration of Congress.

Calvin Coolidge
The White House
December 10, 1925

One year earlier, President Coolidge had signed the Johnson–Reed Act, which restricted immigration from several countries and aimed to exclude immigrants from Asia. America had to remain a country and a culture that was dominated by European immigrants. Because of this law, there was a need for special legislation to invite people from non-European countries. The legislation to hold the congress and invite foreign delegates was approved in April 1926: "…the President was authorized and requested to extend invitations to foreign governments to be represented by delegates at the *International Congress of Soil Science.*" Thirty nations accepted the invitation of the United States Government to send an official delegate.

The organizing committee secured meeting rooms, established a program, and invited every soil researcher they could think of to join the *International Society of Soil Science* and attend the congress. The annual *International Society of Soil Science* subscription was 6.5 Dutch guilders ($50 in 2021) and new members had to pay an entrance fee of 2.5 Dutch guilders ($20 in 2021). All members from the *American Soil Survey Association* were invited. The organizers formed a national committee with leading soil scientist from each state including Dennis Hoagland from California, Merris McCool from Michigan, Percy Brown from Iowa, Harlow Walster from North Dakota, Merritt Miller from Missouri, Frederick Always from Minnesota, William Cobb from North Carolina, Ray Throckmorton from Kansas, Firman Bear from Ohio, and Emil Truog from Wisconsin.

Despite the prosperous economy, the organizing committee had problems financing the congress. Three sources of funding were looked at: contributions by soil researchers, a special grant by the government, and gifts by individuals and organizations. The gifts from the soil researchers did not yield much, and according to Jacob Lipman, most soil scientists were underpaid. Also, a grant from the government was not feasible given the rules and regulations and the possible delay in securing the funds. So, the organizing committee ended up asking for support and contributions from individuals, businesses, and organizations to whom the activities of the congress would be of interest. In total, 140 donors contributed, and most came from individuals, agricultural supply companies, and a few universities. The committee raised $75,000 [5], equivalent to about $1.1 million in 2021. It was used to pay for the congress and for travel expenses of the six commission chairs.

The organizers budgeted the congress and excursion such that participants of the excursion did not have to pay, and they were all guests of the American Organizing Committee.

Curtis Marbut had suggested that the congress should be followed by an extensive excursion to allow increased interaction among participants and to familiarize them with the soil and agricultural conditions of the USA and Canada. He also had higher aims for the excursion and hoped it would increase acceptance of the Russian principles of soil science in the USA which had been slow and problematic. And he hoped that interaction between the American, Russian, and European soil scientists would create consensus on the way soils should be classified and mapped. The excursion was modeled after the Transcontinental Excursion of the *American Geographical Society* that was held in 1912 [6]. The 1927 excursion was a large undertaking that required the cooperation of many people across the country and Canada. Trains were the main mode of transport in 1927, and Curtis Marbut negotiated with the Missouri Pacific Railroad Company in New York the hiring and cost arrangements of train cars and diners. Lunch was to be $1 and dinner $1.25. The organization of the excursion asked so much of Curtis Marbut that a few months before the congress, he was on the verge of collapse and went to a homeopathic hospital in Washington and was sentenced to complete rest. Two weeks later, he was back at work [7].

The *First International Congress of Soil Science* was attended by a little over 500 people which disappointed the organizers as they had expected more than 1,000 participants [8]. There were 366 Americans at the congress, including people from the University of Missouri (William Albrecht, Richard Bradfield, Merritt Miller, and Henry Krusekopf); Michigan (Merris McCool, Georg Bouyoucos, and Jethro Veatch); California (Dennis Hoagland, Walter Kelley, and Charles Shaw); New Jersey (Jacob Lipman, Selman Waksman, Jacob, Joffe, and Robert Starkey); Wisconsin (Emil Truog); North Dakota (Harlow Walster); Iowa (Percy Brown and Bill Pierre); and the *Bureau of Soils* (Wilbert Weir, Hugh Bennett, Curtis Marbut, Arthur McCall, Constantin Nikiforoff, and Thomas Rice).

There were 141 delegates from 31 countries who spoke 25 different languages. Some 29 participants brought their spouses and several of them joined the Transcontinental Excursion after the congress. Russia had sent 20 delegates, Germany 15, and there were six delegates from the UK. The Netherlands, Sweden, Switzerland, Japan, and Hungary each sent four to five delegates. The number of people who were sent by each country probably reflected the soil science activity in that country. There were many countries that sent one or two delegates, like Austria, Finland, Spain, Italy, Norway,

Top photo: Chamber of Commerce in Washington where the *First International Congress of Soil Science* was held from 13th to 22nd June 1927. Photo courtesy of US Chamber of Commerce. Lower photo: Pennsylvania Avenue west from the Old Post Office in Washington. The large building is the Willard Hotel not far from the White House. Many of the international attendees stayed at the Willard Hotel, and the reception after the opening of the congress was held in the hotel. In 1861, the Willard hosted the Peace Convention which was a last effort to avert the Civil War, and the same year President-elect Lincoln stayed at the Willard before his inauguration. Some 100 years later, Dr. Martin Luther King, Jr., finished his "I Have a Dream" speech whilst he stayed at the Willard

Brazil, Chile, Columbia, Cuba, Nicaragua, El Salvador, Australia, Egypt, and South Africa. Japan sent four delegates, India one delegate, but there were no other delegates from Asia. There were also no delegates from Belgium or France except for René Dubos who had emigrated from France to the USA in 1924. Albrecht Penck was, with his 69 years, the oldest participant, followed by Curtis Marbut who was almost 64, Peter Treitz was 61, and Konstantin Glinka had just turned 60. Merritt Miller, John Russell, Alex. de'Sigmond, Eilhard Mitscherlich, David Hissink, Téodor Saidel, and Jacob Lipman were all in their fifties. Hans Jenny and Edward Crowther were 29, Richard Bradfield was 31, and Ivan Tiurin was 35 years old. René Dubos was the youngest participant of the congress; he was 26 years old.

Of the 507 participants, there were only two women. The soil chemist A. N. Goudilina from Russia and Jadwiga Ziemięcka from Poland. Jadwiga Ziemięcka had worked with the soil microbiologist Sergei Vinogradskii at the *Institut Pasteur* in Paris. They studied nitrogen fixation in soils and related density and microbial activity to soil fertility. After Jadwiga Ziemięcka had left the *Institut Pasteur*, she returned to Poland and founded the first laboratory of soil microbiology at the State Scientific Institute of Agriculture in Pulawy. She was head of the laboratory until her death in 1968 [9]. At the first congress, she was the official Polish delegate and presented her work on *Azotobacter* in soil samples collected in 1917 and 1918 [10].

The Russian delegation left Moscow three weeks before the congress and travelled to Berlin to obtain visas. Since the revolution of 1917, the USA and Russia had no diplomatic relationship, so political pressure had to be exerted to allow the Russian delegation to obtain visas. Jacob Lipman had travelled to Moscow in the Spring of 1926 to discuss visas, the organization of the congress, and help securing travel funding for the Russian delegation [11]. It worked out well, and the delegation was given visas in Berlin, after which they visited the city and bought clothes 'in the European fashion' [12]. They boarded the steamer *George Washington* in Bremen on the 1st of June. It was the largest German-built steamship and the third-largest ship in the world when it was built in 1908. Everything in the ship emphasized comfort over speed and it was particularly luxurious in the first-class areas. The steamer travelled to Southampton in the UK, then to Cherbourg in France before it crossed the Atlantic. The journey took ten days, and according to Arseny Yarilov, the whole passage was lively: "...onboard the ship, in conditions where the rock lay at depth of eight kilometers and ground literally slipped from under their feet—what a ground, a ship deck without signs of horizons— the soil scientists held two strategic meetings." The outcome of the

meetings was to suggest commissions on genetic soil science and historical bibliography. The delegation also would like to see that Russian would
become the fourth official language of international soil congresses [13].

The steamer *George Washington* arrived in New York on the 10th of June
1927 [12]. The delegation was met by the organizers and taken to New Jersey
and from there to Washington. On the 11th of June, Washington and most
of the eastern USA celebrated that Charles Lindbergh had flown across the
Atlantic Ocean. There was a reception for the aviation hero hosted by President Calvin Coolidge, but he was keen to get the ceremony over with so that
he could start his three months' vacation in the Black Hills of South Dakota.
President Calvin Coolidge suffered from chronic indigestion and asthma and
was eager to leave Washington. A special train awaited him that included a
small army of reporters along with two collies, Grace Coolidge, and a pet
raccoon that they had named Rebecca [1]. Calvin Coolidge had one task
remaining before he could board the train: the opening speech at the *First
International Congress of Soil Science*.

Konstantin Glinka headed the Russian delegation and was with Curtis
Marbut the pedologic leader of the congress. The delegation included A.
N. Goudilina, Sergei Neustreuv, Boris Polynov, Ivan Tiurin, Dmitri Vilenskii, and 14 others. The Russians were especially welcomed at the congress
and were treated with great respect. The delegation had three commissar-type
functionaries who escorted them throughout the congress and excursion [14].
Several of the Russian scientists expressed their resentment over the surveillance by the functionaries and sometimes quarreled with them. Jacob Joffe
who had emigrated to the USA in 1906 spoke Russian and as a member
of the organizing committee comforted the Russian delegation during the
congress and excursion.

The Russian delegation had brought copies of 13 Bulletins that summarized advances in pedology and soil science in Russia. They were liked by
the congress participants and became influential in the teaching of pedology
in the USA. The bulletins were named *Russian Pedological Investigations*
[15–27]:

   I. Dokuchaev's ideas in the development of pedology and cognate
      sciences. K.D. Glinka.
  II. Achievements of Russian science in morphology of soils. S.A.
      Zakharov.
 III. Genesis of soils. S.S. Neustruev.
  IV. Achievements of Russian science in the province of chemistry of soils.
      I. V. Tiurin.

The bulletins were written by leading soil scientists, and several of them had studied or worked with Vasily Dokuchaev, like Konstantin Glinka, Serghei Zakharov, Jan Afanasiev, Leonid Prasolov, and Sergeï Kravkov. Except for Nikolaï Prokhorov and Boris Keller, all others attended the *First International Congress of Soil Science*.

There was a large group from Germany at the congress and they played an active role in the society. Friedrich Schucht was the editor of the *Internationale Mitteilungen für Bodenkunde* since 1911. Herman Stremme had compiled a soil map for Europe. Otto Lemmerman, Director of the Institute of Soil Chemistry and Bacteriology in Berlin, also attended, and in 1922, he had founded the *Zeitschrift für Pfanzenernährung, Düngung und Bodenkunde* that later became the *Journal of Plant Nutrition and Soil Science*. The geographer Albrecht Penck attended and he was a friend of Curtis Marbut [7]. Geology and physical geography were their common interest and they had first met at the Eight International Geographic Congress in Boston in 1904. If William Davis was the 'founder of American Geography,' Albrecht Penck can be considered as one of the founders of German geography, and his main research was Pleistocene stratigraphy and in particular the four periods of glaciation—Günz, Mindel, Riss, and Würm—named after the river valleys that were the first indication of each glaciation [28]. His son Walther also became a geographer and worked on landscape evolution and was a stern critic of William Davis' theory on the cycle of erosion. Walther Penck worked in Argentina, but died from cancer in 1923, only 35 years old. A year after his son's death, Albrecht Penck finished Walther's book *Die morphologische Analyse. Ein Kapitel der physikalischen Geologie,* which was translated into English in 1953 [29].

In 1926, the German soil community had lost one of its eminent scientists, Emil Ramann, who had died a few months before his retirement. He was of Vasily Dokuchaev's generation and had studied pharmacy, chemistry, and science at the University of Berlin. In 1900, he was appointed professor of forest soil science and chair of Agricultural Chemistry at the University of Munich. In 1905, Emil Ramann had introduced the term *Braunerde* or brown forest soils as a soil type [30]. He wrote soil science textbooks, and one of his books was posthumously published in English [31–33]. Emil Ramann was convinced that advances in scientific knowledge were only possible if there was a close collaboration between all countries of the various climatic and soil zones [34].

There were several Hungarians at the first congress, and it had an active soil science community. The delegation was led by Alex. de'Sigmond who had organized the first international soil meeting in Budapest in 1909. He was born in an aristocratic Transylvanian family and was trained as a chemical engineer [35]. In 1906, he visited the USA, met with Eugene Hilgard, and spent some time with Macy Lapham in the west at the recommendation of Milton Whitney. Macy Lapham remembered him as a: "…courtly, friendly, dapper little gentleman of high scholastic attainment, but by no means devoid of humour. When I last sat with him at the dinner table he laughingly remarked that he felt that if anything were to happen that he should be without other means of support, he could be a jockey. He weighed not much more than 41 kg" [36]. Mark Baldwin and Henry Krusekopf named Alex. de'Sigmond 'the Dean of the soil scientists in Europe,' and Selman Waksman thought Alex. de'Sigmond was an artist and a brilliant scientist [35]. Just before the congress, Alex. de'Sigmond's book on alkali soils was translated and published by the University of California Berkeley [37]. Walter Kelley wrote, somewhat thriftily in the preface of the translation: "This paper is an important one and is worthy of careful study." The Hungarian delegation at the congress also included Peter Treitz, who was at the Agrogeological Department of the Geological Institute, and he made the first soil maps of Hungary based on zonality [38].

Romania was represented by three soil scientists including Téodor Saidel, a pedologist and chemist who worked on salinity, soil pH measurements, and he made one of the first soil maps of Romania [39]. Gheorghe Murgoci had died in Bucharest in 1925 from an albumin disease that affected his lungs [40]. He was 53 years old. Gheorghe Murgoci had introduced the zonal concept of soils in Romania and was eager to attend the 1927 congress and see how soils and their zonal distribution differed in the USA. He was well-connected to Hungarian soil scientists and had helped with founding the

International Society of Soil Science in 1924. With Alex. de'Sigmond, he had supported Curtis Marbut's bid to hold the congress in the USA. Gheorghe Murgoci and Peter Treitz had travelled to Russia in 1907 and 1908, where they conceived the idea to organize the *International Agrogeological Conference* in 1909 [40]. In 1911, Gheorghe Murgoci published *Natural areas of soil in Romania* which was the first pedology book in Romania. His main goal was to attain a universal concept of soil, classification, terminology, and methodology [40].

Hans Jenny, who had emigrated in 1926 from Zürich to New Brunswick, was the official delegate for Switzerland. He attended with the chemist Georg Wiegner from ETH who had supervised his doctoral thesis. Georg Wiegner was a prominent soil chemist and travelled to congress with his new book on quantitative agricultural chemistry, which was partly written by Hans Jenny [41]. In 1921, Georg Wiegner had published a book on the role of colloid chemistry in soil formation, that was praised as: "*...das ganz vortreffliche Werk...Möe es darum eine weite Verbreitung finden*" (a most excellent work which we hope finds wide distribution) [42]. Georg Wiegner was familiar with the studies by Konstantin Gedroiz, who also studied colloid chemistry in relation to soil formation.

The delegation of the UK was led by John Russell and included Bernard Keen, Edward Crowther, Harold Page, and William (Gammie) Ogg [43]. Most of the delegates were from Rothamsted Experimental Station, directed by John Russell since 1912. Bernard Keen had joined the soil physics department at Rothamsted in 1919 and hosted a BBC radio show entitled: *Why and wherefore of farming* [44]. He left Rothamsted in 1947 to become the first director of the *East African Agriculture and Forestry Research Organization* where he was succeeded by Walter Russell in 1954. At the *First International Congress of Soil Science*, there were a few soil scientists who were sent by their British colonial government, including Frederick Hardy from Trinidad, C. H. Knowles of West Africa, and C. L. Whittles from British Guiana, who had translated the work of Emil Ramann [33]. Only two Australians attended the first congress in the same year that the CSIRO Division of Soils was established.

The opening speech of the Congress was given by President Calvin Coolidge. He arrived at the congress with two armed military men, and they were his constant body-guards [45]. Known as 'Silent Cal' he said little and was known to have a dry sense of humour [1], but he delivered a fairly long speech: "The fundamental importance of the soil as a great national and international asset is at once apparent when we reflect upon the extent to which all mankind is dependent upon the soil either directly or indirectly for food,

Group photo (partly) of the *First International Congress of Soil Science* in 1927. Forefront: 1. David Hissink, 2. Jacob Lipman, 3. Konstantin Glinka, 4. Calvin Coolidge (USA President, papers in left hand), 5. Curtis Marbut, 6. William Jardine (Secretary of Agriculture), 7. Oswald Schreiner. In 1927, the USA and Russia had no diplomatic relationships. Photo was taken at the Pan American Union prior to the reception on 14th of June 1927. Photo by M.E. Diemer, Chemist and Photographer of the University of Wisconsin, and friend of Emil Truog

clothing, and shelter. Long after our mines have ceased to give up their treasures of iron, coal, and precious metals, the soil must continue to produce the food necessary for feeding the ever-increasing populations of the world. It is highly proper therefore that representatives of the nations of the earth assemble in groups such as this for the purpose of discussing methods to be employed in the study of the problems of soil conservation and land utilization. The interchange of ideas and the personal contacts made possible by this international gathering cannot fail to be productive of a better understanding between the peoples of the earth and ultimately lead to a more universal desire for peace between all nations. Being a young nation we have not, as yet, been forced to conserve our great natural resources as have some of the older nations where the pressure of population on food supply has long since become acute and has forced them to consider means for conserving the fertility of their soils and at the same time increasing the yield per acre. In the past, with our abundance of fertile acres, we have been able to greatly increase our total production through increased acreage and the use of improved machinery. With practically all of our fertile land now under cultivation, further increases in total production must come from increased acre yields instead of from increased acreage as in the past. The rapid and continued concentration of our people into large industrial centers now makes it necessary to greatly increase the effectiveness of those who remain on the land in order to produce a food supply sufficient to feed our city population." Calvin Coolidge ended his speech with: "I trust that this great gathering of soil scientists from all parts of the world will be productive of great good and that you will return to your respective institutions and countries with your minds enriched with new ideas and that you will carry back with you a renewed enthusiasm for the great work to which you have dedicated your lives."

A response to the opening speech was given by President Jacob Lipman: "On behalf of the *International Society of Soil Science*, under whose auspices this Congress is being held; on behalf of the officers of the Congress; and on behalf of the delegates from many lands, I wish to thank you for the honour you have done us. Your personal interest in our undertaking will lighten our task and will quicken our desire to learn the secrets of soil fertility. We are fully aware of the meaning of soil and land problems in the life of nations. It is our hope that the meetings of this Congress will point the way more clearly and more hopefully toward the effective use of the soil resources of the world, toward prosperity and contentment in all lands, toward friendship and good will among the nations, and toward a more stable material foundation on which mankind may build its spiritual kingdom." A group picture was taken at the Pan American Union prior to the reception on 14th of June, and then Calvin Coolidge hurried to the special train that took him to his vacation address in South Dakota.

Milton Whitney was the honorary chairman of the American Organizing Committee, but he was seriously ill and home in Takoma Park [46]. Just before the congress, he had stepped down from his position at the *Bureau of Soils*, and Arthur McCall was selected to head the soils work, and he became a member of the organizing committee [46]. Arthur McCall had been at the bureau from 1901 to 1904 and was then a professor at the Ohio State

University and the University of Maryland. He obtained a PhD in soil physics from the Johns Hopkins University. The absence of Milton Whitney at the congress gave John Russell the opportunity to propose a resolution expressing: "…the sympathy of the congress to Doctor Whitney in his illness" [46, 47]. Milton Whitney's speech was given by the director of scientific work at the Department of Agriculture who addressed the congress on the origin and work of the *Bureau of Soils*. He warned about soil erosion and the washing away of the fertile soils of the world: "This problem of controlling erosion, both the slow and rapid types, is, I believe, the most vital soil problem we have and the one upon which we are doing the least work" [47].

The congress was held at the Chamber of Commerce building on 1615 H Street. The colossal buildings of the Department of Agriculture on Independence Avenue had yet to be built. The first day was dedicated to the opening speeches and the greetings and roll call of delegates and scientific societies. The day ended with a reception at the Willard Hotel, which was a ten-minute walk across Lafayette Square and Pennsylvania Avenue. At the end of the second day, a reception was hosted by the Secretary of Agriculture, William Jardine, at the Pan American Union. William Jardine had a strong interest in practical farming and wrote several handbooks, such as the *Suggestions for teachers giving practical instruction to city boys* [46]. There was a formal dinner at the Willard Hotel and a reception at the National Gallery of Art—now the Smithsonian American Art Museum. These dinners and receptions were well attended, but no alcohol was served.

Jacob Lipman had made the program for the congress based on the recommendations of the six commissions, which had held meetings in 1925 and 1926 to prepare for the congress. In total, 361 papers were received, with 71 for the commission on Soil Biology and Biochemistry and 78 for the commission Classification, Nomenclature, and Mapping of Soils. The program started with invited papers in the mornings, and the afternoons were used for sessions of the commissions, and these were held in separate rooms. The first commission, The Study of Soil Mechanics and Physics, was chaired by the Czech Véclav Novâk who had introduced the work of Vasily Dokuchaev at the University of Agriculture in Brno [48]. The commission had met in 1926 at Rothamsted in the UK. In total, 34 papers were presented that focused on the preparation of soil samples for mechanical analysis, soil fractions, apparatuses, water holding capacity and moisture equivalents, and how properties behaved under field conditions. Papers were presented by Gilbert Robinson, Charles Shaw, Edward Crowther, Bernard Keen, Merris McCool, Georg Bouyoucos, and Josef Kopecký.

The second commission, The Study of Soil Chemistry, was chaired by Alex. de'Sigmond with the help of Georg Wiegner, Emil Truog, Téodor Saidel, David Hissink, and Merris McCool. The commission had held a meeting at the Soil Science Institute of David Hissink in April 1926, where the members were introduced to the works of Konstantin Gedroiz [49]. The commission agreed that studies should focus on the preparation of soil extracts, particularly from hydrochloric acid, the soil solution, organic matter, and nitrogen in the soil, and chemical determination of the soil fertility. In total, 62 papers were presented, including papers by Richard Bradfield, Hans Jenny, and Georg Wiegner and several by Alex. de'Sigmond, who presented chemical characteristics of soil leaching, and a paper on the classification of alkali and saline soils. The work in the chemistry commission was concerned with soil acidity and base exchange phenomena [5]. According to Curtis Marbut, the study by Georg Wiegner and Hans Jenny on ion exchange was one of the best presented at the congress, but he did not think it was soil science, that he considered much broader than specific chemical reactions [50]. The paper itself, *Ueber Basenaustasch*, was not included in the proceedings. Hans Jenny also presented his studies on soils in the Swiss Alps that he had conducted in 1925 [51].

The commission Soil Biology and Biochemistry was chaired by the Czech Julius Stoklasa but he was ill, and Selman Waksman presided over the session [5]. In 1926, the commission had held a meeting in Berlin, Germany. Papers were presented in seven sessions that included: direct and cultural methods of microbiological analysis of soils, the soil population, nitrogen fixation in the soil, nitrogen transformations, organic matter transformations, mineral transformations, and soil biology from an agronomic standpoint. There were 71 presentations and several papers from Japan. Selman Waksman read the paper by Sergei Vinogradskii that focused on the quantitative aspects of soil bacteriology. He had invited Sergei Vinogradskii to the congress, but Sergei Vinogradskii attended few congresses and regarded them as a waste of time. The discussion was led by John Russell, and he also presented the status of soil biology and showed temporal dynamics in bacteria and protozoa and the fluctuations from day to day and from hour to hour. He had found that soils were more fertile when they have an active bacteria and protozoa population and that those organisms should be fed appropriately.

The fourth commission, Soil Fertility, was chaired by Eilhard Mitscherlich, and the commission had held its preparatory meeting in Düsseldorf, Germany, in 1926. The session at the first congress focused on: nutrient requirements of soils and plants by field, pot and germination experiments, the determination of fertilizer and lime requirements, the effect of cultivation

A corner of the exhibit of soil literature from the USA, and soil profiles from Latvia, where the President of the *International Society of Soil Science*, Jacob Lipman was born in 1874. Photos *International Society of Soil Science* proceedings

on crop yield, and the supply of water and air to plants. In total, 45 papers were presented, and there were studies from Emil Truog, Dennis Hoagland, Merris McCool, Jacob Lipman, Frank Duley, and Andrew Whitson. It had a late start but by the 1920 s American soil fertility research was well-established and dominated the commission on Soil Fertility.

In total, 78 papers were presented in the fifth commission, Classification, Nomenclature, and Mapping of Soils, chaired by Curtis Marbut. The commission had sub-commissions that dealt with soil classification in Europe and the USA, the soil map of Europe chaired by Herman Stremme, and alkali and saline soils. Papers were presented by Konstantin Glinka, Curtis Marbut, Charles Shaw, Merris McCool, Gammie Ogg, Mark Baldwin, Peter Treitz, Gilbert Robinson, and Walter Kelley. Hugh Bennett presented his findings on the soils of Cuba that were published a year later [52]. Gammie Ogg presented the soils of Scotland in relation to climate and vegetation, and Merris McCool presented the relationship between soil colloids and the textural classification of soils. Russian and American soil scientists dominated the session.

The sixth commission was named *The Application of Soil Science to Land Cultivation*, and 37 papers were presented that focused on drainage, irrigation, highway construction, and soil erosion. Georg Wiegner presented some of his works in Switzerland on corrosion in soils, Hugh Bennett on soil erosion, and V. R. Burton on soil science and highway engineering in Michigan.

A wide array of topics was presented for each of the commissions. There were numerous presentations on bases, soil acidity, soil organic matter, and plant nutrition. Harold Page, the soil chemist from Rothamsted, presented a relationship between the base saturation and the pH. He had found that when the soil became more acidic, the exchangeable bases decreased, and at pH 4.0, there was only one-seventh as many exchangeable bases as at pH 5.2. Arao Itano had obtained his BS in Michigan and PhD in Massachusetts and spoke about soil investigations in Japan where most research was directed toward the use of fertilizers and soil improvement. The soil itself was not well studied, and Arao Itano thought that was common in other parts of the world. Emil Truog presented the difference between solid phase feeding and liquid phase feeding. In solid phase feeding, plants directly attack the minerals of the soil, which increased as roots released more carbon dioxide. He found that buckwheat and clover were strong solid phase feeders as opposed to oats and corn.

Curtis Marbut presented a comprehensive scheme of soil classification in which he grouped the units in each soil category on the basis of their properties and characteristics, as opposed to external soil forming factors. The scheme had Pedalfers and Pedocals at the highest level [53]. It was well-received [7]. He also provided a report for the commission and stressed the need for studying soil profiles: "That the committee recommends to all persons in all countries the study, until the meeting of the next International Congress of Soil Scientists, of soil profiles and the accumulation of as much data as possible on the subject, in their respective countries, to the end that a scheme of soil classification, based on these features, may be devised. That soil specialists in all countries be urged to accumulate data, each in his own country, that may serve for the construction of a reconnaissance soil map." These maps were to be presented at the next congress [54].

Jacob Lipman presented a paper, *Soils and Men*, in which he reviewed worldwide problems of plant, animal, and human food. He found that it was in the interest of soil science to familiarize itself with imports and exports of agricultural commodities as they affected soil fertility in different regions, and economic, political, and social trends were based on land and soil resources. He summarized the carbon and nitrogen content of the atmosphere, soils, forests, peat, and coals. The total carbon content of soils was estimated to be 400 billion tonnes, and it was assumed that there was 1.5-times more carbon in the atmosphere than in the soil, and 4-times more carbon in the soil than in forests [55].

Albrecht Penck, the geographer from the University of Berlin, presented a productive capacity map of the world. He calculated calorie production for all regions of the world and compared the calorie production to the number of people who could live from it. Based on his calculations, the carrying capacity of the world was about eight billion people, and the greatest population density was possible in the humid tropics. He thought that Brazil would ultimately be the most populous country in the world [28].

After each presentation, there were five minutes for discussion, and no one was allowed to speak more than twice without special permission. The official languages at the congress were English, French, German, Italian, and Spanish, and the program contained the titles of all papers in those five languages. There were a few delegates who spoke several languages, but most presentations were in English and German. Not every paper was read by its author. The agrogeologist from Wageningen, Jan van Baren, had sent in a paper on limestone soils of the tropics but he was absent [56]. The paper by Konstantin Glinka on developments in soil science in Russia was read by Jacob Lipman.

A paper on bacteria and soil productivity, by Julius Stoklasa from the Technical Institute and Experiment Station in Prague, was read by someone else as he was ill.

The nature of organic matter was an important topic at the first congress, and papers were presented by Harold Page from Rothamsted, Selman Waksman and René Dubos from Rutgers, and Gilbert Robinson from Bangor, Wales. Jacob Lipman presented a paper on the microbial aspects of green manuring and had found that nitrates were lost when green manure high in carbohydrates was plowed into the soil. Richard Bradfield spoke about the use of electrodialysis in soil studies. Jethro Veatch presented a paper on the classification of organic soils [57]. Téodor Saidel presented the 1 to 1.5 million soil map of Romania [58]. There were several papers on soil color, and the topic was of interest in relation to soil organic matter, soil formation, and as an indicator of soil productivity [59–61]. There was no standardized way of observing or recording soil color, and qualitative terms were used. For the clothing industry, Albert Munsell had developed a three-axis system with color chips that had names and numbers. The cards with chips were adapted by Dorothy Nickerson in collaboration with Robert Pendleton, Thomas Rice, and James Thorp to be used in soil survey [62]. Albert Munsell had no connection to soil science and died 30 years before his color charts were introduced in the American soil survey [62].

Each day of the congress, soil exhibits opened at eight o'clock in the Chamber of Commerce Building. A few years before the congress, Jacob Lipman had asked Emil Truog to organize exhibits that would show his soil acidity tester, as well as any other exhibit on soils, with the purpose: "…to give a vision of extensiveness of soil science and the varied nature of research work in different countries" [63]. The exhibit was a large effort, and Emil Truog travelled to Washington a few days before the congress to make the final arrangements, missing a concert by Sergei Rachmaninoff in Madison. Four rooms with a total floor space of 280 m^2 were used for the exhibits, and it became one of the highlights of the congress.

Hermann Stremme had the walls of two rooms covered with draft soil maps from most countries in Europe. A year after the congress, he and the soil survey staff from Danzig published the first soil map of Europe [64]. They used geological maps as the base maps and they had been available since the late 1700 s. The soil map of Europe was published at a scale of 1 to 10 million with a legend in German, French, and Polish, and some years later in English. It had 27 map units and included Chernozems, Rendzinas, Podzols, peat, salty soils, and brown and red soils. Over 30 soil scientists worked on this project including Jan van Baren, Konstantin Glinka, Hans Jenny, Gilbert Robinson, and Georg Wiegner [65].

Wilbert Weir and Arthur McCall of the *Bureau of Soils* were in charge of the American exhibition, which showed soil types, soil maps, instruments for soil analyses, and a soil science literature collection [5]. There were 18 soil monoliths from the USA, and several monoliths from Hungary were collected and prepared by Peter Treitz who had developed a new way to preserve monoliths. The monoliths were treated with gelatin so that the soil particles stuck together, and it also preserved the color of the soil. The monoliths were less than two centimeters thick, and when displayed vertically, the soil did not crumble. It was an improvement over the method that was used by staff of the *Bureau of Soils*; they made 1.5-meter-deep monoliths that were 20 cm wide and 10 cm thick, which were displayed in a box on a table. The soil was kept moist by a layer of sphagnum moss at the bottom of the box [66].

Russia had a large exhibition showing soil maps, periodicals, publications, books, and portraits of Vasily Dokuchaev, Konstantin Glinka, and Konstantin Gedroiz. The exhibit attracted much attention [45]. The Russian delegation had planned the presentation of 50 soil monoliths which were collected by Boris Polynov and Konstantin Glinka along a geographical sequence from St. Petersburg to the Caucasus and in Georgia, Azerbaijan, Kazakhstan, the Amu Darya region, and the Siberian Far East. The monoliths were treated with a sugar solution for preservation, and they were only meant for display and lacked documentation and analytical data [67]. It was a long journey from St. Petersburg to Washington, and the monoliths arrived after the first congress was over. Curtis Marbut offered to store them, and the monoliths ended up in a facility of the Department of Agriculture in Washington, where they remained in the original gray wooden boxes for decades. They were looked after by Constantin Nikiforoff and had been on display in Beltsville once before they were moved to Hyattsville [68]. In 1980, the monoliths were shipped to ISRIC in the Netherlands [4]. Some of the monoliths were returned to Russia, and some of the remaining were subsampled to study soil change [67].

At the exhibit, there were several rooms with novel instruments, equipment, and apparatuses to study soils. The dynamometer of Bernard Keen and William Haines of Rothamsted was on display. It measured the mechanical resistance that must be overcome by the applied force of drawing a tillage implement through the soil [69]. Bernard Keen was at the exhibit and explained how the isodyne maps were made that showed the soil resistance to plowing. They found that the soil resistance at a given location was more or

Russian soil chemist A. N. Goudilina attended the *First International Congress of Soil Science* in Washington, and boarded a ship on the Potomac River, during the excursion to Mt. Vernon; she also participated, along with 205 other delegates, in the Transcontinental Excursion. Right photo: Henry Krusekopf from the University of Missouri in Columbia. When his family came from Central Europe to Missouri, they avoided the prairies because the idea was that if the soil cannot support a forest, it cannot be very good. Between 1909 and 1926, he made soil maps of 22 counties in Missouri. Photos from the Smithsonian Institution Archives

less constant at a normal moisture range, and that a field could be characterized by its isodyne map. Applications of organic matter or lime decreased the soil resistance, whereas the resistance increased slowly when the plow speed was increased. Bernard Keen and William Haines were members of an influential group of English soil physicists, including Earnest Childs, Robert Schofield, and Howard Penman, and their American contemporary Lorenzo Richards [70].

Instruments on display included an electrometric and colorimetric apparatus for the determination of hydrogen-ion concentration, special microscopes, a modified Wiegner tube by Richard Bradfield, and an auto-irrigator, and Georg Bouyoucos showed an apparatus for colloidal analysis. There were soil centrifuges, Pyrex glassware, percolation tubes, motor stirrers, Kjeldahl apparatus, a device that measured the oxygen-supplying power of the soil, and a micro manipulator for single-cell isolations. Nematodes could be viewed under a microscope, and one of the rooms had a miniature plant for the fixation of atmospheric nitrogen. Displays had strain variations in root-nodule bacteria, and there were pots with plants that showed the fertilizer value of manganese and other nutrients.

Journals devoted to soil science and more than 50 influential books were exhibited including Humphry Davy's *Elements of agricultural chemistry* and Edmund Ruffin's *An Essay on Calcareous Manures* [71, 72]. Every attendant was given a copy of *A classified list of soil publications of the United States and Canada*. The bibliography listed soil papers, reports, and books, and they were grouped as follows: General, Soil classification and nomenclature, Soil geography, Soil chemistry, Soil physics and mechanics, Soil biology and biochemistry, Soil ecology, Soil fertility, Soil management, and Fertilizers. It listed all soil publications in the USA up to 1927 and was prepared by Claribel Barnett, librarian of the Department of Agriculture in Washington. She wrote in the Preface: "As it is the first attempt to bring together in a comprehensive way the literature of modern soil science, there were no examples or criteria to guide in the selection and arrangement of the material." There were some limitations: "Many important papers have undoubtedly been left out while others of slight value are included. Articles in farm journals have, with a few exceptions, been omitted, as they are short and popular. Some trivial articles were also deliberately rejected, but on the whole, the list is more inclusive than a selective bibliography" [73]. It included the 1923 paper by Nikolaĭ Tulaikov on the *Drought and the means of overcoming its evil effects in the Volga region of European Russia* that was the first Russian paper published in the *Journal of the American Society of Agronomy* [74].

A classified list of soil publications of the United States and Canada was one of the first soil science bibliographies with over 6,000 references on 500 pages. Freely distributed, it brought American soil science literature to participants from all over the world. It was not the first soil science bibliography. In 1921, Adolf Wulff had published the *Bibliographica Agrogeologica—Essay of a Systematic Bibliography of Agro-Geology*, which listed over 3,300 references [75]. In 1931, a thousand-page German bibliography was published [76], followed by *Bibliography of Soil Science, Fertilizers and General Agronomy* that was published in the 1950 s by the *Imperial Bureau of Soil Science* in the UK.

The *First International Congress of Soil Science* was viewed by many participants as a catalyst for soil science progressions, and it set a standard for future congresses. According to John Russell: "The whole Congress was a brilliant success, a model of what such functions should be" [77]. The congress drew the attention of the public upon the importance of soils, and more importantly, the extension of that knowledge through fundamental soil research [78]. For Leonid Prasolov, it was clear that the congress was dominated by: "...a practical and agronomic trend," and he regretted that the American geographers, geobotanists, and geologists did not participate in the congress since he considered those fields closely related to soil science [12]. Hans Jenny thought the congress was superbly organized and he gave all credit to Jacob Lipman [50]. He did not attend another international soil congress until 1956 in Paris, which was also the last congress that he attended.

Toward the end of the congress, the delegations of Russia and Brazil put in bids to host the second congress in Moscow or Rio de Janeiro in 1929, and Jacob Lipman was in favor of holding the congress in South America [3, 47, 79]. Given the prominence of Russian soil science, it was decided that the *Second International Congress of Soil Science* was to be held in Russia in 1929 [80]. Some contingencies were built in given the political situation: "...that sufficient assurance as to the attendance could be given at an early date. Should this not be possible the General Committee, in conformity with paragraph 12 of the Byelaws will select another country" [81]. Konstantin Glinka was the elected President and charged with organizing the congress. The *Second International Congress of Soil Science* was indeed held in Russia, though not in 1929 as planned, but in 1930. Brazil had to wait another 88 years before it hosted the congress. On the last day of the *First International Congress of Soil Science*, the society elected the following Honorary Members: John Russell, Emil Ramann, Lucien Cayeux, Konstantin Glinka, Sergei Vinogradskii, and Josef Kopecký. Sergei Vinogradskii had been at the *Fourth International Conference of Pedology* in 1924, but he never attended another soil congress; Emil Ramann had died in 1926.

There were two excursion days during the ten days of the congress. On Thursday 16th of June, 320 delegates left the Willard Hotel in 12 motor buses for a trip to Western Maryland. The busses left at eight o'clock in the morning and returned well after midnight [82]. Two police motorcycles accompanied the busses. It visited the University of Maryland—the *alma mater* of Milton Whitney. The excursion crossed the Shenandoah Valley of Virginia, which was the breadbasket of the Confederacy during the Civil War. A second-day excursion was held on Tuesday 21st of June, one day before the congress officially ended. Some 150 participants boarded busses at the Willard hotel and the tour included observations of a soil profile in a road cut. The soil was named a Leonardtown silt loam, which was poorly drained, strongly weathered, and with a root restricting dense clay layer in the subsurface. The excursion guide pointed out the dense clay, which made the soil: "...cold and waterlogged in the spring and droughty in the late summer." Most of the land in the area had been cleared but was lying idle. As the tour guide noted: "It is said that George Washington learned to swear while farming this soil. Do you wonder?" [83]. They then boarded a

Emil Truog (1884–1969) (right) with Walter Kelley (1878–1965) (left) and Téodore Saidel (1874–1967) (center) at the start of the Transcontinental Excursion in 1927. Walter Kelley was born on a tobacco farm in Kentucky and worked most of his life on alkali soils in California, he was elected member of the National Academy of Sciences in 1943. Téodore Saidel produced the soil maps for Romania and was one of the first to quantitatively measure the hydrogen ion exponent using an electrometric method in 1913. Photo by M. E. Diemer, Chemist and Photographer of the University of Wisconsin

boat and visited the harbor and travelled down the Patapsco River. Onboard were several Russian scientists including Konstantin Glinka, Jacob Joffe, A. N. Goudilina, and Vladimir Gemmerling. It was a busy harbor, as Baltimore in 1927 was the largest fertilizer center in the world. They visited the Davison Chemical phosphate plant where raw rock phosphate from Florida was treated with sulfuric acid and converted into superphosphate. The fertilizer industry had rapidly expanded at the beginning of the First World War when America realized its dependence on German fertilizers.

The first congress resulted in a set of proceedings in four volumes, 2,740 pages, that cost only $2 ($30 in 2021) but $4 ($60 in 2021) for non-members of the *International Society of Soil Science*. In total, 197 papers were included, and most of the papers were single authored. Alex. de'Sigmond contributed nine papers. The papers were edited by Jacob Lipman, with the assistance of R.B Deemer, P.R Dawson, and A.R. Metz. Jacob Joffe translated the papers that were in Russian, whereas Selman Waksman assisted in the arranging, editing, translating, and printing the abstracts. He had the help of some of the young researchers in his laboratory including René Dubos, Hans Jenny, and Jacob Blom [84]. At the congress, a book of abstract was available and these were in French, English, and German. René Dubos translated most of the abstracts in French; he was experienced with this and had a similar job at the International Institute of Agriculture in Rome. Hans Jenny and Jacob Blom helped in translating and preparing the abstracts for the printer.

The *First International Congress of Soil Science* impacted the ways in which soils were studied, fostered collaboration, broadened the horizons of many attendees, and inspired others. Some of the participants started to work on aspects of the soil that they had learned at the congress. Gilbert Robinson from the University of Wales in Bangor founded the soil survey of England and Wales [85]. He had seen the works from the *Bureau of Soils* and was convinced that a similar effort should be made in the UK where most soil studies were geared toward soil fertility and crop productivity. Another outcome of the congress was that the *Pochvovedenie* started accepting papers from the USA and in English [86]. The journal *Soil Science,* edited by Jacob Lipman, had already published papers from Russia. At the congress, Jouette (J.C.) Russel from Nebraska heard Konstantin Glinka speak about Vasily Dokuchaev, and after he returned to Nebraska, Jouette Russel started teaching the Russian principles of soil science [87, 88]. He was the first in the USA.

Hans Jenny listened to a presentation relating organic matter content and color of soils. Jouette Russel and Eldron Engle presented a study on soils in Kansas, Nebraska, and the Dakotas which they termed the 'central grassland states.' Samples were taken at 68 sites, and at each site, ten cores were drawn and mixed. They found that organic matter decreased with depth independent of soil horizonation and therefore they sampled by fixed soil depth (0–15 and 15–30 cm). Soils were higher in organic matter and darker in color from south to north and from west to east, which they understood was correlated with temperature and rainfall [89]. It is not unlikely that their work inspired Hans Jenny to collect and analyze soil property and climatic data across the USA and Canada. Following the congress, Hans Jenny published a series of papers in which soil properties such as clay [90–92] organic matter [93], nitrogen [90, 91, 94–97] carbonates [90], cations [90], and soil series [98] were related to climatic factors and in particular rainfall and temperature. He cited the work of J. C. Russel once [97] and in his book of 1941 [99]. At the congress, Hans Jenny also saw a soil formation equation that was presented by Sergei Zakharov based on earlier Russian work. Hans Jenny's equation of 1941 was similar, though he did not cite the book by Sergei Zakharov [100], possibly because it was in Russian.

After the congress, a large number of American scientists became interested in the *International Society of Soil Science*; prior to the congress, it was a European society with a handful of Americans. Membership in the *International Society of Soil Science* increased to 934, with members in 43 countries [101]. Alex. de'Sigmond was pleased that the Russian delegation had the opportunity to share their genetic school of soil science, as much of that work had been overlooked in the USA. Some years after the congress, Alex. de'Sigmond wrote a textbook in Hungarian, and following conversations with John Russell, the book was translated in English and entitled *The principles of soil science*. The book was edited by Graham Jacks, and John Russell wrote the preface [102, 103]. Like many Hungarians, Alex. de'Sigmond was an early adopter of the Russian principles on soils and familiar with soil studies across Europe and the USA. In *The principles of soil science*, he distinguished the following soil forming factors in a section named 'genetics': climate, orographical and hydrographical, animals, micro-organisms, age of the soil, and man. He expanded the factors as they had been formulated by Vasily Dokuchaev some 50 years earlier.

Alex. de'Sigmond was part of the second wave of soil researchers that followed Vasily Dokuchaev, Eugene Hilgard, Franklin King, and Emil Ramann, and he saw the need for independent soil studies: "The fact that pedology has risen to the dignity of an independent natural science has in no way impaired its connection with the other natural sciences. On the contrary, all kindred sciences have gained by the concentration in a scientific system of all our pedological knowledge." *The principles of soil science* was a superb and distinctive synthesis, but not a widely read book. Alex. de'Sigmond wrote a few more papers after his textbook, but he died in 1939, at a time when the Kingdom of Hungary had become stridently nationalistic and joined the Axis powers.

The *First International Congress of Soil Science* was followed by a month-long excursion through the USA and Canada. In 1912, the American Geographical Society had organized an excursion for geographers across the USA called the Transcontinental Excursion. The idea was based on the tours that William Davis from Harvard University had experienced in Italy and France in 1908 [104]. The 1912 geographical excursion was organized by the geologist Albert Brigham, a friend of Curtis Marbut. In total, 43 European geographers and 12 Americans toured for eight weeks across the USA in two Pullman railcars, two observation cars, and a diner and baggage car. Several physical geographers were present among the participants and they carried maps and geological and geographic libraries, with the aim: "…to get as much first-hand knowledge about the United States as possible" [6]. They made frequent stops, travelled by car, and visited the Chicago Stock Yards and the 'bonanza farms' in North Dakota.

The European geographers in 1912 were particularly interested in American agriculture and dryland farming, which they had read about but not seen before [105]. The Europeans had a new treat in the form of sliced watermelons, and John Muir, the father of the national parks, showed them the redwoods near the Golden Gate Bridge of San Francisco. They visited the University of Wisconsin where Emil Truog had just started his position at the Department of Soils. They traversed across the terminal moraine, the Driftless Area, and the Baraboo range where the pink quartzite rock was over 1.5 billion years old [105]. The participants were offered dinners by companies and local governments, and in Minnesota, they were welcomed by its Governor, who observed that he had: "…never expected to see so many people who knew so much about the earth and owned so little of it" [105]. A daily bulletin was issued on the train giving an outline of the stops and visits. The excursion ended in New York, with a dinner at the Waldorf Astoria where every European participant was presented with a personalized brass plaque.

Curtis Marbut did not attend the 1912 excursion organized by the American Geographical Society. He was a member of the much younger Association of American Geographers and had been its president in 1924. Curtis Marbut modeled the 1927 excursion after the 1912 excursion, and he designed a train journey across the country with stops that had geological and pedological significance. In 1925, Emil Truog saw a draft itinerary that showed that the excursion would go through the SouthWestern part of the country but would not pass through Wisconsin.

Emil Truog thought it would be a missed opportunity and wrote to Jacob Lipman: "In talking over with various people the proposed transcontinental trip which Dr. Marbut has outlined, there seems to be a feeling that too much of the time of the trip will be spent in the arid and semi-arid section of the country. Since this trip will be taken during June and July, I doubt the advisability of spending so much time in the hot, dry sections of the country. I am wondering if Dr. Marbut is not stressing the point of seeing different soil profiles and soil types much more than should be the case and is thus lessening the emphasis on different kinds of agriculture as found in this country.

Konstantin Glinka (1867–1927) (center) holding David Hissink (1874–1956) (left) and Curtis Marbut (1863–1935) (right) at several photo occasions during the Transcontinental Excursion in 1927. Photos by M.E. Diemer, Chemist and Photographer of the University of Wisconsin

I know a great many people feel that it will be too bad not to have the trip include the great wheat sections of the North-west and also the dairy sections of Minnesota and Wisconsin. If it is possible, from a financial standpoint and also from the railroad standpoint, I believe it would be much better to have the trip go northward from California up to Oregon and Washington and then eastward through Montana, North Dakota, Minnesota, Wisconsin and then through the corn belt of Illinois and Indiana and then eastward as planned. This, I believe, would give a trip which would be much more pleasant and fully as profitable as the one going through the arid and semi-arid section of the country from California to Iowa. As already stated, I have the feeling that Dr. Marbut is placing too much emphasis on trying to have the people see different soil profiles and types. The greater percentage of these people will probably be interested more in the great agricultural sections of the country and the different kinds of agriculture which have been developed. As regards the matter of financing the trip, I believe the route which I have suggested would have certain material advantages. I feel much more hopeful that we would be able to secure considerable contributions from the milling and other industries of Minneapolis and St. Paul if the trip were to include those." He concluded his long plea with: "I also have the feeling that if the trip were to go through Madison that it would be much easier to secure contributions here." It did not help; the excursion route went as planned by Curtis Marbut.

Emil Truog only attended the first part of the Transcontinental Excursion and when the train headed into the dry heat of the southwest, he returned home to Madison. He had obtained one of the congress handbooks bound in leather that was presented by the fertilizer company Chilean Nitrate of Soda Educational Bureau of the United States of America. It contained the congress program, an introduction from Jacob Lipman [8], the new concept of soils explained by Wilbert Weir [106], and short biographies of all members of the organizing committee. Of the 51 members, 31 members had been: "…reared on farm" and eight members held degrees from the University of Wisconsin. Emil Truog used the last pages, named Memoranda, to collect names, signatures, positions, and addresses from excursion participants. Several participants wrote their names in pencil, including Gammie Ogg, Edward Crowther, Gilbert Robinson, Peter Treitz, David Hissink, Téodor Saidel, Arao Itana, Herman Stremme, Boris Polynov, and Konstantin Glinka. They wrote down their title as 'professor of agricultural chemistry,' Hans Jenny pencilled down 'colloid chemistry and climatic soil types,' where Frederick Hardy from Trinidad listed himself as 'Professor of Soil Science and Chemistry.' No one wrote 'pedology.'

Curtis Marbut designed a tour that reflected his knowledge of the soils across the country. "The regions visited included: (1) the northern Piedmont and Coastal Plain, representing the forested or woodland soils of the mid latitudes of the United States, all of which are Podsolic but not true Podsols and all of which have been developed under a cover of deciduous forest and a high rainfall; (2) the Red soils of the southern Piedmont; (3) the Yellow soils of the southern Allegheny Plateau, although no stop was made to examine them; (4) the Prairie soils of southwestern Missouri and southeastern Kansas; (5) the true Prairie soils; (6) the belt of Black Earths or their equivalent in central-western Kansas; (7) the Brown Grassland soils in western Kansas and eastern Colorado; (8) the Desert soils of southern Nevada; (9) the Old and Young soils of the Pacific Coast; (10) the Humid soils of the Northwest; (11) the Dark-colored soils of the Canadian Northwest, the Dakotas, and Minnesota; (12) the Prairie soils of Iowa; and (13) the western end of the belt of Deciduous Forest soils in Indiana, the eastern end of which had been examined in the local excursions around Washington. By selecting a route of this kind, some of the great north-south belts of soils were seen in two places. This is especially true of the belts of prairie soils and that of the Black Earths. The latter were seen in Kansas and also in several places in Canada. The belt of Brown Forest soils was seen both at its western and eastern ends" [2].

First Intern. Congress of Soil Science, U.S.A. 1927 – Excursion – Fargo.

Marbut showing the mature profile.
(see Proceedings, I Int. Congress (1927), V, page 175)
"That really mature Prof. Ile".

"That really mature Profile", Curtis Marbut discussing a road cut soil profile near Fargo in North Dakota and handing a sample to Herman Stremme. The soils were Chernozems, characterized by level topography and the site was in the basin of former glacial Lake Agassiz. Photo by M. E. Diemer, Chemist and Photographer of the University of Wisconsin. Handwriting by David Hissink

The tour guide was 170 pages and had descriptions of all stops, physical and chemical data of the soil profiles, as well as discussions and interpretations [107]. An appendix had descriptions of soils and agriculture in Kansas, the cotton states (North and South Carolina, Georgia, Alabama, Mississippi), Iowa, Saskatchewan, Alberta, Manitoba, and Minnesota and a report on the soils of California by Charles Shaw. The guide started with a description of the soils from New York to Washington, because for: "The European visitors the excursion will begin at New York City" [107]. It explained that the soils between New York and Washington: "…are podsolic but not podsols" and that they seem to correspond to the soils of parts of Germany, France, and: "…probably the brown soils of Ramann." For most Europeans, it was their first visit to the USA, and these words at the start of the congress and Transcontinental Excursion must have felt like a warm pedological welcome.

The Transcontinental Tour left on the 10th day of the congress at 10 o'clock at night from Washington Union Station. The organizers had hired nine Pullman sleepers, two dining cars, and a club car that sold food and drinks. There were 206 participants and about half of them were from overseas. Only official delegates participated, and they were representatives of national, state, provincial or municipal governments, or educational or research institutions [108]. Every participant was assigned a Pullman car, with a name tag on the door. Traveling was done by night, breakfast was on the train, and in the morning, there were 70 to 80 excursion cars, mostly Model T-Fords, lined up to bring the participants to a research station, soil pits, or geologic points of interest [50]. At the end of the day, participants could take a bath or shower in pools, local clubs, or student gymnasia [45]. In the evenings, there were banquets accompanied by speeches, mostly on agriculture and comparisons between agricultural practices and productivity in the USA and other countries. When there was a piano, Alex. de'Sigmond would play improvisations, and Selman Waksman found him a brilliant pianist [35].

From Washington, the tour went south, as far as Georgia, then north and west to southern California. From there, the tour traversed the length of California, passed through Oregon and Washington, went as far north as Edmonton, then east to Winnipeg, south to Des Moines, Iowa, east to Lafayette, Indiana, and back to Washington. The distance covered by train was 19,000 km and the cars for the local excursions added another 2,500 km. The trip lasted thirty days and the train passed through 23 states and four Canadian Provinces. Thirty-four stops were made, and twelve agricultural stations were visited. Curtis Marbut led all discussions at the soil profiles that were especially dug for the excursion or observed along road cuts.

The days through Canada were particularly welcomed by the participants who fancied an alcoholic beverage, or two, as the USA had been dry since 1920. When the group was shown the wineries in California, there was some puzzlement, among the delegates, why there was no tasting: "...we were treated in the most cruel manner." When the group visited the University of California Berkeley, there was excitement among the Hungarian delegates. They visited the home of Eugene Hilgard, whose work on alkali soils was well-known in Hungary where such soils were called *szik* soils. Peter Treitz had translated some of Eugene Hilgard's work in the 1890 s and Alex. de'Sigmond had visited him in 1906. Eugene Hilgard was well-known in Europe, and according to Hans Jenny, very few Americans on the excursion had heard of Eugene Hilgard [50]. In part, that was due to the disagreements he had with Milton Whitney, who continuously referred to Eugene Hilgard as the 'old school' of soil science [109, 110]. There were others, however, like Charles

Téodore Saidel (1874–1967) (left) from Romania, Konstantin Glinka (1867–1927) (centre) from Russia, and Albrecht Penck (1858–1945) (right) from Germany in Kansas on the 28th June 1927. Photo by M. E. Diemer, Chemist and Photographer of the University of Wisconsin. Handwriting by David Hissink

Lipman from the University of California Berkeley, who realized the greatness of Eugene Hilgard, as he wrote in 1916: "Dr. Hilgard was a blazer of trails, a pathfinder in the vast and uncharted realm of soil science... whatever he touched he illuminated" [111]. But in 1927 Eugene Hilgard had been dead for 11 years, and somewhat forgotten.

Agriculture was changing in the USA in the 1920s. Tractors had replaced horses and mules, and between 1920 and 1925 over 600,000 tractors were sold. The number of people who worked in agriculture decreased, whereas productivity per hectare slowly increased. Over 80% of the people in the USA were farmers in the mid-1800s, but by the 1920s, it was less than 25%. In the mid-1800s, an average agriculture worker cultivated about five hectares, and in 1927, it was 15 to 40 ha. The costs of agricultural machinery and labor took up some 60% of the cost of the product. Nikolaĭ Tulaikov, who had visited Eugene Hilgard in 1908, noted during the 1927 excursion that American farmers were becoming businessmen and in his view, they were little distinguished from other commercial people: "...perhaps only by the fact that his own profession is much more complicated and more unsettled than many of the city undertakings" [45].

For many of the participants, a striking feature was the abundance of different kinds of products and the potential for agricultural production. Fertilizers were little used, as it was assumed that the inherent fertility of the soils was high enough that they were not needed [112]. But that changed in the 1920s. In 1900, organic manures supplied more than 90% of the nitrogen added to the soil, but by 1950, fertilizers were widely used and organic manures supplied less than 4% of the nitrogen. Bernard Keen from Rothamsted compared agriculture in Europe to the USA. In Europe, agriculture was on land that had been cultivated for generations and it had yielded a conservative resistance to any change. In the USA, farming methods had not changed much since the initial settlement [113].

Despite the pedological emphasis of Curtis Marbut, agriculture was an important subject during the Transcontinental Excursion. Many of the participants had grown up on farms and were interested in American agriculture. They were pleased to visit the plants of John Deere in Moline, Illinois, and the International Harvester Company in Chicago, where 120 tractors were being produced every day. They also visited Swift & Company at the Stockyards of Chicago, where a living pig entered at one end of the conveyor and left at the other end in the form of bacon, Frankfurters, and various conserves [45]. Chicago in the 1920s was the hog butcher for the world. A few years later, the stockyards burned, destroying 25 ha of buildings, barns, and cattle pens. The fire had started with a carelessly tossed cigarette, and it was the worst fire in Chicago since the Great Fire of 1871, which had left 100,000 homeless and killed 300 people. Excursion participants also visited the markets of New York, a dairy farm with 1,750 cows in Plainsboro, the Mormon temple in Salt Lake City, and the film studios of Hollywood. None of these stops were chosen by Curtis Marbut.

The Transcontinental Excursion participants included Konstantin Glinka, Richard Bradfield, Hans Jenny, Frederick Hardy, René Dubos, John Russell, David Hissink, Jadwiga Ziemięcka, Téodor Saidel, Alex. de'Sigmond, and Albrecht Penck. Jadwiga Ziemięcka was the only female soil scientist on the excursion, but 29 participants, including Jacob Joffe, had brought their spouses for who there was a special program at each stop [114]. The excursion was a distinctive and memorable feature and the total costs, some $50,000 (equivalent to one million dollars in 2021), were paid for by the organizers. Most of the conversations were in English, but some people provided translations, such as David Hissink. He wore a straight beard which he had the habit of combing whenever he was excited, and as Secretary, he was ever anxious to preserve his dignity [115]. Konstantin Glinka and John Russell spoke

577 Soil examination and sampling, near Athens, Georgia

Dr. Glinka collects his own samples

#579 Cecil soil showing relation to underlying neck.

Photos from soil observations and sampling during the Transcontinental Excursion in 1927. Photos by Merritt Miller (1875–1965) from Missouri State University, and courtesy Missouri State Archives, and Senator Mark Miller

French together, whereas Sergei Neustreuv spoke English. Besides the scientific members of the Russian delegation, there were a few Russians, whose function was to keep watch on their fellows.

There was some quarrelling among the people on the excursion; some of it came from having an intense program for two hundred people from across the world in trains and cars for a month, and some of it was because of an age difference. Younger scientists talked about pH, exchangeable hydrogen ions, and zeta potentials, whereas the older soil scientists discussed textures, mechanical analyses, total analyses, and fertilizer responses [50]. Some people in the group would talk to each other, and some would not. Discussions were held on the mechanical analyses of soil by participants of countries where rain was sufficient, whereas soil moisture was discussed by participants from drier parts of the word. Participants learned that in some states, soil surveyors were concerned with laboratory work whereas, in other states, there were groups that would not get to the field or were never in the laboratory. For example, in California, the soil surveyors worked in the field all year round and were seldom in the laboratory. In the midwest, they spent the summer in the field and the winter in the laboratory [50].

Many of the participants saw for the first time subtropical soils, such as the yellow and red soils in Virginia, Georgia, and California. There were several discussions about those soils, and some thought yellow soils should not be distinguished separately from red soils. They were the same soil and in the same subtropical zone, but the red soil had been further developed. Some Russian participants had problems classifying the soils according to their own classification scheme because the temperature and rainfall gradients were not parallel on the American continent [50].

Curtis Marbut used the excursion to discard the system of soil classification of provinces and to promote a classification system that was closer to the Russian system. It included terms such as Podzol, Chernozem, and Rendzina, which were generally unknown to Americans but were familiar to most of the European attendees. Curtis Marbut was in a difficult position for he was in a government agency that struggled with change. Milton Whitney was opposed to the Russian principles of soils, as was Arthur McCall, who was in charge of all the soils work. Curtis Marbut had fewer problems with Arthur McCall than with Milton Whitney, as Arthur McCall did not interfere and: "…has never developed a particular line of work of his own and will have no special predilections for any branch of work." [11] Curtis Marbut used the excursion and the soil sites to bring a different view to the foreground. He had to do this carefully, as Arthur McCall was on the excursion along with others from the *Bureau of Soils*. He had asked Wilbert Weir of the *Bureau of Soils* to write

#617   In the Nevada desert

A typical caravan on the tour except that there are more Folds in sight than usual.

rnegel

Photos from a stop during the Transcontinental Excursion. Photos by Merritt Miller (1875–1965) from Missouri State University, and courtesy Missouri State Archives, and Senator Mark Miller

about the new concept of soils for a handbook that was especially prepared for the congress and excursion [106]. Not everybody was impressed by the new approach to soil studies and its classification. For example, Richard Bradfield was interested, but the chair of the Department of Soils at the University of Missouri, Merritt Miller, was not. The scheme that Curtis Marbut laid out during the excursion met resistance and it did not work everywhere. For example, in California, Charles Shaw was in charge of the soil surveys and pointed out several cases where the scheme failed. Curtis Marbut had success with the Europeans, but not so much with the Americans.

On the 19th of July, a few days before the Transcontinental Excursion ended in New York, Curtis Marbut celebrated his 64th birthday. He was presented with an album that contained the autographs of all participants and some photographs: "Presented to Dr. C. F. Marbut on his birthday, July 19, 1927, by members of the *International Society of Soil Science* as a token of esteem and affection, with best wishes for long life and prosperity and with thanks for many helpful and kindly acts during the Transcontinental Tour." Curtis Marbut received daily accolades and applause [50], and Frederick Hardy wrote a poem for him:

In Quest of Prof. Ile.
From Memphis to Portland we sought him
In journeys of mile upon mile
Oh where, oh where shall we find him
That true undegraded Prof. Ile.

Some scratched every blue "C" horizon
Some looked over meadow and stile
And yet never once we have seen him,
That really mature Prof. Ile.

Many think he was never created,
A myth, a snare and a guile
Yet in hole and boring we seek him;
We seek that elusive Prof. Ile.

Perhaps at the end of our journey
When past the barrier we file,
We shall awake from our dreams to discover
The cause of that Marbutian smile.

Macy Lapham, who was the soil inspector for the *Bureau of Soils* in the western USA since the early 1900 s, guided the excursion through Nevada,

California, and Oregon. Several stops were made for examining soil profiles and like many other American soil scientists, Macy Lapham was pleased to meet the eminent soil scientists that he knew only vaguely by name [36]. For Wilbert Weir, the excursion was an unforgettable experience and he took over 350 pictures. The tour was also a confirmation of the Spirit telling him to leave his job with the *Bureau of Soils* [116]. He was a deeply religious man.

Selman Waksman and Georg Wiegner were critical during the excursion, and according to Hans Jenny, it was because they knew nothing beyond their own specialty. For example, Georg Wiegner did not know much about soil physics, which was often debated, and according to Hans Jenny: "...what he didn't understand, he thought was no good." They also made fun of the people at the experiment stations that were visited or thought some of the work was stupid [50]. At times there was confusion about units, as the Americans used inches and Fahrenheit which made some European participants grossly under-appreciate the rainfall and over-appreciate the temperature.

A mimeographed newsletter came out every night that described what had happened during the day. It was called the *Boden Bull* and aimed to be semi-informative and semi-humorous in nature. The idea came from the staff of the New Jersey Agricultural Experiment station; they had considered the titles 'The dirt farmer' or 'The dirt scientist,' but decided that *Boden Bull* would be more international [117]. A mimeographing machine, a stenographer, paper, and pencils were brought on the train. The newsletter was prepared by people from a different country every day and there was some worry that jokes would not be understood. In the end, the *Boden Bull* was characterized by unsavory humor and included several poems by Frederick Hardy from Trinidad:

> Chanson
> To C. F. Marbut with homage
> Sing a song or profiles, sing with heart and soil,
> Four and twenty Russians crowded in a hole.
> Trowels, bags, and acid bottles, filling every chink,
> Stremme, Treitz, and Marbut dancing on the brink

When it was Hans Jenny's and some of the younger participants' turn to describe the day, they commented on John Russell, who never missed an opportunity to talk about Rothamsted. So, they put in the newsletter: "Why did we come on this transcontinental excursion? Was it to hear about Rothamsted?" It caused a little outburst and Selman Waksman warned them that they had the English against them. All of the newsletter items were published in a book edited by Selman Waksman and R. Deemer [2], but the contribution of Hans Jenny was not included.

The transcontinental tour returned to New York on Friday 23rd of July 1927, after it had travelled for a month and had induced endless discussions and debates that deepened friendships and differences. For Arthur McCall, who was the executive secretary of the organizing committee, it was clear that the personal contact had created friendships and provided opportunities for discussions that had not been possible during the days of the congress [114]. For most participants, it was a unique and important event in their scientific lives. Nonetheless, there were critical notes despite the fact that everything was paid for, the excursion was well-organized, and all efforts were made to show the great diversity in soils and agriculture across the USA and parts of Canada. For many, it was their first visit to the USA which in the 1920 s was staggeringly well off [1]. About 40% of the households had a car, and most homes had refrigerators, radios, telephones, electric fans, and electric razors. Such appliances would not become available in other countries until well after the Second World War.

Paolo Bignami from Italy greatly enjoyed the excursion, which: "…was marked by discipline and organization so characteristic of the Americans and by a great spirit of hospitality." He thought the month-long excursion was not long enough for detailed studies, but he considered it sufficient: "…for giving a general idea of this part of the American continent which passed before our eyes like a cinematographic film." Overall, he thought that agricultural production was more about quantity than quality, but he believed that great advantages might develop from the application in Italy of what is done in America [112]. The Norwegian geologist and sedimentologist Knut Björlykke was not so impressed with the whole excursion, and at the congress, he had presented the climatic soil regions of Norway [118]. Visits to the agricultural colleges and stations proved to him that Americans were well equipped but: "…they are very young and have not as yet obtained very great results…and everything in America is the biggest or largest in the world, but we cannot deny that this is often actually the truth." He enjoyed meeting the Norwegians, although for them America had also become the best in the world [119].

The soil chemist Georg Wiegner from Switzerland compared America to Europe: "…ideals live in Europe, whereas in America blooms The *Dollarismus*. In Europe people are free, individual, and thoughtful; in America, it is narrowed down by mass suggestion and uncritical. There is art in Europe, kitsch in America; in Europe, there is a deep, theoretical science; in America, something is chemically synthesized and measured physically, especially for industry. According to these reports, clean, intensive, thorough-going European agriculture is countered by the lazy, extensive American predatory economy." But he was also impressed with some of the equality and the way people treated each other: "What strikes you immediately on a working day in America is the generous, benevolent assessment of every job and every worker without distinction" [120].

The Polish microbiologist Jadwiga Ziemięcka wondered why farmers were not so wealthy given the efforts of the government to assist and support agriculture. For her: "…the farm owners work hard—as hard as their laborers—but in comparison to the enterprising elements in the cities, do not accumulate great wealth." She found the excursion: "…thought out deeply, organized great, without taking into account the cost and difficulty, with a great generosity of the feeling of American hospitality, which allowed us to understand how real and important on this earth is the meaning of the words, we read at the University of Oregon: The glory of the home is hospitality." Jadwiga Ziemięcka visited a beach in California, where she observed: "I go to the beach with one English scholar. Empty. Nobody comes here, they all drive cars or sit in traffic lights and crowds. Nature doesn't kidnap an American. In Portland, I heard how it was decided that the largest and most beautiful of the local waterfalls and the Columbia River would be exploited for water-power, which would bring the city millions of dollars. Our persuasion that the current landscape worth more than any billionaires was not convinced, and we only seemed unreal. America is still living only practical life" [121].

For Bernard Keen: "…the traditional hospitality and generosity of the American citizen and his readiness to organize official receptions, the great welcome and assistance we received all through our tour can only be explained by the fact that the importance of soil research is becoming under-stood both in the rural districts and—what is even more important—among the great centers of population." [113] John Russell reflected on the excursion and its participants in his 1956 autobiography: "These men and women of widely different origin and background, who at the outset were eyeing each other with some suspicion, were before long prepared to pool their stocks of scientific and technical knowledge and help each other in the solution of their technical problems." He thought that new methods and techniques should help some of the poorer countries to increase their food production [115]. Similar thoughts were expressed by Selman Waksman: "…this excursion illus-trated better than it would ever be possible that all nations of the world may unite and work together for a purpose which advances the common good." The geographer Albrecht Penck enjoyed the excursion and returned a year later to teach at the University of California Berkeley and work with his old friend, the historical geographer Carl Sauer. For Arthur McCall, the excursion "…proved to be a fitting climax" [114].

René Dubos travelled back to France after the congress and excursion to see his family for the first time in three years. He boarded the *Leviathan*, which ferried many other soil scientists returning home, several of whom carried bags of soil samples collected during the excursion for analysis in their laboratories [122]. René Dubos felt that the Europeans had enriched America with the long history of their soil research, and he predicted it would improve the science because: "...it is the shock of ideas that causes sparks to fly." The excursion had given him new ideas that soils differed widely and that microbes play an important role in the soil. He returned to the USA in September 1927 and started working at the Rockefeller Institute for Medical Research in New York where he studied medical microbiology and human diseases. The contract for that position was signed in the Willard Hotel while he was at the congress [122].

For Jacob Joffe, the outstanding feature of the congress was the 'invasion' of the Russian viewpoint of soils. He found that the discussions at the congress influenced the way soils were studied and that: "...it opened new avenues of approach to the many complex problems of the science" [123]. Jacob Joffe believed that: "...the illustrious delegation from the Soviet Union had the key to this new school of Soil Science" [124]. Three Russian soil scientists attempted to fit their classification schemes to the soils as they had seen them during the Transcontinental Excursion. Jacob Joffe summarized their findings [123]. Firstly, Dmitrii Vilenskii, professor and head of the botany chair at the Charkov Agricultural Institute, made a map of the soil regions in North America based on the zonal principle. He noted, like Curtis Marbut, that: "...in America, there is not the zonation of soils in our sense of the word, neither latitudinal nor longitudinal, but there is a checkerboard pattern." The geographer Leonid Prasolov from the Soil Institute of the Academy of Sciences in Moscow found errors in the maps of the *Bureau of Soils* [123]. He reviewed the proposed system of Curtis Marbut and how it compared to soils in Russia [53]. Ivan Tiurin, who studied soils under forest at Kazan University, had sampled several soils during the excursion and compared the morphological descriptions with some of his own chemical analyses. All in all, the three Russian investigators had similar findings and used the zonal division of soils. For Jacob Joffe, it was clear that the genetic approach of Vasily Dokuchaev was the proper method for studying soils, for if it worked in Russia, it should work in America [123].

Jacob Lipman found that the congress and the excursion did much to promote international good will: "...as we take stock of the work accomplished, we are led to conclude that the congress has been the means of broadening our outlook on the entire field of soil science. There is reason to expect that soil research will receive as a result of the activities of the congress better support from governments, educational institutions, and individuals of means interested in soil research" [79].

Curtis Marbut returned to Washington, after he had brought Albrecht Penck, Herman Stremme, and Téodor Saidel to the boat that would take them across the Atlantic and back to Europe. Konstantin Glinka had already left. Back in his Washington apartment, Curtis Marbut wrote a long letter to his eldest daughter Louise. She was 17 when her mother Florence died and had a good relationship with her father. He wrote her regularly [11]. Louise lived in North Dakota and was married to Leroy Moomaw, who was at the Dickinson Agricultural Experiment station. Curtis Marbut regretted that Louise had not come to Fargo when the Transcontinental Excursion passed through there one week earlier: "Wish you could have come to Fargo. I was on the go all day, but could have taken you along. It would have been a hard trip though and I comfort myself with the thought that your decision not to come was wise." Curtis Marbut was satisfied with the excursion, as he wrote to Louise: "The trip was much more than I dreamed it could be. Everything went off without a hitch. I made hosts of friends I think. I have gotten many suggestions (scientific) from the excursionists and I am convinced they have had some from the trip. It was fully worthwhile from the scientific standpoint and from the personal standpoint, pleasant to everyone." He signed his letter: "With love. Dad" and then he took the train to Chicago where he stayed with the plant physiologist Charles Shull until early September. They were friends and Charles Shull had attended the first congress [78]. With his brothers George Shull, professor of botany and genetics at Princeton, and Aaron Shull, professor of zoology at Michigan, they were important in the biological sciences for more than a generation [125].

The first issue of the journal *Soil Science* in 1928 was dedicated to the congress and contained reviews and reports from the six commissions. Shortly after it was published, Jacob Lipman wrote to Curtis Marbut and hoped that he had looked at the *Soil Science* issue, and he was "…rather interested to know" what Curtis thought of it. Curtis Marbut responded a week later: "I have received and enjoyed the January issue of *Soil Science*. It is a noble contribution to soil literature in this country, and I hope it will be widely read. I am very glad indeed that such a number has been published. It will do a great deal of good to the soil men throughout the United States." It did more than that—it founded a world of soil science.

References

1. Bryson, B. (2013). *One summer: America 1927*. London: Transworld Publishers.
2. Waksman SA, Deemer RB, editors. Transcontinental excursion and impressions of the congress and of America. Washington D.C.: The American Organizing Committee of the First International Congress of Soil Science; 1928.
3. Hissink DJ. Jacob Goodale Lipman and the international society of soil science. Trans Third Commission Int Soc Soil Sci 1939; A:1–3
4. Moon, D., & Landa, E. R. (2017). The centenary of the journal soil science: reflections on the discipline in the united states and russia around a hundred years ago. *Soil Sci, 182*(6), 203–15.
5. Keen, B. A. (1927). An international congress of soil science. *Nature, 3019*, 385–86.
6. Brigham, A. P. (1913). The transcontinental excursion of the american geographical society. *Science, 37*(945), 210–13.
7. Krusekopf HH, editors. Life and works of C.F. Marbut. Madison: Soil Science Society of America; 1936.
8. Lipman JG. The first international congress of soil science. A handbook especially prepared for the meeting & tour of the First International Soil Congress. New York: Chilean Nitrate of Soda Educational Bureau of the United States of America; 1927. p. 3–5.
9. Królikowski L. Professor Dr. Jadwiga Ziemiecka (1891–1968). Bullet Int Soc Soil Sci 1986; 33:51.
10. Ackert L. Sergei Vinogradskii and the cycle of life. From the thermodynamics of life to ecological microbiology, 1850–1950. Dordrecht: Springer; 2013.
11. The, Moon D., & Steppes, American. (2020). *The unexpected russian roots of great plains agriculture, 1870s–1930s*. Cambridge: Cambridge University Press.
12. Prassolov L. International congress of soil science in Washington and soil excursion through North America in 1927. In: Waksman S, Deemer R, editors. Proceedings and paper first international congress of soil science. Transcontinental excursion and impressions of the congress and of America. Washington D.C.: The American Organizing Committee of the First International Congress of Soil Science; 1928. p. 147–51
13. Krupenikov, I. A. (1992). *History of soil science from its inception to the present*. New Delhi: Oxonian Press.
14. Joffe JS, editors. Russian contribution to soil science. In: Christman RC, editors. Soviet science; a symposium presented on December 27, 1951, at the Philadelphia meeting of the American Association for the Advancement of Science. Washington: American Association for the Advancement of Science; 1952.

15. Glinka, K. D. (1927). *Dokuchaev's ideas in the development of pedology and cognate sciences*. USSR: Publishing Office of the Academy, Leningrad.

16. Zakharov, S. A. (1927). *Achievements of Russian science in morphology of soils*. USSR: Publishing Office of the Academy, Leningrad.

17. Neustreuv, S. S. (1927). *Genesis of soils*. USSR: Publishing Office of the Academy, Leningrad.

18. Tiurin, I. V. (1927). *Achievements of Russian science in the province of chemistry of soils*. USSR: Publishing Office of the Academy, Leningrad.

19. Afanasiev, J. N. (1927). *The classification problem in Russian soil science*. USSR: Publishing Office of the Academy, Leningrad.

20. Prasolov, I. L. (1927). *Cartography of soils*. USSR: Publishing Office of the Academy, Leningrad.

21. Polynov, B. B. (1927). *Contributions of Russian scientists to paleopedology*. USSR: Publishing Office of the Academy, Leningrad.

22. Gemmerling, V. V. (1927). *Russian investigations concerning the dynamics of natural soils*. USSR: Publishing Office of the Academy, Leningrad.

23. Kravkov, S. P. (1927). *Achievements of Russian science in the field of agricultural Pedology*. USSR: Publishing Office of the Academy, Leningrad.

24. Tulaikov, N. M. (1927). *Russian Pedology in agricultural experimental work*. USSR: Publishing Office of the Academy, Leningrad.

25. Yarilov, A. A. (1927). *Brief review of the progress of applied soil science in USSR*. USSR: Publishing Office of the Academy, Leningrad.

26. Prokhorov, N. I. (1927). *Soil science in the construction of highways in USSR*. USSR: Publishing Office of the Academy, Leningrad.

27. Keller, B. A. (1927). *Russian progress in geobotany as based upon the study of soils*. USSR: Publishing Office of the Academy, Leningrad.

28. Penck A. Das Hauptproblem der physischen Anthropogeographie. Paper presented at first international congress of soil science, Washington DC; 1928.

29. Penck W. Morphological analysis of land forms. Translated by H. Czech and K.C. Boswel. London: MacMillan & Co., Limited; 1953.

30. Vernier RT, Smith GD. The concept of Braunerde (brown forest soil) in Europe and the United States. Adv Agron 1957; 9:217–289.

31. Ramann, E. (1893). *Forstliche Bodenkunde und Standortslehre*. Berlin: Verlag von Julius Springer.

32. Ramann, E. (1911). *Bodenkunde (Dritte, umgearbeitete und verbesserte Auflage)*. Berlin: Verlag von Julius Springer.

33. Ramann E. The evolution and classification of soils. Translated by C.L. Whittles. Cambridge: W. Heffer & Sons Ltd; 1928.

34. Krauss G. General notices. E. Ramman; 1926.

35. Waksman, S. A., & Alexius, A. (1940). J de'Sigmond 1873–1939. *Soil Sci, 49*(4), 251–52.

36. Lapham, M. C. (1949). *Crisscross trails: narrative of a soil surveyor*. Berkeley: W. E. Berg.

37. de'Sigmond AAJ. Hungarian alkali soils and methods of their reclamation. Special publication issued by the California agricultural experiment station. Berkeley, California: University of California Printing Office; 1927.

38. Kele GZ, Hernadi G, Mako A, editors. Bridging the centuries 1909–2009. Historical gallery. Veszprém, Hungary: Ook-Press Ltd; 2009.

39. Davidescu, D. (1967). Théodore Saidel (1874–1967). *Int Soc Soil Sci Bull, 32,* 20–21.

40. Florea, N. (1997). The contribution of Gheorghe Munteanu-Murgoci (1872–1925) and his Romanian colleagues to soil science. In D. H. Yaalon & S. Berkowicz (Eds.), *History of soil science—international perspectives* (pp. 365–75). Reiskirchen: Catena Verlag.

41. Wiegner, G. (1926). *Anleitung zum quantitativen agrikultuchemischen Praktikum.* Berlin: Verlag von Gebrüder Borntraeger.

42. Wiegner, G. (1931). *Boden und Bodenbildung in Kolloidchemischer Betrachtung* (Sechste ed.). Dresden und Leipzig: Verlag von Theordor Steinkopff.

43. Mitchell RL. Sir William Gammie Ogg. Year Book R.S.E. 1980; 67–71.

44. Pereira C. Bernard Augustus Keen 5 September 1890–5 August 1981. Elected F.R.S. 1935. Biographical Memoirs. Royal Society; 1981.

45. Tulaikov MM. Some impressions of a trip to the United States and Canada. In: Waksman S, Deemer R, editors. Proceedings and paper first international congress of soil science. Transcontinental excursion and impressions of the congress and of America. Washington D.C.: The American Organizing Committee of the First International Congress of Soil Science; 1928. p. 121–33

46. Anon. Prof. Milton Whitney, soil scientist, dead. The Official Record. United States Department of Agriculture. 1927; VI(46):1 and 5

47. Anon. Five hundred attend world soil congress. The official record. United States Department of Agriculture. 1927; VI(25):1–2

48. Valek B. Prof. Dr. Véclav Novâk (1888—1967). Bull Int Soc Soil Sci 1967; 30:17–18.

49. Page HJ. The investigations of K.K. Gedroiz on base exchange and absorption. vol A. Groningen, Holland: Transactions of the Second Commission of the International Society of Soil Science; 1926.

50. Maher D, Stuart K, editors. Hans Jenny—soil scientist, teacher, and scholar. Berkeley: University of California; 1989.

51. Jenny H. Soil investigations in the Swiss Alps. Vol Commission V. Commission VI. Miscellaneous papers. Washington D.C.: The American Organizing Committee of the First International Congress of Soil Science; 1928.

52. Bennet, H. H., & Allison, R. V. (1928). *The soils of Cuba.* Washington DC: Tropical Plant Research Foundation.

53. Marbut CF. A scheme for soil classification. Paper presented at first international congress of soil science: Washington DC; 1928.

54. Marbut CF. General report of the fifth commission. Vol Part I. Proceedings. Washington D.C.: The American Organizing Committee of the First International Congress of Soil Science; 1928.

55. Lipman JG. Soils and men. Vol Part I. Proceedings. Washington D.C.: The American Organizing Committee of the First International Congress of Soil Science; 1928.

56. van Baren J. Profiles of limestone-soils from the tropics. Vol Commission V. Commission VI. Miscellaneous papers. Washington D.C.: The American Organizing Committee of the First International Congress of Soil Science; 1928.

57. Veatch JO. The classification of organic soils. Vol Commission V. Commission VI. Miscellaneous papers. Washington D.C.: The American Organizing Committee of the First International Congress of Soil Science; 1928.

58. Saidel T. Die Bodenkarte von Romanian, 1:1,500,000. Vol Commission V. Commission VI. Miscellaneous papers. Washington D.C.: The American Organizing Committee of the First International Congress of Soil Science; 1928.

59. Bushnell TM. The soil color field. Vol Commission V. Commission VI. Miscellaneous papers. Washington D.C.: The American Organizing Committee of the First International Congress of Soil Science; 1928.

60. O'Neal AM. The effect of moisture on soil color. Vol Commission V. Commission VI. Miscellaneous papers. Washington D.C.: The American Organizing Committee of the First International Congress of Soil Science; 1928.

61. Hutton JG. Soil colors: their nomenclature and description. Vol Commission V. Commission VI. Miscellaneous papers. Washington D.C.: The American Organizing Committee of the First International Congress of Soil Science; 1928.

62. Landa, E. R., & Albert, H. (2004). Munsell: a sense of color at the interface of art and science. *Soil Sci, 169*(2), 83–9.

63. Truog, E. (1928). General exhibits. *Soil Sci, 25*, 89–5.

64. Stremme, H. (1928). *General map of the soils of Europe (Ogolna Mapa Gleb Europy)*. Warszawa: International Society of Soil Science.

65. Hartemink AE. Soil maps of Europe. Book review of: soil Atlas of Europe, by European Soil Bureau Network of the European Commission, 2005. J Environ Qual. 2006; 35:952–955.

66. Anon. Preserving soil samples with gelatin and glycerin method. The Official Record. United States Department of Agriculture. 1927; VI(27):1.

67. Muggler CC, Spaargaren O, Hartemink AE. The Glinka Memorial soil monolith collection: a treasure of soil science. Paper presented at: EGU, Vienna; 2012.

68. Fanning DS. Niki stories by Roy W. Simonson. Pedologue. Mid-Atlantic association of professional soil scientists, Winter 2017. p. 11–2.

69. McBratney, A., & Minasny, B. (2010). The sun has shone here antecedently. In R. Viscarra Rossel, A. B. McBatney, & B. Minasny (Eds.), *Progress in soil science*. Dordrecht: Springer.

70. Gardner WH. Early soil physics into the mid-20th century, New York, NY; 1986.

71. Davy, H. (1815). *Elements of agricultural chemistry*. New York: Eastburn, Kirk & Co.

72. Ruffin E. An essay on calcareous manures. Petersburg Va.: J.W. Campbell; 1832.

73. Barnett, C. R. (1927). *A classified list of soil publication of the United States and Canada*. Washington D.C.: United States Department of Agriculture.

74. Tulaikov, N. (1923). Drought and the means of overcoming its evil effects in the Volga region of European Russia. *J Am Soc Agron, 15*, 6–15.

75. van Baren J. Agrogeology as a science. Mededeelingen van de Land-bouwhogeschool en van de daaraan verbonden instituten. In: Wulff A, editors. Bibliographia Agrogeologica, vol. 20. Wageningen; 1921. p. 5–9.

76. Niklas, H., Czibulka, F., & Hock, A. (1931). *Literautirsammlung aus dem Gesamtgebiet der Agrikulturchemie* (Vol. I). Bodenkunde. München: Verlag des Agrikultuchemischen Instituts Weihenstephan der Technische Hochschule.

77. Russell, E. J. (1939). Dr. J. G. Lipman. *Nature, 143*(3633), 1012.

78. Shull, C. A., & Thone, F. (1927). The first international congress of soil science. *Plant Physiol, II*, 369–83.

79. Waksman SA. Jacob G. Lipman. Agricultural scientist and humanitarian. New Brunswick, New Jersey: Rutgers University Press; 1966.

80. Anon. Next soils congress to be held in Russia. The Official Record. United States Department of Agriculture. 1927; VI(26):8.

81. Anon. The second international congress of soil science held in the Soviet Union, 1930. Proceedings and papers of the second international congress of soil science, vol VII. Moscow: State Publishing House of Agricultural, Cooperative and Collective Farm Literature (Selkolkhozgis); 1932. p. 11–8.

82. Bailey EH. Trip through western Maryland, vol. Part I. Proceedings. Washington D.C.: The American Organizing Committee of the First International Congress of Soil Science; 1928.

83. Bailey EH. Trip to Baltimore and Baltimore harbor, vol Part I. Proceedings. Washington D.C.: The American Organizing Committee of the First International Congress of Soil Science; 1928.

84. Waksman, S. A. (1954). *My life with the microbes*. New York: Simon and Schuster, Inc.

85. Young A. Thin on the ground. Soil science in the tropics. Second edition. Norwich: Land Resources Books; 2017.

86. Joffe, J. S. (1928). Soil science publications in Russia. *Science, 67*, 105.

87. Glinka KD. Dokuchaeiv's ideas in the development of pedology and cognate sciences, vol Part I. Proceedings. Washington D.C.: The American Organizing Committee of the First International Congress of Soil Science; 1928.

88. Simonson, R. W. (1997). Early teaching in USA of Dokuchaiev factors of soil formation. *Soil Sci Soc Am J, 61*(1), 11–6.

89. Russel JC, Engle EB. The organic matter content and color of soils in the central grassland states, vol Commission V. Commission VI. Miscellaneous papers. Washington D.C.: The American Organizing Committee of the First International Congress of Soil Science; 1928.

90. Jenny, H., Gessel, S. P., & Bingham, F. T. (1949). Comparative study of decomposition rates of organic matter in temperate and tropical regions. *Soil Sci, 68,* 419–32.

91. Harradine F, Jenny H. Influence of parent material and climate on texture and nitrogen and carbon contents of virgin California soils: I. Texture and nitrogen contents of soils. Soil Sci. 1958; 85(5):235–43.

92. Jenny, H. (1935). The clay content of the soil as related to climatic factors, particularly temperature. *Soil Sci, 40*(2), 111–28.

93. Jenny, H. (1930). Soil organic matter-temperature relationship in the Eastern United States. *Soil Sci, 31,* 247–52.

94. Jenny, H. (1950). Causes of the high nitrogen and organic matter content of certain tropical forest soils. *Soil Sci, 69*(1), 63–70.

95. Klemmedson, J. O., & Jenny, H. (1966). Nitrogen availability in California soils in relation to precipitation and parent material. *Soil Sci, 102*(4), 215–22.

96. Jenny, H. (1929). Relation of temperature to the amount of nitrogen in soils. *Soil Sci, 27*(3), 169–88.

97. Jenny, H. (1930). The nitrogen content of the soil as related to the precipitation-evaporation ratio. *Soil Sci, 29*(3), 193–206.

98. Jenny, H. (1946). Arrangement of soil series and types according to functions of soil forming factors. *Soil Sci, 61,* 375–92.

99. Jenny, H. (1941). *Factors of soil formation. A system of quantitative pedology.* New York: McGraw-Hill.

100. Florinsky, I. V. (2012). The Dokuchaev hypothesis as a basis for predictive digital soil mapping (on the 125th anniversary of its publication). *Eurasian Soil Sci, 45,* 445–51.

101. de'Sigmond AJ. Developments of soil science. Soil Sci. 1935; 40(1):77–88.

102. de'Sigmond AAJ. The principles of soil science. London: Thomas Murby & Co.; 1938.

103. Szabolcs I. The 1st international conference of agrogeology, April 14–24, 1909, Budapest, Hungary. In: Yaalon DH, Berkowicz S, editors. History of soil science international perspectives. Armelgasse 11/35447 Reiskirchen/Germany: Catena Verlag; 1997:67–78.

104. Davis W. The development of the transcontinental excursion of 1912. Memorial volume of the transcontinental excursion of 1912. New York: American Geographical Society of New York; 1915:3–7

105. Brigham AP. History of the excursion. Memorial volume of the transcontinental excursion of 1912. New York: American Geographical Society of New York; 1915. p. 9–45.

106. Weir WW. The new concept of soils. A handbook especially prepared for the meeting & tour of the first international soil congress. New York: Chilean Nitrate of Soda Educational Bureau of the United States of America; 1927. p. 62–67.

107. Marbut CF. The transcontinental excursion. Descriptions, discussions and interpretations of soils and soil relationships along the route of the excursion. The American Soil Survey Association; 1927.

108. Science ISoS. Official program of the first international congress of soil science. Washington DC: The American Organization Committee and the International Society of Soil Science; 1927.

109. Amundson R. Philosophical developments in pedology in the United States: Eugene Hilgard and Milton Whitney. In: Warkentin BP, editors. Footprints in the soils. People and ideas in soil history. Amsterdam: Elsevier; 2006. p. 149–66.

110. Tandarich JP, Sprecher SW. The intellectual background for the factors of soil formation. Factors of soil formation: a fiftieth anniversary retrospective. Madison, WI: Soil Science Society of America; 1994. p. 1–13.

111. Lipman, C. B. (1916). Eugene Woldemar Hilgard. *J Am Soc Agron, 8,* 160–62.

112. Bignami P. Excursions of scientists and agriculturists though North America. In: Waksman S, Deemer R, editors. Proceedings and Paper first international congress of soil science. Transcontinental excursion and impressions of the congress and of America. Washington D.C.: The American Organizing Committee of the First International Congress of Soil Science; 1928. p. 134–43.

113. Keen BA. The American agricultural research and advisory system. In: Waksman S, Deemer R, editors. Proceedings and paper first international congress of soil science. Transcontinental excursion and impressions of the congress and of America. Washington D.C.: The American Organizing Committee of the First International Congress of Soil Science; 1928. p. 157–62.

114. McCall, A. G. (1928). The transcontinental excursion. *Soil Sci, 25,* 105–106.

115. Russell EJ. The land called me. An autobiography. London: George Allen & Unwi; 1956.

116. Weir, W. W. (1956). *How real is religion?.* New York: Vantage Press.

117. Anon. The story of the "Boden Bull". In: Waksman S, Deemer R, editors. Proceedings and paper first international congress of soil science. Transcontinental excursion and impressions of the congress and of America. Washington D.C.: The American Organizing Committee of the First International Congress of Soil Science; 1928. p. 163–178.

118. Björlykke KO. Die klimatischen Bodenregionen in Norwegen. vol Part I. Proceedings. Washington D.C.: The American Organizing Committee of the First International Congress of Soil Science; 1928.

119. Björlykke KO. The big excursion. In: Waksman S, Deemer R, editors. Proceedings and paper first international congress of soil science. Transcontinental excursion and impressions of the congress and of America. Washington D.C.: The American Organizing Committee of the First International Congress of Soil Science; 1928. p. 144–146.

120. Wiegner G. Reiseeindrücke aus Nordamerika. In: Waksman S, Deemer R, editors. Proceedings and paper first international congress of soil science. Transcontinental excursion and impressions of the congress and of America. Washington D.C.: The American Organizing Committee of the First International Congress of Soil Science; 1928. p. 89–120.

121. Ziemięcka J. Some impressions from a journey in the United States. In: Waksman S, Deemer R, editors. Proceedings and paper first international congress of soil science. Transcontinental excursion and impressions of the congress and of America. Washington D.C.: The American Organizing Committee of the First International Congress of Soil Science; 1928. p. 152–54.

122. Moberg, C. L. (2005). *René Dubos, friend of the good earth: microbiologist, medical scientist, environmentalist*. Washington DC: American Society for Microbiology.

123. Joffe, J. S., & Antipov-Karataev, I. (1929). American soils as seen by Russian investigators. *Soil Sci, 27,* 159–66.

124. Joffe JS. A pedologist reflects on the third international congress of soil science. History and Present State of Soil Science; 1935. p. 427–29.

125. Loomis WE, Hartt CE. Charles A. Shull 1879–1962. Plant Physiol. 1964; 39:137–38 (Charles Albert Shull memorial issue).

12

From 1927 to 1960, and a Favor Returned

"Soil is a very broad term, like plant or animal."

Charles Kellogg, 1948

"The problems which are growing out of overspecialization are not limited to soil science but are common to many sciences. Physicists, engineers, chemists, biologists all see the limitations of too narrow specialization and the need for more generalists."

Richard Bradfield, 1960

There were some remarkable international soil activities in the interbellum. Charles Shaw, from the University of California, worked in China in the early 1930s and his work was published by the Academy of Sciences of Russia as part of the soil map of Asia project [1]. Curtis Marbut had never been in Africa, but made the first continental soil map in 1923 [2]. It was based on vegetation and geology and the map was prompted by the Peace Conference in Versailles in 1919. Books by Emil Ramann and Paul Vageler were translated from German into English in the 1930s [3]. Alex. de'Sigmond was an authority on salinity and alkali soils, and his 1923 book was translated from Hungarian into English and published by the University of California in 1927 [4]. Boris Polynov's book *The Cycle of Weathering* was translated by Alexander (Sandy) Muir with a foreword by Gammie Ogg [5], both at the Macaulay Institute for Soil Research in Scotland. These translations disseminated knowledge across a growing global soil science community. There was genuine fear of another war and international collaboration was seen as a way to avoid it.

Numerous textbooks were published that showed the diverse aspects of the discipline, such as five editions of *Soil Conditions and Plant Growth* by John Russell [6], *Principles of soil microbiology* in 1928 by Selman Waksman, *Micropedology* in 1938 by Kubiëna [7], and the voluminous *Soils & Men* yearbook from 1938 [8]. Soil science was still young, but it had accumulated much knowledge and in the late 1920s, Ernst Blanck of the University

© The Author(s), under exclusive license to Springer Nature
Switzerland AG 2021
A. E. Hartemink, *Soil Science Americana*
https://doi.org/10.1007/978-3-030-71135-1_12

of Göttingen published the *Handbuch der Bodenlehre,* which was a work of 10 volumes written by many German soil scientists including Ernst Blanck, Fritz Giesecke, Herman Stremme, and Eilhard Mitscherlich [9]. Hans Jenny contributed a chapter on soils in mountainous regions. The *Handbuch der Bodenlehre* covered all aspects of soil science including soil genesis and cartography, weathering and climate, zonal soils, soil physics, chemistry and biology, soil fertility and management. It was monumental for soil science.

After a successful *First International Congress of Soil Science* held in 1927, soil science prospered from international soil congresses in Russia in 1930 and the UK in 1935. The *Second International Congress of Soil Science* was held in St. Petersburg in 1930. In 1927, Konstantin Glinka had been elected the President of the *International Society of Soil Science* and placed in charge of organizing the congress. He died within three months after his return from the first congress. He was 60 years old. Konstantin Glinka was well-known and much liked in Europe, and after 1927, in the USA. He had dreamt of visiting America to see how the classification of soils in Russia could be applied in the USA: a dream fulfilled by the 1927 congress and the Transcontinental Excursion. Konstantin Glinka became the first soil scientist to be elected Member of the Russian Academy of Sciences. According to John Russell, he was among the foremost pioneers in soil science, and he admired that Konstantin Glinka studied the soil as a distinct natural object [10]. Gammie Ogg found Konstantin Glinka the master in the field study of soils, and he felt privileged to have accompanied him in an excursion across Hungary in 1926, and the USA in 1927 [11]. According to Gammie Ogg, Konstantin Glinka brought Russian soil science to the world: "Although we must not forget the pioneer work of Dokuchaev and Sibirtsev, nor the work of the other present-day members of the Russian school, it would be difficult to over-estimate the contributions of Glinka, not only in developing the subject, but in making it known throughout the world. He was a man of strong and most attractive personality, and his modesty and unfailing kindliness and good humor endeared him to all those with whom he came in contact." [11]. Selman Waksman regarded him as an indefatigable investigator, a brilliant teacher, and one of the best pupils from Vasily Dokuchaev [12].

The cause for his sudden death was unclear. John Russell noted that he was fine during the 1927 congress and excursion, and that no one knew he had failing health. Others alleged he was not well at the time of the congress [13]. Selman Waksman thought that the travel from Russia to the USA, the congress, and the one-month excursion proved too great an exertion for Konstantin Glinka [12]. According to Jacob Joffe, Konstantin Glinka had talked too freely at the *First International Congress of Soil Science,* and although he had died "undoubtedly for other sins," his actions were regarded as non-loyal to the Russian government. As Jacob Joffe summarized it: "He was accused of such 'short comings as fostering the cult of Russian pedology, Russian investigators, Russian spirit, Russian school, etc.' Such utterances were considered then as reactionary, counter-revolutionary and a stab in the back of internationalism." [14]. The official reason for his death was cancer of the stomach [15].

Konstantin Glinka was born in 1867, in an aristocratic family that included the composer Mikhail Glinka, whose compositions and opera were influenced by his stay in Italy, Austria, and Germany in the 1830s. Mikhail Glinka was the first composer to gain wide recognition in Russia and brought European influence into Russian music [16]. Incidentally, some 60 years later Konstantin Glinka brought Russian soil science to Europe. Konstantin Glinka studied at the University of St. Petersburg and obtained his PhD degree at the University of Moscow. In 1900, he was appointed as professor of mineralogy at the Agricultural Institute in Novo Alexandria that became the State Scientific Institute of Agriculture in Pulawy near Lublin in Poland. At the institute, he succeeded Nikolai Sibirtsev, who had died of tuberculosis in 1900. Konstantin Glinka trained notable students including Constantin Nikiforoff. Inspired by the work of Vasily Dokuchaev and Nikolai Sibirtsev, he wrote a book that expanded their legacy: *Die Typen der Bodenbildung* [17]. In the early 1900s, soil researchers in Europe became familiar with the work of Konstantin Glinka, but for others, particularly those in the USA, the English translation of *Die Typen der Bodenbildung* in particular, opened up a new understanding of pedology.

Nikolai Sibirtsev had stressed the climatic importance in the origin of the soil and that soils should be classified based on climatic zones. Initially, Konstantin Glinka took soil moisture as the basis for distinguishing soils, and he defined soils groups of optimum, medium, deficient and over-abundant moisture content, but dissatisfied with these groupings, he sought a more pedological basis for classifying soils. Trained as a mineralogist and glauconite weathering in particular, Konstantin Glinka became a pedologist and studied the soil from the surface to over two meters depth. He had a sharp eye for features in the soil profile, which became apparent during the excursion following the congress in 1927, as he observed details in the soil profile and was able to relate them like no one else did.

With the death of Konstantin Glinka, the *International Society of Soil Science* had to seek a new president. David Hissink was in Rome for a conference at the International Institute of Agriculture when he received a

Tweede Intern. Commissie – Groningen 1926.

Glinka by het bekende profiel Hondsrug.

Konstantin Glinka in Drenthe, The Netherlands, kneeling down for a soil at the medial moraine Hondsrug (caption: "Glinka at the well-known profile Hondsrug"). Konstantin Glinka was in Groningen attending the meeting of the Commission *The Study of Soil Chemistry* in which the session for the *First International of Congress of Soil Science* was prepared. It was in Drenthe that he remarked to David Hissink: "Sie haben in Holland doch gar keine Böden," to which David Hissink replied: "Aber wir haben Ernten." Photo and handwriting from David Hissink

telegram from Boris Polynov informing him of Konstantin Glinka's death. David Hissink instantly called a meeting with some other society members who were in Rome, including Friedrich Schucht. They agreed that Russia should submit nominations for a new President, and that the *International Society of Soil Science* would make the choice for the next President. After some time, Arseny Yarilov, replied that Konstantin Gedroiz was unanimously selected as the candidate. He was a chemist, director of the Dokuchaev Soil Science Institute, and had worked on base exchange capacity and colloidal properties of soil [18]. He was a friend of David Hissink, as their research overlapped. The society accepted the Russian nomination and Konstantin Gedroiz was appointed as President of the *International Society of Soil Science* and organizer of the *Second International Congress of Soil Science.*

In 1929, conferences of the six commissions were held in Prague, Budapest, Stockholm, Gdańsk, and Kaliningrad. At a meeting of the presidium in Budapest in the summer of 1929, the dates for the second congress were decided, as well as its duration, and what excursions should be organized [19]. The organizing committee was chaired by Arseny Yarilov, and the committee of 100 people met 27 times before the second congress. At each of these meetings, committee members were presented with a collection of soil samples, neatly boxed, from a location of the congress excursion; descriptions and analytical data of the samples were included. The organizing committee decided that the congress should not only highlight Vasily Dokuchaev's work, but also new developments in Russian soil science, with studies that dispelled the idea that Russian soil science was merely theoretical [19]. Secondly, it was decided that the activities of the congress should be directed toward the application of soil science and not only focus on general and theoretical problems. Therefore, all plenary speeches and addresses should have an applied agronomic character. Lastly, the organizing committee decided to show "...the brilliant successes of socialist construction in the whole people's economy, in industry, especially heavy industry, in agriculture, education, science, art, in questions of national relations, and in social and professional organizations." The organizers assumed that the foreign members of the *International Society of Soil Science* were "...poorly informed of the USSR, but, much worse, had received quite erroneous ideas about it from newspaper and periodical articles inspired by the Russian emigres."

The *Second International Congress of Soil Science* was held in St. Petersburg and Moscow from 20th until 31st of July 1930. The President of the *International Society of Soil Science,* Konstantin Gedroiz, was to open the congress, but he was ill and unable to attend. Instead, the opening address was given by the geographer and polar explorer Rudolf Samoilovich, who had studied mining engineering in Freiberg, Germany, and law in St. Petersburg. In 1908, he had been convicted of revolutionary activities, escaped from prison and joined expeditions to Spitsbergen, where he discovered high-quality anthracite. In 1918, he founded the Institute for Arctic Research.

Deuxième Congrès Intern. de la Sc. du Sol, Leningrad - 1930.

Séance d'Ouverture, Académie des Science, lundi 21/7 '30.
Jarilov, Hissink, Samoïlowitsch, Lemmermann (Lenin).

Opening speech (in German) of David Hissink at the *Second International Congress of Soil Science* in 1930. Left to right: Arseniy Yarilov, the editor of *Pochvovedenie,* David Hissink, Rudolf Samoilovich and Otto Lemmerman. The polar explorer Rudolf Samoilovich gave the speech on behalf of Konstantin Gedroiz. Photo and handwriting from David Hissink

At the second congress, Rudolf Samoilovich read the opening speech of Konstantin Gedroiz who apologized for his absence but noted that: "…our country with great attention, as never before, listens to the voice of science, awaits and seeks its aid. And undoubtedly our country will get this aid from the International Congress. At the same time, it is quite certain that members of the Congress will also return home after the Congress and the excursion are over, enriched by the scientific achievements and the experience accumulated by five generations of soil scientists of our Union. They built up and are continuing to build a theoretical pedology and during recent years, have begun to work closely on its practical application to life."

Rudolf Samoilovich had become an international celebrity when he led an expedition in 1928 to rescue the survivors of Umberto Nobile's Arctic flight [20]. Famous as he was, it did not end well for Rudolf Samoilovich. In 1938, he was arrested, committed to a sanatorium, and accused of "… being an agent of German and French intelligence services, creating an anti-Soviet wrecking organization at the institute." On the 4th of March 1939, he was sentenced to death and executed. His name was erased from all publications [20].

The Moscow part of the second congress was held in the State Conservatory. The congress was structured similarly to the first congress, with presentations, discussions, exhibitions, social events, and after the congress an excursion by train. The second congress was attended by 460 soil scientists, and there were 123 people from 20 countries. In 1930, the *International Society of Soil Science* had 1,116 members, compared to 934 members in 1927. The USA and Russia each had over 230 members. In total, 37 Americans attended the congress despite the economic depression, and it included Curtis Marbut, Charles Shaw, Robert Starkey, Jacob Joffe, Selman Waksman, Homer Shantz, Richard Bradfield, and William Albrecht. The geographer Elmer Ekblaw attended from Clark University. Emil Truog did not attend the second congress. He had a major role in the first congress, but he was not charmed by Russia, and had suffered some major financial losses following the stock market crash of October 1929. Charles Kellogg also did not attend; he was only eight months into his assistant professorship at *North Dakota Agricultural College* which was struggling to keep its people on the payroll. Charles Shaw, Selman Waksman, Robert Starkey, and Jacob Joffe were accompanied by their spouses.

At the time of the second congress, there was no diplomatic relationship between the USA and Russia. American scientists who worked for the federal government were not allowed to travel to Russia. For Curtis Marbut, attending the congress presented difficulties. He had to take unpaid leave and attend as a special representative of the American Geographical Society that paid for part of his travel [21, 22]. The congress and excursion was expensive and cost him $725 ($11,400 in 2021) [22]. In Russia, the American delegates were restricted in their travel and subjected to surveillance [22]. Several of the Russians who had fled to the USA in 1917 were worried about their safety if they were to attend the congress [22]. Such fear was justified, as the communist rulers declared the emigrants to be traitors and enemies. In the 1920s and 1930s, leading scientists who did not return to the Soviet Russia after traveling abroad were persecuted. Elected members of the Russian Academy of Sciences who emigrated were expelled from the academy [23]. Selman Waksman and Jacob Joffe, who had emigrated to the USA from Ukraine and Russia in 1910 and 1906, attended the *Second International Congress of Soil Science*, but Constantin Nikiforoff and Vladimir Ignatieff, who had both fled in 1917, did not attend. Jacob Lipman did also not attend.

The UK was represented by Bernard Keen, Edward Crowther, John Russell, and Gilbert Robinson. Japan had sent seven delegates. There were 27 Germans at the congress—the largest group after the Americans and Russians, and it included Friederich Schucht, Eilhard Mitscherlich, Otto Lemmerman, and Herman Stremme who was accompanied by his wife. Georg Wiegner from ETH in Switzerland attended, as did David Hissink, Albert Demolon, and Václav Novak. There were no participants from Hungary, which in 1930 had a strongly anti-Soviet government.

The congress started with a week of presentations and exhibitions, followed by an extensive excursion. Subjects that were discussed in the soil physics commission included mechanical analyses, soil capillary phenomena, photoelectric measurement of soil albedo, particle size fraction classes, the use of electro-dialyses for the pretreatment of soil samples, soil color, microstructure, and soil aggerates and structure relationships. Reports from various countries were presented on the determination of clay content and what size fractions should be preferred. Two methods of mechanical analysis were used: the elutriation method that used currents of water of varying velocity, and the sedimentation and pipette methods in which soil particles settle in a stationary column of water. The elutriation method was favored for fractions of 0.01 or 0.05 mm, and the pipette method was favored for the finer fractions [24]. New methods were presented for measuring horizontal and vertical soil wetting fronts, and water permeability at different soil depths. A penetrometer was presented as was a metal device for gaging swelling in clayey soils [25].

The soil chemistry commission had 31 papers that dealt with the pH of soil, phenomena of absorption in soils, soil colloids, liming, phosphates, soil organic matter, and analytical methods. Victor Kovda presented his studies on the charging of soil suspension particles and cataphoresis which was the motion of charged soil particles in an electric field. George Wiegner attended his presentation and invited Victor Kovda to join his laboratory

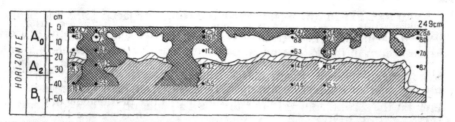

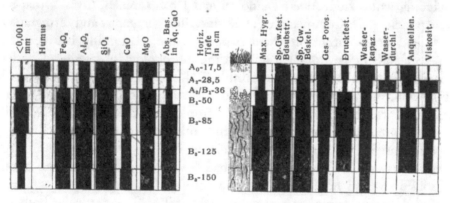

Upper diagram: Wetting front in a Podzol up to 50 cm soil depth across a 250 cm wide soil profile with moisture content ranging from 6 to 28%. Middle photos: measuring water permeability at different depths. Bottom diagrams: chemical and physical properties of a loamy podsol sampled by horizon (six in total). All figures from N. A. Kachinsky: *Nue Methoden zur Bestimmung eniger Physicalischen Eigenschaften des Bodens*, in the Proceedings of the *Second International Congress of Soil Science*

for a year at ETH in Switzerland after the congress. Jacob Joffe, the pedologist from Rutgers University, presented a paper on the movement of iron and aluminum in the soil. Some 50 papers were presented in the soil biology commission chaired by Selman Waksman. Jadwiga Ziemięcka from Poland presented work on nitrification and nitrogen-fixation. Hans Jenny was not present at the second congress but submitted a paper on the relation between soil humus and climate for the USA [19]. In desert regions, the soil nitrogen content was very low, and no difference was found between the northern or southern zones. Soil nitrogen increased logarithmically with an increasing humidity factor, and the rate of increase was highest in the north and lowest in the south. With increasing temperature soil nitrogen decreased exponentially, and the rate of decrease was greatest in humid regions and smallest in arid regions. Hans Jenny had been working on such relationship since he heard J. C. Russel speak at the congress in 1927 [26] and while attending the Transcontinental Excursion. Hans Jenny had published several papers on the relationship between soil properties and climate [27–29], and the findings formed the building blocks for his 1941 book *Factors of Soil Formation* [30].

In the symposium of the soil fertility committee, 34 papers were presented, including works from Albert Demolon and Jacob Lipman. Over 50 papers were presented for Commission V: Classification, geography and cartography of soils. Curtis Marbut presented papers on the relationship between the soil type and the environment and on the morphology of laterites. Charles Shaw presented a soil formation formula [31], and Edward Crowther related climate, clay composition and soil type. Soil studies of all parts of the world were presented, except for the soils of the tropics and Antarctica [32]. Dmitrii Vilenskii presented an overview of saline and alkali soils [33], which was a topic of interest to many soil researchers in Europe, Russia, and the USA. Some years earlier, Dmitrii Vilenskii had developed a genetic classification of soils in which he stressed the importance of the age of the soil. He coined the term *pedosphere* and defined the soil as "…a particular body of nature, extending like a fine epithelium over the surface of the lithosphere and forming the pedosphere." [34]. Throughout the presentations at the congress, contributions of Vasily Dokuchaev's genetic school were mentioned and validated. The sixth commission, named Application of Soil Science to Agricultural Technology, had 18 papers read that dealt with soil drainage, groundwater, irrigation, land reclamation, swelling of soils, and their influence on the destruction of buildings. Like the first congress, pedology was the main topic of the second congress.

The proceedings of the second congress were published in seven volumes in 1932, covering over 2,100 pages. In total, 235 papers were included, with

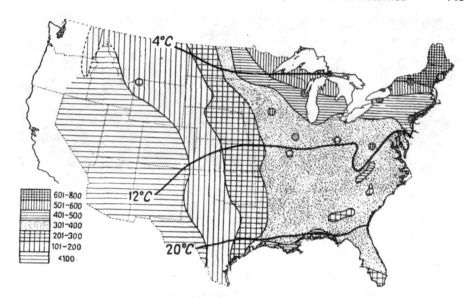

Map of annual NS-quotients (precipitation: saturation deficit) and the annual isothem of 4, 12 and 20 °C in the USA. From Hans Jenny's paper *Gesetzmässige beziehungen zwischen Bodenhums und Klima* presented at the *Second International Congress of Soil Science* in 1930. He realised that the temperature versus the precipitation in the USA formed a checkerboard for investigating soil-climate relationships. The 'checkerboard' term was coined by Curtis Marbut in 1927 during the Transcontinental Excursion: "...the checker-board distribution of soil character produced by the two factors of difference in rainfall and difference in temperature, it is evident that we have east-west soil belts across the arid and semiarid region of the United States in which soil differences are due entirely to differences in rainfall, the temperature in these belts remaining constant."

26 papers from the six sub-commissions on peat soils, forest soils, alkali soils, Mediterranean soils, the soil map of Europe, and the soil map of Asia. Most of the papers were edited by Konstantin Gedroiz, Arseny Yarilov and Dmitrii Vilenskii, and they had help from others and the commission chairs. Commission reports and papers were in either English, French and German and some papers had abstracts in another language.

There were soil exhibitions in St. Petersburg and Moscow. The Dokuchaev Soil Institute had a newly enlarged and reorganized museum, with sections on cartography of soils, genesis and systematics, and the geography of soils. There was an exhibit on 'road soil science' from the Auto Road Institute, and numerous specimens of soils, pictures, photographs, and maps based on the climatic soil zones. A collection of soil maps from other countries was on display. A guide to the exhibition was published in English and Russian. The exhibition in Moscow was held in the foyers and halls of the four-story

Exhibition in the Conservatory Mansion, Moscow, during the *Second International Congress of Soil Science* showing exhibits of the Agro-Soil Institute (left) and of the Timiryazev Agricultural Academy (right). Photos from the Proceedings of the *Second International Congress of Soil Science*

Conservatory where the congress took place. It focused on the practical application of soil science for agriculture, forestry, and construction. There were soil profiles exhibited as monolithic blocks and the important soil types were represented [35]. The monoliths were several meters high, and collected using the Russian box method. Also on display was a collection of the nodule bacteria exhibited in test tubes and Petri dishes, and various cultures of bacteria and plants [36]. Literature on soil science and related disciplines was on display including Russian soil books and journals from 1896 to 1930 on 15,000 bibliographical cards [36]. According to the proceedings: "It was a brilliant example of the energetic work of Russian scientific organizations."

The tour after the congress lasted 24 days and covered the steppe, the Volga region, the Ukraine, and the Caucasus region including Armenia. The tour took participants through the center of European Russia, toward the southeastern section, through the Caucasus, then by boat across the Black Sea, and from there through Ukraine. Kiev was the last stop after more than 7,000 km of travel [14]. Established and new experimental stations were visited; some of these had communist directors with little or no knowledge of science or agriculture. At every stop, soil pits had been dug, and the profile walls gave the Russian scientists an opportunity to demonstrate their method of soil profile description and analyses. Morphological descriptions and a genetic analysis of the soil profile were given [35]. Participants were divided into groups speaking English, German, and Russian. Many participants saw for the first time the Chernozems developed in loess.

The excursion took the participants to a tractor factory in Volgograd, an agricultural implements factory in Rostov-on-Don, tea plantations in Georgia, and the massive Dnieper hydroelectric station that was under construction and completed in 1932 but destroyed during the Second World War. At each of the excursion stops, the participants listened to speeches on Lenin-Marxism, and scientific papers were often started with a quote from Lenin, Marx, or Engels [37]. The organizers did not miss an opportunity to highlight the success of the socialist experiment. According to Georg Wiegner from ETH, the *Pyatiletka* (Five-Year Plan) and the tractor were at the center of interest in Soviet Russia; he would never forget: "...the tremendous propaganda for the *Pyatiletka* which was being carried on so successfully everywhere - at railway stations, in museums, schools, hotels." Georg Wiegner noted that: "...you can address to the hotel doorkeeper the question: How many tractors are there in your country? With the complete certainty he will immediately and unhesitatingly answer: At present, in 1930, there are 75,000, in 1931 there will be 100,000, and in 1932, 200,000. You will get the same prompt and definite answer to the question: How many of the peasant farms are collectivized?" [35]. For Edward Crowther of Rothamsted, the excursion was a once-in-a-lifetime pilgrimage. Curtis Marbut appreciated the Russian work on soils, and a few months after the congress, he started to study Russian so that he could read books and journals in Russian [22]. But during the excursion, Curtis Marbut developed stomach problems and canceled his plans to travel to Turkestan [21]. Instead, he traveled to Berlin, where he recovered, and then made a trip to the Black Forest: the 'Ozarks of Germany'.

During the excursion there were discussions on the international character of science and whether science has boundaries and united people. Russian workers responded with: "...a view of science is quite foreign to the scientific workers of the socialist parts of the world. For them science, economics and politics are all one. They are bound together by the class struggle, socialist construction and Marx-Lenin methodology. The Soviet citizen and the Soviet scientist make no distinction between science and life, between science and socialist construction and the class struggle." [35]. There was discussion about these views but most attendees focused on aspects that united the soil community rather than divided it. At one of the banquets, Curtis Marbut was asked to speak, and he summarized it: "...there are reasons why Americans are interested in the U.S.S.R. There are two ideals in Russia which are also American: universal education and a universally high standard of living. Americans have achieved much in both these respects. They have established universal education and accepted the principle that the better the environment, the higher the standard of living, the finer man will be." He did not say whether that was achieved in Russia.

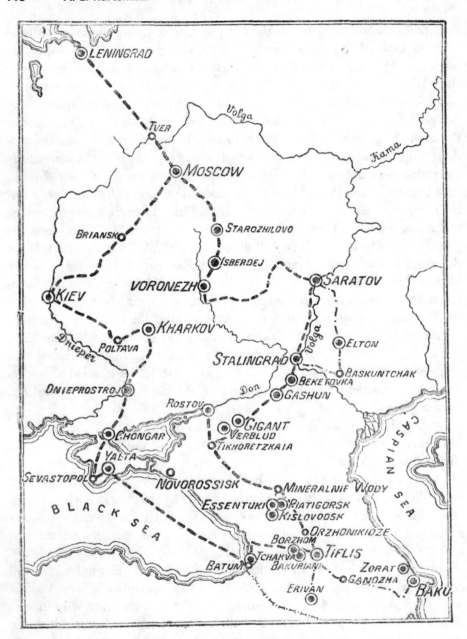

Excursion after the *Second International Congress of Soil Science* in Russia in 1930. Some 7,000 km were traveled from St. Petersburg to Moscow, to Saratov, Volgograd, Rostov, and then to Azerbaijan, Georgia and by boat across the Black Sea and onto Kiev, the last stop. The excursion took 24 days. Map from Proceedings of the *Second International Congress of Soil Science*

Excursion participants at the Georgian Military Road (top photo, left); Soil profile examination at Chongar, Ukraine and soil profile examination at Gashun, Northern Caucasus (top photos right). Participants at a giant loess profile with fossils at Chongar in the Ukraine (left bottom), and Chernozem profile examination at the Agricultural Experiment Station in Kharkov in the Ukraine (right bottom photo). Photos from the Proceedings of the *Second International Congress of Soil Science*

An *International Society of Soil Science* council meeting was held on the last day of the congress. Russian was adopted as an official language of the Congress, in addition to German, French, English, Spanish, and Italian. The organization of a third international congress was discussed and some suggested that the third congress should be held in the sub-tropics as those soils were understudied [32]. Konstantin Glinka had noted in 1914 that laterites and soils of subtropical latitudes were not studied by Russian investigators and that little was known about soils of the sub-tropics [17]. It would be novel and offer new perspectives to soil scientists across the world, just as the first two congresses had done. But a congress in the sub-tropics did not happen. John Russell convinced the council to hold the third international congress in the UK.

The *Third International Congress of Soil Science* was held in Oxford, UK, during the first week of August in 1935. The congress was presided by John Russell, who had been knighted by King George V in 1922. John Russell had participated in the first and second congress and was a well-known soil scientist. He had been the director of Rothamsted Experimental Station since 1912, and the station largely expanded during his time, from 140 staff in 1912 to 471 staff by the time he left in 1958 [38]. He had been succeeded by Gammie Ogg at the Rothamsted centennial in 1943. John Russell wrote the first seven editions of *Soil Conditions and Plant Growth* [6], and his son, Walter, a soil physicist, authored the eighth, ninth and tenth edition.

The third congress was to be opened by the Prince of Wales, who would become King Edward VIII half a year later. But he could not come, so the congress was opened by the vice-chancellor of Oxford University, who spoke the words: "It was the production from the soil of what is needed for the maintenance and enrichment of human life which evoked the earliest exercise of man's intellectual powers, and it is still the most important of all his enterprises." This was followed by the presidential address of John Russell and the report of Secretary General David Hissink on the work of the society since 1930. There were 968 members in 1935, down from 1,116 members in 1930. The Great Depression that had hit the USA at the end of 1929 had spread across Europe and it was clear that the world's economies were interconnected.

There were six plenary sessions, and the congress was organized around the six commissions. Sub-commissions met on problems related to alkali soils and peat soils. There was a large session on soil physics, and several researchers from Rothamsted presented their work, including Herbert Greene and Walter Russell. Robert Schofield presented the hysteresis effect in the pF curve. The relation between the water content and the soil water potential was established by early 1900 and the curve was characteristic for different soils. It was used to predict soil water storage and supply to plants. Water filling and draining the pores gave different curves, and the effect of that was named hysteresis. It was one of the discoveries of Robert Schofield who had been a student of Bernard Keen and succeeded him as head of the Physics department in Rothamsted in 1939. In 1956, Robert Schofield became Reader in Soil Science at Oxford; he died after a short illness in 1960 at 59, a brilliant soil physicist [39]. There were others from Rothamsted and the UK who died young; Edward Crowther, who was the Head of the Chemistry Department and founder of the British Society of Soil Science, died suddenly in 1954 at the age of 56 [40]. The pedologist Alexander Muir, also at Rothamsted and the Director of the Soil Survey of England and Wales, died in 1962 at only 56 years old [41]. He had studied two years in Russia and translated *The cycle of weathering* by the Russian pedologist Boris Polynov [5]. The tropical soil scientist Cecil Charter, trained by Gilbert Robinson in Wales, died in 1956,

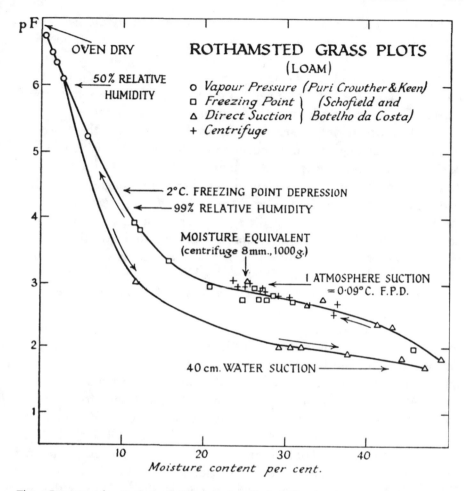

The pF curve of soils from Rothamsted and their hysteresis presented by Robert Schofield at the third congress in Oxford in the summer of 1935. From the proceedings of the *Third International Congress of Soil Science*

at 51 years old. Geoffrey Milne died when he was 43. They were all British and born at the turn of the century.

At the third congress Emil Truog presented several of his recent findings, including the mineral compounds of soil colloids and a test for determining available phosphorus that he had developed with Adolph Mehlich. Charles Kellogg presented his system of land classification based on the work with Ken Ableiter in North Dakota [42]. In their view, natural land classification should be based on inherent qualities of the land that included climate, soil, relief, stoniness, native vegetation, and their mutual relationships. Land classification had two purposes: as a planning tool for land utilization, and secondly, for rural zoning and tax assessment. Economic conditions were not part of land classification, and Charles Kellogg thought that no additional field work was needed if the economic conditions changed. A heated debate followed on the usefulness of mapping for land-use planning [32]. There were those for whom the soil survey had no other purpose than a soil inventory, whereas for others the soil survey had a definite agrarian purpose.

Over 150 papers were presented at the third congress and there were 427 participants, three-quarters from outside the UK. The *International Society of Soil Science* mourned the loss of Konstantin Gedroiz who had died in 1933, and Peter Treitz who had died a few months before the congress. Peter Treitz had organized the *First International Conference of Agrogeology* in Budapest in 1909 and helped to establish the society. The group picture taken at the congress showed Emil Truog and Charles Kellogg standing in the third row close to each other, not far from Geoffrey Milne. The third congress ended with a positive balance of over £500 [43], which was credited to *The* British Empire Section of the *International Society of Soil Science*, as there was no British soil science society.

Like the previous congresses, there was an exhibit with soil monoliths, maps, literature, and instruments. It was held in the newly established Soil Science Museum and Laboratory of Oxford University. A collection of soil monoliths from England and Wales, and from Sudan, Nigeria, Ghana, and South Africa was on display and the soil monoliths from Africa had been collected by Geoffrey Milne, C. R. van der Merwe, and C. G. T Morison. Soil maps and monoliths from China were prepared by the Geological Department and included soil horizon samples and full-sized colored photographs of 'rice-paddy soils.' There were five attendees from China at the congress. Curtis Marbut had sent monoliths from the Great Soil Groups that were donated to the Soil Science Museum in Oxford. He presented the new Soil Map of the United States at a scale of 1–2.5 million that was published in the *Atlas of American Agriculture* [44]. The first Soil Map of Europe at a scale of 1–2.5 million was presented; it was compiled by Herman Stremme with the help of many of his colleagues throughout Europe. The map had been planned since the 1924 congress in Rome, and a draft version was presented in 1927. Other exhibits included Alex de'Sigmond's system of soil classification, numerous soil specimens from David Hissink, and Charles Shaw exhibited a method for estimating soil color.

Geoffrey Milne, who was born in the same year as Hans Jenny, attended the congress. He was on leave from the East African Agricultural Research Station in Tanzania which he had joined at its establishment in 1928. He visited his brother Edward Milne, who was professor of mathematics at the University of Oxford [45]. Prior to the congress he spent some time at Leeds University and worked in the laboratory with Edmund Marshall, who became fascinated by Geoffrey Milne's knowledge on the soils of the tropics. At the congress, Geoffrey Milne presented a preliminary soil map of East Africa that he had just completed, including the soil catena concept [46]. The plan was to produce a soil map for the whole of British Africa and to present it at the next international congress in 1940 [47]. In 1938, he traveled for half a year through the West Indies and visited soil conservation stations in the USA. He realized the importance of green manures and cover crops: "…if then the soil continues to grow plants for us, in turn we must grow plants for the soil." The soil map for the whole of British Africa was never completed; the world went to war again, and Geoffrey Milne died at the beginning of 1942.

The Russians had dominated the first congress in Washington in 1927 [32], and organized the second congress in 1930 where 350 Russian soil scientists had participated. There were only six Russian at the third congress in 1935, including Boris Polynov, Leonid Prasolov, and Arseny Yarilov; they had all attended the first congress. Also present was Viktor Kovda, and the agrochemists Dmitriy Pryanishnikov and Oscar Kedrov-Zikhman. It was Victor Kovda's second international soil congress. For his PhD, he had studied Solonetzes and Solonchaks and his thesis was examined by Leonid Prasolov [48]. Victor Kovda had studied one year at the laboratory of George Wiegner in Switzerland [49]. After the 1935 congress, he became involved in international soil science. In 1945, he attended the meeting that aimed to resurrect the *International Society of Soil Science*. In 1968, Victor Kovda was elected President of the society and organized the *Tenth International Congress of Soil Science* in Moscow—six years after the ninth congress in Adelaide. The Russians wanted to celebrate the Golden Jubilee or 50th Anniversary of the *International Society of Soil Science*, and for that reason the tenth congress was in 1974, deviating from the four-year cycle.

In the 1920s, Joseph Stalin was part of a collective leadership, but by the 1930s, the country was under his dictatorship. Two years after the congress in Oxford, Boris Polynov was arrested and spent several years in prison, as he was suspected to be an English spy. He joined the Communist party in 1951 and died a year later. The third congress was the last international congress that Russian soil scientists attended until the sixth congress in Paris in 1956 [14]. By that time, Joseph Stalin had been dead for four years.

Victor Kovda in Oxford (left photo) and during the post-congress excursion across the UK of the *Third International Congress of Soil Science* in 1935 (right photo). It was his second international congress. He became the President of the *International Society of Soil Science* in 1968, and organized the *Tenth International Congress of Soil Science* in Moscow in 1974

A special issue of the *Transactions of the International Society* was published with the theme *Pedology in the USSR* that contained 22 papers from Russian soil scientists who could not attend the third congress in Oxford. The transactions included a paper by Sergei Zakharov who had been a student of Vasily Dokuchaev, on the fertility of deep horizons in the soil [50]. After the third congress, a special issue of *Pochvovedenie* was published, with contributions by David Hissink, Georg Wiegner, Richard Bradfield, Albert Demolon, Selman Waksman, Jacob Joffe, Jacob Lipman, Jadwiga Ziemięcka, and several others.

The third congress in Oxford lasted from Wednesday 30th of July to Thursday 7th of August 1935. During the weekend, the Rothamsted Experimental Station and the Imperial Chemical Industries station were visited.

There were short excursions around Oxford that showed Podzols, brown forest soils, and rendzinas under beech and conifers on an escarpment of oolitic limestone. Examples were shown of how soils under forest had been brought under cultivation and made productive. The post-congress excursion lasted two weeks and was organized by Gammie Ogg from Scotland and Gilbert Robinson from Wales. In total, 146 delegates participated, most of them from outside the UK including Victor Kovda and Charles Shaw [51]. The excursion went from Wellington, to Bangor in North Wales, to Perth, Aberdeen, and Edinburgh, and then to Newcastle, York, and Cambridge. A guide book for the excursion included detailed descriptions of each of the regions visited [52].

One to three days was spent at each location, and soil profiles were visited that had been dug in advance or to exposures in quarries and road cuts. Soils under agriculture and forest were compared, but the soil pits were small, certainly when compared to the pits observed in Russia in 1930, and it was not always possible to hear the discussions that mostly focused on the genesis and nomenclature of the soils [53]. Terminology differed, although there was some agreement on Podzols [53]. The brown earth group of Emil Ramann was used as a group of soils for anything that was not podzolised [53, 54]. For Charles Shaw from the University of California, who had seen great diversity of soils in Pennsylvania and California, the most striking pedological feature were the young soils and the time since the recession of glacial ice. Some of the recently drained soils were less than 20 years old [51]. There was little variation in the soils, and for Charles Shaw it was an opportunity to study the minor soil modifications under a relative uniform climate and similar land use [51]. Although several soil profiles were examined during the excursion, considerable attention was given to the subjects of soil fertility, drainage, and methods of reclamation—subjects that were close to the heart of John Russell [51].

The *International Society of Soil Science* had established a peat subcommission in 1930, as peat soils were found in all parts of the world including tropical regions. They had not received the same research attention as alkali soils, which had been studied in Hungary, USA, and Russia. In the USA, peat soils were widespread in Florida, along the Atlantic coast and in northern Michigan, Wisconsin and Minnesota. The peat soils had been ignored by American settlers because of their wet conditions and the prevalence of malaria along the Atlantic coast. But now there was interest in developing peat soils for agriculture, or to restore lands from which the peat was removed for heating or burning. During the excursion, participants went by air from Aberdeen to the Island of Lewis to study peat reclamation in the

Charles Kellogg and Cees Edelman in Richmond Hill Plantations on the coastal plains of Georgia in 1957, looking for acid-sulphate soils. Charles Kellogg was born in 1902 and Cees Edelman was born in 1903 and both led the soil survey in their countries. They met for the first time at the *Third International Congress of Soil Science* in Oxford in 1935. In 1947, they resurrected the *International Society of Soil Science* and decided that *Fourth International Congress of Soil Science* was to be held in Amsterdam in 1950

Hebrides [51]. Peat covered more than one-third of the island, and it was one of the largest blanket bogs in the world.

The third congress lacked the brilliance and vitality of the first two congresses. There were discussions about the fate of soil science led by Albert Demolon, who found that soil science had regressed [55]. According to Jacob Joffe, who had attended the first two congresses, the third congress did not have the same excitement as the first two congresses: the diminished novelty of the gatherings, the lack of big things and the achievements since the first congress in 1927, made it all less interesting. As he summarized it: "…at Oxford the congress worked under the shadow of the first congresses which were colorful, youthful, spurting with enthusiasm. In other words we must remember that the congress at Oxford was the *Third International Congress of Soil Science* – with the accent on the 'Third'." [32].

After the third congress in Oxford, Emil Truog traveled to Switzerland and Curtis Marbut started his long train journey to China. Emil Truog wanted to visit the birth places of his parents and he had asked Charles Kellogg to accompany him, but Charles Kellogg was too busy, for: "...there is an unprecedented demand for soil survey work and information regarding soils in connection with a great many of the federal programs." The third congress was over, and it took 15 years before another international soil congress was held.

In the late 1930s, a grim and hostile atmosphere descended upon the world. With the Marco Polo Bridge Incident between Japan and China and the German invasion of Poland, a new world war began. It was a little over twenty years after a war in which 37 million people had died. The memory of that calamity could not avert a tragedy that affected most nations and most people. The war would also affect the international community of soil science, which after successful international congresses, had hoped to expand and advance the activities and recognition of soil science. In the late 1930s, the *International Society of Soil Science* had about a thousand members [56].

At the third congress in 1935, the German delegation had extended an invitation to hold the next congress in Heidelberg in 1940, but the proposal was met with protests at the council meeting. Shortly after Hitler had obtained power in March 1933, a process known as *Gleichschaltung* was implemented, in which politically-suspect and Jewish civil servants were dismissed, trade unions were replaced by the *Deutsche Arbeitsfront,* and other political parties forbidden. From July 1933 onwards, everything 'un-German' had to disappear, and books written by Jewish, left-wing, or pacifist writers were burned. The protests against holding the international soil congress in Germany were from Selman Waksman and Jacob Lipman who was the chair of the American delegation [57]. Jacob Lipman feared both Fascism and Communism and regarded them, in their methods and psychologies, as not far apart.

Jacob Lipman and Selman Waksman mildly condemned the German government and said that the congress should not be held in such a country [58]. They saw it as serving the Nazis and warned that the congress would weaken the science. It created unrest at the council meeting and prompted the President of the *International Society of Soil Science*, John Russell, to take Selman Waksman aside for a private conversation. They had been friends, and Selman Waksman had written, in one of his first books: "The book is dedicated to Sir John Russell - investigator and writer, whose books on soil fertility and plant growth have disseminated widely the knowledge of the soil and its practical application." [59]. John Russell said to Selman Waksman that: "...scientists must not be punished for the action of their government, that there was no other invitation received by the society other than the Germans, and that a new president must be elected and he had to come from a country where the congress was to be held." [58]. Selman Waksman suggested that the congress should be held in Switzerland and that Georg Wiegner from ETH should become the *International Society of Soil Science* President. But Switzerland had not extended an invitation, and so that idea was not adopted. For the final vote on the 1940 congress to be held in Germany, Jacob Lipman abstained, and Selman Waksman voted against—all other countries voted in favor of holding the fourth international soil congress in Germany in 1940. Disappointed by that outcome, Selman Waksman instantly resigned from the commission on Soil Biology that he had chaired since 1927 [58]. The *Third International Congress of Soil Science* ended on a bad note. John Russell omitted the entire matter in his autobiography of 1956 [37]. When Selman Waksman was invited to write for the Rothamsted centennial in 1943, he barely mentioned the role of John Russell in the growth of the station and the expansion of soil microbiology [60].

The fourth congress was to be held in Heidelberg in German in 1940, and the *International Society of Soil Science* council approved Friedrich Schucht as its President. Friedrich Schucht had participated at the *First International Conference of Agrogeology* in Budapest in 1909, had served as secretary when the society was formed in 1924, and had been the editor of the journal *Soil Research* since 1928. Friedrich Schucht was also the President of the German Soil Science Society and a member of the NSDAP—the *Nationalsozialistische Deutsche Arbeiterpartei* or Nazi party. Some council members of the society had preferred Eilhard Mitscherlich as president, but that was not possible because he was not the president of the German Soil Science Society [61]. With that position, Friedrich Schucht automatically became the *International Society of Soil Science* President.

Shortly after the German invasion of Poland in September 1939, it was decided that the *Fourth International Congress of Soil Science* planned for 1940 had to be postponed. One of the last acts of the *International Society of Soil Science* was the meeting of the commission on Soil Biology held at the end of August 1939 at Rutgers University in New Jersey. After the meeting, most European participants had problems returning home [61]. Besides the start of the Second World War, there was more sadness. Just before the biology commission meeting, Jacob Lipman had died from heart failure. He was much liked in Europe, Russia, and across the USA. As President of the *International Society of Soil Science,* he had brought the world of soil science together. He was a great scientist and teacher, and had expanded the New Jersey Agricultural Experiment Station [62]. Selman Waksman was hired by him, and would write Jacob Lipman's biography 27 years after his death in which he called Jacob Lipman an agricultural scientist and humanitarian [63]. John Russel found that: "...Doctor Lipman's work has always been characterized by originality of outlook." [64].

When the Second World War broke out, the co-founder of the French society of soil science (AFES), Albert Demolon, recommended to suspend all efforts by the *International Society of Soil Science*. That did not happen. The Secretary, David Hissink, though retired from the Soil Science Institute in Groningen, maintained private and scientific correspondence from his home. He continued to administer activities of the society, although by 1942, there were only 220 paying members left, and the number of members dwindled further as war raged across the world. In Nazi-occupied Europe, official scientific communications were only possible through the *Deutsche Wissenschaftliche Institute*—the German Scientific Institute which had the task of promoting Nazi Germany by means of research, art, and culture. David Hissink had to work through this institute in order to communicate with members of the *International Society of Soil Science*.

President Friedrich Schucht died in 1941, and the German society of soil science appointed the chemist Fritz Giesecke as the new President. Through this, he also became the President of the *International Society of Soil Science*. In the 1920s, he had worked at the University of Göttingen as a scientific assistant to Edwin Blanck, who had signed a confession to the Nazis in 1933 as was required for university professors and school teachers [65]. During the Second World War, Fritz Giesecke held various Nazi party positions including head of training at the SS Race and Settlement Office, and chairman of the Agricultural Chemistry Working Group [65]. None of the *International Society of Soil Science* council members were informed about the appointment of Fritz Giesecke as the society President. In 1943, Fritz Giesecke became the President of *Deutsche Wissenschaftliche Institute* and was stationed in Sweden. He corresponded with David Hissink through the Swedish embassy and the *Germanisches Forschungsinstitut* (German Research Institute) in the Netherlands. Towards the end of the war, the *International Society of Soil Science* activities of David Hissink and Fritz Giesecke became impossible and ceased.

When the Second World War was over, 75 million people had died and the economies in Europe had collapsed and their industries ruined and bombed. The Nazi war machine was defeated, and Germany had surrendered. In May 1945, Russian, American, French, and British soil scientists were invited for a meeting in Moscow. The meeting was intended to resurrect the *International Society of Soil Science* and to celebrate the 200th anniversary of the Russian Academy of Sciences, which traced its origin back to the Emperor Peter the

Great. Another purpose was to prepare for the celebrations of 100th Anniversary of Vasily Dokuchaev's birth in 1846 [66]. Only soil scientists from the allied countries were invited to the meeting in Moscow, and Secretary David Hissink and the German President Fritz Giesecke were excluded.

On the 18th of June 1945, Russian, American, French, and British soil scientists met at the Dokuchaev Institute in Moscow. Leonid Prasolov, Boris Polynov, and Arseny Yarilov participated, and they had attended the three international soil congresses. The meeting was also attended by Dmitri Vilenskii, Sergei Zakharov, Innokenti Gerasimov and Viktor Kovda. Charles Kellogg, Edward Crowther, and Gammie Ogg were present. At a banquet during the meeting, the hosts served mastodon soup. The mastodon had been found frozen in a glacier, thawed out, and made into soup [67]. It was Charles Kellogg's first visit to Russia and he saw the devastation and burned-out tanks on his way to Moscow. The Nazis had been a few hundred kilometers from Moscow in 1941, and over 1.5 million people had died in the battle of Moscow.

Edward Crowther and Gammie Ogg had been to Russia for the *Second International Congress of Soil Science* in 1930, and Gammie Ogg had made a field trip with Konstantin Glinka in 1926. The French soil scientist Albert Demolon, who had just been part of the formation of INRA, also participated. In 1939, he had urged a halt to all society activities and at the meeting in Moscow, Albert Demolon stated that all participants should distance themselves from the *International Society of Soil Science* activities during the war. A motion was accepted to abolish the *International Society of Soil Science* and form a new organization that was to be named the *Association of Soil Scientists of the United Nations*. It was decided that the celebration of the 100th anniversary of Vasily Dokuchaev's birthday, 1st of March 1846, would be used to launch the new association. A temporary executive committee was elected that included Boris Polynov and Dmitri Vilenskii, Charles Kellogg, Richard Bradfield, Gammie Ogg, Edward Crowther, Albert Demolon and Auguste Oudin. They adopted the following resolutions [66]:

1. It is necessary to completely dissociate the new organization from the wartime activities of the Presidium of the I.S.S.S. All decisions taken by that body, including the appointment of Giesecke as the new President, are considered null and void.
2. The need is recognized for the renewal of scientific cooperation of the national soil science organizations of the U.N. in the form of an Association of Soil Science of the United Nations.

3. A Temporary Executive Committee should be formed to include USSR—
   Drs. B. B. Polynov, D. G. Vilenskii; USA—Drs. Ch. E. Kellogg,
   R. Bradfield; Great Britain—Drs. W. G. Ogg, E. M. Crowther; France—
   Drs. A. Demolon, A. Oudin. The Secretary of the Committee will be
   Prof. D. Vilenskii, Prof. A. Yarilov as one of the oldest pedologists will be
   included in the Committee and charged with calling the first meeting of
   the Executive Committee.

All seemed fine among these Russian, American, British, and French soil
scientists—fascism was defeated, and it was time for the allies to rebuild an
organization that would foster international soil science. There was a great
need to rebuild the world, hunger was to be eliminated, and no more war was
ever to be allowed. Russia had re-established diplomatic relationships with the
USA in 1933, thanks to President Franklin Roosevelt who had written a letter
to Joseph Stalin in October 1933: "Since the beginning of my administration,
I have contemplated the desirability of an effort to end the present abnormal
relations between the hundred and twenty-five million people of the United
States and the hundred and sixty-million people of Russia." [68].

The USA and Russia had held a joint conference in New York in 1943
that focused on science. At the conference, Selman Waksman had presented
research on bacteriology in the USA and Russia, and the conference seemed
to bode well for cooperation between the two nations. The USA had secured
the war's end by dropping atomic bombs on the Japanese cities of Hiroshima
and Nagasaki. By some estimate, the cost of the war for the USA was over
400,000 people and $300 billion. With that price paid and the strength
earned, came self-esteem and responsibility. President Franklin Roosevelt
emphasized it as follows: "...we all have to recognize—no matter how great
our strength—that we must deny ourselves the license to do always as we
please." There grew a great desire to lead.

President Franklin Roosevelt, who had led the country for 12 years, died
just before the war was over. He was replaced by Harry Truman, and he
invited Winston Churchill, whom he greatly admired and who had lost re-
election in the UK in July 1945. Winston Churchill accepted the invitation
and came in March 1946 to speak at the all-male and Presbyterian West-
minster College in Fulton, Missouri, where Putnam silt loams were common
soils. In 1933, President Roosevelt had hoped that the relation between the
USA and Russia would "...forever remain normal and friendly" but the
Fulton speech ruined the relationship and affected the organization of a new
international soil science society.

In Fulton, President Harry Truman joined Winston Churchill on the platform. Winston Churchill began his speech by praising the USA, which he declared stood "…at the pinnacle of world power" and with that power came responsibility to future generations. Winston Churchill argued for an even closer relationship between the USA and the UK—the great powers of the 'English-speaking world' as he called it—in organizing and policing the world. He warned against the expansion of Soviet Russia and used the term 'iron curtain' which was originally coined by the Nazi Minister of Propaganda Joseph Goebbels in reference to Soviet Russia. Winston Churchill warned that in dealing with the Russians, there was "…nothing which they admire so much as strength, and there is nothing for which they have less respect than for military weakness." Harry Truman warmly received the speech.

Meanwhile, in the Soviet Russia, Joseph Stalin condemned it as warmongering and referred to the comments about the English-speaking world as imperialist racism. It was the beginning of the Cold War that did little good to the world. Contacts between Russian and American soil scientists were frozen, and all plans made in June 1945 for the formation of the *Association of Soil Scientists of the United Nations* were tabled.

The years directly after the Second World War saw the formation of various soil organizations, such as the Dutch soil survey (*Stichting voor Bodem Kartering*—StiBoKa) [69], and the *Institut National de la Recherche Agronomique* (INRA) in France. Some of these had been planned before the war and were part of the rebuilding or establishment of soil institutions. Several countries had established soil science societies before the Second World War, such as Germany, Denmark, France, India and the USA, but many new soil science societies were started after the war, including China (1945), Brazil, Spain, UK (1947), Philippines (1948), Belgium, Ethiopia, Israel (1950), Italy, New Zealand (1952), Zambia (1953), and South Africa, Venezuela (1954). Sweden started its national society in the middle of the war in 1943. Between 1940 and 1980, an average of one national soil science society was established every year [70], reflecting the growth of soil science.

Despite the lack of an international soil science organization, there were several international meetings in the years following the Second World War. In May 1947, a pedology conference was organized by Albert Demolon in France, and it was planned as continuation of what was presented at the third congress in 1935 [71, 72]. There was uncertainty about the dates for the 1947 conference and the dates changed shortly before its start, which prevented many from attending, including most American attendees. The conference focused on the red soils of Mediterranean regions that were termed *terra rossa*. The conference included 10-day excursions in southern France and across Algeria. The conference had a pedologic focus with examination and discussions of soil profiles. Besides the French delegation, there were 22 delegates from 10 countries, mostly from Europe, with the UK sending eight delegates, including Walter Russell, Edward Crowther, and Gammie Ogg. Russian pedologists were invited but did not participate. Jacob Joffe from Rutgers participated, and he described the bus tour as "*En route* from Montpellier to Nimes, the color of the prevailing soil landscape as it appeared through the window of a speeding autobus was brown to red, reminding one that this must be the beginning of the much-described but little-known *terra rossa*." [71]. Jacob Joffe was critical about the way French pedologists studied soil profiles, and found that: "…one of their weaknesses lies in not fully appreciating pedologic methods. Recognition of this shortcoming will serve as an incentive to return to the problem, dig up more typical profiles, subject the soils to a more critical pedologic analysis, and follow the more advanced chemical and mineralogical methods being used in soil studies in the United States and Russia." [71].

In June 1948, the Commonwealth Bureau of Soil Science organized a conference on tropical and subtropical soils at Rothamsted that focused on tropical and subtropical soils, soil classification, fertility problems, soil erosion, and miscellaneous problems [73]. It was attended by British soil scientists, including Edward Crowther, Peter Nye and Gilbert Robinson, and there were several Belgians, French, South Africans, Dutch and two Americans—Charles Kellogg and Robert Pendleton. There were no German attendees. Reports and papers were from British, Dutch, and French workers in the colonies, but no one from the colonized countries attended. Several of the participants had met at the *Second International Congress of Soil Science* in Oxford. Cees Edelman presented research from Indonesia where he had spent some time in the 1930s. He had compiled a bibliography of soil studies that was partly burned in the Second World War, but was published in 1947 [74]. Charles Kellogg presented the classification and nomenclature of Great Soil Groups in tropical and equatorial regions. He stressed that systems developed in the temperate regions cannot be extended into the tropics, although the fundamental principles would be useful. In his view, there was good evidence that the guiding principles used in the soil surveys of the USA could be applied in tropical regions. Charles Kellogg stressed the need for scientific research, and that the increase in agricultural efficiency in Europe and the USA should inspire similar efforts in the tropics. There was debate about agricultural production in relation to soil productivity, and Albert Demolon reviewed the difference between native agriculture which he considered destructive, and 'intensive capitalistic' agriculture which introduced modern methods such irrigation, mechanization, and the use of inorganic fertilizers.

There was discussion on those soils of the tropics that were named lateritic soils, allitic soils or *sol dur alitic* or *ferralsol* in French. The need to distinguish between the soils with laterite that hardened upon drying compared to the well-structured red soils of the tropics was debated [75]. Charles Kellogg had ideas about these soils and coined the term Latosol: "We should like to suggest that some new term be adopted to comprehend all the zonal soils in tropical and equatorial regions having their dominant characteristics associated with low silica-sesquioxide ratios of clay fractions, low base-exchange capacities, low activities of the clay, low content of most primary minerals, low content of soluble constituents, a high degree of aggregate stability, and (perhaps) some red color. The word *Latosol* has been proposed as the name of the group at the categorical level of suborder. Perhaps some other term would be better. Chromosol has been suggested." It took another 10 years before the Oxisol order was launched in the *7th Approximation* [76]. Latosols or Oxisols were not well-researched and little information on these soils was available. For some soil scientists, the deep red soils were considered featureless and properties used for classifying soils of other orders were meaningless in Oxisols. There were few Oxisols in the USA, as the diagnostic criteria were made such that they would be found only in Hawaii and Puerto Rico [76]. The strongly weathered soils of the southeastern USA around the Piedmont that were called Red Yellow Podzolic soils became Ultisols.

At the 1948 conference in Harpenden, Charles Kellogg, Edward Crowther, Albert Demolon, and Cees Edelman discussed the status of *International Society of Soil Science* and the upcoming international congress in 1950. Charles Kellogg and Cees Edelman had met in Paris in 1947 for an international meeting on food aid for Poland. After that meeting, they met with some representatives from some national soil societies and from countries belonging to the United Nations. Cees Edelman, on behalf of the Dutch government, called for resurrecting the *International Society of Soil Science* and to organize an international congress in 1950. He had the support of Charles Kellogg. The 1945 idea of a new soil association was no longer pursued, and it was decided to resurrect the *International Society of Soil Science*. It was determined that the *Fourth International Congress of Soil Science* was to be held at the Royal Tropical Institute in Amsterdam, the Netherlands. Hungary, Romania, Russia, and Germany had been part of the formation of the society in 1924 but they had no participants at the meetings in Paris and Harpenden. Some of that had to with travel restrictions, though politics also played a role.

The *Dutch Society of Soil Science* (NBV), which had been established in 1935, was in charge of organizing the *Fourth International Congress of Soil Science* to be held in 1950 [77]. Cees Edelman was elected as President of the society and in charge of the meeting and the excursion after the congress. He spoke reasonably good French, English and German, was a good friend of Charles Kellogg, and was liked by the American members of the society [78]. That was important after the Second World War. It had been fifteen years since the last international congress, and there was optimism and a desire to rebuild and recover from the devastation of the Second World War. Food shortages in combination with the post-war baby boom required a considerable increase in agricultural production [69]. John Russell had predicted in the middle of the Second World War that: "…we must face the fact that post-war problems will be at least as difficult as those of the War itself." [79]. Science came out of the war with high status [80], and the call for a new international soil science congress was warmly welcomed.

The fourth congress was held at the Tropical Institute in Amsterdam and lasted from Monday 24th of July until Tuesday 1st of August 1950. There were almost 500 attendees, with about 120 from the Netherlands, including the doyen of tropical soil science, Jules Mohr. It was his first, and last, *International Congress of Soil Science*. Over 40 participants came from the USA and many brought their spouses, including Charles Kellogg, Ken Ableiter, William Albrecht, Firman Bear, Hugh Bennett, Henry Krusekopf, Lorenzo Richards, and James Thorp. Some spouses had been on the Transcontinental Excursion in 1927, and at the congress in Amsterdam, there was a special program for spouses organized by the spouses of the congress organizers. The 'Ladies Trips' as it was named, included excursions to the flower auction in Aalsmeer, Frans Hals Museum in Haarlem, cheese-market in Alkmaar, and a drive through the dunes along the North Sea.

There were close connections and friendship among soil scientists in the USA and from the Netherlands. Some of these originated from contacts and travel before the Second World War, but on the whole, the relationships became stronger and more intensive after the war. The Netherlands came plundered out of the war and received aid from the USA as part of the Marshall Plan under the motto: "Whatever the weather we only reach welfare together." Much of the aid was used for the rebuilding of industry, but there was also aid for agricultural equipment, the reclamation of land, and the rebuilding of Hotel de Wageningse Berg, accidentally bombed by allied forces in September 1944. Two showcase villages were modernized with Marshall aid. Dutch war crimes in Indonesia from 1945 to 1949 and the refusal to send soldiers to fight communism in Korea in 1950 almost ended the Marshall aid for the Netherlands, but effective diplomacy continued aid, and more than one billion dollars was received between 1948 and 1952.

David Hissink was retired as Secretary General of the *International Society of Soil Science*. The tropical soil scientist Ferdinand van Baren became acting Secretary and Treasurer. He held that position until 1955, when it was changed to Secretary General.

Ferdinand van Baren had graduated in 1934 on a study of potassium minerals in Dutch soils, and he was the first PhD student of Cees Edelman. Ferdinand van Baren had worked in Indonesia with Jules Mohr, who had conducted pioneering research on the soils of the tropics. Together they wrote a book on the pedogenesis of tropical soils [81, 82]. Ferdinand van Baren was only two years younger than Cees Edelman, and they were of the same generation as Charles Kellogg. Cees Edelman had struggled with his health for most of his career. It was thought to be the aftermath of the tuberculosis that he suffered in the 1920s. He was given regular injections with gold that were considered anti-rheumatic and anti-inflammatory. In fact, he suffered from leukemia which was not diagnosed until shortly before he died in 1964 at the age of 61.

At the fourth congress in Amsterdam in 1950, were no participants from Russia, Poland, Hungary, or Romania; soil scientists from these countries had actively participated in the first three congresses and the establishment of the society but travel and visa restrictions now precluded their participation. The exclusion of soil science from communist-led countries and its effect on the exchange of ideas has not been investigated. The 1950 congress was the first for the Portuguese soil scientist Luis Bramão, and for M. S. Swaminathan from India, who was on a UNESCO fellowship to study potato breeding in Wageningen. He was a plant geneticist, interested in soils, and eventually led the Green Revolution in India. Father and son John and Walter Russell attended the conference, and so did Peter LeMare and René Tavernier. There were only 12 German soil scientists at the congress in Amsterdam. Edwin Blanck led the delegation; Fritz Scheffer attended but Fritz Giesecke was absent.

The congress was opened on Monday 24th of July 1950 by Cees Edelman, who reflected: "It is fifteen years since the *Third International Soil Science Congress* was held at Oxford. During the intervening period much has happened which tends to make us sad. Many who were reckoned among the leading spirits of soil science in 1935 are no longer among the living. You came to Amsterdam to attend the *Fourth International Soil Science Congress*. You are in for a week of hard work. Not only have we the scientific side of the Congress in connection with which numerous lectures will be delivered, but we also expect to deliberate on the future organisation of the international contact in the field of soil science. The old *International Soil Science Society* will return to life in the coming days, and it will have to serve as the international forum of science for a long time to come."

Cees Edelman then asked David Hissink, whom he called the most famous Dutch pedologist, to speak to the congress participants. David Hissink had an important role in the formation and the expansion of the society and corresponded with an extensive group of soil scientists, and in Groningen, he welcomed and entertained visitors from all over Europe and the world. David Hissink spoke in French and reflected on his involvement with the *International Society of Soil Science* since 1909 and the resurrection of the society in 1947. He listed the achievements of the society but noted: "...we must never lose sight of the fact that organization in itself is not the goal: organization can only be one of the means to achieve this goal. And the goal, which we will pursue using this organization, will remain the same: to contribute to the development of Soil Science, but above all, to form a universal consciousness and thus help in the organization of a truly peaceful and prosperous world. The members of the Organization Committee do not look upon this Congress as their own. We have only been able to pave the way. The Congress is your Congress. The inspiration and wisdom to make the Congress a real success will have to come from you. But among you there are so many outstanding personalities, so many talented, enterprising, resourceful and willing men that our Committee is certain that the Congress will attain its two-fold objective." [83].

The last speaker at the opening session was Sicco Mansholt, who had been a farmer in a Dutch polder before joining the government as Minister of Agriculture in 1945. He was the first European Commissioner responsible for Agriculture and laid the basis for the Common Agricultural Policy. He was a friend of Cees Edelman and helped him to secure funds for establishing the Dutch soil survey. Some 20,000 people had died during the famine in the war winter of 1944 in the Netherlands and Sicco Mansholt stated that a stable supply of affordable food should be guaranteed for all. He emphasized that Europe needed to become self-sufficient: " ... although at the moment it would seem as though the balance between the productiveness of the soil and the food requirements has been restored, we all are well aware of the fact that the potential food requirements are far in excess of the quantity man to-day is in a position to supply. It will now be clear that if the standard of living of a nation will rise, food requirements will increase and the production will have to be pushed ahead. This is the problem with which every country and every people are faced, and which is the more important since it is almost universal." As a farmer and minister of agriculture, he was well aware of the importance of soil: "…it is always the nature of the soil, which in the first place calls for our attention. The aspects of the problem, however, are not the same in all countries. Densely populated lands will feel the need of paying attention to soil problems sooner than those where there is an abundance of arable land." [84].

Cees Edelman thanked Sicco Mansholt for his speech, and then gave his keynote address on some unusual aspects of soil science that he felt would not be dealt with during the congress. History, archaeology and toponomy had his interest, and he reviewed plaggen formation, the riverine clay areas, and the naming of soils. He discussed age-old agricultural systems and how they affected soil conditions. Some of that was in his book *Soils of the Netherlands,* written especially for the congress [85]. He saw no need to draw lines between sciences and offered a holistic view of the soil science discipline: "What is soil science? It is science developed by people who call themselves soil scientists. What do these soil scientists do? To find this out one should consult a number of textbooks or the register of an important periodical or the programme of an international congress. They show what the soil scientists themselves consider the most important objectives of their scientific activity. It is up to the soil scientists to look for an explanation of striking soil characteristics, isn't it? Why should soil scientists be allowed to borrow methods from chemistry, physics, biology, climatology, geology, geography and since our last Congress, also from social sciences and not from archaeology, history and philology? It should be left to the discretion of the soil scientists themselves to judge how far they want to go and how broad they want to make the theoretical foundation of their knowledge and understanding." [86]. From then on, he often publicly stated his credo: "Soil science is everything a soil scientist is interested in."

Cees Edelman also congratulated Charles Kellogg, who had just received the Distinguished Service Award and gold medal of the Department of Agriculture in the USA for his: "…outstanding leadership in the field of soil science; for unique effectiveness in interpreting soil uses to better serve human welfare, and for outstanding contributions to public understanding of world food potentials." It was quite the recognition for the 48-year-old Charles Kellogg, who had joined the department 15 years earlier.

The first day of the congress ended with a reception at Hotel Krasnapolsky, situated opposite the Royal Palace in the center of Amsterdam. David Hissink had contributed to the *International Society of Soil Science* since 1909 and was made Honorary Member. The war years in which David Hissink played a role were not mentioned, and the society has never made an effort to investigate its activities during the Second World War [87]. Besides David Hissink, the *International Society of Soil Science* elected Walter Kelley and Albert Demolon as Honorary Members. Albert Demolon had been critical of the society during the Second World War, and that was directed towards David Hissink, and Fritz Giesecke who had been appointed President of the *International Society of Soil Science* but held Nazi party positions. Walter Kelley was in 1950 the first American to become Honorary Member. Although they both had died in the 1930s, it is remarkable that Jacob Lipman and Curtis Marbut who had done much for establishing the society and organizing its first congress were never elected as Honorary Members. The society also failed to elect Selman Waksman as Honorary member; he had chaired the commission on Soil Biology from 1927 to 1935 and was after all the only soil scientist awarded the Nobel prize.

At the reception for David Hissink, there were several speeches, and Richard Bradfield recalled visiting Groningen when he was on sabbatical at ETH. They had first met at the soil congress in 1927 and both worked on acid clays that yield protons when the clay is in suspension. Richard Bradfield wrote: "In 1928, I visited him at Groningen for the first time. I was anxious to see with my own eyes the polders of which he had written so interestingly. Dr Hissink arranged such a tour and escorted me personally to see several successive dikes and polders. This cordial reception to a young, unknown American was in very sharp contrast to the reception in a few other well known continental laboratories, laboratories whose directors had not yet learned the truth so aptly expressed by Kettering, of the General Motors Research Laboratory: 'He who works in his Laboratory behind closed doors locks out much more than he locks in.'"

David Hissink replied to the speeches: "Forty years ago, in 1910, I had the privilege of sitting in the chair at the final meeting of the Second Conference in Stockholm, and so I had to close this Conference. In doing this, I pointed out that a considerable quantity of work was lying before us and that it would require much labour and much patience." To stimulate the work and patience, he quoted the last stanza of the American Henry Longfellow's *Psalm of Life*:

> Let us then be up and do
>   With a heart for any fate
>   Still achieving, still pursuing,
>   Learn to labour and to wait.

The organizational structure of the congress was somewhat different in 1950. At previous congresses, the meeting was centered around the commissions but in 1950, there were eight sections: soil physics, clay minerals, soil chemistry, soil biology, soil fertility, tropical and subtropical soils, soil conservation and management, land classification and evaluation, and saline soils. The discipline of soil science had broadened, and there was less focus on pedology, which had dominated the first two congresses.

The congress started on Tuesday 25th of July 1950, with a plenary lecture by Richard Bradfield on soil structure, and by the successor of Hans Jenny in Missouri, Edmund Marshall, on the electrochemistry of clay minerals in relation to pedology. There were sectional meetings and a discussion on the future of the *International Society of Soil Science*. In the evening, a reception was held at the Rijksmuseum. The next day a plenary lecture was given on soil fertility and visual crop symptoms, and Firman Bear spoke about soil organic matter. Other plenary lectures were given by Hugh Bennett on modern soil conservation, Robert Schofield on soil moisture and evaporation, and on the saline and alkali soils of the western USA by Lorenzo Richards. Almost half of the plenary lectures were given by American soil scientists.

During the congress, there were excursions to the dune region along the North Sea and to the sandy soil region of the Veluwe in the center of the Netherlands. There was a boat trip through the canals of Amsterdam enjoyed by over 400 participants. After the boat trip, they were entertained by the Associate Bulb Growers of Holland and learned how tracts of dune sand had been converted into bulb fields. For many participants, it was a sensation to hear that a large part of the route was about five meters below sea level. Like previous congresses, there was an exhibition at the congress which was named *Soils and People* and showed the relationship between land reclamation and the high population density in the Netherlands.

The fourth congress had a special session on the soils of tropics, and more was known about the soils of the tropics compared to previous congresses. Charles Kellogg's interest in tropical soils was awakened by his visit to Congo in 1947 [88]. At the session, he predicted that, as detailed soil maps would become available, more distinct soil types would be found within the tropics than in the other parts of the world. He thought that weathering masked all influence of parent rock on soil properties [88]. The Dutch presented colonial research in Indonesia and Surinam, which was new to many of the attendees, as most of the research had been published in the Dutch language. The tropical soils book by Jules Mohr had been translated into English by Robert Pendleton in the 1930s while he worked in Thailand, but the translation was not published until 1944 [89]. Interestingly, it was published by J.W.

Edwards in Ann Arbor, Michigan, which had also published the translation of Konstantin Glinka's book by Curtis Marbut in 1928 [90].

On the last day of the congress in Amsterdam, Tuesday 1st of August, a new foundation was laid for the *International Society of Soil Science*. The plans made at the Dokuchaev Institute in Moscow in 1945 to form an *Association of Soil Scientists of the United Nations* had been abandoned in 1947. At the congress in Amsterdam, there were no participants from Russia, Poland, Hungary or Romania, so the society was restarted without those members. The *International Society of Soil Science* had rules and procedures established at its foundation in 1924, and in 1950, revisions were deemed necessary. Charles Kellogg organized and chaired the session for the reconstruction of the *International Society of Soil Science* rules, with help from Walter Russell and Auguste Oudin. Given the problems during and directly after the Second World War, there was a need for committees on rules, nominations, and the international congress. No member from Germany was on any of these committees, but a resolution was passed declaring that German delegates should be invited to participate in subsequent work, and that they should be permitted to become members of the reorganized *International Society of Soil Science* like delegates from any other country [91].

The statutes for the reconstituted *International Society of Soil Science* included: Name, objects, and seat of the society; Membership; Structure of the Society; The Congress; The general meeting of the society; The officers; The Executive Committee; The council; The commissions; The Bulletin; Voting; Finance of the society; Other regulations; Change of rules; and Temporary provisions. It was decided that the *International Society of Soil Science* should publish a regular bulletin to inform its members about society activities, reports of meetings, new meetings, and publications. The Bulletin would give information on 'Who's Who' in soil research, keep up-to-date listings of members of the Society, publish new legislative measures related to soil science, information on new research centers, and addresses of directors when they take up other appointments. The first *International Society of Soil Science* bulletin appeared in 1952. It was edited by Hendrika (Kiek) de Nobel, who was married to the Secretary General, Ferdinand van Baren.

In 1950, information on the soils of the Netherlands was summarized in a book by Cees Edelman published in Dutch and English [85, 92]. The Dutch soil survey had unofficially started in 1943 in the riverine clay areas of the Bommelerwaard, where students from Wageningen hid to avoid being forced to work in German war factories. Among them were Kees Hoeksema, Pieter Buringh, Hendrik Jan Henkes, and Leen Pons. The soil survey in the Bommelerwaard was led by Cees Edelman, who used it to develop a soil mapping approach that was termed physiographic but in essence had a geomorphic base with subdivisions based on soil profile characteristics. The soil survey work in the Bommelerwaard led to the formation of the Dutch soil survey in August 1945. The post-congress excursion in 1950 was used to highlight the physiographic soil survey approaches and the successes in agriculture. It was organized by Pieter Buringh and Cees Edelman and covered the major soil regions of the Netherlands [93]. There were excursions for three, six or nine days, and more than one-third of the congress participants attended. From the three centers (Groningen-Zwolle, Wageningen, and Breda) visits were made by bus to land reclamation works that focused on soil ripening, river-clay soils in relation to fruit tree growing, and agriculture and horticulture on sandy soils. The nine-day excursion went partly through Belgium. There was a tour guide, and each participant received a copy of the *Soils of the Netherlands* that included a colored Provisional Soil Map of the Netherlands at a scale 1 to 400,000. According to the organizers, the excursions went according to plan, and for one rare moment in the Netherlands, the weather was fine, most of the time.

Charles Kellogg was active during the congress; he organized a symposium on tropical soils and re-drafted the statutes of the *International Society of Soil Science*. After the congress, he wrote a note to Cees Edelman: "If all the soils in the world with at least as good potentialities as those in Holland were reclaimed and managed with similar efficiency, world food production would be truly astronomical. What they (Dutch engineers and soil scientists) have done in the Rhine delta with a lot of water and a lot of work and a little soil material is amazing" [94]. Charles Kellogg had been in soil science for twenty years, and the congress in Amsterdam was his second *International Congress of Soil Science*. After that, he became active in the society and attended every congress until the eleventh congress in Edmonton in 1978.

The friendship between Charles Kellogg and Cees Edelman resulted in an invitation by the Department of Agriculture to a Dutch delegation to study acid sulfate and marsh soils along the Gulf of Mexico and the Atlantic Coast [95]. Cees Edelman and Koos van Staveren, a drainage engineer with the International Land Reclamation Institute in Wageningen, were interested in studying the 'cat clays' along the coast. The invitation made Cees Edelman proud and he remarked afterwards: "See, even the USA came to us for help." Although the invitation came from the USA, the trip was paid for by the Dutch government as a gesture of thanks to American troops who helped after the 1953 floods that killed over 1,800 people in southwest of the Netherlands.

Dutch soil scientists working in Indonesia visiting Hugh Bennett (center) in his office at the *Soil Conservation Service* in Washington in June 1946. They made an extensive tour through the southern states and visited the *Tennessee Valley Authority*. On the right (with bow-tie) Ferdinand van Baren, the delegation leader. Photo from Hans van Baren

Acid sulfate soils had been named 'cat clays' by the Dutch since the seventeenth century, and they has been studied by the chemist Jakob van Bemmelen [96]. Some soils of reclaimed areas turned extremely acidic, and developed yellowish mottles composed of an iron-potassium sulfate mineral named jarosite. It resembled cat excreta, and also smelled like that [97]. The soils had low agricultural productivity. The purpose of studying the acid sulfate soils along the Atlantic Coast was to compare them with similar soils in the Netherlands, and to evaluate the potential for agriculture or other uses. Cees Edelman and Koos van Staveren found that many of the marsh soils along the Atlantic Coast had 'cat clay' characteristics, whereas the marsh soils along the Gulf Coast did not acidify enough to qualify as acid sulfate soils. The marsh soils did not contain calcium carbonate, unlike those in the Netherlands, and they recommended that these soils should not be reclaimed but be left as wildlife areas [95].

In the first half of the twentieth century, there had been several visits by Dutch soils scientists to the USA. In 1905, Jules Mohr went to the USA to study the soil survey methods developed by the *Bureau of Soils* [98]. He was head of the Laboratory of Agrogeology and Soil Research at Bogor in Indonesia, and wondered whether the soil survey methods developed in the

USA could be used in Indonesia. Jules Mohr was familiar with Russian studies and had corresponded with Eugene Hilgard. In Washington, he visited with Milton Whitney and spent time in the field with some soil surveyors [99]. According to Jules Mohr, the American soil survey ignored the origin of the soils or its methods of formation and there was no attention for the effect of climate and vegetation on soil formation. He was critical of the quality of the soil surveyors, their scientific level, and all they did in 1905 was "...see well with their eyes and feel with the hands." It was clear to him that the 'American method' had little to offer for the soil survey work in Indonesia.

In 1946, Ferdinand van Baren headed a delegation of nine Dutch soil researchers based in Indonesia for an soil erosion study tour through the USA [100]. The tour was financed by the Ministry of Economic Affairs of the Dutch East Indies and the initiative came from Lourens Baas Becking, the director of the Botanical garden in Bogor who had studied with Martinus Beijerinck. The participants of the tour had survived the Japanese war camps where one out of six people died. The tour through the USA was meant for 'mental rehabilitation.' Ferdinand van Baren led the delegation, and he had succeeded Jules Mohr as director of the soil research institute in Bogor, that was housed in a colonial building near the botanical garden. From June to November 1946, they traveled through North Carolina, Tennessee, South Carolina, Alabama and Georgia and visited the Tennessee Valley Authority. They wrote a 500-page report that focused on the causes, effects and control of soil erosion, and how it should be handled in Indonesia where soil erosion was a problem on various islands. During the tour, the group was hosted by the *Soil Conservation Service* and, in particular by its director, Hugh Bennett. Ferdinand van Baren thought Hugh Bennett was 'the genius of conservation.' [100].

In 1951, several Dutch soil scientists made a tour across the USA that was considered as technical cooperation and funded by the Marshall Plan. The team included Jaap Schelling of the soil survey and Albert Jan Zuur, head of the pedological department of the State Authority for the reclamation of the central polders. They traveled for three months and visited Charles Kellogg, Emil Truog, Francis Hole, and many others. They realized that there was much more experience with soil and land classification in the USA than in the Netherlands, and that better land assessment and classification was needed to help increase crop yields [101]. During the tour, the Great Soil Group classification was discussed, but no agreement was reached on a preference for a natural system of classification or a system that had practical application for agriculture. The physiographic or geomorphic soil mapping approach developed by Cees Edelman was found less suitable for soil classification, as it

Plaggen soil photographed by Roy Simonson using a Contaflex SLR in August 1956. The profile was near Raalte in the Netherlands. The plaggen soil formed from long term addition of heather sod and bedding sand covering the humus Podzol (Spodosol). Cees Edelman highlighted these soils in his opening speech at the *Fourth International Congress of Soil Science* in 1950. Such soils are absent in the USA but were included in the *7th Approximation* soil classification system of 1960. Note the lamellae below the spodic horizon, and plain white measuring tape in feet. Such simple tape was necessary, as Roy would say "...the soil withstands little competition."

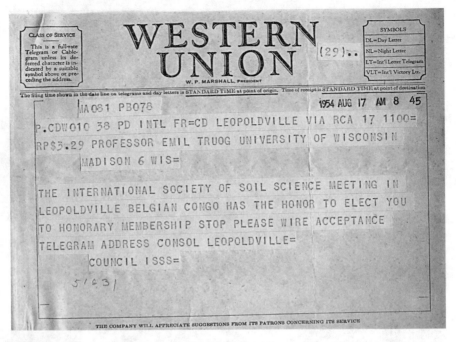

Telegram that was sent to Emil Truog on 17th August 1954 inviting him to become a Honorary Member of the *International Society of Soil Science*. He was the second American; Walter Kelley had been elected in 1950. Charles Kellogg had nominated Emil Truog for the Honorary Membership. Richard Bradfield and Charles Kellogg became Honorary Member in 1974. Hans Jenny became Honorary member in 1968, and Roy Simonson in 1990

provided more insight in geogenesis than in pedogenesis [102]. In the end, influenced by the American approach, the Dutch developed a soil classification system that had parent material and hydromorphic properties at the highest level [102].

At the *Fourth International Congress of Soil Science* in 1950, the society members expressed a clear interest in having a congress in a tropical country. Many soil scientists had read much but seen little or nothing of tropical soils while tropical soils cover about one-third of the earth's land surface. But it was not the first call for such a congress. In 1914, Konstantin Glinka had noted that laterites and soils of subtropical latitudes were not studied by Russian investigators and that little was known about such soils [17]. In 1927, he stressed that there was a lack of knowledge on the distribution of soil types in the tropical zone, and the relation of soils to the topography [103, 104]. In 1930, it was suggested that a congress should be held in the sub-tropics as soils from the sub-tropics were understudied [32]. Belgium had organized in 1948 the *First African Soil Conference* in Congo with participants from several African countries, the UK and two observers from FAO [105]. The conference focused on the low fertility soils under tropical rainforest. The conference included a session on social and economic problems in relation to soil conservation. Over 180 papers were presented, and the success of the conference prompted Belgium to extend an invitation to host the *Fifth International Congress of Soil Science* in Belgian Congo, in 1954. The invitation was accepted by the society, and the fifth congress was *Sous le haut patronage de sa Majesté le Roi des Belges.*

The *Fifth International Congress of Soil Science* was a small congress, with attendance by some 200 soil scientists, of which 48 worked in Congo, mostly Belgians [106]. There were 12 attendees from the USA including Richard Bradfield, Charles Kellogg, and Earl Storie. Rudi Dudal, Peter Nye and Walter Russell participated, as did Vladimir Ignatieff from FAO. The Netherlands was represented by Cees Edelman, Ferdinand van Baren, and Pieter Buringh. Edward Crowther, the chair of the Chemistry Department at Rothamsted, had died suddenly a few months before the congress; he had helped in organizing the *Third International Congress of Soil Science* in 1935, and with Herbert Greene translated the book of Paul Vageler, one of the first books on tropical soils [107]. The congress in Congo was organized by Floribert Jurion and René Tavernier, who had become the President of the *International Society of Soil Science*. He was trained as geologist and mineralogist and had founded, in 1950, the Soil Science Society of Belgium, as well as the Center of Soil Cartography.

The congress in Congo started on Sunday 15th of August 1954 with a morning reception in the Hôtel de Ville. The next day was the official opening by the Governor General of Congo and Floribert Jurion, and each day began with general lectures, followed by commission meetings. On the second day, Charles Kellogg spoke about soil conservation and noted that: "…no civilization has been challenged more sternly than ours." The Frenchman Georges Aubert spoke on laterite soils and some of the soil laterite profiles that would be observed during the excursion to Lubumbashi. Cees Edelman presented, in his best French, the importance of pedology for agricultural production based on his work in the riverine areas of the Netherlands. Richard Bradfield spoke about soil structure as he had done at the 1950 congress in Amsterdam. He was introduced by Charles Kellogg: "…our speaker is the ideal specialist with a broad point of view. He tells me that he spent 10 years studying the B-horizon of one local soil type. At that early stage of our science such concentration was necessary. I give you my dear friend, an outstanding example of the best in American soil science, Dr. Richard Bradfield." Charles Kellogg was referring to the Putnam silt loam, a cracking clay soil with a bleached layer that was a common soil north of Columbia, Missouri, and thus well-studied by Richard Bradfield.

Herbert Greene spoke about fertilizer prospects in Africa, and the Portuguese Joaquim Bothelo da Costa, who had worked in Angola, spoke about the relationship between soils, water, and plant growth. All presentations were followed by a discussion, and they were published in the congress proceedings, a tradition that gradually disappeared from proceedings. On Thursday evening, there was a visit to the Leopoldville Brewery, and Friday and Saturday were used for commission meetings. Like previous congresses, there was an exhibition that focused on the development of agriculture and institutions that studied the soils in the Congo.

At the council meeting, Charles Kellogg suggested that Emil Truog be invited to accept Honorary Membership of the Society: "Prof. Truog has attained world-wide prominence along two main lines. Firstly he has contributed fundamentally to our knowledge (1) of soil acidity and liming, (2) of the nature of cation exchange, (3) of the chemistry of phosphorus compounds, (4) of the relation of soil conditions to plant nutrition, and (5) of the proper methods for useful soil analysis. Secondly, he has been an inspiring teacher of soil science to hundreds of graduate students from all parts of the world. Prof. Truog has long taken a deep interest in soil research in other countries besides his own. His plans were well advanced for attending the Congress in Leopoldville, but because of a serious illness during a part of last winter his physician recommended against the long trip at this time. Prof. Truog is still actively engaged in soil science as the President of the Soil Science Society of America." The vote in favor was unanimous, and a telegram was sent to Emil Truog, and to the Swedish soil chemist Sante Mattson, requesting them to accept the nomination as newly elected Honorary Member. The Swedish chemist Sante Mattson had studied soils under Eugene Hilgard, and back in Uppsala, he studied soil colloids and isoelectric precipitates of iron and aluminum, from which he derived ideas on how soil horizons developed.

Peter Stobbe from Canada, who was trained by Joseph Ellis, summarized the congress in Congo as follows: "This congress offered the first opportunity for soil scientists interested in tropical soils to get together, to discuss research findings and to share their views. There were no startling announcements made in connection with new discoveries nor were there any profound treatises presented dealing with new fundamentals in soil science. However,

The congress location of the *Fifth International Congress of Soil Science* in Kinshasa, Congo in August 1954, and congress participants on a field excursion along the Lower Congo River. Photos by Peter LeMare

many factual results of soil investigations were presented. The sharing of such results and the discussion of common problems and of the approaches to their solution should provide a stimulus to further research and should give guidance in soil investigations in tropical countries. The success of the congress in my opinion was also due to a large extent to the organizing ability of its president, Mr. F. Jurion, the very able director of I.N.E.A.C. and the committee associated with him."

The congress was officially closed by Ferdinand van Baren, and on Sunday members of the congress visited Brazzaville, whereafter excursions were held to the Lower Congo River, Lubumbashi, and Yangambi. The excursion included a flight from Kinshasa to Kisangani in the heart of Congo, and a boat trip on the Congo River to the I.N.E.A.C. Research Station in Yangambi on the banks of the Congo River. Research at Yangambi focused on oil palms, soybeans, and dessert banana. It had a large Belgian staff, including Jules d'Hoore, who ten years later would make a soil map for Africa [108]. The participants of the excursion were indulged with an extensive soil trench that was dug across a hill. The soil profile at Mayumbe was some 200 m long and 1.2–2.5 m deep, it ran down the slope of one hill and up on another. Charles Kellogg called the trench 'a pedological extravaganza.'

The soils of Congo were new to many of the participants, and seeing these soils had, for some participants, a long-life influence. It was the second congress for Pieter Buringh from the Netherlands, who presented his ideas about the analysis of pedological elements in aerial photographs. During the Second World War, he had conducted detailed soil surveys in the Bommelerwaard, and his PhD research on soils around Wageningen was supervised by Cees Edelman. In 1950, he had helped with the excursion following the fourth congress in Amsterdam, and in 1955, he was appointed by the Government of Iraq as a consultant for soil survey and land evaluation [109]. There he met J. C. Russel, who had worked in Missouri and Nebraska and had been appointed professor in soil physics at the Agricultural College in Iraq after his retirement [85, 92]. Pieter Buringh became the first professor of tropical soil science at Wageningen University and, in 1964, wrote an introductory and widely-used textbook on the soils of the tropics and sub-tropics [110, 111].

The soils of Congo were new to the pedologist Peter Stobbe from the Central Experimental Farm in Ottawa, Canada. He observed: "Not being familiar with tropical soils, the featureless nature of their profiles and the lack of distinct pedogenetic horizons was rather disappointing to me. Some of the important distinguishing morphological soil features of temperate regions are very feebly expressed in many of the tropical soils. This makes one wonder how much stress should be placed in these soils on some of the commonly accepted morphological characteristics. Would closer studies reveal some other, perhaps less striking, characteristics which would be more significant in the interpretation of the genesis of these soils? It would seem

The extensive soil trench that was dug across a hill at Mayumbe in Congo as part of the excursion after the *Fifth International Congress of Soil Science* in 1954. The trench was some 200 m long and 1.2–2.5 m deep. Photo by Peter LeMare

that closer chemical and mineralogical analysis are a 'must' in the study of tropical soils." [112]. Peter Stobbe's unfamiliarity of soils of the tropics was no exception. Charles Kellogg often stated that Americans think of tropical soils as 'peculiar' and soils in the temperate regions as 'ordinary.' [113].

Left to right: Ferdinand van Baren, René Tavernier, Vladimir Ignatieff and Walter Russell (with sunglasses) at a boat on the Congo River in 1954. Photo by Charles Kellogg

For Charles Kellogg, the congress was his second visit to Congo. He traveled through West Africa after the congress, and in Ghana he met with Cecil Charter who had studied with Gilbert Robinson [114, 115]. From 1937 to 1951, Cecil Charter had worked on soil surveys in Trinidad, Honduras and Tanzania and developed a system of soil classification for the soils of the tropics [116]. The system interested Charles Kellogg, who needed knowledge on soils of the tropics for a new system that his office was developing. While traveling through Ghana, he noted how little soil erosion there was "...all the accelerated erosion I saw in over 200 miles could be found on a small Georgia farm." [114]. While he visited the soil research institute in Kumasi, he did not visit the university in Accra, and therefore did not meet with Peter Nye, who had been in Ghana and Nigeria since 1947, or with Dennis Greenland, who had just arrived [117, 118].

Peter Nye and Dennis Greenland would write a seminal book on shifting cultivation, based on their studies in Ghana, that analyzed and unraveled the system and showed how it sustained production if fallow periods were long enough [119]. Much of the book, *The Soil under Shifting Cultivation*,

Opening session of the *Sixth International Congress of Soil Science* in Paris, France, in August 1956. Left to right: René Tavernier, Auguste Oudin, Charles Kellogg, Ferdinand van Baren and Stéphane Hénin. Photo from Proceedings of the *International Society of Soil Science*

was written during evenings under a petroleum lamp. If population pressure combined with land hunger reduced the fallow period or increased the cropping period, soil fertility dwindles, and nutrient inputs were deemed essential. *The Soil under Shifting Cultivation* was influential, particularly with ecologists, and can be seen as a quantitative effort in soil fertility and nutrient cycling, and in its approach and impact equivalent to Hans Jenny's *Factors of Soil Formation* in pedology.

Charles Kellogg reviewed *The Soil under Shifting Cultivation* in 1962, and it influenced his thinking about the system [120]. He thought it was a welcome and timely book and stated: "...so far as this reviewer knows this is the first sound book about soil management under shifting cultivation in the English language." [120]. He also called it: "..a good beginning...this little book." [120]. A year later, he published his own review on shifting cultivation based on his visit and data from *Institut National pour l'Etude Agronomique du Congo Belge* in 1947. Charles Kellogg emphasized that research should start with a good study of how the system works, and he had no doubts that soil scientists could find easier ways than shifting cultivation to get the same or better results [121]. Such was Charles Kellogg's thinking in everything he did: understand the system, then improve it.

At the fifth congress in Congo, it was decided that the sixth congress was to be held in 1956 in Paris, France. Several people recommended to Charles

Kellogg that the seventh congress should be held in the USA in 1959 or 1960. It had been more than 30 years since the first congress and the success of that congress was legendary. Charles Kellogg was asked to extend such an invitation at the *Sixth International Congress of Soil Science* in Paris in 1956 [106]. Back in the USA, Charles Kellogg presented the suggestion to the *Soil Science Society of America,* and the President of the society at that time was not a stranger to him: Emil Truog. Both were aware that the USA government had to agree on organizing the congress, and that considerable funds would be needed. Charles Kellogg doubted whether any funds would come from government sources, given the costs of the war and the support that had been given for the rebuilding of Europe through the Marshall plan.

At the *Soil Science Society of America* meeting in Minneapolis in November 1954, the American delegation was endorsed to invite the *International Society of Soil Science* to hold the *Seventh International Congress of Soil Science* in the USA. A location was not decided, but Emil Truog advocated that it should be held at the University of Wisconsin [122]. Emil Truog had worked with Jacob Lipman and Curtis Marbut in 1927, was in charge of the exhibits, and he saw the organization of an international congress as the *grande finale* of his long career in soil science. In 1954, he had officially retired from the University of Wisconsin in Madison and so he had time for the organization of the congress. But no decision was made on whether the congress would be in the USA, not to mention whether it would be in Madison, his hometown.

The *Sixth International Congress of Soil Science* was held in Paris at the end of August 1956. Roy Simonson traveled to the Netherlands before traveling to Paris. In Amsterdam, he was met at the airport by Leen Pons, who took him to Wageningen, where the agricultural university and soil survey center was located. He stayed at the Hotel de Wageningse Berg and the next day he met with Cees Edelman and the 27 staff of the Dutch soil survey. They presented their experiences and reflected on the *4th Approximation* that was published in 1955 and its usefulness for classifying Dutch soils. The *4th Approximation* included, for the first time, diagnostic horizons to distinguish soil orders. The approximations were part of a soil classification system that Guy Smith and Charles Kellogg were developing since 1950 and that culminated in the *7th Approximation* in 1960, and its final product *Soil Taxonomy* in 1975.

To his chagrin, Cees Edelman had been asked to resign from the directorship of the Dutch soil survey that he had started in 1945. According to Roy Simonson, he also resented the fact that he had a small country rather than a continent to exercise his talents as a soil scientist. The Dutch soil surveyors took Roy Simonson on a tour across the country and in his diary, he sketched some aspects of Dutch life. He found the people hospitable and friendly, but they want to get their money's worth. Like in his own youth at the farm in North Dakota, the Dutch only changed clothes after a bath on Saturday, or on special occasions. He learned to eat with a knife and fork, and lunch was pieces of ham on slices of white bread with two or three eggs on top of that. The Dutch drank beer with lunch and had coffee afterwards, and the coffee was better than at home. They called greenhouses 'glass houses.' He noted that respect for authority was greater than in the USA, that the pace was more leisurely, and that in the Netherlands many women look like their queen. After his tour with the soil surveyors, he returned to Amsterdam where he took a Convair plane to Paris.

It was customary for the organizing country to elect the President of the society, and so the pedologist Auguste Oudin presided over the *Sixth International Congress of Soil Science* in Paris. He had been severely wounded on the battlefield of Verdun in 1916. In the 1930s, he taught pedology, and contributed to the first soil classification system that was based on genetic principles to which morphological and physico-chemical properties were added to define taxonomic units. Auguste Oudin assisted with the first soil map of France, which was presented at the fourth congress in 1950 [123]. The fifth congress in Paris had almost 800 attendees, of which 200 were from France, and over 40 American attendees.

About 1,000 people attended the opening session on 28th of August 1956. Auguste Oudin brought greetings from the *Haute Patronage* of the congress—President René Coty, who had political turmoil and the beginnings of a war of independence in Algeria to contend with. Symposia were held in different buildings, and the congress participants had to take the metro or bus to travel between symposia. Georges Aubert was the vice-president and organized the field excursion through the northwestern part of France [124]. Years earlier, Charles Kellogg had characterized him as "…an enthusiastic and very capable young soil scientist; however he is a bit shy but nevertheless he will become a great guy." [125, 126]. Georges Aubert conducted research with Stéphane Henin in the laboratory of Albert Demolon in Versailles near Paris, and had worked in the French colonies in West Africa. In 1952, he made a soil map of France at a scale of 1 to 1 million [124].

Several Russians attended the congress, including Victor Kovda and Innokenti Gerasimov; both were members in good standing with the Communist Party [49]. There had not been any Russian participation since the congress in Oxford in 1935. In Paris, the Russian delegation arrived ahead of the congress but failed to show up for several of the presentations as they had decided to celebrate for a change. Roy Simonson observed that most Russians were nice to the Americans and especially to Charles Kellogg. Discussions were held on a series of monographs on the major soils of the world that were to be published by the FAO. But as Russia, China, Romania, and several other countries were not FAO members, Victor Kovda suggested that the project should be carried out by the *International Society of Soil Science*. In separate meeting, Roy Simonson and Guy Smith met with Luis Bramão and Vladimir Ignatieff from the FAO, and with René Tavernier and Cees Edelman to discuss maps and monographs on the major soils of the world. Not much progress was made on how that should be organized, and it was not until the seventh congress in 1960 that plans for a soil map of the world were formalized.

At the congress and its subsequent excursion, French pedology was presented by Philippe Duchaufour, Jean Lozet, Jean Boulaine, and Yvon Dommergues. The world soil scientists learned about the process of *lessivage* and the benefits of morphogenetic soil classification that linked genetic evolution and knowledge of processes with morphological characteristics of soil profiles. The congress was the first for Rudi Dudal who had obtained his PhD in 1955 studying the loess soils of Central Belgium. He had joined the FAO as technical assistant on soil resources appraisal in Indonesia and taught soil science at the Faculty of Agricultural Sciences in Bogor, where he used mimeographed lecture notes from Roy Simonson. It was also the first congress for Dennis Greenland, who presented his findings on denitrification in tropical soils. Roy Simonson presented a paper with Luis Bramão on a new Great Soil Group that they named Rubrozems. These soils were observed in Brazil in 1954, and appeared to have Chernozems over the B and C horizons of Red Yellow Podzolic soils [127].

As the *International Society of Soil Science* was growing there was a need for more help to the Secretary General, Ferdinand van Baren. At the council meeting, it was decided that a Deputy Secretary General would be needed, and at the recommendation of Cees Edelman, Pieter Buringh from the Netherlands was appointed. At the council meeting, Charles Kellogg presented a proposal to host the seventh congress in the USA. While most countries had recovered from the war, they had no capacity or resources to organize a large congress, and so the council voted unanimously to accept the American proposal. The council also approved the recommendation of the USA delegation to nominate Richard Bradfield as President of the *International Society of Soil Science*, and Charles Kellogg as vice-president.

At the last council meeting in Paris, Auguste Oudin as retiring-President handed over the presidency to Charles Kellogg, as Richard Bradfield had already returned home to Cornell University. Charles Kellogg closed the meeting: "I had intended, Mr President, to speak in your language, but I fear that the result would have been too droll for this serious occasion. First of all, prof. Oudin, we, from the United States, wish to thank you, and through you, all the members of your Committee for the splendid arrangements that have made this Congress such a success. The resolutions just accepted by the Congress should assure you of our thankfulness. Obviously, you have worked very hard in order to make our meeting so comfortable and to provide such a profitable exchange of views among scientists using different languages. Then too, we ask that you extend our hearty thanks to the Government of France and to the Council of Paris. We should also like to express our thanks to Professor van Baren for his great activity on behalf of the Society between the Congresses as well as during this Congress. Secondly, we are very happy in the United States to be trusted to arrange for the next Congress in our country. We shall do our best, I promise you. Yet certainly the soil scientists of the Netherlands, of Belgium and the Belgian Congo and of France have set very high standards. All of us will look back to this Congress with the greatest pleasure. It is said that every man prefers his own country; but that all men outside of France call France their second country. We can see why this is so. And especially Paris is the heart and pulse of France. So much is concentrated here - in science, art, history and politics. We have all been thrilled, Mr President, in fact nearly overwhelmed with the power and beauty of it all. We hope to see all of you and many of your colleagues in the United States in 1960."

The meeting location of the *Seventh International Congress of Soil Science* in 1960 was not yet determined. There were two contenders—Cornell University and the University of Wisconsin. The proposal from Cornell was led by Richard Bradfield and the Wisconsin proposal by Emil Truog. Richard Bradfield had credentials in the society as he had attended all six congresses and had been part of the resurrection of the society after the Second World War. At the *Soil Science Society of America* meeting in Cincinnati in November 1956, it was decided that the congress would be held at the University of Wisconsin in Madison, and not at Cornell University.

A few months later, Norman Volk from Purdue University wrote to Richard Bradfield and expressed his disappointment with the decision to hold the congress in Madison. He asked whether other locations could be considered and in particular Purdue University. A follow-up letter was sent by Tom Bushnell from Purdue: "…personally, I have no desire to harass the committee with futile criticism of their failure to openly and thoroughly make extensive investigation of all possible alternatives for locating the Congress site." Richard Bradfield had indeed mentioned at the meeting in Cincinnati that more proposals for the congress were allowed but shortly after that it was announced the congress was to be held in Madison. Emil Truog then responded to Richard Bradfield in the hope that the decision was not to be reversed, as he wrote, tongue in cheek: "…as you may know, I did not initiate the idea of the meeting in Madison but was urged by others to present the invitation largely because of the impression we made in connection with the Mineral Nutrition Symposium held at Madison in 1949." He summed up all the advantages and found that Purdue had better meeting rooms, but that they could not match the lakeshore setting of Madison.

In the end, the decision was not changed, and the *Seventh International Congress of Soil Science* was held in Madison. Charles Kellogg, as an influential soil scientist in the government, had weighed in on the decision. The congress at the University of Wisconsin in Madison fulfilled a longtime wish of Emil Truog's. It also returned a favor. In exchange for a suit given freely in the depths of the Great Depression, Emil Truog received from Charles Kellogg an international soil congress.

References

1. Shaw CF. A preliminary field study of the soils of China (from the "contributions to the knowledge of the soils of Asia, 2" compiled by the Bureau for the Soil Map of Asia. Leningrad: Publishing office of the Academy of Sciences of the USSR; 1933.
2. Marbut CF. The vegetation and soils of Africa. New York: American Geographical Society; 1923.
3. Ramann E. The evolution and classification of soils. Translated by C.L. Whittles. Cambridge: W. Heffer & Sons Ltd; 1928.
4. de'Sigmond AAJ. The principles of soil science. London: Thomas Murby & Co.; 1938.
5. Polynov BB. The cycle of weathering (translated by Alexander Muir, Foreword by Gammie Ogg). London: Thomas Murby & Co.; 1937.
6. Russell EJ. Soil conditions and plant growth. London: Longmans; 1912.
7. Kubiëna WL. Micropedology. Ames, Iowa: Collegiate Press; 1938.
8. United States Department of Agriculture. Soils & men. The year book of agriculture. House Document No. 398. Washington United States Department of Agriculture. United States Government Printing Office; 1938.
9. Blanck E. Handbuch der Bodenlehre (vol. I-X). Berlin: Verlag von Julius Springer; 1929–1932.
10. Russell EJ. Obituary. Prof. K.D. Glinka. Nature. 1927;120:887–888.
11. Ogg WG. The contributions of Glinka and the Russian school to the study of soils. Scottish Geogr Mag. 1928;44:100–6.

12. Waksman SA. Professor K.D. Glinka. Soil Sci. 1928;25(1):3.
13. Shull CA, Thone F. The first international congress of soil science. Plant Physiol. 1927;II:369–383.
14. Joffe JS, editor. Russian contribution to soil science. Washington: American Association for the Advancement of Science; 1952. Christman RC, editor. Soviet science; a symposium presented on December 27, 1951, at the Philadelphia meeting of the American Association for the Advancement of Science.
15. Joffe JS. Pedology (with a foreword by C.F. Marbut). New Brunswick: Rutgers University Press; 1936.
16. Brown D. Mikhail Glinka, a biographical and critical study. Oxford: Oxford University Press; 1974.
17. Glinka K. Die Typen der Bodenbildung - Ihre Klassification und Geographsiche Verbreitung. Berlin: Verlag von Gebrüder Borntraeger; 1914.
18. Waksman SA. Professor K.K. Gedroiz. Soil Sci. 1932;34:403–416.
19. Anon. The second international congress of soil science held in the Soviet Union, 1930. In: Proceedings and papers of the second international congress of soil science, vol. VII. Moscow: State Publishing House of Agricultural, Cooperative and Collective Farm Literature (Selkolkhozgis); 1932. p. 11–8.
20. Koryakin VS. Rudolph Lazarevic Samoilovich, 1881–1939. Nauchno-biograficheskaya literatura: Nauka-M; 2007.
21. Rice TD. C.F. Marbut. Life and works of C.F. Marbut. Madison: Soil Science Society of America; 1936. p. 36–48.
22. The MD, Steppes A. The unexpected Russian roots of great plains agriculture, 1870s–1930s. Cambridge: Cambridge University Press; 2020.
23. Soyfer VN. Setting the record straight. Nature. 2002;419(6910):880–1.
24. Robinson GW. The division of the soil into fractions in mechanical analysis. In: Proceedings and papers of the second international congress of soil science, vol. I. Moscow: State Publishing House of Agricultural, Cooperative and Collective Farm Literature (Selkolkhozgis); 1932. p. 281–4.
25. Kachinsky NA. Nue methoden zur bestimmung eniger physicalischen eigenschaften des bodens. In: Proceedings and papers of the second international congress of soil science, vol. I. Moscow: State Publishing House of Agricultural, Cooperative and Collective Farm Literature (Selkolkhozgis); 1932. p. 129–160.
26. Russel JC, Engle EB. The organic matter content and color of soils in the central grassland states. Vol Commission V. Commission VI. Miscellaneous papers. Washington D.C.: The American Organizing Committee of the First International Congress of Soil Science; 1928.
27. Jenny H. Relation of climatic factors to the amount of nitrogen in soils. J Am Soc Agron. 1928;20:900–12.
28. Jenny H. Klima und Klimabodentypen in Europa und in den Vereinigten Staaten von Nordamerika. Soil Res. 1929;3:139–89.

29. Jenny H. Soil organic matter-temperature relationship in the Eastern United States. Soil Sci. 1930;31:247–52.

30. Jenny H. Factors of soil formation. A system of quantitative pedology. New York: McGraw-Hill; 1941.

31. Shaw CF. A soil formulation formula. In: Proceedings second international congress of soil science. ISSS; 1932.

32. Joffe JS. A pedologist reflects on the third international congress of soil science. History and present state of soil science; 1935. p. 427–9.

33. Vilenskii DG. Saline and alkali soils of the Union of Socialist Soviet Republics. Pedology. 1930;4:32–86.

34. Vilenskii DG. The analogical series in soil-formation and a new genetic classification of soils. Moscow; 1925.

35. Anon. Notes and jottings from the excursion. In: Proceedings and papers of the second international congress of soil science, vol. VII. Moscow: State Publishing House of Agricultural, Cooperative and Collective Farm Literature (Selkolkhozgis); 1932. p. 87–98.

36. Anon. Exhibition. In: Proceedings and papers of the second international congress of soil science, vol. VII. Moscow: State Publishing House of Agricultural, Cooperative and Collective Farm Literature (Selkolkhozgis); 1932. p. 147–152.

37. Russell EJ. The land called me. An autobiography. London: George Allen & Unwi; 1956.

38. Mitchell RL. Sir William Gammie Ogg. Year book R.S.E.; 1980. p. 67–71.

39. Anon. Obituary Robert Kenworthy Schofield. Géotechnique. 1960;10:127.

40. Russell EW. Obituary Dr. E. M. Crowther. Bull Int Soc Soil Sci. 1954;5:24–5.

41. Stewart BA. Dr. Alexander Muir (1906–1962). Int Soc Soil Sci Bull. 1962;21:29–30.

42. Kellogg CE, Ableiter JK. A method of rural land classification. Washington D.C.: United States Department of Agriculture; 1935.

43. Russell EW. The early history of the British Society of Soil Science. J Soil Sci. 1974;25:399–407.

44. Marbut CF. Soils of the United States. Part III. In: Baker OE, editor. Atlas of American agriculture. Washington: United Sates Department of Agriculture. Bureau of Chemistry and Soils; 1935. p. 1–98.

45. Anon. Obituaries. Mr. Geoffrey Milne. Nature. 1942;149:188.

46. Borden RW, Baillie IC, Hallett SH. The East African contribution to the formalisation of the soil catena concept. CATENA. 2020;185.

47. Milne G. Some suggested units of classification and mapping, particularly for East African soils. Soil Res. 1935;4:183–98.

48. Krupenikov IA. Memorable meetings with great pedologists that have passed away. Eurasian Soil Sci. 2012;45:895–9.

49. Yaalon DHVA. Kovda—meeting with a great and unique man. Newsl IUSS Comm Hist Philos Sociol Soil Sci. 2004;11:4–9.

50. Zakharov SA. A study of the fertility of the deep horizons of the soil of the USSR. Transactions of the international society. Pedology in the USSR. 1935. p. 156–166.

51. Shaw CF. Third international congress of soil science. Geogr Rev. 1936;26(1):139–41.

52. Ogg WG, Robinson GW, et al. Guidebook for the excursion round Britain of the third international congress of soil science. Oxford: Clarendon Press; 1935.

53. Muir A. The post-congress excursion round Britain. Paper presented at: third international congress of soil science. Oxford; 1935.

54. Ramann E. Bodenkunde (Dritte, umgearbeitete und verbesserte Auflage). Berlin: Verlag von Julius Springer; 1911.

55. Krupenikov IA. History of soil science from its inception to the present. New Delhi: Oxonian Press; 1992.

56. van Baren H, Hartemink AE, Tinker PB. 75 years the international society of soil science. Geoderma. 2000;96(1–2):1–18.

57. McCall AG. Jacob Goodale Lipman. Science. 1939;89(2313):378–9.

58. Waksman SA. My life with the microbes. New York: Simon and Schuster, Inc.; 1954.

59. Waksman S, Starkey R. The soils and the microbe. An introduction to the study of the microscopic population of the soil and its role in soil processes and plant growth. New York: Wiley; 1931.

60. Waksman SA. The men who made Rothamsted. Proc Soil Sci Soc Am. 1944;8:5–5.

61. Blume HP. Some aspects of the history of German soil science. J Plant Nutrition Soil Sci-Zeitschrift fur Pflanzenernahrung und Bodenkunde. 2002;165(4):377–81.

62. Allison RV, Jacob G. Lipman as a teacher and director of research. Soil Sci. 1935;40(1):31–8.

63. Waksman SA. Jacob G. Lipman. Agricultural scientist and humanitarian. New Brunswick, New Jersey: Rutgers University Press; 1966.

64. Russell EJ, Jacob G. Lipman and soil science. Soil Sci. 1935;40(1):3–10.

65. Klee E. Das Personen Lexikon zum Dritten Reich. Koblenz: Edition Kramer; 2003.

66. Anon. Minutes of the conference of the activities of the I.S.S.S. held at the Dokuchaev Soil Institute in Moscow on 18-th June 1945. Pochvovedeniye. 1946;5:328.

67. Kellogg RL. Remembering Charles Kellogg. Soil Surv Horiz. 2003:86–92.

68. Roosevelt FD. Letter to Joseph Stalin, October 20, 1933. The Franklin D. Roosevelt "Day by Day" Project. FDR Library. 1933.

69. Bouma J, Hartemink AE. Soil science and society in the Dutch context. Netherlands J Agric Sci. 2002;50(2):133–40.

70. Larson WE. Soil science societies and their publishing influence. In: McDonald P, editor. The literature of soil science. Ithaca: Cornell University Press; 1994. p. 123–42.

71. Joffe JS. Notes on the international conference of Pedology in the Mediterranean region. Soil Sci. 1948;65(5).

72. Kellogg CE. Soil genesis, classification, and cartography: 1924–1974. Geoderma. 1974;12:347–62.

73. Commonwealth Bureau of Soil Science. In: Proceedings of the First Commonwealth conference on tropical and sub-tropical soils, 1948. Technical Communication No. 46. Harpenden: CAB; 1949.

74. Edelman CH. Studies over de bodemkunde van Nederlandsch-Indië. 's-Gravenhage: Publicatie no. 24 van de Stichting Fonds Landbouw Export Bureau; 1947.

75. Dudal R. Evolving concepts in tropical soil science: the humid tropics. Evolution of tropical soil science: past and future. Brussels: Royal Academy of Overseas Sciences; 2003. p. 15–38.

76. Soil Survey Staff. Soil classification. A comprehensive system, 7th approximation. Washington D.C.: Soil Conservation Service. United States Department of Agriculture; 1960.

77. Bouma J, Hartemink AE, Jellema HWF, Grinsven JJMv, Verbauwen EC, editors. Profiel van de Nederlandse bodemkunde. 75 jaar Nederlandse Bodemkundige Vereniging (1935–2010). Wageningen: NBV; 2010.

78. Vink APA. Vijf en veertig jaar uit het blote hoofd. Bodemkartering, Landschapsecologie, Landevaluatie. Bussum: unpublished biography; 1989.

79. Russell EJ. Restarting agiculture in devastated Europe. Nature. 1943;3833:433–8.

80. Tinker PB. Soil science in a changing world. J Soil Sci. 1985;36:1–8.

81. Mohr ECJ, Van Baren FA. Tropical soils—a critical study of soil genesis as related to climate, rock and vegetation. The Hague: N.V. Uitgeverij W. van Hoeve; 1959.

82. Mohr ECJ; Van Baren FA, van Schuylenborgh J. Tropical soils—a comprehensive study of their genesis. The Hague: Mouton - Ichtiar Baru - van Hoeve; 1972.

83. Hissink DJ. II. Inaugural session. Address by Dr D.J. Hissink. Paper presented at: fourth international congress of soil science, Amsterdam; 1950.

84. Mansholt SL. II. Inaugural session. Address by his excellency S.L. Mansholt. Paper presented at: fourth international congress of soil science, Amsterdam; 1950.

85. Edelman CH. Soils of the Netherlands. Amsterdam: North-Holland; 1950.

86. Edelman CH. Some unusual aspects of soil science. Paper presented at: fourth international congress of soil science, Amsterdam; 1950.

87. ISSS. The history of the International Society of Soil Science. Bulletin. 1974;45.

88. Kellogg C, Davol F. An exploratory study of the soil groups in the Belgian Congo. Congo: Institut National pour l'etude agronomique du Congo Belge; 1949.
89. Mohr ECJ. The soils of equatorial regions with special reference to the Netherlands East Indies. Ann Arbor: J. W. Edwards; 1944.
90. Glinka KD. The great soil groups of the world and their development (translated from German by C.F. Marbut in 1917). Ann Arbor, Michigan: Mimeographed and Printed by Edward Brothers; 1928.
91. Giesecke F. Geschichtlicher Uberblick under die Entwicklung der Bodenkunde bis zur Wende des 20. Jahrhunderts. Berlin: Verlag von Julius Springer; 1929.
92. Edelman CH. Inleiding tot de Bodemkunde van Nederland. Amsterdam: North-Holland; 1950.
93. Hartemink AE, Sonneveld MPW. Soil maps of The Netherlands. Geoderma. 2013;204:1–9.
94. Anon. 25 years ago: the 4th congress of the international society of soil science. Bull Int Soc Soil Sci. 1975;47:5–7.
95. Fanning DS, editor. Leen Pons. Father of the International Acid Sulphate Soils Symposia/Conferences. University of Maryland; 2008; No. College Park MD.
96. Dent DL, Pons LJ. A world perspective on acid sulphate soils. Geoderma. 1995;67(3–4):263–76.
97. Prokopovich NP. Cat clays. General geology. Boston, MA: Springer US; 1988:65–69.
98. van Baren FA. In memoriam Prof. Dr. E. C. Julius Mohr. Bull Int Soc Soil Sci. 1970;36:18.
99. Shaw CF. The American soil survey. A rebuttal of criticisms by Dr E.C.J. Mohr. Landbouw 1933;4:169–175.
100. van Baren FA. Erosie, oorzaak, gevolgen en bestrijding. Mededeelingen van het Departement van Economische Zaken in Nederlandsch-Indie. No. 8; 1947.
101. Contactgroep Opvoering Productiviteit (COP). De bodem- en landclassificatie in de Ver. Staten van Amerika. 's-Gravenhage: Studierapport Landbouw; 1951.
102. Hartemink AE, de Bakker H. Classification systems: Netherlands. In: Lal R, editor. Encyclopedia of soil science. 2nd ed. New York: Taylor & Francis; 2006. p. 265–8.
103. Glinka K. Treatise on soil science (Pochvovedenie), fourth posthumous edition. Translated from Russian in 1963 by the Israel Program for Scientific Translations. Washington DC: National Science Foundation; 1931.
104. Wilde SA. Glinka's later ideas on soil classification. Soil Sci. 1949;67(5):411–4.
105. Pendleton RL. African conference on soils at Goma, Belgian Congo. Soil Sci. 67:481–486.
106. Kellogg CE. The fifth international congress of soil science Leopoldville, Belgian Congo, August 1954. Soil Sci Soc Am Proc. 1955;19:115–118.
107. Vageler P. Grundriss de Tropischen und Subtropischen Bodenkunde für Pflanzer und Studierende. Berlin: Verlagsgesellschaft für Ackerbau M.B.H; 1930.

108. d'Hoore J. Soil map of Africa scale 1 to 5 000 000. Explanatory monograph. Lagos: Commission for Technical Co-operation in Africa; 1964.

109. Blokhuis W, de Meester T. In memoriam Pieter Buringh (1918–2009). Bull Int Union Soil Sci. 2009;115:26–7.

110. Buringh P. Introduction to the study of soils in tropical and subtropical regions. Wageningen: PUDOC; 1970.

111. Buringh P. Introduction to the study of soils in tropical and subtropical regions, 3rd ed. Wageningen: PUDOC; 1979.

112. Stobbe PC. Some observations on the fifth international congress of soil science held in the Belgian Congo, 16.8-5.9.1954. ISSS Bull. 1955;7:29.

113. Kellogg CE. Soil use for abundance in the newly developing countries. Food for peace. ASA Special Publication Number 1; 1963. p. 28–35.

114. Mount HR. Retracing Charles Kellogg's path in Ghana. Soil Surv Horiz. 2004;45:52–4.

115. Effland W, Asiamah R, Adjei-Gyapong T, Dela-Dedzoe E, Boateng E. Discovering soils in the tropics: soil classification in Ghana. Soil Surv Horiz. 2009;50:39–46.

116. Charter CF. The aims and objects of tropical soil surveys. Soils Fertil. 1957;XX:127–158.

117. Greenland DJ, Tinker PB, Kirk GJD. Peter Hague Nye (1921–2009). Biogr Mems Fell R Soc. 2011;57:315–26.

118. Gregory PJ. Dennis James Greenland (1930–2012). Biogr Mems Fell R Soc. 2019:1–17. https://doi.org/10.1098/rsbm.2018.0030.

119. Nye PH, Greenland DJ. The soil under shifting cultivation. Harpenden: Commonwealth Bureau of Soils; 1960.

120. Kellogg CE. Book review of: the soil under shifting cultivation, by P.H. Nye and D.J. Greenland. Agron J. 1962;54(3):279–280.

121. Kellogg CE. Shifting cultivation. Soil Sci. 1963;95:221–30.

122. Truog E. Reflections of a professor of soil science. Soil Sci. 1965;99:143–6.

123. Begon J. In Memoriam. Auguste Oudin (1886–1979), Membre honoraire de l'AISS. Int Soc Soil Sci Bull. 1979;61:49.

124. Nortcliff S. Georges Aubert (1913–2006). Bull Int Soc Soil Sci. 2008;112:31–2.

125. Feller C. Georges Aubert et les soils (1919–2006). Marseilles: IRD; 2006.

126. Tavernier R. Georges Aubert et son rôle international. Pedologie. 1981;XVIII:183–185.

127. Bramão DL, Simonson RW. Rubrozem—a proposed Great Soil Group. In: Proceedings of the sixth international congress of soil science, vol. V; 1956. p. 25–30.

13

Seventh International Congress of Soil Science 1960

"…people from foreign countries may learn. just how our research and educational institutions working with farmers. have provided for our abundant food production."

Emil Truog in a letter to President Dwight Eisenhower, 1960.

The first half of the twentieth century was turbulent. There was panic in the early 1900s with a stock market crash, followed by the First World War in 1914, and the revolution in Russia in 1917. The economic recession that started in 1929 was followed by the totalitarianism of communists and fascists, and tens of millions of people lost their lives, countries were ruined, and very few people were untouched by the devastation of war and the tyranny of ideologies. After the Second World War, there was a spirit of turning over a new leaf, massive reconstruction, and a sense of 'never again.' Several international soil organizations were formed, and older ones were resurrected. The *International Society of Soil Science* had a laborious restart, its war years were disremembered, and international soil congresses were held in Amsterdam in 1950, Kinshasa in 1954, and Paris in 1956. By the time the international congress was held in Madison in 1960, the world had recovered, but the optimism of global peace and cooperation had begun to dwindle, upon facing the reality of international dissimilarities and the brewing of yet another war, a cold one.

Soil science had become a global science with defined principles and practices, a burgeoning number of scientists, and some early signs of maturation. The Russian-American pedologist Constantin Nikiforoff summarized the situation in 1959 as: "Soil science is passing through a difficult stage in its development. After a promising start at the end of the last century and the beginning of the present one, it has been plagued by formalism and has lost much of the original impetus. The aftermath of the two world wars and widespread social upheavals put severe stresses on world agriculture, and soil science, which is still oriented toward agronomy, was called on to concentrate

© The Author(s), under exclusive license to Springer Nature Switzerland AG 2021
A. E. Hartemink, *Soil Science Americana*
https://doi.org/10.1007/978-3-030-71135-1_13

on purely practical problems dealing with the betterment of crops. Research in basic science was forced to surrender priority to sheer technology and soon found itself in the strait jacket of bureaucratic supervision, and its methods of scientific inquiry were largely replaced by sadly unimaginative empiricism" [1]. It is against major scientific advances, a degree of soil science maturity, and the background of international tension that the *Seventh International Congress of Soil Science* was held in Madison in 1960.

In November 1956, it had been decided that the seventh congress would be held at the University of Wisconsin in Madison. A committee was established in 1957 to plan for the tours, programs, publications, and finances. Richard Bradfield became the committee chair; Charles Kellogg vice chair, and Emil Truog was appointed as the finance chair and congress manager. Financing the congress was a concern, and Charles Kellogg knew that not much support would come from the federal government. Emil Truog was familiar with the financial struggles of the congress in 1927, for which Jacob Lipman eventually raised $75,000 from private donors and industry. Emil Truog thought that what could be done in 1927 could be done in 1957. He had a large network in the industry, academia, and the public sector, and asked donations and support from colleagues, industry partners, and former students; in 1960, and the Department of Soil Science at the University of Wisconsin had graduated almost 900 soil science students.

Emil Truog, seeking funds for the congress, wrote to the National Science Foundation (NSF) a two-page document entitled *Soil Science is a Distinct Science in its own Right*. He stressed the importance of soil science and its contributions to other disciplines: "…much of the information the geologist now uses regarding the weathering of primary minerals, and formation of secondary minerals was provided by soil scientists. The same holds true in the case of chemistry, particularly analytical chemistry, in physics, particularly movement of water, heat, and air in porous materials, and in biology, particularly nitrogen fixation and nutrition of plants." He concluded: "Soil science has well-earned its position among the sciences and deserves full recognition in the National Science Foundation and other agencies or organizations supporting or sponsoring scientific endeavors." He requested $5,000 ($43,000 in 2021) from NSF and sent Charles Kellogg on: "…a reconnaissance mission to the NSF director." Emil Truog was keen to get the funds, as he wrote to Richard Bradfield in May 1960: "…this is the strategic moment to get soil science recognized in NSF for the all-time future regardless of the soil congresses. I feel we have them over the barrel right now as may never occur again." Indeed, he had them over the barrel and NSF provide $5,000 for the congress. He also received a donation from the Rockefeller family and raised $175,000, which is equivalent to about $1.5 million in 2021.

Emil Truog arranged for faculty and staff of the Department of Soil Science in Madison to be involved in the organization of the congress. He had been at the department since 1912 and was its Head for 15 years. Faculty, staff, and some 50 soil science graduate students helped with field trips, planning of the congress, and during the congress dinner. They did not see any of the presentations and were the worker bees of the congress [2]. Early in 1959, Emil Truog asked the President of the *Soil Science Society of America*, Guy Smith, to investigate issuing commemorative postage stamps for the congress, since the *Soil Conservation Service* had such stamps printed. Guy Smith responded that it was too late, too expensive, and suggested using the soil conservation stamp of 1959, which emphasized the importance of contour cropping. The stamp was issued in panes of fifty each, and 120 million stamps were printed.

In April 1960, Emil Truog suggested to the organizing committee to invite President Dwight Eisenhower as the opening speaker. President Eisenhower was well-respected across Europe having led the allied invasion of Normandy in June 1944. Emil Truog liked his Republican viewpoint. Charles Kellogg contacted the President's brother, Milton Eisenhower, who suggested writing to the President directly, so Emil Truog wrote the President a letter: "Because of your often expressed concern about adequate supplies of food, and the health and welfare of peoples in the underdeveloped countries, I believe you will especially be interested in an important event that will take place this fall. It is the 7th International Soil Science Congress which will be held on the Campus of the University of Wisconsin." Emil Truog mentioned that there would be 50 attendees from Russia and that: "…40 papers will be presented by the Soviets." Lastly, he wrote that the 1927 soil congress was opened by President Calvin Coolidge and: "…that it would be most appropriate if you could find time to address the congress." Emil Truog estimated that some 15,000 people would attend a speech by the President and that a large radio and TV audience could be expected. What worked in 1927 did not work out in 1960; President Eisenhower declined to come to Madison.

The *Seventh World Congress of Soil Science* started on a Monday and lasted from the 15th to the 23rd of August 1960. It was a warm and humid summer in Madison. Before and during the congress, delegates strolled across the Ho-Chunk land and the drumlins on which the university campus was built and back to the Memorial Union Terrace to enjoy the lake view underneath old bur oak trees. Small sailboats flocked over Lake Mendota, but there was little wind and the air shimmered across the lake. The Memorial Union had been opened thirty years earlier, designed by a university architect who drew

The theme of the *Seventh International Congress of Soil Science* in Madison in 1960, and Emil Truog's depiction of the wealth of Wisconsin soils in relation to the world's gold; a clear imbalance

inspiration from the palaces of Venice and Padua. The building was made of limestone from Indiana, with panels of sandstone from Madison and steps of Winona travertine from Minnesota. The Wisconsin architect Frank Lloyd Wright summed it up as: "Yes it speaks Italian, extremely bad Italian, and very difficult to understand" [3].

The congress was opened in the adjacent union theater which had been built in 1938; all 1,300 seats were occupied. The morning was filled with welcoming speeches from the Governor's representative, the University of Wisconsin President, and the Secretary General of the *International Society of Soil Science*. The three foremost presentations were on the congress theme that had been coined by Emil Truog: *To Promote Peace and Health by Alleviating Hunger through Soil Science*. The opening presentation was given by the Director General of the Food and Agricultural Organization of the United Nations (FAO). The FAO had increasing influence in the world since its establishment in Quebec in 1945; its Land and Water Division had about 180 experts working in the field program as technical assistants, and a handful of them were soil surveyors [4]. With increased decolonization, FAO and other UN agencies oversaw, and occasionally overtook, developmental issues of newly independent governments [5].

The Director General spoke about the Freedom from Hunger Campaign which had been launched six weeks earlier, in Rome. He stressed that where there is poverty, there is hunger. The situation in Europe 200 years earlier was similar to the current situation in some regions of the world but, now, Europe enjoyed a level of socio-economic growth promising a full life for all. What has been possible for Europe should be possible for the rest of the world. The two crucial factors in soil management were increased use of chemical fertilizers and conservation of water use; newly cropped soils in undeveloped countries were perceived to be low in nutrients and needing lime. With his concluding point, the FAO Director General stressed how hunger remained a bigger threat to humankind than nuclear weapons and, though he was not a soil scientist, he invited all soil scientists to contribute to the Freedom From Hunger Campaign.

The next presentation was by the President of the society, Richard Bradfield, who had attended every international soil congress since 1927. He spoke about the opportunities in freeing the world from hunger and could think of no single group of scientists who have more to contribute to feeding the world than soil scientists. In his view, soil scientists had more experience and, in general, more success in increasing food production than population experts had in population control [6]. As a professor of soil science, Richard Bradfield was concerned about the training of the next generation of soil scientists; in 1960, there were several hundred students in soil science enrolled in undergraduate degrees across the USA [7]. Richard Bradfield, who taught at Cornell University, summarized his concerns: "Are we in soil science, for example, teaching our students to see the whole patient or are we turning

Left photo: Emil Truog (with hat), Vladimir Ignatieff, and Ivan Tiurin who headed the Russian delegation to the *Seventh International Congress* of *Soil Science* at Truax Municipal Airport in Madison in 1960. Ivan Tiurin was the Director of the Dokuchaev Institute, President of the Russian Soil Science Society, and editor of *Pochvovedenie*. He suddenly died two years after the congress. Right photo: Jadwiga Ziemięcka from Poland and Alfred Aslander from Sweden had both attended the first congress in 1927. There were only a few who attended the first and seventh congress: Ivan Tiurin, Jacob Joffe, Emil Truog and Richard Bradfield. Jadwiga Ziemięcka and Alfred Aslander had also attended *Sixth International Soil Science Congress* in Paris in 1956, and Alfred Aslander attended the congress in Amsterdam in 1950. Photos from the Department of Soil Science UW-Madison, and *International Society of Soil Science*

out men who can see only his eyes, his heart, or his liver? I fear we are graduating many PhDs in Soil Science who have acquired a great deal of specialized knowledge of some phase of the subject but who have had but little contact with the whole patient, the soil as it exists in its natural setting, the dynamic, living system in which crops grow. Are we turning out men who are qualified and willing to forsake the ease of their offices and laboratories when necessary and go out to the fields, the only place where many of these problems can be solved?" Richard Bradfield was trained as a soil chemist, but also worked on soil structure, soil fertility, and international agriculture. The lack of a broader view and the inevitable specialization and fragmentation in soil science worried him.

Charles Kellogg was the next speaker. He had just recovered from kidney stones and was glad to be back in Madison where had conducted his PhD research in the late 1920s. He spoke about the productivity of arable soils in the USA and how agricultural production has increased since the *First*

International Soil Congress. In 1927, the congress emphasized the hazards of soil exhaustion through erosion, loss of organic matter, and plant nutrients. At that time, the prevailing idea was that there should be a natural balance between humans, somewhat similar to other emerging concepts in ecology. Charles Kellogg thought that 'natural balance' was a vague concept and discounted what could be achieved by science. Between 1927 and 1959, science and technology had developed a concept of soil and water management termed 'soil conservation' [8]. It stood for the effective use and conservation of each acre of soil so that it attained an optimum level of productivity. Charts of increasing crop productivity reflected the impact of science and, based on those trends, Charles Kellogg was optimistic in the ability of soils to feed the world and eliminate hunger. Between the congresses of 1927 and 1960, the global population had increased from two to three billion people and, as the population continued to grow, there was a need for rational food production and the widespread adoption of soil conservation measures.

As the last speaker of the opening session, Vladimir Ignatieff of FAO presented the revised and enlarged edition of the *Multilingual vocabulary of soil science*, and the first edition had been presented in 1954 [9, 10]. Since its inception, a rich soil terminology had been developed in several languages, and dissemination in soil science was held back by language barriers. This vocabulary was an important contribution toward harmonized terminology and consistent use of a soil science lexicon. The collection was started by Herbert Greene in 1949, and the vocabulary was finalized by Graham Jacks. The vocabulary languages were English, French, German, Spanish, Portuguese, Italian, Dutch, and Swedish and, for the 1960 edition, Russian vocabulary was added. Like the 1954 edition, the title of the book was in English, French, and Spanish, but not in German. The first copy was presented to the *International Society of Soil Science* President Richard Bradfield. With the presentation of the vocabulary, the opening session of the seventh congress ended. Delegates poured out of the union theater for a cigarette, lunch, and, perhaps, a beer.

The seventh congress was attended by 1,260 people, of which 467 delegates were from overseas. It was the largest gathering of soil scientists to date and included almost 200 graduate students. Still, the number of participants was disappointing to Emil Truog. In April 1960, he had anticipated 1,500 to 2,000 participants, with 500 attendees from overseas. The *Soil Science Society of America* had about 1,600 members in 1960, and participation in the congress was below expectation by the congress organizers. The congress was attended by 71 participants from Canada, 44 from the UK, 30 from

Russia, 29 from Germany, 22 from the Netherlands, 21 from Belgium, and 18 from India whose travel were funded through the Ford Foundation at the instigation of Charles Kellogg and Richard Bradfield. The international delegates were housed in the newly built Chadbourne Hall, an eleven-story student dormitory.

The *International Society of Soil Science* had about 3,300 members in June 1960, compared to 924 members in 1927, when half of the members attended the first congress. In 1960, the *International Society of Soil Science* had 1,022 members from the USA, 218 from New Zealand, 212 from the UK, 155 from Belgium, and 118 from Germany. Since the international soil congress in Paris in 1956, eight national soil science societies had been formed: Argentina, Bulgaria, Denmark, Hungary, Ireland, Peru, Sweden, and U.A.R. (Egypt).

There were a few participants from East Germany, which had been under communist rule since 1949. More had applied and assumed that the USA would give visas to all member countries of the *International Society of Soil Science*. But they did not get visas. Charles Kellogg, as a government official in the Department of Agriculture, was criticized by members from Hungary and Russia for the lack of visas given to soil scientists from East Germany. A week before the congress, Charles Kellogg wrote to Emil Truog: "...the United States will not place a United States visa in an East German passport. The Government of the United States does not recognize East Germany as a sovereign state." He was told by the Department of State that the number of visas was greatly reduced: "...because they have found that some East Germans have abused these privileges, violated our hospitality, and engaged in propaganda." Emil Truog, who had little sympathy for communism, was unsurprised by that finding. Charles Kellogg, however, was accused of playing politics and refusing participation of East German soil scientists in the seventh congress. This was not the only political issue. Before the congress, Emil Truog wrote to Richard Bradfield and Charles Kellogg complaining about the payments for the tours and congress by the Russian delegation whom he called: "our good friends the Soviets." Checks arrived with no details and Emil Truog complained: "...no this is not an isolated case. Just recently got checks totaling $950 with little or no information. Do you think they could set up an effective organization for a war? I am somewhat inclined to think Nikita is whistling in the dark. Of course, maybe only their Soil Scientists are dumb." Richard Bradfield and Charles Kellogg, who had been to Russia several times, held their tongues.

The Russian delegation was headed by Ivan Tiurin and Victor Kovda, who had been Director of the Department for Natural Sciences at the United

Stamp and envelopes produced for the *Seventh International Congress of Soil Science* in 1960. Subsequent international soil congresses had special stamps printed like Romania in 1964 with Gheorghe Murgoci; Australia 1968, Russia in 1974 (not shown), and India in 1982. There is a book on stamps in soil science by the German pedologist Hans-Peter Blume, entitled *Ein philatelistischer Streifzug durch die Bodenkunde*, published in 2018. It includes over 2,000 stamps from about 100 countries

Nations Educational, Scientific and Cultural Organization (UNESCO) since 1958 [11]. Victor Kovda was in good standing with the Communist Party [12]. His work was influenced by the geochemist Vladimir Vernadskii. As Charles Kellogg discovered during the congress, Victor Kovda was a man with a temper and not afraid of making enemies [11–13] Russian pedologists had been treated as celebrities at the congress in 1927, but the ambiance in 1960 was different. The general feeling was that the pedological and soil chemical leadership that Russian scientists had provided in the first decades of the twentieth century was now outshone by American work and confidence.

Except for Victor Kovda, very few of the Russian attendees spoke English. A woman police officer was in charge of the Russian group who spoke English, paid the bills, and each Russian participant was given two dollars, so that they could buy lunch but not travel anywhere. Several of them talked to Constantin Nikiforoff, who had emigrated from Russia to the USA in 1921 and asked him about political asylum or whether there were any jobs available. Some of the American attendees wondered if there were spies among the Russian delegation [2].

At the congresses of the *International of Soil Science*, a tradition had formed whereby distinguished soil scientists who had died since the last congress were memorialized. Just before the 1960 congress, the tropical soil scientist Robert Pendleton, the pedologist Dmitrii Vilenskii, the soil conservationist Hugh Bennett, and the soil physicist Robert Schofield had passed away. They had all made contributions to soil science in the first half of the twentieth century. It was also noted that there were six attendees in 1960 who had attended the first congress in 1927: Ivan Tiurin from Russia; Jadwiga Ziemięcka from Poland; Alfred Aslander from Sweden; and Jacob Joffe, Emil Truog, and Richard Bradfield from the USA. Ivan Tiurin studied soils under forest and soil organic matter and he developed a carbon oxidation method that became known in Russia as the 'Tiurin method.' He had described a zonal pattern of organic matter composition in soils and, in 1960, he was the director of the Dokuchaev Institute and Editor-in-Chief of *Pochvovedenie*. Jadwiga Ziemięcka had founded the first laboratory of soil microbiology in Poland [14]. Alfred Aslander had been a student at Cornell University and studied the effectiveness of sulfuric acid and solutions of iron sulfate as herbicides [15].

Charles Kellogg noted the absence of his friends from the Congo due to turmoil following its independence from Belgium in June 1960. Sir John Russell was also absent; according to Charles Kellogg, he was: "…too old and ill to come, or too busy writing." Indeed, a year after the congress, he published a book with Norman Wright entitled *World hunger: can it be averted?* [16] followed by the third edition of *The World of Soil* [17] and *A history of agricultural science in Great Britain* [18], which was published after his death in 1965.

A congress dinner was held in the Edgewater Hotel in Madison, and Charles and Lucille Kellogg were joined by Richard Bradfield, Emil and Lucy Truog and Cees Edelman, René Tavernier, José Fripiat, Norman Taylor, Ivan Tiurin, Charles Stephens, Georges Aubert, Boris Johannesson, and three FAO people: Vladimir Ignatieff, Rudi Dudal, and Luis Bramão. It was an international group of soil scientists, several of whom were leaders in their own country. According to Charles Kellogg it was a pleasant dinner, and everyone seemed to have a good time. The next evening, the Russian delegation had a party in the same room and used the table flowers again. The official congress

Some of the American and international leaders of the *Seventh International Congress of Soil Science* posing at the Memorial Union terrace in Madison, with a sign that had the name of the society wrong. From left to right: Cees Edelman from the Netherlands, Ivan Tiurin from Russia, the President of the *Soil Science Society of America* James Fitts, the Secretary General of the *International Society of Soil Science* Ferdinand van Baren from the Netherlands, Charles Kellogg, Emil Truog from the USA, and René Tavernier from Belgium. Seemingly, they had a jolly good time. Photo *International Society of Soil Science*

conference dinner was held in the Shell built in 1954 next to Camp Randall. Some 1,600 people attended; they watched a dance and pageantry by the Menominee tribe, Swiss Yodelers, and precision maneuvering with music. Sweet corn, brats, and beer were served. For many, it was an unforgettable Wisconsin evening and meal.

During the congress, some 400 presentations were given, grouped into the six commissions of the *International Society of Soil Science*. All were in English. Before the congress and on orders from the French government, Georges Aubert had made a formal protest to the organizers about the lack of simultaneous translations into French and German. The organizing committee had considered it and found out that the costs for translators would have been over $100,000 and decided to use that money to provide travel grants to soil scientists overseas who otherwise would not have been able to attend. So, the congress was all in English, breaking with earlier congress tradition.

Science had come out of the Second World War with high status, and universities and research centers had grown rapidly [19, 20]. Soil science benefited from this, and soil knowledge was seen as important to feed the growing population. There were major advances in the 1950s, such as improved theory and instrumentation following investments in soil research, as well a growing number of soil scientists. Radioactive and heavy isotopes had become available and these were accompanied by the development of instrumentation like flame emission and atomic absorption spectrometers, mass spectrographs, X-ray diffractometers and fluorescence, colorimeters, spectrophotometers, column and gas chromatographs, and the first computers [21]. Other developments that aided soil research were advances in statistical theory and designs of field experiments, theories on ion transport from the solid phase to the root surface, and the increased understanding of soil chemical and biological properties and processes [22]. New instruments and technology allowed for improved soil and plant tissue testing so that better fertilizer recommendations could be made.

In the early 1900s, Fritz Haber and Carl Bosch had discovered how to produce ammonia from atmospheric nitrogen [23]. The Haber–Bosch process made nitrogen fertilizers more widely available, and fertilizer use was much increased after the Second World War. At the same time, nitrification inhibitors, new nitrogen compounds, coated fertilizers and synthetic chelates were developed, all of which increased crop production [21, 24]. Inequalities in food supplies had long been accepted as part of the natural order of things, and during and immediately after the war, Russia and some European countries had suffered millions of deaths through the famine [25]. Now, there was unity across the globe that famine and starvation should be eradicated [26].

In the 1950s, a significant part of the increase in crop and farm productivity was attributed to nitrogen fertilizers, improvements in soil management, and the application of soil science knowledge [27]. While this had benefitted consumers, with increased yield leading to relatively cheap and abundant food, it did not benefit farmers [20]. In the early 1960s, 45% of the farm families in the USA were poor, defined as families earning less than $3,000 per year, compared with 18% of non-farm families who were poor [20]. The farm population formed less than 10% of the total population in 1960 and had been halved between 1932 and 1960, from 30 to 15 million [20]. In 1960, the USA had about four million farms, and most of these were highly mechanized, and increasing amounts of fertilizers and biocides were used.

The environment and protecting water, air, and soil from excess nutrients and contaminants were not discussed at the seventh congress in 1960. These issues had received some research attention and there were worries about the quality of food [28], but the dominant view in 1960 was that more food needed to be produced. A few years after the soil congress, environmental awareness grew and the limitations of some agricultural practices and the earth's finite soil resources were slowly recognized [29, 30]. Victor Kovda played an important role in international meetings related to the environment and its protection [31]. He recognized the significance of soil degradation and was less confident in scientific and technical solutions for solving the emerging agricultural and environmental difficulties.

The congress manager Emil Truog, at the age of 76, presented an overview of 50 years of soil testing. He recounted how he was raised on a farm in Wisconsin and that virgin land was being cleared of trees in the 1880s. His father had told him that, on new land, crops were always good but that yields declined after continuous cropping. These observations sparked his interest in soil fertility, plant nutrients, and the need for soil testing. In high school, he read agricultural bulletins and the works of George Washington and Edmund Ruffin. At the University of Wisconsin, he studied soils and chemistry, and after he was hired in the Department of Soils in 1912, the state legislature made an appropriation to finance a soil testing program. A state laboratory was erected, and Emil Truog was in charge of developing soil tests and fertilizer recommendations for oats, barley, and alfalfa. In 1915, he developed a litmus paper test for soil acidity and subsequently developed pH and nutrient deficiency testing kits [32]. Soil testing and improved fertilizer recommendations were successful and, by the 1950s, Wisconsin was leading the nation in alfalfa production. Improved fertilizer use, planting of corn hybrids, and weed control resulted in corn yields of 100 bushels per acre, compared to 45 bushels per acre in the absence of soil testing. Although Wisconsin was not considered to be in the corn belt, the state had some of the highest corn yields in the country.

The presentation at the seventh congress was Emil Truog's last paper, after which he left the University of Wisconsin which had been his professional home for almost 50 years. He helped in making the congress proceedings publishable and the 2,500-page proceedings came in four volumes ($25 for society members) and were mostly in English. A few papers had summaries in other languages, and a handful of papers were in German, with one or two in French. In total, 1,500 copies of the hardbound proceedings were printed. One of Emil Truog's last activities was the selection of a site for the new headquarters of the *Soil Science Society of America*.

There was one tour before and two tours after the congress. The pre-congress tour from New York to Madison lasted ten days, and the 3,000-km tour passed through the states of New York, Pennsylvania, Ohio, Indiana, Illinois, and Wisconsin. It had 191 participants from 45 countries and costs $110 per person. A tour guidebook was prepared by the staff of the soil survey, and Charles Kellogg's office had paid for the printing. Marlin Cline from Cornell University was in charge of the tour and its financial administration. In his role, he discovered that one of the other administrators was untrustworthy, leading to concerns about the collected funds. The Head of

The official conference dinner of the *Seventh Congress of Soil Science* in the Shell next to Camp Randall in Madison, Wisconsin. Some 1,600 people attended the buffet dinner that largely consisted of sweet corn, brats and beer. They watched a dance and pageantry by the Menominee tribe dressed in full regalia, a performance by Swiss Yodelers and Flag Throwers of New Glarus, Wisconsin, and precision maneuvering with music by the famed Four Lakes Bugle Corps. Photo *International Society of Soil Science*

the Department of Agronomy at Cornell University, Nyle Brady, visited the start of the tour to investigate the matter. There was a search for the administrator and, when they finally found him, he was dead. He had killed himself and the money was gone.

Charles Kellogg wanted someone on the tour who could speak on behalf of the soil survey, and Roy Simonson was sent. He joined the French-speaking bus and his explanation on soils, landscapes, and history was translated by Georges Aubert and soil scientists from Quebec. Between New York and Madison, they made stops in every state, and local soil experts joined the group. At some stops, two or three of the participants were late and left behind but, after that, everyone was on the bus when Roy Simonson blew the whistle. Roy Simonson explained to the participants why he and Marlin Cline were so skinny; unsurprisingly he had a pedological enlightenment. Marlin Cline worked in New York state where rather poor soils such as Podzols and *Sols Brun Acides* were common whereas in Maryland, where Roy Simonson lived, the soils were even poorer and classified as Red-Yellow Podzolic [33]. The connection between the inherent fertility of soils and the health and prosperity of humankind was noticeable, even in the USA where in 1960, less than 10% of the people lived on farms. With increased mobility, wealth, urbanization, and a very small proportion of the population living directly from the land, that connection faded.

Roy Simonson gave the tour participants a free copy of the 7th *Approximation* soil classification system, much to the chagrin of Victor Kovda, who was the only Russian on the tour. Roy Simonson wrote in his diary: "I got the feeling he believed the 'damned Americans' had stolen a March on the Russians by getting out a system of soil classification and passing out a monograph to every soil scientist attending the Congress. Apparently, the monograph came as a complete surprise to all Russians, including Kovda." Victor Kovda also challenged Roy's interpretations of some soil profiles in loess; Victor Kovda was familiar with loess and he had seen the giant loess profile with fossils at Chongar in Ukraine after the second congress in 1930. Roy Simonson was four years younger than Victor Kovda; they both had studied saline soils, Victor Kovda in Uzbekistan, Roy Simonson in North Dakota. They did not get along. The last stop of the tour was in Illinois, and the police escorted the buses to Wisconsin at full speed and with the sirens on. In Wisconsin, Francis Hole, with degrees in both geology and French, came on the bus and spoke in French about the geology and soils of southern Wisconsin that was only partly covered with ice during the last glaciation. The participants were tired from the travel but they perked up at having an American speak to them in French.

In the middle of the congress week, there were tours through the state of Wisconsin. The tours had an agrarian focus and visited research stations and some large farms. Many participants had grown up on farms and appreciated the herds of Holstein (Friesian) cattle and farm machinery. In 1960, the average farm size was 60 ha, and the principal crops were oats, corn, alfalfa hay, clover hay, canning peas and corn, barley wheat, potatoes, and soybeans for grain. Wisconsin had over four million cows, almost two million pigs, and four million people.

There were two tours after the congress: one to the southeast from Madison to Washington, and one to the west, ending in Berkeley, California. The tour from Madison to Washington had 69 participants from 26 countries. They

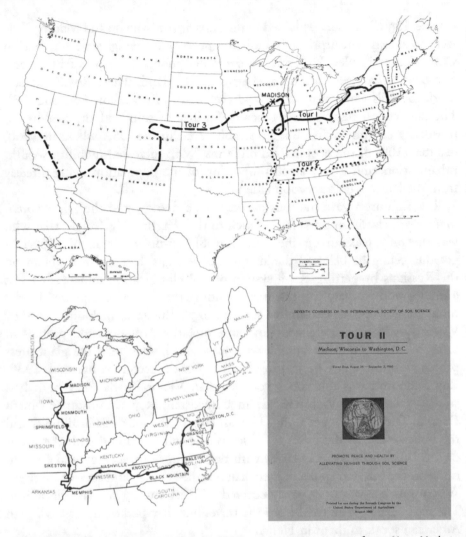

There were three congress tours in 1960; a pre-congress tour from New York to Madison (tour 1) and two post congress tours: from Madison to Washington DC, and from Madison to Berkeley. The post congress tour from Madison to Washington, traversed the mid-western loess plains to the Mississippi delta, and across the half-billion-year-old Appalachians to the District of Columbia

traveled in two buses and were on the road for nine days. The route was from Madison south to Memphis, Tennessee, from there it went east to Raleigh, North Carolina, and then north to Washington where the tour ended. It was led by Eric Winters who had studied soils at the University of Wisconsin at the same time as Roy Simonson. In each state, local soil scientists joined the tour and showed soil profiles. Francis Hole was in charge of preparing information for a segment of the Madison–Washington tour. As there was segregation in the southern USA and black people had to sleep in different hotels, Francis Hole, a Quaker, refused to cross the Mason–Dixon line and did not participate in the southern part of the tour.

The tour from Madison to Berkeley was over-subscribed and, according to Emil Truog, it was: "...especially by people of far-way places. Seems they all want to see our Golden West." The tour had 144 participants from 30 countries, including Victor Kovda and Ferdinand van Baren. Andy Aandahl, who had been a student of Charles Kellogg in North Dakota in the early 1930s, led the tour. The tour went through Iowa, Nebraska, Colorado, New Mexico, Arizona, and ended in California. The tour visited Disneyland and Joshua Tree National Park in the Mojave desert. The buses were equipped with a toilet, air conditioning, and a loudspeaker system. The tour book stressed that it was forbidden to take photographs of military objects, fuel storage depots, seaports and tunnels, radio and television facilities, and from airplanes. It was a long tour, and all went well except for a participant who ate some mushrooms in Southern California. It would have killed him if not for emergency hospital treatment.

The *Seventh International Congress of Soil Science* was the first soil congress for the pedologist Dan Yaalon, who much enjoyed the pre- and post-congress field trips because: "...the more soil profiles you examine and discuss with others, the better you understand soil formation and distribution" [12]. He had just finished a post-doctoral position at Rothamsted and was appointed as a lecturer at the Hebrew University of Jerusalem. He had received $500 from the organizers to attend the seventh congress [34]. At the congress, he presented a paper on fundamental concepts of pedology in soil classification [35]. For him, the decision to produce a world soil map and the introduction of the *7th Approximation* were memorable parts of the congress.

Soil mapping and classification were important topics at the congress. To be precise, it was not all about soil mapping or classification, but about who would lead the creation of the world soil map, and what soil classification system would be used across the world. Soil science had reached a stage where some of the debates shifted from novelty of content to practicality and plain geopolitics. The lack of debate about scientific development in the soil survey was described by Constantin Nikiforoff just before the 1960 congress: "...a mere scratching of old cliffs and road cuts for the study of soil anatomy or collecting of samples for chemical analysis served the purpose in the pioneering stage of soil science. It still satisfies the requirement of a routine soil survey, but it is no longer adequate in scientific investigation. Chemical and physical analyses of soil are adequate insofar as the procedure

Congress tours after the *Seventh International Congress of Soil Science* in 1960—by bus (not train), and with a boxed lunch on strawbales at the Aurora Research farm of Cornell University. Photos *International Society of Soil Science*

is concerned, but methods for interpretation of the results and especially for coordination of various analyses and selection of proper material for laboratory study are in urgent need of radical improvement. Careful study of soil morphology is an essential prerequisite for the identification of the soil and the collection of samples for analysis. Morphology as such, however, is just an empty shell if its origins are unknown. Hence, the need for a shift of attention in soil study from a descriptive morphology to study of the genesis and dynamics of soils is clearly indicated. Such an orientation of soil science was foreseen by the pioneers (Dokuchaev, Shaler, and others) but later on was overshadowed by practical considerations" [1]. Constantin Nikiforoff's urge for a more detailed study of soil formation and the quantification of soil dynamics passed as a ship in the night.

Left picture: Participants studying a fragipan (Orthic Fragaquepts) on the tour from New York to Madison. Person on the right is Koos van Staveren, who was Director of the International Institute for Land Reclamation and Improvement (ILRI) in Wageningen, The Netherlands. Right picture: Andy Aandahl receiving a gift from Ferdinand van Baren after the completion of the Madison-Berkeley tour where he was the excursion leader. Photos *International Society of Soil Science*, and Pieter Buringh

The decision to create a soil map of the world was made at the *Seventh International Congress of Soil Science*. The idea was not new and continental maps were available for the USA, Europe, Africa, South America, and parts of Asia. The first soil map of the world, at a scale of 1 to 80 million, had been made by Konstantin Glinka in the early 1900s and showed that the soil pattern followed global climatic zones [36]. The map legend included 18 units such as Podzolic soils, Chernozem, Chestnut soils, and Krasnozem (*terra rossa*). The first soil map covering the USA was published by Milton Whitney in 1909 and depicted soil provinces that were largely based on the geology of an area [37]. In 1913, George Coffey published the first country-wide map based on soil properties, and in that same year, Curtis Marbut published a map with 13 physiographic units that were subdivided into soil series, soil classes, and soil types [38–40]. In 1935, Curtis Marbut made a Great Soil Group map at a scale of 1 to 8 million that introduced the Pedocals and Pedalfers and Russian soil names such as Chernozems and Podzols. Curtis Marbut also made the first soil map of Africa in 1923 [41]. The first European soil map was published in 1928 at a scale of 1 to 10 million and had 27 map units and was based on the geological map. A 1 to 2.5 million soil map of Europe was published in 1937 and 1956 [42, 43]. At the *Sixth International Congress of Soil Science* in Paris in 1956, there was interest for these soil maps [44], and at the seventh congress in Madison, soil maps were exhibited from 29 countries.

The grand idea behind a soil map of the entire world was that it would bring all information together and enhance standardization in methods and terminology. It would also reveal where the need for soil surveys was largest, and the inventory of global soil resources could be used for a wide range of applications. Noble as that idea was, the decision to make a new soil map of the world became political and territorial. Charles Kellogg and his World Soil Geography Unit had some of the best soil information and maps, but the Americans were tactically surpassed.

Two months before the congress, Charles Kellogg had written to the Ford Foundation, Rockefeller Foundation, and the FAO about funding the publication of a world soil map to be made by his office in Washington. He had launched the idea of the world soil map in 1943 when he wrote the *Proposal for the Department of Soil Science and Management of a World Organization on Food and Agriculture* [45]. The Ford and Rockefeller Foundations pledged financial support, but Charles Kellogg did not receive a response from the FAO. A few weeks prior to the congress, Charles Kellogg met with Luis Bramão, the head of the FAO soil survey who pretended that FAO, and in particular their soil scientist Vladimir Ignatieff, had no interest in a soil map of the world. This did not go down well with Charles Kellogg, who knew Vladimir Ignatieff from the FAO foundational meeting in 1945 and he had worked with him on the study *Efficient Use of Fertilizers* [46]. "What a damn fool he was to talk to me like that about the most dedicated and highly motivated soil scientist in the world," wrote Charles Kellogg about Luis Bramão in his diary. Certainly, Charles Kellogg was disappointed that the FAO was not interested in a world soil map project.

At the seventh congress, continental and regional soil maps were presented from all parts of the world: [47]

Soil Map of South America, presented by Luis Bramão and Petezval Lemos of FAO.

The Soil Map of Africa South of the Sahara by Jules D'Hoore of the CCTA (Commission for Technical Cooperation in Africa South of the Sahara).

The Australian Soil Landscape by Charles Stephens of CSIRO (Commonwealth Scientific and Industrial Research Organization).

La Carte Des Sols de L'Asie by E.V. Loboba and Victor Kovda of the Soils Institute of the Soviet Academy of Sciences.

International Soil Map of Eastern Part of Europe by Ivan Tiurin, N.N. Rozov, and E.N. Rudneva of the Dokuchaev Soil Institute.

Classification of Soils and the Soil Map of the USSR by E.N. Ivanova and N.N. Rozov

The Soil Map of Western Europe on Scale of 1:2.5 Million by René Tavernier of Centre for Soil Survey, Ghent and Eduard Mückenhausen of Institute for Soil Science, Bonn.

Half the soil maps presented were made in Russia, and Russian pedologists had been mapping soils in Asia, particular in India and China, since the 1920s [5, 48]. Various soil survey organizations were interested in copies of the maps but the FAO had no funds to reproduce them. Victor Kovda, Charles Kellogg, and Vladimir Ignatieff held a meeting, and Victor Kovda offered to reproduce the maps. The three drafted a resolution for the *International Society of Soil Science* council in which it was stated that their organizations would reproduce all soil maps that were exhibited and, importantly, any future soil maps. They would work together on a new *Soil Map of the World*.

The council recommended that an attempt be made to harmonize and synthesize the knowledge acquired on the soils of the world [36, 49]. It was widely discussed and favorably received by the council members. Victor Kovda of UNESCO and Luis Bramão of FAO had, however, convinced the Secretary General, Ferdinand van Baren, to make the resolution as an endorsement of the *Soil Map of the World* project by two organizations: FAO and UNESCO. That resolution was approved by the council. The original resolution had been twisted: Charles Kellogg and the Americans were excluded from the effort to produce the *Soil Map of the World*. It was a strategic decision propelled by Victor Kovda; Russia had been a UNESCO member since 1954 but had not joined the FAO. Charles Kellogg was furious and wrote in his diary: "I made what turned out a colossal error," and called the secret deal to produce the *Soil Map of the World* by FAO and UNESCO a fraud.

In 1961, the *Soil Map of the World* became an official joint project between two UN organizations: the Food and Agricultural Organization (FAO) and the United Nations Educational, Scientific and Cultural Organization (UNESCO). The FAO had some soil surveyors on its staff and the organization was fiercely territorial [5], but soil survey and mapping were alien to UNESCO. The *Soil Map of the World* would be coordinated by a World Soil Resources Office established at FAO in Rome [5]. An advisory panel met in 1961 in Rome; it included Georges Aubert, Jules D'Hoore, Guy Smith, Charles Stephens, René Tavernier, Ivan Tiurin, and Ferdinand van Baren, and was chaired by Luis Bramão with help from Rudi Dudal and Victor Kovda. They discussed the international cooperation that was needed to publish the maps. It required coordination of national or regional efforts and activities, and it was unclear how all soil maps and information could

Victor Kovda and Ferdinand van Baren enjoying an all-American liquid in Colorado at the post-congress tour from Madison to Berkeley in 1960. They were good friends. Victor Kovda was a Russian pedologist and organizer of the Unesco biosphere conference. As he used to say: "My father was from North Caucasus, so I have a big temper, my mother was Ukrainian, so I have two big tempers." Ferdinand van Baren, son of Jan van Baren, was Secretary General of the *International Society of Soil Science* from 1950 to 1974, co-author of *Tropical Soils* with Jules Mohr, and founder of ISRIC and *Geoderma*. Photo *International Society of Soil Science*

be obtained from the FAO member nations, or from nations that were not members of the FAO. Other issues were the scientific basis of the map, the choice of mapping units, legend, and financing of the project. FAO had no budget for this undertaking, but Victor Kovda provided UNESCO funds to help FAO undertake the work. When that news reached Charles Kellogg, he was even more upset, as the USA had been instrumental in founding UNESCO in 1946 and had contributed to its budget since. And more than 25% of FAO's annual budget came from the USA.

The Portuguese soil surveyor, Luis Bramão, was appointed project coordinator and the pedologist Rudi Dudal was appointed soil correlator and technical secretary. Both had attended the seventh congress in Madison and were aware of the arrangements made between the FAO, UNESCO, and the *International Society of Soil Science*. Rudi Dudal had worked for four years in Indonesia as a technical assistant in soil survey and had produced, with M. Supraptohardjo and D.Z. Sahertian, a 1 to 1 million soil map of Java and Madura. Luis Bramão had studied differential thermal analysis of kaolinite and obtained his MS degree at Cornell University in 1938 under Richard Bradfield, and then worked in Portugal and for the Rockefeller Foundation

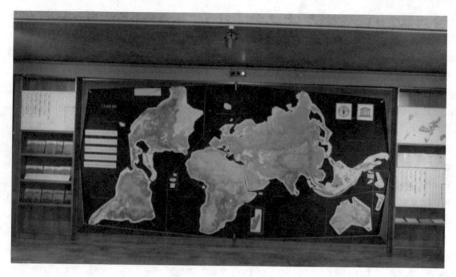

The 1 to 5 million *Soil Map of the World* as it was displayed for many years at the FAO headquarters in Rome, Italy. It included maps for all continents except Antarctica

on a soil map of Brazil [44]. He had joined the FAO office in Rome in 1955. Luis Bramão was a friend of Victor Kovda and an avid promotor of international cooperation [44].

The newly established FAO World Soil Resources Office was assisted by recent soil science graduates through a technical assistance program. In 1955, Luis Bramão had toured soil survey institutes in Europe and found interest in international work in Belgium and Netherlands, some of which came from decolonization and the need for expertise on soils in tropical and arid regions [5]. Rudi Dudal, Hans van Baren, Wim Sombroek, and Klaas Jan Beek were part of this technical assistance scheme and played important roles in the soil offices of FAO for extended periods. No Americans were hired in this program.

The advisory panel worked on the preparation of an international legend and the selection of the topographic base for the map [36]. The American Geographical Society map became the topographic base, and the map scale was set at 1 to 5 million. The following aims were set for the *Soil Map of The World*: the first appraisal of the world's soil resources, supplying a scientific base for the transfer of experience between areas with similar environments, and promoting the establishment of a generally accepted soil classification system and nomenclature. The panel hoped that the *Soil Map of The World* would establish a common framework for more detailed investigations in developing countries and that it would strengthen international soil science.

The map was seen as an important layer for planning agricultural development. The project was estimated to cost $176,000 and was to be completed by the eighth congress in 1964, but by 1968, the cost was over $600,000, and about half of the mapping had been completed [5].

The *Soil Map of The World* was made at a time when the world was submerged in a cold war, and political differences trickled down into the discussions and decisions. For example, there were disagreements about the use of Russian terminologies such as Solonetz, Podzols, Solonchaks, and the most iconic soils of all: the Chernozems. Correlation meetings were organized that included technical discussions and field tours. Compromises were being sought for the soil names of the Major Soil Groupings but in the end, all Russian terms were retained. Guy Smith represented the USA in most of these meetings; he was well-regarded by many of the European participants but the USA and Russia were not overly active in the project and were more focused on their own soil survey and mapping [5]. Both countries had more detailed soil maps than the proposed 1 to 5 million map and, probably, they did not support the global compromise that the FAO's World Soil Resources Office tried to establish. None of the correlation meetings was held in Africa, but that would change in the 1970s. By that time, however, the system was developed, and African soil scientists had little chance to bring in their expertise.

The legend of the *Soil Map of the World* became, in effect, a soil classification system in 1974, with 26 soil units at the highest level. The soil units were called Major Soil Groupings, reference soil groups, or soil groups, and were approximately equivalent to Great Soil Groups in the American soil classification system [39]. The FAO–UNESCO system was a mono-categorical map legend, not a taxonomic system, and FAO noted upfront that the system could be seen as a 'correlating medium' [5]. It had borrowed little from the hierarchical classification system that Guy Smith and Charles Kellogg had been working on. FAO assiduously promoted the classification as the international standard [5], and it became a parallel effort to the *7th Approximation* in developing a global system of soil classification.

A first draft of the *Soil Map of the World* was presented at the *Ninth International Congress of Soil Science* in Australia in 1968 but, after that, the project took another ten years to complete. The final maps were published on 18 sheets at a scale of 1 to 5 million. There were extensive areas where the soil distribution had to be inferred from other maps such as climatic, vegetation, or geological maps. Antarctica was not included. Some 600 soil maps were used as primary sources, and over 10,000 soil maps were collected by FAO's World Soil Resources office. The *Soil Map of the World* became an important activity by the international soil science community [50], but, according to Charles Kellogg, it was executed by an organization that did not have the capacity or the knowledge and it wasted hundreds of thousands of dollars. Nonetheless, his office cooperated with FAO in the making of the North American map. Guy Smith and Roy Simonson worked well with Luis Bramão and Rudi Dudal [51]. Many of the experts that worked on the map were more worried about users mistaking the map for reality than the actual soil map [5]. Its usefulness, as well as its accuracy, was questionable at a scale of 1 to 5 million but it became part of the global knowledge infrastructure on which the environmental modeling from the 1970s was built [5]. It did enhance global cooperation but, also, created division and disagreements—in particular the FAO–UNESCO soil classification system that came with it.

A major event at the seventh congress in Madison was the presentation of *Soil Classification—A Comprehensive System*, which became known as the *7th Approximation* [52]. It was freely distributed at the congress and on the tours, and presented with aplomb. The *7th Approximation* was a conceptual change to the earlier soil classification systems and relied on diagnostic horizons and measured properties. In 1950, after Charles Kellogg had finished the second edition of the *Soil Survey Manual*, he became convinced that a reexamination of the system of soil classification was needed [53]. Soil classification in the USA had been geologic in the beginning, then was based on soil physical properties, and factorial-genetic concepts dominated soil classification approaches during the 1920s to 1950s [54, 55]. Given the changes in soil classification, Charles Kellogg realized that a new system would not be received with great excitement: "…all the natural sciences have the same problem. A classification is good to the extent that it serves the purpose of remembering characteristics, seeing relationships, and developing principles. A classification is bad to the extent that scientists become slaves to it and twist their data and ideas to fit the classification. It improves as our knowledge grows. Some wonder when soil classification will 'settle down'—when names and definitions will no longer be changed. This will happen when soil science has ceased to discover anything new—in other words, when it dies" [56].

There were essentially two systems for classifying soils in the USA: one for the Great Soil Groups (Pedalfers with podzols and laterite soils; Pedocals with Chernozems and brown soils); one for the soil series, for example, Miami silt loam; the two systems did not blend. There was a problem with the Great Soil Group classification system which at the highest level was: zonal, azonal, and intrazonal soils—inherited from Konstantin Glinka and Nikolai Sibirtsev and adopted in the USA by Curtis Marbut and Mark Baldwin.

Zonal soils were based on climatic factors, whereas azonal soils were not well-developed because they were young, or some condition of relief or parent material inhibited soil development. Intrazonal soils had profiles that reflected the dominant influence of relief or parent material over the zonal effects of climate and vegetation such as Rendzinas, gley soils, or peat. The problem was that many soils could not be put in any of these three groups. In 1950, Charles Kellogg charged Guy Smith, who had worked for the soil survey since 1946, to develop a new system [57]. Guy Smith placed all established soil series in the existing classification system and it became apparent that it would be impossible to find places for all soil series without substantially changing the system [52]. Numerous soils in a given Great Soil Group were not closely related. Changing the definition of one Great Soil Group merely shifted the problem to others, and the limits of any Great Soil Group were also limiting other groups.

Charles Kellogg and Guy Smith put together a group of soil scientists to devise a system that could accommodate all soils, not just the soils of the USA. They decided to develop a system in which soils were solely classified on morphology and not on genetic processes. Guy Smith found that the concepts of the zonal soils could be maintained if soil moisture and temperature were introduced at a high level, and Charles Kellogg agreed that such approach would maintain some continuity with previous classifications [58]. Soils in arid regions, the Aridisols, were the only soil order defined by its soil moisture regime. The reason hereto was according to Guy Smith: "One of the most important boundaries on soil maps is the limit between the sown and the unsown. The land that can be cultivated and the land that can only be grazed, and so it seemed to us that it would be useful to have an order that included the bulk of the soils that were too dry to be cultivated and that did have some horizons" [58]. It became clear that the new soil classification system had an utilitarian, if not an agrarian, start.

A series of approximations was developed and tested to investigate its defects and, thus, gradually approach a workable system [52]. Charles Kellogg believed in the evolution of thinking, sharpening predictions, and including more alternatives. As he wrote in 1950: "It is through a series of approximations that we progress towards the truth" [59]. The goal was to develop a universal soil classification system designed to encompass all known soils. The first approximations were sent to a few soil scientists, but subsequent versions were widely circulated for comment, correction, and addition. The soil classification project was started in 1950 and a *3rd Approximation* was published in 1953, the *4th Approximation* in 1955, *5th Approximation* in 1956, and the *6th Approximation* was sent around in 1957. The system was tested by putting soil series in the newly established soil orders and it seemed to work well. In 1956, a new nomenclature was developed so that each class had a name that was mnemonic that connoted some properties of the soils [60]. The system of nomenclature was primarily developed by classical linguists, and the class names were coined from Greek and Latin roots [58].

The *7th Approximation* was incomplete in several respects. All the definitions were considered tentative and definitions of the classes of soil that did not occur in the USA were not tested. Oxisols and Histosols required much more work, as Charles Kellogg wrote in the Foreword: "A final system for adoption has not yet been achieved. But we think that we have arrived again at a place where what has been done needs to be set down in detailed form for study, criticism, and testing. Some gaps and inconsistencies are pointed out. Perhaps there are others. But regardless of these, we believe that the system is a forward step made by many soil scientists working together toward a common end."

For all these good intentions, the *7th Approximation* drew some stern criticism [61–65]. Its presentation at the congress in 1960 overshadowed several national efforts in soil classification and it received instant disapproval from the Russian and French delegates; their classification system was derived from genetic principles but also used morphological and physico-chemical properties to define taxonomic units [66]. For many, the *7th Approximation* was a key rather than a classification system [67]. And although attempts had been made to explain the rationale for the nomenclature [68], the soil names were considered bizarre, incomprehensible, barbarous in formation, and conspicuously lacking in euphony [60]. The soil names were all new, and terms that had been used for decades—such as Podzols and Chernozems—were abandoned. James Raeside from the New Zealand Soil Bureau summarized it as follows: "Although pedology is rapidly becoming accepted as an established discipline, that is as an independent earth science, and becoming less and less looked on as an appendage to agriculture, it is still not sufficiently mature to risk its credit by such bizarre adventures in philology" [69]. The criticism was directed toward Guy Smith who, in the end, worked for over twenty years on the system [57]. A lot of it was anti-American sentiment, repeating clichés from the 1920s and 1930s when Americans were accused of superficiality [70]; some of it was envy which might have been induced by the insistent presentation and free distribution of the work.

The *7th Approximation* was meant as strictly morphometric, as opposed to the genetic system or the morphogenetic hybrid as developed by George Aubert in France. After he studied the *7th Approximation*, Cees Edelman thought that soil classification should be named 'soil systematics' [71]. Hans

Jenny saw no guiding principles in the system, why there were ten orders and not nine or twelve, and he did not see the scientific background; for him, it was very specific but not very scientific. He considered it a bureaucrat's delight but he had nothing better to offer and realized it was better than the previous soil classification systems. Some perceived that the system was biased toward soils from the prairies [72]. Charles Kellogg and Guy Smith defended the *7th Approximation* and knew that the complexity of soil classification and the pedological understanding at that time would have required a new approach and hundreds of workers.

In 1962, Roy Simonson published an eight-page paper in *Science* on soil classification in the USA in which he reviewed progress in soil classification and the *7th Approximation* and noted that no scheme can be any better than the state of knowledge in the soil science of its day [73]. His paper was introduced by a soil profile picture on the cover of the *Science* issue. In 1963, a special issue of the journal *Soil Science* focused on the *7th Approximation*, with contributions from Firman Bear, Charles Kellogg, Guy Smith, Marlin Cline, Roy Simonson, Bill Johnson, Frank Riecken, and from René Tavernier and Charles Stephens, who had tested the system in Europe and Australia.

In 1959, Guy Smith spent some time in Australia and examined 220 soils and studied how the soils could be classified using the *7th Approximation*. It worked fairly well but the Australian pedologist Charles Stephen found that it separated soils more on the basis of their degree of development than on the kind of development and that the system violated soil–climate relationships. He thought that the system could become useful for Australia if it were revised, definitions and criteria were simplified, closely related soils were in the same class, and a new nomenclature was devised [67]. René Tavernier, a friend of Charles Kellogg and Guy Smith, was less critical and expected that in the future systems similar to the *7th Approximation* would be developed across Europe. He also thought that the *7th Approximation* was the greatest contribution to soil classification [71]. Charles Kellogg reviewed the disparagement of the *7th Approximation* and saw the development of a soil classification system as: "…the best classification that can be devised is still only a tool to help us understand soils better and to apply soil science to achieve better systems of soil use. No system of classification should ever become so sacred or so classical that the system becomes an end in itself. It should always remain a tool for use, to be sharpened or to be replaced as the attainment of our objectives in applying soil science demands" [74].

After the introduction of the *7th Approximation* in 1960, eleven international committees focused on particular soil orders or properties and aimed to improve the system. Some of the committee members had fundamental objections to the system and favored the FAO–UNESCO system. The use of soil classification peaked in the decades after the introduction of the *7th Approximation* [54], and the number of soil series in the USA increased from less than 5,000 in 1960 to 10,000 in 1980 [39]. The *7th Approximation* was modified, taking into account some of the critiques, and published in 1975 as *Soil Taxonomy: a Basic System of Soil Classification for Making and Interpreting Soil Surveys* [75].

The world was politically divided in 1960 and, on a smaller scale, the soil science community was divided by soil classification. Both would remain divided for decades. The effect of this divide on the prosperity of the soil science community and on the advancement of its science has not been studied. One could debate the progress in soil classification in all directions, but the soil is a four-dimensional continuum with diffuse boundaries between millions of individuals, and any classification is a generalization and reflects current knowledge. One of the effects of the *7th Approximation* was that soil science and classification became strongly data driven and, within the soil survey, this attitude flourished. The idea of more data, better information, and better management of resources was unwritten but ubiquitously entrenched. Charles Kellogg held on to the emerging system until it was so large, it could not be abolished. According to some soil scientists, the system, wide-ranging and hierarchical as it was, had become dinosauric [76].

At the *Seventh International Congress of Soil Science*, a new journal of soil science was proposed for research papers and reviews on diverse aspects of soil science [77]. The journal was approved four years later at the eighth soil congress in Bucharest. The *International Society of Soil Science* could not accept responsibility for the implementation of a new journal, but the publisher Elsevier was interested in adding a soil science journal to their portfolio of earth-science periodicals. They also published the proceedings of the seventh soil congress. The Secretary General, Ferdinand van Baren, was a friend of the geologist who headed Elsevier's earth-science division. *Geoderma*—the earth's skin—was chosen as the journal's name, and the first issue was published in 1967 with a journal cover that in its color matching, font and design exceeded any cover design of any other soil science journal. The Editorial Board was headed by Ferdinand van Baren and included 32 soil scientists from across the globe, among them Rudi Dudal, Charles Kellogg, Victor Kovda, Roy Simonson, Guy Smith, René Tavernier, and Dan Yaalon.

Two years before his retirement, Roy Simonson was invited by Elsevier to become the Editor-in-Chief of *Geoderma*, and he would remain in charge until 1989. During those 18 years, 823 papers were published [77]. The journal greatly expanded, and papers from many different parts of the world were published. He encouraged manuscript submissions from authors in countries where English is not spoken and authors who were unfamiliar with publishing in international journals. A helpful editor Roy Simonson put considerable effort into the editing and improvement of manuscripts. He would write a seven page-review report with recommendations for a paper which he considered relevant but poorly written [77], and would often enclose a document, *How to Write in the English Language*, with his reports, including reviews directed to native English speakers. In his farewell editorial in 1989, there was some Norwegian modesty about his dedicated editorial activities where he provided the following motivation: "A revision, ranging from little to much, has been required for a majority of the manuscripts received. The intent has been to improve the manuscripts on several counts, including the validity of the data, their interpretations, and their presentations. Revising manuscripts provides practice in writing, which is the only way to improve the skill. There is no substitute for such practice." Roy Simonson had much practice himself; he learned to write clearly and succinctly from Emil Truog and had assisted in the editing of the *Soils & Men* yearbook in the 1930s.

Charles Kellogg had tense meetings with lots of people at the seventh congress, and in his diary, he provided vibrant descriptions of some of these meetings. He noted that Ferdinand van Baren had done almost nothing for the council meetings, and the reports were untyped and contained many errors. Besides the betrayal over the *International Society of Soil Science* resolution on the *Soil Map of the World*, he considered the election of Ferdinand van Baren as Secretary General a mistake. In 1950, the Dutch had pressed for continuity in having the *International Society of Soil Science* secretariat in the Netherlands, as David Hissink had served in that role since 1924. Ferdinand van Baren had visited the USA in 1946 and had not met with Charles Kellogg but only with his rival Hugh Bennett [78]. Charles Kellogg had little regard for Ferdinand van Baren, but at his retirement in 1975, Charles Kellogg wished him a pleasant retirement and congratulated him on all the work done for the society.

Charles Kellogg had a row with another Dutchman who was his friend: Cees Edelman. He had established the Dutch Soil Survey and had developed the physiographic soil survey approach in which landscape units and soil types were interconnected [79]. Like soil surveys in other countries, the approach was derived from methods described in the *Soil Survey Manual* [53],—a manual often praised by Cees Edelman. However, the Dutch Soil Survey used the term 'land classification' when, according to Charles Kellogg, they meant 'soil interpretation.' Charles Kellogg had explained to Dutch visitors that land classification involved many other items, especially economic appraisal [80]. Lee Schoenmann had taught him that in Michigan, and Charles Kellogg had applied it in North Dakota and everywhere else since. For Charles Kellogg, the precise use of words and terms was important, in order to avoid confusion and provide direction. Some years later, Cees Edelman wrote to him that the Dutch Soil Survey had agreed to make the change and complimented him on his persistence.

Charles Kellogg and Cees Edelman had much in common, they were of the same generation and both believed that soil science was strengthened when it worked with other scientific disciplines. In the 1950s, Cees Edelman summarized this as follows: "…defining a science is a precarious matter, even more precarious is defining the boundary between two sciences" [81]. Both Charles Kellogg and Cees Edelman were interdisciplinary in soil science, and they were interdisciplinary in a time when specialization had begun to fragment the soil science discipline. The term interdisciplinary was used for the first time in agriculture in the 1940s [20].

The last issue that the council of the *International Society of Soil Science* had to resolve in 1960 was the location of the next congress in 1964. The council had received two letters of invitation, one from the Romanian Academy of Science to hold the congress in Bucharest, Romania, and one from the German society of soil science. The *International Society of Soil Science* President, Richard Bradfield, talked with the Russian Ivan Tiurin who recommended Romania as a veritable 'Soils Museum.' Romania had a communist and pro-Russian government. Richard Bradfield had met with the Romanian soil scientist Téodore Saidel at the first congress and was familiar with soil research in Romania. After discussion and some balloting, the German invitation was withdrawn. Hence, it was decided to hold the *Eighth International Congress of Soil Science* in 1964 in Bucharest. Germany had to wait another 26 years before it would host an international congress of soil science.

Charles Kellogg found that the Russians, whom he consistently named *the Soviets* in his diary, had some beautiful soil maps on display at the congress. He asked Ivan Tiurin whether he could have them copied, and Ivan Tiurin agreed provided he could take them back to Russia. Charles Kellogg took the train to Chicago and then to Washington with his luggage and the soil maps. At the soil survey office, the maps were copied and given back to Ivan Tiurin, who had just returned from the southern tour after the congress. As a sign of appreciation, Lucille Kellogg presented the Russian spouses with

Cees Edelman with Charles Kellogg on 31st May 1957 holding the 1950 soil map of the Netherlands that had been prepared for the *Fourth International Congress of Soil Science* in Amsterdam
"See Charles, this is where Wageningen is at the intersection of ice-pushed pre-glacial sands, riverine clays and Holocene peat. For that reason, the university...."
"I know Cees, I know, been there and it's all fine. Let's sit down and talk about land classification. You should know, it is wrongly used by your office....."
"I know, Charles, I know, and it's all fine."
Photo National Agricultural Libraries

stretch gloves. René Tavernier, who had been the President of the *International Society of Soil Science* for the congress in Congo, stayed with the Kelloggs for ten days after the congress. It was the end of August 1960, and it was so hot in Washington that even the Kellogg's old gray cat, Finnogan, was willing to listen to the roar of the air-conditioner.

Despite the betrayal over the *Soil map of the World* and the criticism of the *7th Approximation*, Charles Kellogg was satisfied with the achievements of the *Seventh International Congress of Soil Science*. The congress proved to him that there was fellowship among soil scientists. On the last day of the congress, he had proposed the following resolution: "Since the theme of the *Seventh Congress of the International Society of Soil Science* was *To Promote Peace and Health by Alleviating Hunger through Soil Science* and since the Food and Agriculture Organization launched a Freedom from Hunger Campaign in July of 1960, the Congress expressed its full support of this campaign and appealed to soil scientists throughout the world to assist the campaign through the full application of their knowledge and experience for sustained and efficient food production." On soils and peace. A better and more agrarian ending of the congress could not have been wished for.

References

1. Nikiforoff, C. C. (1959). Reappraisal of the soil. *Science, 129,* 186–96.
2. Keeney DR. The Keeney place: a life in the Heartland. Levins Publishing; 2015.
3. Feldman, J. (1997). *The buildings of the University of Wisconsin Madison, Wisconsin.* Madison: The University Archive.
4. Phillips, R. W. (1981). *FAO: its origin, formation and evolution.* Rome: Food and Agricultural Organization of the United Nations.
5. Selcer, P. (2015). Fabricating unity: the FAO-UNESCO soil map of the world. *Hist Soc Res/Historische Sozialforschung, 40,* 174–201.
6. Bradfield R. Opportunities for soil scientists in freeing the world from hunger. In: Transactions 7th international congress of soil science, vol. I; 1960. p. xxix–xxxviii.
7. Blanck E. Handbuch der Bodenlehre (vol. I–X). Berlin: Verlag von Julius Springer; 1929–1932.
8. Kellogg CE. Productivity of the arable soils of the United States. Paper presented at 7th International Congress of Soil Science, Madison; 1960.
9. FAO. Multilingual vocabulary of soil science. In: Jacks GV, Tavernier R, Boalch DH, editors. Vocabulaire multilingue de la science du sol. Vocabulario multilingue de la ciencia del suelo. Rome: Land & Water Development Division; 1960.
10. FAO. Multilingual vocabulary of soil science. In: Jacks GV, editors. Vocabulaire multilingue de la science du sol. Vocabulario multilingue de la ciencia del suelo. Rome: Agriculture Division; 1954.
11. Dmitrieva, V. A., & Polunin, N. (1992). Victor Abramovich Kovda 1904–91. *Environ Conserv, 4,* 364–65.
12. Yaalon, D. H. V. A. (2004). Kovda—Meeting with a great and unique man. *Newsletter IUSS Commiss Hist Philos Sociol Soil Sci, 11,* 4–9.
13. Dobrovolsky GVK, V. I. In memoriam Prof. Dr. A. Kovda (1904–1991). Bull Int Soc Soil Sci. 1991; 80:77–78.
14. Królikowski L. Professor Dr. Jadwiga Ziemiecka (1891–1968). Bull. Int. Soc. Soil Sci. 1986; 33:51.
15. Aslander, A. (1927). Sulphuric acid as a weed spray. *Journal of Agricultural Research., 34,* 1065–091.
16. Russell, E. J., & Wright, N. C. (Eds.). (1961). *Hunger—Can it be averted?.* London: British Association for the Advancement of Science.
17. Russell, E. J. (1963). *The world of the soil* (3rd ed.). London: Collins.

18. Russell, E. J. (1966). *A history of agricultural science in Great Britain, 1620–1954*. London: Alien and Unwin.
19. Tinker, P. B. (1985). Soil science in a changing world. *J Soil Sci, 36,* 1–8.
20. Kellogg CE, Knapp DC. The college of agriculture. Science in the public service. New York: McGraw-Hill Book Company; 1966.
21. Viets, F. G. (1977). A perspective on two centuries of progress in soil fertility and plant nutrition. *Soil Sci Soc Am J, 41,* 242–49.
22. Hartemink, A. E. (2002). Soil science in tropical and temperate regions—some differences and similarities. *Adv Agron, 77,* 269–92.
23. Smil, V. (1999). Detonator of the population explosion. *Nature, 400*(6743), 415.
24. Bouma, J., & Hartemink, A. E. (2002). Soil science and society in the Dutch context. *Netherlands J Agric Sci, 50*(2), 133–40.
25. Russell, E. J. (1954). *World population and world food supplies*. London: George Allen & Unwin Ltd.
26. Collingham, L. (2011). *The taste of war: World War Two and the battle for food*. London: Allan Lane.
27. Greenland DJ. The responsibilities of soil science. Paper presented at 11th international congress of soil science, Edmonton; 1978.
28. Albrecht, W. A., & Smith, C. E. (1952). *Soil acidity as calcium (fertility) deficiency*. College of Agriculture: University of Missouri.
29. Carson, R. (1962). *Silent spring*. USA: Houghton Mifflin.
30. Ehrlich, P. (1968). *The population bomb*. New York: Ballantine Book, Inc.
31. Elie, M. (2015). Formulating the global environment: Soviet soil scientists and the international desertification discussion, 1968–91. *Slavonic East Eur Rev, 93,* 181–204.
32. Truog E. A new test for soil acidity. Bulletin 249. Madison, Wisconsin: Agricultural Experiment Station of the University of Wisconsin; 1915.
33. Simonson, R. W. (1950). Genesis and classification of red-yellow podzolic soils. *Soil Sci Soc Am J, 14,* 316–19.
34. Yaalon, D. H. (2012). *A passion for science and zion*. Jerusalem: Maor Wallach Press.
35. Yaalon DH. Some implications of fundamental concepts of pedology in soil classification. Trans 7th Int Congress Soil Sci. 1960; IV:119–23.
36. Hartemink, A. E., Krasilnikov, P., & Bockheim, J. G. (2013). Soil maps of the world. *Geoderma, 207,* 256–67.
37. Whitney M. Soils of the United States. Bureau of Soils Bulletin No. 55. Washington DC: Government Printing Office; 1909.
38. Coffey GN. A study on the soils of the United States. Bureau of Soils—Bulletin no. 85. Washington: Government Printing Office; 1913.
39. Brevik, E. C., & Hartemink, A. E. (2013). Soil maps of the United States of America. *Soil Sci Soc Am J, 77*(4), 1117–132.
40. Marbut, C. F., Bennet, H. H., Lapham, J. E., & Lapham, M. H. (1913). *Soils of the United States (edition, 1913)*. Washington: Government Press Office.

41. Shantz, H. L., & Marbut, C. F. (1923). *The vegetation and soils of Africa.* New York: National Research Council and the American Geographical Society.

42. Stremme HE. Preparation of the collaborative soil maps of Europe, 1927 and 1937. In: Yaalon DH, Berkowicz S, editors. History of soil science international perspectives. Armelgasse 11/35447 Reiskirchen/Germany: Catena Verlag; 1997. p. 145–58.

43. Stremme, H. (1937). *International soil map of Europe, 1:2,500,000.* Berlin: Gea Verlag.

44. Dudal R. Dom Luis Bramão (1909–2007). Bull Int Union Soil Sci. 2008; 112(32–33).

45. Kellogg CE. Proposal for Department of soil science and management of a world organization on food and agriculture. Unpublished Confidential Report 9/25/43; 1943.

46. Ignatieff V, editor. Efficient use of fertilizers. Washington DC: Food and Agricultural Organization of the United Nations; 1949.

47. Selcer, P. (2011). Patterns of science: developing knowledge for a world community at Unesco. Ph.D. Thesis. University of Pennsylvania.

48. Polynov, B. B. (1932). *Contributions to the knowledge of the soils of Asia. Academy of Sciences of the USSR.* Leningrad: Dokuchaev Institure of Soil Science.

49. Dudal, R., & Batisse, M. (1978). The soil map of the world. *Nat Resour, 14,* 2–6.

50. van Baren, H., Hartemink, A. E., & Tinker, P. B. (2000). 75 years the international society of soil science. *Geoderma, 96* (1–2), 1–18.

51. Deckers, J. (2014). Raoul (Rudi) Dual. *Bull Int Soc Soil Sci, 124,* 44–6.

52. Soil Survey Staff. Soil classification. A comprehensive system, 7th Approximation. Washington D.C.: Soil Conservation Service. United States Department of Agriculture; 1960.

53. Soil Survey Staff. (1951). *Soil survey manual.* Washington DC: USDA.

54. Hartemink, A. E. (2015). The use of soil classification in journal papers between 1975 and 2014. *Geoderma Reg, 5,* 127–39.

55. Bockheim, J. G., Gennadiyev, A. N., Hartemink, A. E., & Brevik, E. C. (2014). Soil-forming factors and Soil Taxonomy. *Geoderma, 226–227,* 231–37.

56. Kellogg, C. E. (1948). Modern soil science. *Am Sci, 36,* 517–35.

57. Brasfield J. In Memoriam Guy D. Smith. Soil Horizons. 1981; 22.

58. Forbes TR, editors. The Guy Smith interviews: Rationale for concepts in soil Taxonomy. SMSS Technical Monograph no. 11. Washington DC: USDA Soil Conservation Service; 1986.

59. Kellogg, C. E. (1950). The future of the soil survey. *Soil Sci Soc Am Proc, 14,* 8–13.

60. Heller, J. L. (1963). The nomenclature of soils, or What's in a name? *Soil Sc Soc Am J, 27* (2), 216–20.

61. Webster, R. (1960). Fundamental objections to the 7th approximation. *J Soil Sci, 19,* 354–66.

62. Leeper, G. W. (1952). On classifying soils. *J Aust Inst Agric Sci, 18,* 77–80.

63. Thorp, J. (1948). Practical problems in soil taxonomy and soil mapping in great plains states. *Soil Sci Soc Am J, 12,* 445–48.

64. Manil, G. (1959). General considerations on the problem of soil classification. *J Soil Sci, 10,* 5–13.

65. Pierre, W. H. (1958). Relationship of soil classification to other branches of soil science. *Soil Sci Soc Am Proc, 22,* 167–70.

66. Aubert G, Duchaufour P. Projet de classification des sols. In: Transaction of the 6th International Congress of Soil Science; 1956. p. 597–604.

67. Stephens, C. G. (1963). The 7th approximation: Its application in Australia. *Soil Sci, 96,* 40–8.

68. Smith GD. The Guy Smith interviews: Rationale for concepts in Soil Taxonomy. SMSS Technical Monographs no. 11. Washington DC: Soil Conservation Service; 1986.

69. Raeside, J. D. (1960). Letter to the editor. *ISSS Bull, 19,* 20–2.

70. Klautke, E. (2011). Anti-Americanism in twentieth-century Europe. *Hist J, 54,* 1125–139.

71. Tavernier, R. (1963). The 7th approximation: its application in Western Europe. *Soil Sci, 96* (1), 35–9.

72. Maher D, Stuart K, editors. Hans Jenny—soil scientist, teacher, and scholar. Berkeley: University of California; 1989.

73. Simonson, R. W. (1962). Soil classification in the United States. *Science, 137* (3535), 1027–034.

74. Kellogg, C. E. (1963). Why a new system of soil classification. *Soil Sci, 96,* 1–5.

75. Soil Survey Staff. Soil taxonomy. A basic system of soil classification for making and interpreting soil surveys. Agricultural Handbook no. 436. Washington DC; 1975.

76. Gray, B. (1980). Popper and the 7th approximation: the problem of taxonomy. *Dialectica, 34,* 129–53.

77. Hartemink, A. E., McBratney, A. B., & Cattle, J. A. (2001). Developments and trends in soil science: 100 volumes of Geoderma (1967–2001). *Geoderma, 100* (3–4), 217–68.

78. van Baren FA. Erosie, oorzaak, gevolgen en bestrijding. Mededeelingen van het Departement van Economische Zaken in Nederlandsch-Indie. No. 8; 1947.

79. Hartemink, A. E., & Sonneveld, M. P. W. (2013). Soil maps of the Netherlands. *Geoderma, 204,* 1–9.

80. Kellogg, C. E. (1951). Soil and land classification. *J Farm Econ, 33,* 499–513.

81. Heslinga MW, Wiggers AJ. Over de betekenis van C.H. Edelman voor de geografie. Tijdschrift van het Koninklijk Nederlandsch Aardrijkskundig Genootschap. 1966; LXXXIII:4–14.

14

Chronicles and Progressions

"My desire and wish is that the things I start with should be so obvious that you wonder why I spend my time stating them. This is what I aim at because the point of philosophy is to start with something so simple as not to seem worth stating, and to end with something so paradoxical that no one will believe it."

Bertrand Russell, 1918

Chronicles and progressions are made daily and can be dissected and aggregated in diverse ways. They can be presented as series of sequential happenings (and then… and then… and then), or as a meandering flux of events—mostly backward and linear but at times curving or directionless. In all efforts, chronicles and progressions are about connections using knowledge of the past to relate events that led to the present. Progressions are more than a timeline of happenings and discoveries, as they harbor hidden and convoluted social processes, tensions and conflicts [1]. Some of that is driven by personal or geopolitical interests, some of it might be the inherent nature of all human activities. Unraveling soil science progression is probably as complex as the soil itself and, in both fields, much remains to be discovered. In the previous chapters, chronicles and progressions have been described and portrayed in the spirit of Charles Kellogg, who wrote: "It is rather a knowledge of relationships that the general reader seeks, not facts per se." In this final chapter, some deeper and wider excavations are attempted in the search for the American soul in soil science.

Soil science had a late and uncertain start in the USA. For years, Americans had drawn on Europe and, for years, it was assumed that the findings from Europe could be applied in the USA [2]. Ignorance and carelessness caused some severe soil degradation but, for a long time, settlers could move on. That ended in the early 1900s, when there was no more new land for farm settlement [3]. More soil knowledge was needed. Experimental stations were founded, soil research and teaching were started at universities, and a national

© The Author(s), under exclusive license to Springer Nature
Switzerland AG 2021
A. E. Hartemink, *Soil Science Americana*
https://doi.org/10.1007/978-3-030-71135-1_14

soil survey was established. Soil research shed its agricultural chemistry, bacteriology and geology ancestries, but efforts remained practical and preserved an agrarian emphasis. The *First International Congress of Soil Science* boosted confidence in soil science in the USA and across the globe. Economies were roaring; there was optimism about the future of the land and the prosperity it could bring. Then came the Great Depression; prolonged drought; ecological disaster; and mass migration. Early on in that Great Depression, a successful soil chemist gave a new suit, shirt, and pair of shoes to an aspiring young pedologist who was to be interviewed for a job in North Dakota. Several decades later, the pedologist had become highly influential, and returned a favor. Emil Truog and Charles Kellogg were friends for as long as they lived, and both profoundly influenced soil science. By the 1950s, there was no trace of its uncertain start in the USA.

American soil science was started by chemists and geologists that had studied or had spent time in Europe, including Samuel Johnson, Charles Dabney, Eugene Hilgard, Cyril Hopkins, Curtis Marbut, Frederick Alway and, in later years, Richard Bradfield and Merritt Miller. They brought back academic practices and ideas that were tried, tested and adopted on American soil. Soil science in the USA also benefited from the migration of people within the country. In the *Bureau of Soils*, field crews worked in southern states during the winter and in northern states during the summer, which helped in standardization of practices and unraveling the country's soil geography. Migration also took place through the educational system. Most students attended a university close to their home but, for advanced degrees, they often enrolled in a university in neighboring state or another part of the country. Both the jobs and universities brought about mixing of people and ideas.

Soil science in the USA has benefited from immigration. Immigrants such as Eugene Hilgard, Jacob Lipman, Selman Waksman, Jacob Joffe, Sergei Wilde, Constantin Nikiforoff, Hans Jenny, and Edmund Marshall came from Germany, Latvia, Lithuania, Ukraine, UK, Switzerland, and Russia. They brought with them skills, ambitions, and aspirations. There were second-generation immigrants like Emil Truog and Roy Simonson who were born shortly after their parents had emigrated to the USA, and there were others like Milton Whitney and Charles Kellogg whose families had been in the USA for generations [4]. The mingling of the European training, migration, and new and old immigrants into vast areas of unknown soils developed into a research approach that was down-to-earth, and proved to be a fruitful breeding ground for the development of the soil science discipline. Although the formative years of American soil science involved a mixture of people, they were all men, and they were all white; this would remain the norm for many years.

A characteristic of early soil science was that people stayed long-term in leadership positions. Except for the economic crisis of the 1930s, there was little turnover, few layoffs, and no fixed terms of office. For example, Andrew Whitson was head of the Department of Soils at the University of Wisconsin for 38 years, Frederick Alway was head of the Divisions of Soils at the University of Minnesota for 28 years, Charles Kellogg was in the soil survey for 37 years and most of that time was in a leadership position, and Jacob Lipman was the editor of *Soil Science* from 1916 to his death in 1939. Emil Truog spent 50 years at the University of Wisconsin, George Bouyoucos was at Michigan State University for 53 years, Charles Black for 46 years at Iowa State college, and Henry Krusekopf, Merritt Miller and Edmund Marshall were at the University of Missouri for over 40 years. There are many such examples and it was not unique to the USA; John Russell was director of Rothamsted Experimental Station for 31 years, Rudi Dudal worked for over 30 years at the FAO, and Ferdinand van Baren was Secretary General of the *International Society of Soil Science* for 24 years. They established patterns of soil research, trained generations of soil scientists, and maintained the norm.

The founding soil science community in the USA was a generation who grew up on the farm. Samuel Johnson, Nathaniel Shaler, Thomas Chamberlin, Andrew Whitson, Franklin King, Curtis Marbut, Richard Bradfield, Cyril Hopkins, Merritt Miller, Emil Truog, Charles Kellogg, Hugh Bennett, Walter Kelley, Roy Simonson, Percy Brown, Ray Throckmorton, Arthur McCall, Merris McCool, Firman Bear, William Albrecht, Bill Pierre and Henry Krusekopf were all raised on farms, and most of them in the Midwest.

Milton Whitney did not grow up on a farm; his father was an attorney who practiced in Baltimore [4]. Eugene Hilgard's father was a lawyer but he grew up on a farm east of St. Louis in Illinois. Of the 51 members of the organizing committee of the *First International Congress of Soil Science*, 31 members were from the farm. Even in 1970, it was noted that a disproportionately high portion of men listed in *Who's Who in Science* had a rural background. A reason for the disproportion was the distance from the farm to plumbers, electricians, repair shops and all those service available in towns and cities [5]. Farmers must solve most of their own problems, and they learned the joy and independence that comes from solving these problems. Those traits made for good soil scientists.

Hans Jenny, who emigrated to the USA in 1926, felt he was often criticized and disadvantaged as he had not grown up on a farm, and that he was not good in translating his work to the farmer—which was considered important. He found those who grew up on a farm were good with their hands and explored anything that suited their research [6]. The connection between being raised on a farm and the soil was seen as natural. Merritt Miller, who grew up on a farm in Ohio and established the soils department at the University of Missouri, put it in 1950 as "Every farm-reared youth knows something about the soil. People who live in the open country work with the soil…they know their welfare depends on it. The farm boy is brought up with a respect for it. The respect is part of the inheritance of country people which most town people do not have" [7].

The agrarian background of the early soil scientists had some practical advantages but also limited soil research and caused some bias. For example, Selman Waksman was refused for graduate studies at the University of Illinois, as Cyril Hopkins only admitted students that came from the farm. Cyril Hopkins was dead by the time Selman Waksman received the Nobel Prize. Emil Truog did not offer a position to Hans Jenny in 1933 for administrative reasons, but Hans Jenny had not grown up on a farm and Emil Truog had a preference for hiring men from the farm.

Much of the early soil research in the USA focused on agricultural development which was needed as in the mid-1800s over 80% of the population in the USA were farmers. The farm population decreased to 25% by the 1920s [8], and to less than 10% of the total population in 1960 [3]. But much of the soil research kept an agrarian focus, and as a result the discoveries of the soil profile, soil-climate relationships and the founding of pedology were not American achievements, even though it was the only other country that had similar conditions and vastness to Russia where such discoveries could possibly have been made.

In the USA in the early 1900s, Land-Grant Universities were open for those that had grown on a farm and were transitioning into occupations other than farming [3]. This was mostly the case for men of European descent. Curtis Marbut, Charles Kellogg, Emil Truog, and Roy Simonson exemplify access to university education. By contrast, most of the early Russian and European soil researchers did not come from the farm. Konstantin Glinka, Alex. de'Sigmond and Vladimir Ignatieff were born in aristocratic families, Martinus Beijerinck was the son of a railway clerk, Sergei Vinogradskii came from a banking family, Konstantin Gedroiz was the son of a medical doctor, Eugene Hilgard's father was a lawyer, Nikolai Sibirtsev's father was a high-ranking priest in the Spiritual Academy, David Hissink was the son of a municipal secretary, Eilhard Mitscherlich grew up in an academic family,

Emil Ramann was the son of a manufacturer and his mother was a great cultivator of the mind, Hermann Stremme was the son of a merchant, and John
Russell's father was a Reverend. Several of the immigrants who came to the
USA were not raised on farms, including Jacob Lipman, Selman Waksman,
Sergei Wilde, Jacob Joffe, and Constantin Nikiforoff. Both Selman Waksman
and Jacob Lipman worked on a farm for some years before they studied
agriculture and soils at Rutgers College in New Jersey.

The farm background of American soil scientists burst through in
manuscripts, speeches, research focus, and in their conversations and letters.
Sergei Wilde, who joined the Department of Soils at the University of
Wisconsin in 1934, wrote in his autobiography: "...when I joined the
department it had only six professors and about ten graduate students. My
professional colleagues were non-drinking, non-swearing fragments of the
Victorian era, dedicated primarily to the production of crops. A conversation usually started with alfalfa and invariably culminated with fertilizers, the
omega of the discussions" [9]. The agrarian connection was deeply engrained
and it basically determined what was researched, what was being taught, and
what was talked about. For example, in August 1944, Emil Truog wrote to
Roy Simonson, who had just moved from Iowa to Tennessee: "Crops in
Wisconsin in general will be good or at least considerably above average.
In general, hay is a good crop, small grains average plus and corn, while
slightly behind schedule in many cases, may make a very good crop. Our
rainfall has been quite favorable. Truck crops are in general very good. All in
all, I think Wisconsin will come through with another bumper crop." Some
days later, Roy responded: "I am glad to know that there are good crops in
some parts of the country. East Tennessee has had the driest summer since
recording was started about 75 years ago. Corn and hay crops are extremely
poor, although the pastures have been revived a little by the rains of the past
two weeks. Feed is short enough generally in this part of the state, and in
most parts of Tennessee, so that hay shipments are now coming in—some
of them from Nebraska. The recent rains do give promise that the drought is
broken, however, and they will help the pastures a good deal." It was not until
the 1960s that soil science broadened its horizons and embraced environmental aspects that conversations changed, and a farm background became
less important for a career in soil science.

The early soil scientists like Eugene Hilgard and Samuel Johnson had PhD
degrees in geology or chemistry. In 1903, Jacob Lipman and James Bizzell
were the first awarded a PhD in soil science in the USA [10]. Thereafter, most
leading soil scientists obtained a PhD in soil science, but not everyone. For
example, Emil Truog mentored over 100 PhD students, but he had no PhD

himself. He finished all of his course requirements in chemistry for a PhD except a dissertation [11]. Neither Henry Krusekopf nor Hugh Bennett had PhD degrees. Merritt Miller of the University of Missouri had no PhD but was awarded honorary doctorates from Kansas State University and the Ohio State University. Curtis Marbut never sat for his PhD exam at Harvard but was awarded two honorary doctorates. Dennis Hoagland from the University of California Berkeley had no PhD and he brought it up often enough that it must have bothered him [6]. At the *Bureau of Soils* in the 1930s, only Constantin Nikiforoff, Thomas Rice, and Mark Baldwin had PhD degrees; most soil surveyors had B.S. degrees, and there were a few who had an M.S.

Emil Truog was a chemist by training but was more of an edaphologist, that is to say an agronomist who studies the soil part of growing crops for human or animal consumption. Emil Truog's science was to help the soil to grow plants with better quality and greater yields than ever before. For Charles Kellogg, science should inform where the soils differed and how that should be observed and mapped. Both proclaimed that a deeper understanding of soils was needed, whether it concerned fundamental properties or an understanding of how those properties varied across the landscape and world. Their science was guided by the development of ideas and theory, and they were reductionists who condensed complex interactions to the sum of their parts for easier study. As Charles Kellogg wrote in 1941: "In the first place, there are two things necessary to science - facts and ideas. Simple facts or observations can only be useful to us if there are some connecting ideas; and ideas must be illustrated and supported by facts, or else they may lead us in the wrong way. Sometimes we complain that the man who is all idea doesn't get things done - that it is the practical man who really goes places. But unless the man who does things has correct ideas as well as facts, we will find that he has gone, to be sure, but to the wrong place. Thus ideas without facts or facts without ideas accomplish nothing. Not only must we have facts and ideas, but our facts must be plentiful, else our ideas will be too narrow" [12].

Emil Truog and Charles Kellogg grew up on farms under pioneering conditions. With university educations and years of dedicated work, they became leaders in their field and advocated science, and soil science in particular, to help make this world a better place. Hans Jenny reckoned Franklin King, Curtis Marbut, Emil Truog, and Charles Kellogg influential leaders in the development of soil science in the USA [6]. There were connections between all of them: Emil Truog arrived at the University of Wisconsin two years before Franklin King died, and he was trained by Andrew Whitson. Emil Truog became the Head of Department of Soils after Franklin King

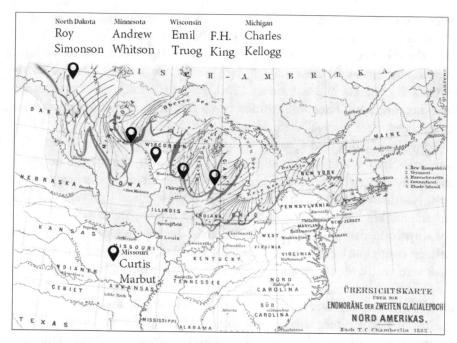

Some pioneering American soil scientists and their place of birth: Curtis Marbut (1863–1935) Spring River Valley, Missouri; Roy Simonson (1908–2008) Agate, North Dakota; Andrew Whitson (1870–1945) Stanton, Minnesota, Emil Truog (1884–1969) Independence, Wisconsin; Franklin King (1848–1911) LaGrange, Wisconsin; and Charles Kellogg (1902–1980) Palo, Michigan. They were all raised on the farm. Map is the glacial extent map produced by the geologist Thomas Chamberlin in 1882

and Andrew Whitson had retired. Charles Kellogg was trained by Merris McCool in Michigan who had taken a geology course from Curtis Marbut in Missouri. For his PhD studies, Charles Kellogg worked for a year under Andrew Whitson, and Emil Truog became his mentor and helped him get his first job in North Dakota. When Charles Kellogg worked in North Dakota, Emil Truog continued to mentor him, and Curtis Marbut came to visit Charles Kellogg in the field. When Curtis Marbut was about to retire, he hand-picked Charles Kellogg as his assistant and as his successor as chief of the soil survey. Charles Kellogg had an excellent undergraduate student, Roy Simonson, who he sent over to Emil Truog for graduate studies.

Progressions in the discipline of soil science are historically connected and influenced by a small number of people: from the friendship between Konstantin Glinka and Curtis Marbut to a pedological approach in the American soil survey; from the sabbatical of Richard Bradfield that brought Hans Jenny to the University of Missouri to the writing of *Factors of Soil Formation*; from the immigrant Jacob Lipmann to the establishment of the journal *Soil*

Science and the naming of the soil science discipline; from Sergei Vinograd-skii investigating nitrifying bacteria to the Nobel prize by Selman Waksman; from the soil survey in Iowa with Guy Smith and Roy Simonson, to the hiring of both of them by Charles Kellogg; and from a new suit for Charles Kellogg to the soil congress in Madison in 1960. These chronicles and progressions were not random. There are numerous connections, and once established, connections were not well-defined but deeply nurtured.

Both Emil Truog and Charles Kellogg helped their former students and colleagues that maintained and expanded their own thinking. The loyalty and hard work were rewarded. Charles Kellogg hired his former students from North Dakota, and they stayed with him in the soil survey until their retirement. He had strong opinions and Roy Simonson, noted that it was not surprising that so many people disliked Charles Kellogg. When Charles Kellogg started at the *Division of Soil Survey* in 1934, he was younger than most staff in the division, and he was seen as a bit brash, had a high regard for his ability, wanted things his way, was not challenged much, and domi-nated most conversations. He was good in promoting soil science, creating teams but had a direct way of providing leadership. Internationally, he had his struggles too, and did not get along with the soil science leadership in the UN agencies and in the *International Society of Soil Science*. There was some resentment of American supremacy, particularly after the Second World War. American soil science spread across its borders, and was welcomed and led to activity and inspiration, just as much as it led to resistance and, perhaps, envy. It was a strong force in the world due to the 1927 and 1960s congresses, as well as the *Division of Soil Survey* machinery that provided free assistance and publications.

Charles Kellogg was aware of the resistance and animosity he at times created. In 1950, he started a diary and wrote about events in an abbrevi-ated and candid way: "A few of the people I worked with, in the universities, in the Department of Agriculture, and overseas, lied to me and to others on numerous occasions. The most difficult were those with narrowly specialized training, without the basic subjects, and who were both uncertain of them-selves and eager for money, rank, and power. They ran scared much of the time and tended to be increasingly arrogant, unhappy, and untruthful as they grew older. Many men took their religion for granted until they reached 60 or more. Then it occurred to them that possibly their only monuments would be in this world, yet they had none. Several men I've known in both univer-sities and government became very bitter at this stage." Charles Kellogg was a man of little doubt and regret. His main interest was to advance soil survey for the benefit of humankind, and he had little patience for different ideas. Progressions were to come through conventional wisdom.

Hans Jenny considered Charles Kellogg a pioneer of American soil science. They met for the first time at the funeral service for Curtis Marbut, and Hans Jenny remembered that Charles Kellogg kept the pipe in his mouth while talking. He made an arrogant and conceited impression on Hans Jenny, and they ended up having a formal and stiff relationship [6]. Despite Charles Kellogg's encouragement to his students to take more mathematics and the basic sciences, he showed no interest in the quantitative pedology that Hans Jenny developed. Hans Jenny's work was not brought into the soil survey, just as the work of Eugene Hilgard had been ignored some decades earlier. The reasons for disregarding that work in the soil survey can only be guessed, but the fact that the two had poor relations might be part of it. Another possibility is that Charles Kellogg's education in mathematics and chemistry was insufficient to grasp its innovative value. Nonetheless, Charles Kellogg included Hans Jenny's book in the various reading lists for soil scientists [13–16].

Some of the connections between soil scientists came from professional organizations, such as the *American Association of Soil Survey Workers* and its successors, or they knew each other through their college years, or meetings of the *International Society of Soil Science* or FAO. There were exclusive groupings such as the Cosmos Club in Washington, where Milton Whitney, Curtis Marbut and Charles Kellogg were members. According to Charles Kellogg, it was at the club where: "…the real decisions on scientific plans were made." There were connections through fraternal organizations like the Freemasons to which Richard Bradfield, Charles Kellogg, and Harlow Walster belonged. Charles Kellogg had joined the Masons in 1924 and became a member of the Masonic order to please the Reasoners—his parents-in-law. Although he had joined the Freemasons, Charles Kellogg believed more in the influence of the Cosmos Club members than in religion.

Several soil scientists studied at religious college such as Brigham Young College in Utah which was founded by the Latter-Day Saints, better known as Mormons, and Champ Tanner and Nyle Brady received degrees from Brigham Young. Many soil physicists studied at neighboring Utah State Agricultural College, such as Lorenzo Richards, Walter Gardner, Wilford Gardner, Arthur Wallace (Dr. Iron), and Don Kirkham. There was also Earlham College, a Quaker College in Indiana, where Mark Baldwin, Earl Fowler, Ralph McCracken, Francis Hole, and James Thorp were trained [17].

Some soil scientists were deeply religious such as Wilbert Weir, a farm boy from southern Wisconsin. After he had obtained a soils degree, he moved back to his parent farm but, after some years, he accepted a graduate student position at the Department of Soils at the University of Wisconsin. In 1923, he was hired by Curtis Marbut to work for the *Bureau of Soils*. With Earl Storie, he worked on soil rating and mapping in California [18, 19], and was then hired as soil technologist at the Southwestern Cemeteries in Arizona. He retired and wrote the book *How real is religion*, in which he sought to find scientific answers to religious questions [20]. After he had finished his M.S. degree in soils in 1917, he felt that he: "…had unwittingly allowed scientific studies and false interpretations of the Scriptures to destroy a simple religious faith…I was truly an infidel. The word 'science' had come to mean far more to me than the word 'God'" [20]. He resigned from the *Bureau of Soils*, as the Spirit spoke to him during the Transcontinental Excursion in 1927. There are very few chronicles in soil science in which a religious influence was so outspoken.

Charles Kellogg was an optimist, trusted in the ability of science to help humanity and believed in the role of government in advancing the public good. Charles Kellogg admired the Democratic President, Franklin Roosevelt, who was raised in an aristocratic family—quite different from the Kellogg family. A supporting government for the individual was central to Charles Kellogg's political views and, in several aspects, he may have modeled himself after President Franklin Roosevelt [21]. He was a Democrat his entire life but he did not come from a progressive and democratic family or region. When he was born in 1902, there were more Republicans in Ionia County of Michigan than Democrats and the Republican Party has carried Ionia County in most elections ever since.

Emil Truog was a rock-ribbed Republican, an anti-communist, no supporter of the New Deal or President Roosevelt, and he rarely missed an opportunity to hold forth on politics [22]. The soil physicist Champ Tanner was a PhD student with Emil Truog and, one day, his mother visited him. She was an English teacher at Brigham Young College in Utah and used a text by Emil Truog as an example of excellent writing. Emil Truog was pleased to hear that. When the conversation turned into politics, she was under the impression that Emil Truog was of the same political side as she was—Emil Truog was very quiet, straining to be polite after the compliments he had received over his text. Champ Tanner carefully interrupted: "Mother, you need to know that Professor Truog is a very conservative Republican." It surprised his mother who exclaimed: "But he's so bright!" [23].

Emil Truog was a Republican and Charles Kellogg a Democrat. They knew their political differences but it did not affect their relationship, which was initially one of mentoring, then grew into friendship and collegiality.

Politically they had different views, scientifically they were in different sub-disciplines, but both were convinced that science was the best way to develop the nation, if not the world. It would bring peace and prosperity in the American tradition. They lived through two world wars, the global rise of communism and fascism, a severe economic depression, and both came from humble beginnings. There are gestures of gratitude and selflessness in their story, and they had a relentless determination to make contributions to the welfare of humankind. Since they met for the first time in 1929, Emil Truog was a dedicated mentor, and Charles Kellogg, 18 years younger, was always appreciative of his advice. In May 1931, some 17 months into his first job at *North Dakota Agricultural College*, Charles Kellogg wrote to Emil Truog: "I can't thank you enough for all you've done for me. Your last letter was gold to me. I certainly hope the chance comes some day for me to return something to you."

Despite their differences, Emil Truog and Charles Kellogg got along well but some people just did not, such as Milton Whitney and Eugene Hilgard, Milton Whitney and Cyril Hopkins, Jacob Joffe and Selman Waksman, Emil Truog and Andrew Whitson, Hans Jenny and Charles Shaw, Charles Shaw and Dennis Hoagland, and James Thorp and Constantin Nikiforoff. Charles Kellogg did not get on with Hugh Bennett, Victor Kovda, Ferdinand van Baren, or William Albrecht. They argued, they clashed, did not talk to each other, or wrote letters to their superiors complaining about the other. Some of the differences and disagreements were scientific, others were personal, but most were intertwined. As soil science was a small community, these disputes affected progress; a clear example was the conflict between Milton Whitney and several of his contemporaries. Some have argued that the soil survey progressed slowly because of Milton Whitney's stubbornness and the conflicts that he had with leading soil scientists in the country [24].

The soil science community had also members that did not conform to the general type, such as, for example, Sergei Wilde, Constantin Nikiforoff, Hans Jenny, and William Albrecht. They had different backgrounds, ideas, and in a different time and place, their ideas may have been more influential than in their own times. On the other hand, the community was probably tolerant of eccentrics, provided they worked hard and produced results to the benefit of soil science. As Bertrand Russell stated: "…a community needs, if it is to prosper, a certain number of individuals who do not wholly conform to the general type. Practically all progress, artistic, moral, and intellectual, has depended upon such individuals" [25].

Very few women appear in the chronicles and portraits recounted in this book. Vasily Dokuchaev acknowledged the help of his wife Anna Sinkler,

Charles and Lucille Kellogg with John F. Kennedy in 1963. Two democrats shaking hands. Photo special Collections, USDA National Agricultural Library

who had worked with him for 20 years [26]. She had become interested in the earth sciences and mineralogy, and Vasily Dokuchaev stated that she deserved the title of the first Russian female pedologist. Her soil science contributions have never been studied and highlighted. In the USA, the first women were hired by the soil survey in the late 1890s, but they could not join the field crews, and were banned from field work until the 1940s [27, 28]. Some women participated unofficially, such as Mary Baldwin, who joined her husband Mark Baldwin in surveys in Wisconsin and Minnesota during the 1920s. She was not allowed to work for a federal agency, as married couples were discouraged from working for the same agency [27].

Many women made contributions to soil science. In 1927, Claribel Barnett of the United State Department of Agriculture library prepared the voluminous *A classified list of soil publications of the United States and Canada* [29]. It listed all soil studies conducted up to 1927. Lillian Wiland and Lois Olson contributed to the *Soil Conservation Service* through the compilation of a bibliography on erosion. Dorothy Nickerson developed the soil color standards that led to the Munsell color chart system, introduced in the soil survey in 1949 [30]. Other women contributed to the American soil survey through editing and laboratory work. In 1946, Mary Baltz graduated from Cornell University and was hired as junior soil surveyor. She was the first female soil surveyor, and the opportunity arose because of labor shortages after the war. Female participation in the *International Society of Soil Science* congresses was very low. There was one female soil scientist who attended both the 1927 and 1960 congress, Jadwiga Ziemięcka, who founded the first laboratory of soil microbiology in Poland [31]. It may be that the gender inequity in soil science was particularly problematic in the USA, and the balance was better in other countries, but no comparative studies have been conducted.

Sexism in soil science was common. The first sentence of the widely used textbook *Nature and Properties of Soil* was "Man is dependent on soil." In 1962, Charles Kellogg held a recruitment speech in Ithaca, New York, and expressed his view for good candidates: "...especially of well-trained, broadly educated young men who can develop rapidly" [27]. He also stated: "Our service offers a scientific career with opportunities for research to men in the field of soil science. Our staff includes some of the outstanding men in the field of soil science." Lucille Kellogg was introduced to President John F. Kennedy as "Mrs. Charles E. Kellogg." Many of the wives of the soil scientists moved all over the country and might have had little to no say in those moves. For example, the soil scientist Wilbert Weir moved with his wife and daughter from Wisconsin to Washington to New Jersey to Arizona. He was indebted to his wife for all the moves and wrote in 1956 about her: "...never asking the reason why, even though the changes and moves seemed unreasonable" [20]

There were no female soil scientists who worked with Charles Kellogg or Emil Truog other than those who provided office or laboratory assistance. They both referred to students as "the boys." Roy Simonson called undergraduate soil courses a "boy's course" and courses in the 500 series that were generally taken by graduate students "man's courses." Emil Truog was a mentor to 175 graduate students, but never mentored a female student. The first female student who graduated with a PhD in soil science at the University of Wisconsin was Jaya Iyer in 1968—that was 14 years after Emil

Truog had retired. Jaya Iyer then became the first female professor of soil science at the University of Wisconsin. The first female that graduated with a PhD in soil science was Ester Perry at the University of California Berkeley in 1939. She was mentored by Charles Shaw [27]. The first female president of the *Soil Science Society of America* was Mary Collins in 2005, and by that time the society was almost 70 years old. In the late 1980s, women moved into leadership roles in soil survey: Carol Wettstein became in 1988 the first female state soil scientist in Maryland, Carole Jett in California in 1991, and Carol Franks in Arizona in 1994. Since 1999, the soil survey has had women in national leadership positions like Maxine Levin, Carolyn Olson, and Pam Thomas. Although the American soil survey is now over 120 years old, it has not had a female director.

The soil survey excluded all minorities during its first 60 years [32]. Bill Shelton from Virginia fought in Italy during the Second World War, and in 1949, he earned a degree in agronomy at Virginia State University through the G.I. bill that provided funds for education. As there was apartheid in the south, African Americans could not enroll for a graduate education at Virginia Polytechnic Institute, so Bill Shelton moved north to Michigan State University where he earned a M.S. degree in soil science in 1953. After finishing his degree, he had no job offers, and the *Soil Conservation Service* did not hire African Americans when they had to work with farmers. Bill Shelton was hired as a soil surveyor in North Dakota [32]. The passage of civil rights legislation in 1960 forced the soil survey to become more inclusive.

Some soil research took place at the historically black Land-Grant institutions in the early 1900s but, compared to other universities, their capacity and funding were limited. Teaching programs in soil science began at the North Carolina Agricultural and Technical University in 1936 and at Virginia State University in the 1940s. The black Land-Grant institutions and their graduates have played an important but understudied role in the soil survey. Garland Lipscomb became in 1981 in Pennsylvania the first African American state soil scientist. In 1996, Horace Smith was the first African-American director of the *Soil Survey Division*; by that time, the soil survey was almost 100 years old [32]. Horace Smith had obtained a M.S. in soil science from the Ohio State University in 1971. In 1999, Birl Lowery was the first African American to Chair the Department of Soil Science at the University of Wisconsin; the department was founded in 1889. The *Soil Science Society of America*, 85 years old in 2021, has never had an African-American President.

Emil Truog and Charles Kellogg flourished in an academic era that was less managerial and less dependent on external funding. Attractive as that may sound, it was a time of the "old-boys" network, somewhat feudal, and lacked diversity of any kind. There was a degree of academic freedom and most saw the sciences not as job but as a vocation [33]. It was an era of discoveries and progressions, but future generations will have to determine how the rate of discoveries in the first half of the twentieth century compares to later periods, when there was more diversity. With the lack of diversity, the amount of bias, idiosyncratic behavior, and the agrarian focus, one wonders whether soil science was unique in these aspects, and whether other related sciences had similar issues. For example, would soil science had advanced more had it been more inclusive and less focused on agriculture? Contemporary research shows that diversity matters, and that science benefits from diversity, suggesting that soil science would have had more frequent paradigm shifts, had the discipline been more diverse. However, the recipe for making breakthroughs and discoveries in soil science has yet to be written. A prerequisite seems to be autonomy to explore new pathways; there was a degree of autonomy during the first half of the twentieth century, but it was mostly for white men.

Studying soils has rarely been seen as a cultured profession, and in the USA, the study of soils was dominated by men from the farm. The appreciation of soil closely followed societal needs, in particular, the need for affordable food. Beyond that, society has shown indifference to the soil and that has increased with urbanization and greater food security. This indifference by society to soil is no new thing. The geologist Nathaniel Shaler had an interest in soils and wrote in 1892: "Now and then a poetic spirit, anticipating with the imagination the revelations of science, has spoken of the earth as the mother of all; but the greater part of mankind, those who are well instructed as well as the ignorant, look upon the soil as something essentially unclean, or at least as a mere disorder of fragmentary things from which seeds manage in some occult way to draw the sustenance necessary for their growth. Any chance contact with this material fills them with disgust, and they regard their repugnance as a sign of culture" [34]. It was in that climate that the soil science discipline arose, and the ignorance for soil was in 1894 worded as "The phenomena exhibited by the soil coating of the earth are so familiar that they are often contemptuously overlooked" [35].

Some 35 years later, Bernard Keen from Rothamsted attended the *First International Congress of Soil Science* in 1927 and had the following observation: "Agriculture differs in this respect from say, the medical and electrical sciences. The average educated citizen approves of active research in these sciences for he cannot avoid knowing how closely they concern his material comfort and health. But the equal urgency of studies on, for example, the physiology of a fungus or base exchange phenomena in soil, is not evident to him. I experienced an amusing example of this fact. Members of the Congress who stayed at the Willard Hotel in Washington will remember that there was also in progress at the Hotel the 'Convention' of another Society, whose members were well educated professional men. One of them approached a group of us in the hotel lobby and began to speak about his own subject. Then he noticed that our badges were different from his own and realizing his mistake politely asked who we were. I explained we were members of the *International Society of Soil Science*, whereupon he asked with a somewhat puzzled tone what we did. I replied that the physical, chemical and biological properties of soil were the basis of all agriculture and that our members were engaged in research on these important matters. He appeared still more puzzled and asked: And what is the use of all that, anyway?" [36].

In the 1930s, Gilbert Robinson wrote: "I am not sure that the general public is really interested in soil. I notice two types of reaction when I am introduced to people as a soil expert (a title I should never claim). Either the word 'soil' suggests some ribald joke as to my occupation, or else my interlocutor asks me if I can tell him what is good for his garden! I have rarely met anyone interested in the soil for its own sake" [37]. Those are not the observations of some contemporary defeatist, but they were written in 1937 by a great pedologist, and a friend of Charles Kellogg. Gilbert Robinson thought that everyone should become "soil-conscious." He also found that much of the failure in applying science to soil problems has been due to the elementary mistake of confusing soil material with soil as an individual in the field. That was problematic in the 1930s and continues to be so today.

Education was seen as a way to promote soil science and in 1950, Emil Truog had wondered why so little was taught about soils in schools: "In elementary geography classes, pupils learn the names of the tallest mountains, the highest waterfalls, the largest lakes, the longest rivers, and the locations of the greatest deserts and most productive gold mines. What do they learn about the character and location of the soils which feed and shelter all of us for the most part? Surprisingly little. The first step to take in correcting this situation is, I believe, the preparation of soil maps which are suitable for various kinds and levels of classroom work. An introduction to soil maps and their use at an early stage will without question help greatly in promoting usage and understanding of these maps in later years" [38]. Surprisingly perhaps, this can still be heard decades later [39–41].

In 1960, Emil Truog wrote to the National Science Foundation in the hope to get some funding for the *Seventh International Congress of Soil Science*. He noted that: "...the public preoccupation with successful practical application has led to erroneous conclusions about the content and nature of soil science, and to underemphasis of its basic aspects, notable soils per se and their complex, mineral, organic, gaseous and microbiologic systems." Funding was received and soil science flourished in what we can now see as a Golden Age of the 1950s and 1960s.

Emil Truog was trained in analytical, organic and inorganic chemistry and made contributions to the understanding of soil acidity, liming, phosphorus, and cation exchange properties of soils as well fertilizer requirements [6]. His eagerness for improving the productivity of the land brought him to the university and characterized his career. One of his final contributions was the organization of *Seventh International Congress of Soil Science* in 1960, for which he had chosen the theme: *To Promote Peace and Health by Alleviating Hunger through Soil Science*. The program book contained the quote: "It is sometimes said that soil is Wisconsin's most valuable natural resource. Its worth in terms of dollars may be forcibly expressed by comparing the cash value of what it produces to that of some other natural resource. For example, if all of the world's gold mines were concentrated within the borders of Wisconsin, how might the production value of this resource compare with that of her present soil? Startling as it may seem, during recent years the annual value of the agricultural products produced from the soil of Wisconsin greatly exceeded the annual market value of the world's gold production. Moreover, gold mines peter out, never to recover, while soil when properly managed according to the modern teachings of soil science lasts indefinitely, and in many cases even gains in productivity." Soil degradation was not acceptable to Emil Truog; he held a strong conviction of the continued rise in agricultural production, provided the soils were properly managed, and that there was modern teaching in soil science.

Emil Truog had clear-cut and well-defined opinions, and he was fearless in defending them [42]. His opinions stretched across all areas that affected his work, including the use of soil survey maps and reports [38]. As part of his B.S. degree training, he had participated in a soil survey in Wisconsin, and he later advocated that different maps of the same area should be made for different users [43]. He had profound ideas on a range of issues related to farming, and alternative views were contested. Emil Truog was convinced that Russia could not feed itself [44]; argued against the sole dependence on organic inputs to maintain soil fertility [45]; called the organic agriculture movement "a cult" [46]; testified against William Albrecht and Jerome Rodale in the 1950 Congressional hearings on Chemicals in Food [47], and wrote numerous letters to organizations expressing his views. Emil Truog refuted the *Plowman's Folly* that advocated less plowing [48], and wrote in *Harper's Magazine*: "...it is nonsense to maintain that the moldboard plow has sapped the soil of its fertility, raided the nation's food basket, fostered crop pests, and even paved the way for current vitamin-pill fad" [48]. In many of these opinions, he was not alone; Charles Kellogg agreed with him on the work on William Albrecht, whom he called "a quack," and on the *Plowman's Folly*— as did Robert Schofield, the soil physicist from Rothamsted, who wrote in *Nature*: "Mr Faulkner's folly in committing himself to print on so slender a pretext is infinitely greater than that of any ploughman" [49].

Soil research has national characteristics that includes a strong focus or neglect of particular research topics, and a higher level of self-citations in some countries, including the USA [50, 51]. The acceptance of new ideas and approaches is different in different nations. The beginning of this book reviewed how the genetic principles of soil science developed by Russian workers were ignored; also, within the USA, work of some soil scientists was disregarded. The late acceptance of ideas not only reduced progress in the late 1800s and early 1900s but continued for decades. In the 1960s, many in higher education in agriculture held the view: "Why change? Our only need is to explain what has been done", which implies that conventional thinking will eventually lead to a conservative attitude.

Charles Kellogg and David Knapp summarized the status of university education in the 1960s as follows: "We Americans have some curious notions about scientists. We want men and women in scientific research who are unconventional within their areas of scientific responsibility. We want them to question the traditional and the current textbooks and authorities. We want them to look for deeper truth and to find new principles that can lead to better technology, better institutions, and the better life. Still we insist that they be conventional in everything else" [3]. They were concerned about intellectual provincialism in which there was a climate lacking in scholarship, imagination, and curiosity. Geographic provincialism referred to faculty that have degrees from the university at which they worked, whereas solving local problems is often done by those with wide geographic experience and a more cosmopolitan experience [3]. Geographic provincialism was more widespread in agricultural colleges than elsewhere, and as soil science often resides within agricultural colleges it may explain some of the slow progression patterns in the USA.

American soil science has many exceptions to narrow intellectualism. Emil Truog, born and raised on a farm in Wisconsin, never left the university where he received his B.S. and M.S. degrees. He rarely traveled abroad and did not have the interest in soils of the tropics that came naturally to Charles Kellogg. But he liked to work on soil problems that affected crop production and increased crop yields, which he viewed as essential for a peaceful world.

He was a geographic intellect; his research was far from provincial and he stimulated curiosity, scholarship, and imagination in all his graduate students. Charles Kellogg grew up in a small village, had his education in the Midwest but became an internationalist who firmly believed in global cooperation.

Charles Kellogg was in favor of the SI units or metric system, as he wrote in 1968: "For many years I have favored adoption of metric units in the United States, both in science and in industry. Thus, I heartily approve the action taken by the *Soil Science Society of America* Editorial Board making use of the metric system mandatory in all society papers published after January 1, 1967. Advantages of using a uniform system throughout the world are obvious. In a scientific journal our principal purpose is to communicate as clearly and briefly as possible to a large scientific audience that reads papers from many countries and that is already fully familiar with the metric system. The authors converting inches, pounds, and the like to their metric equivalent should be warned against the saving of small fractions that give an obviously false impression of accuracy" [52] The proposal to adopt the metric system received quite some criticism: "The exclusive use of the metric system in reporting data makes no contribution to the quality or fundamental importance of the research but does seriously impair the understanding and comprehension of the results by a large and important segment of readers of these journals" [53]. A conventional attitude dominated the use of units, why change!

Despite their difference in international activities, both Emil Truog and Charles Kellogg became Honorary Members of the *International Society of Soil Science* which was internationally the highest recognition for soil science and service. Emil Truog became an Honorary Member in 1954 and he was the second American; Charles Kellogg became Honorary Member in 1974. Emil Truog attended the first and third international soil congress and led the organization of the seventh congress. Charles Kellogg attended the third international congress in Oxford in 1935 and every congress thereafter until the eleventh congress, in Edmonton in 1978. By that time, he was 76 and not in good health. According to Tony Young, he was hoisted onto the platform in Edmonton, and Victor Kovda thought he looked "like a walking cadaver." Victor Kovda and Charles Kellogg were not friends. Roy Simonson thought he should not have come to Edmonton, and Charles and Lucille Kellogg left before the congress was over.

Charles Kellogg grew from international travel and cooperation and became a world citizen with views that he brought home in the soil survey. When he first traveled through the Congo and Ghana he must have reminisced about his own upbringings, the hardship on a farm with poor soils in

Michigan. He felt a connection to farmers across the world, and found that: "…every farmer should have at his command the knowledge of the whole world about his particular type of soil – knowledge gained through scientific researchers and from the experience of farmers everywhere on the same kind of soil" [54]. He observed how farming in the USA changed, and somewhat apprehensively, he wrote in the early 1940s: "The *business of farming* has replaced the *art of agriculture*. *Soil* becomes *land* or real estate" [12]. In the 1960s, he also realized that the social aspirations of farm families had changed: "…they need good housing, tasteful furniture, and pretty gardens as much as other people. They want well-equipped libraries, fine paintings, and good music" [3].

Farming and agriculture changed greatly between the late 1800s and 1960 in the USA. The farm population was 32 million in 1910, but less than 15 million in 1960, or from 35 to 9% of the total population. The decrease in farm population was accompanied by an increase in higher education in agriculture and expanding Colleges of Agriculture. In 1960, there were 240 colleges in the USA that taught agriculture, and 50 of those were in Land-Grant universities [3]. These colleges were often named the "people's college" and much of their research, education and extension focused on soil, water, plants, and animals with the unstated assumption that if these were managed well, the welfare of farm families would grow. There was the widely accepted belief that knowledge would serve the people. Charles Kellogg and David Knapp addressed this in the mid-1960s: "In a way this philosophy flows from the basic American dream - the opportunity to get ahead under a fair set of rules" [3]. As agriculture changed, so did the colleges. The student population in agriculture increased, but their proportional share decreased. Between 1910 and 1960 the number of students in the Colleges of Agriculture increased from 5,400 to 35,200 but as percentage of the total university student population, the number of students in Colleges of Agriculture declined from 20% to less than 8% [3].

The Colleges of Agriculture put much emphasis on plants and animals, and Charles Kellogg and David Knapp became concerned that the sole focus was on production and efficiency. They foresaw more attention paid towards the environmental sciences that would concentrate on the protection of air, water, and soil from contaminants. Besides the environmental aspects, they urged more sociological studies, and in their view, the Colleges of Agriculture had neglected the responsibility for people and families in rural areas, many of them having a low-income. In the early 1960s, 45% of the farm families were defined as poor whereas 18% of the non-farm families were poor [3]. There was work to do, a generation to be educated, scientific advancements to be made, and in all that, the people should not be forgotten, and the environment preserved.

Up to 1960, there was little attention given to the environmental impact of agriculture and alternative methods or ideas such as those of William Albrecht, Edward Faulkner [55, 56], Lionel Picton [57], Jerome Rodale, or Louis Bromfield were ignored, if not ridiculed. Productivity, profitability, and rational agriculture was the way forward [58]. In 1961, this was worded by the *Successful Farming's Soils Book*, published in Iowa: "…new ideas in soil management…money-making keys to big crop yields *at lowest cost*. How to understand and use the latest fertilizers, weed killers, and insecticides on your farm" [59]. The book stressed that the: "…soil is the farmer's lifeblood, and the lifeblood of civilization and must pass on to sustain the civilizations which follow." Many soils required a "build-up" of plant nutrients and more fertilizers should be used; similar recommendations were made for insecticides and herbicides.

Charles Kellogg had a philosophical and classical side; he enjoyed literary discussions while smoking his pipe and was always well-dressed, if not a little vain. Let's assume that was because he grew up in Ionia—those Ionians with their love for philosophy, art, democracy, and even pleasure. As a child, he was embarrassed by his shabby clothes and, when he had more means, he dressed well, even when he went to the field. John Arno was a soil surveyor in Maine and, in the late 1930s Charles Kellogg came for visit. John Arno was too inexperienced to know what all the discussion was about but realized that: "If I could not understand the classification of soils, I could appreciate the value of Dr. Charles Kellogg's highly polished riding boots and pants and his gold watch chain" [60].

More notable than how he looked, Charles Kellogg was a bibliophile and gathered first editions of books by James Joyce, Virginia Woolf, and Dorothy Richardson [61]. After he had arrived in Fargo in January 1930, he complained to Emil Truog: "…there isn't a single book store; everyone in Fargo has a book!" He adored James Joyce and writers influenced by Joyce, especially Marcel Proust, Dorothy Richardson, and D.H. Lawrence. Charles Kellogg moved his books from East Lansing to Madison, from Madison to Fargo, and Fargo to Washington, and in the 1940s, he bought a bookcase with three shelves: for Marcel Proust, D.H. Lawrence, and James Joyce. His James Joyce collection was started in 1930 with a reproduced copy of *Ulysses*, and he might have read the interview with James Joyce in *Harper's Magazine* in 1929, in which he said: "The demand that I make of my reader is that he should devote his whole life to reading my works." He surely did. Charles Kellogg corresponded with famous James Joyce scholars and led a colloquium on James Joyce at Texas A&M in the English Department [61]. He named their cats after Joycean characters.

Charles Kellogg did not grow up with much literature but became literary, erudite, and widely read. He was a book collector but, in the beginning, money was seldom spent except for a few lucky bargains. The year 1930 was warm and dry, and Charles conducted a soil survey of the estate of the President of the Davison Chemical Company in Baltimore. It paid handsomely. On weekends, he visited the Peabody Bookshop and Beer Stube in Baltimore. It was a crowded, dusty bookshop in front, and in the backroom customers could get beer and a sandwich. There was a piano in the shop, mounted animal heads, and wooden tables carved with the names of patrons. The bookshop was started by the brothers Hugo and Siegfried Weisberger, Austrians who had immigrated in 1912, and started a bookshop in 1922 with its own beer bar two years into the prohibition. Siegfried Weisberger sold the Peabody Bookshop in 1954, convinced that people were no longer interested in: "…books and ideals and culture. They only want dollars." He foresaw dark days for America.

In 1930, Siegfried Weisberger persuaded Charles to buy a copy of *Ulysses*. That was his first James Joyce book that grew over the years to a large collection. Later on, Charles Kellogg was often asked where his fascination came from and explained that by saying: "…James Joyce influenced good writing in English more than any other author after William Shakespeare and the compilation of the King James' version of the Bible." Charles Kellogg admired James Joyce's unusually retentive memory and education by Jesuits at the University College in Dublin. James Joyce taught himself Norwegian so he could read Henrik Ibsen's plays. According to Charles Kellogg, the only test for a good book, poem, or essay is whether it answers the question: "Is this the way the world is?" And reading James Joyce was for Charles Kellogg a confirmation of the way the world was.

Charles Kellogg was a serious man, and his son Robert recalled him as stern, focused entirely on his work and reading [61]. His need to read was encouraged by his grandmother, who had a Victorian intellectualism, in which reading was seen as virtue. The 1920s was the peak decade for reading in America [62]. There were about 120 million people in 1927 and each year about 100 million books were published, and over 36 million newspapers were printed every day [62]. That was all before the radio. Charles Kellogg also liked teaching and, after his retirement, taught at the graduate school which was an institute for continuing education by several government agencies. He taught the history of classical liberalism from John Stuart Mill that justified the freedom of the individual in opposition to state and social control. When he became older, students would come to their home in Hyattsville, and he would talk about the importance of soil and, of course, about James Joyce [63]. The Kelloggs were welcoming and Lucille would serve drinks.

Their son, Robert Kellogg, became a professor of English literature at the University of Virginia in Charlottesville [64]. Robert had his father's passion for literature inborn. His M.A. and PhD from Harvard University were in medieval English literature but, also, became interested in Icelandic literature through the friendship between Charles Kellogg and the Icelandic soil scientist Björn Johannesson [63]. Robert Kellogg edited the *Soils of Iceland* that Bjorn Johannesson published in 1960 [65]. Robert's position at the University of Virginia might have been a position to which his father, in a different life, would have aspired. Literature meant much to Charles Kellogg but he grew up differently. He was an only child and his farm background and scientific education rooted in him the urge to help people understand the soil that supported them.

Charles Kellogg directed and expanded one of the world's largest soil surveys. Bureaucracy in the federal government, the clashes about direction and funding, and the daily mendacities and disputes were mollified by vanishing into the great books of literature. It had comforted him ever since he could read, and when he first found contentment in books after fights with his father. Given his steadfast character, reading might not have changed his mind on daily matters, but it certainly soothed his soul. Besides reading and collecting books, Charles Kellogg grew azaleas and found gardening relaxing. Combining his hobby and work, he wrote a book on garden soils that showed gardeners the importance of the soil [66].

In 1940, Charles Kellogg prepared a paper about reading for soil scientists, together with a library for a soil scientist, that listed books on Soil science; Related sciences; Early agriculture; Science history, meaning, method, and philosophy; Philosophy, conduct of life, history, etc.; Novels and stories; Biography; and Drama and Poetry [13]. This list was published in his 1941 book *The soils that support us*, and was updated in 1964 and 1971 [12–16]. He noted upfront: "Most people will not read much unless they derive a certain satisfaction from it. Some get satisfaction from the very music of word combinations. Others enjoy reading about places or experiences that recall to themselves pleasant places they have seen or pleasant experiences they have had." He listed some of his favorite Russian books: *The Brothers Karamazov* by Fyodor Dostoevsky, *Father and Sons* by Ivan Turgenev, *War and Peace* and *Anna Karenina* by Leo Tolstoy and Leon Trotsky's *History of the Russian Revolution* from 1930 [61].

In 1968, Charles Kellogg published a poem entitled *A Lament for B* that he wrote for Guy Smith in 1957 on the train home from Monticello where he had visited the plantation that had been owned by Thomas Jefferson. He reflected on soil-horizon nomenclature and assured Guy Smith that: "It will definitely NOT be helpful." The B horizon became more specifically defined, and Charles Kellogg was worried about the loss of simplicity. He was a generalist and disagreed with Guy Smith, who advocated a quantitative approach, which was an assault on Charles Kellogg and his authority [67]. A fragment of *A Lament for B*:

> The old podzolic texture B
> Has clay-skins now to make it be,
> And if there are no skins to see
> It still may be a clayey B;
> Or better still, try hard to see.

And the poems end with a plea:

> One sigh for Mother B
> And in her name a plea:
> Don't lose her major connotation
> By unneeded mutilation.

His protest had little effect and over the years a fashionable alphabet of suffixes was added to magnify the B horizon.

After Emil Truog had finished his M.S. degree, he was offered a position in the Department of English at the same time he was offered a position in the Department of Soils. He had a Swiss precision about language, and became known as a rigorous editor among his students and colleagues. He insisted on good writing, and much of his own writing was well-balanced, with a degree of starkness that left no room for ambiguity, not that different from Charles Kellogg's, who was more widely read and a more prolific writer.

Given the background and career passages of Charles Kellogg and Emil Truog, it is tempting to ponder a little about *William Stoner*—the campus novel by John Williams about the farm boy entering the University of Missouri in the early 1900s [68]. William Stoner was an only child and raised on a small farm west of Columbia, Missouri. An extension agent visited their farm one day and recommended to his father that William attend the new course in agriculture at the university. During his second year, William had a course in soil chemistry and a course that surveyed the English literature,

Robert L. Kellogg (left), son of Charles Kellogg, taught at the University of Virginia for 42 years before retiring in 1999. He served as chairman of the English department and Dean of the College of Arts and Sciences. His specialty was medieval English literature and Icelandic sagas. John Williams (right) was born in Texas in 1922 and fell in love with literature in high school. His grandparents had been farmers. He served in the Air Force from 1942 to 1945 and the GI Bill enabled him to go to college. John Williams obtained a PhD degree at the University of Missouri, where the book *Stoner* was set early 1900s

as John Williams wrote: "The course in soil chemistry caught his interest in a general way; it had not occurred to him that the brownish clods with which he had worked for most of his life were anything other than what they appeared to be, and he began vaguely to see that this growing knowledge of them might be useful when he returned to his father's farm." But the course in English literature transformed William Stoner, and the English writers, particularly William Shakespeare, opened a new world for him. He fell in love with literary studies, and it was a world very different from the toiling of the land that William Stoner had done since he was a small boy. Without telling his parents, he dropped out of the agriculture program and at graduation, he told them he will not be returning to the farm and complete his M.A. in English. After his degree, he began teaching and developed a passion for love and learning but failed at both. His parents sold their farm, and the book did not have a happy ending. The academic path of William Stoner was not walked by Charles Kellogg or Emil Truog who resided within the territories of their ancestry.

America is a big country and its soil science is geographically and academically interwoven, with several centers in the Midwest. Over time, the universities and centers at the east and west coast gained prominence; the southern states created some excellent universities as well but did not become the center of soil science. There are connections in time and space between all members of the soil science community described in the previous chapters. There are unique aspects to the American chronicles and progressions, and in

Charles Kellogg's view: "…there are a great variety of soils, of landscapes. No other country has greater contrasts, and in no other country have the roads and other conditions permitted such a large percentage of the people to travel. Some hold that the peculiar characteristics of the Americans are due in part to this movement from landscape to landscape. It may account for some of our nervousness, tension, and hurry - whether there is any reason to hurry or not" [12].

The world of soil science in the USA is tightly intertwined, and there may be something peculiar about American soil science due to the movement from landscape to landscape. The cradle and history traces back to a few personalities. Its future is wide open, and as diverse and complex as the soil itself.

References

1. Swidler E-M. The social production of soil. Soil Sci. 2009;174(1):2–8.
2. Kellogg CE. A challenge to American soil scientists on the occasion of the 25th anniversary of the Soil Science Society of America. Soil Sci Soc Am Proc. 1961;25:419–23.
3. Kellogg CE, Knapp DC. The college of agriculture. Science in the public service. New York: McGraw-Hill Book Company; 1966.
4. Fanning DS, Fanning MCB. Milton Whitney: soil survey pioneer. Soil Horiz. 2001;42:83–9.
5. Kemper WD. Environment and its role in the development of a scientist. Soil Sci Soc Am Proc. 1970;34:365–8.
6. Maher D, Stuart K, editors. Hans Jenny—soil scientist, teacher, and scholar Berkeley: University of California; 1989.
7. Miller MF. The soil—its improvement and conservation. Columbia, Missouri: Lucas Brothers; 1950.
8. Marbut CF. Translator's preface. In: The great soil groups of the world and their development, by K.D. Glinka. Ann Arbor, Michigan: Edward Brothers; 1928.
9. Wilde SA. Dr. Werner's facts of life. New Delhi: Oxford & IBH Publishing Co.; 1973.
10. Cline MG. Agronomy at Cornell 1868 to 1980. Agronomy Mimeo No. 82-16. Ithaca, NY: Cornell University; 1982.
11. Anon. Resolution in memory of Victor Lenher. Science. 1927;LXVI:76.
12. Kellogg CE. The soils that support us. New York: The Macmillan Company; 1941.
13. Kellogg CE. Reading for soil scientists, together with a library. J Am Soc Agron. 1940;32:868–76.

14. Kellogg CE. Reading for soil scientists, together with a library. Washington: United States Department of Agriculture; 1971.

15. Jenny HEW. Hilgard and the birth of modern soil science. Pisa: Collana della revista agrochemica; 1961.

16. Jenny H. Factors of soil formation. A system of quantitative pedology. New York: McGraw-Hill; 1941.

17. Helms D. Early leaders of the soil survey. In: Helms D, Effland ABW, Durana PJ, editors. Profiles in the history of the U.S. soil survey. Ames: Iowa State Press; 2002. p. 19–64.

18. Weir WW, Storie RE. A rating of California soils. Bulletin 599. Berkeley, California: University of California; 1936.

19. Storie RE, Weir WW. Generalized soil map of California. Berkeley: University of California, Division of Agricultural Sciences; 1953.

20. Weir WW. How real is religion?. New York: Vantage Press; 1956.

21. Grossman RB. Note on C.E. Kellogg by a junior staff member. Soil Surv Horiz. 2004;45:144–148.

22. Beatty MT. Soil science at the University of Wisconsin-Madison. A history of the department 1889–1989. Madison: University of Wisconsin-Madison; 1991.

23. Tanner CB. Letter to Walter Gardner. Department of Soil Science Archive; 1976.

24. Moon D. The American steppes. The unexpected Russian roots of Great Plains agriculture, 1870s–1930s. Cambridge: Cambridge University Press; 2020.

25. Russell B. Authority and the individual. New York: Simon and Schuster; 1949.

26. Dokuchaev VV. The place and significance of modern pedology in science and life (Translated from Russia by the Israel Program for Scientific Translations). Izbrannye Sochineniya. 1949;III:330–338.

27. Levin MJ. Women in soil science (USA). In: Hillel D, Hatfield JL, Powlson DS, et al., editors. Encyclopedia of soils in the environment, vol. 4. Amsterdam: Elsevier Academic; 2005. p. 345–352.

28. Lapham MC. Crisscross trails: narrative of a soil surveyor. Berkeley: W.E. Berg; 1949.

29. Barnett CR. A classified list of soil publication of the United States and Canada. Washington, DC: United States Department of Agriculture; 1927.

30. Simonson RW. Soil color standards and terms for field use—history of their development. In: Bigham JM, Ciolkosz EJ, editors. Soil color. Madison, WI: Soil Science Society for America; 1993. p. 1–20.

31. Królikowski L. Professor Dr. Jadwiga Ziemiecka (1891–1968). Bull Int Soc Soil Sci. 1986;33:51.

32. Dewayne Mays M, Smith H, Helms D. Contributions of African-Americans and the 1890 Land-Grant Universities to soil science and the soil survey. In: Helms D, Effland ABW, Durana PJ, editors. Profiles in the history of the U.S. soil survey. Ames: Iowa State Press; 2002. p. 169–189.

33. Philip JR. Soils, natural science, and models. Soil Sci. 1991;151:91–8.

34. Shaler NS. The origin and nature of soils. Department of the Interior. U.S. Geological Survey. Washington: Government Printing Office; 1892.

35. Shaler NS, editor. The United States of America. A study of the American Commonwealth, its natural resources, people, industries, manufactures, commerce, and its work in literature, science, education, and self government (3 volumes). New York: D. Appleton and Company; 1894.

36. Keen BA. The American agricultural research and advisory system. In: Waksman S, Deemer R, editors. Proceedings and paper first international Congress of Soil Science. Transcontinental excursion and impressions of the Congress and of America. Washington, DC.: The American Organizing Committee of the First International Congress of Soil Science; 1928. p. 157–162.

37. Robinson GW. Mother earth. Being letters on soil addressed to Sir R. George Stapledon C.B.E., M.A., F.R.S. London: Thomas Murby & Co.; 1937.

38. Truog E. Enlarging the use of soil survey maps and reports. Soil Sci Soc Am Proc. 1950;14:5–7.

39. Hartemink AE, Balks MR, Chen ZS, et al. The joy of teaching soil science. Geoderma. 2014;217:1–9.

40. Herrmann L. Soil education: a public need developments in Germany since the mid 1990s. J Plant Nutr Soil Sci. 2006;169(3):464–71.

41. Dobrovol'skii GV, Orlov DS, Rozanova MS. Soil science in secondary school and popular books on soils. Eurasian Soil Sci. 2002;35(2):210–5.

42. Jackson ML, Attoe OJ. In memoriam Emil Truog 1884–1969. Soil Sci. 1971;112(379–380).

43. Whitson AR, Geib WJ, Dunnewald TJ, Truog E, Lounsbury C. Soil survey of Iowa County, Wisconsin. Madison, Wisconsin: Wisconsin Geological and Natural History Survey; 1914.

44. Truog E, Pronin DT. A great myth: the Russian granary. Land Econ. 1953;29:200–8.

45. Truog E. "Organics only?"—Bunkun! Land. 1946;5:317–21.

46. Truog E. The organic gardening myth. Soil Surv Horiz. 1963;4:12–8.

47. Peters S. Organic farmers celebrate organic research: a sociology of popular science. In: Nowotny H, Rose H, editors. Counter-movements in the sciences. Sociology of the sciences. Volume III. Dordrecht: D. Reidel Publishing Company; 1979.

48. Truog E. Plowman's folly refuted. Harper's Mag. 1944(July Issue):173–177.

49. Schofield RK. Disk-harrowing versus ploughing. Nature. 1944;3883:391.

50. Yaalon DH. Has soil research national characteristics? Soils Ferti. 1964;27:89–93.

51. Minasny B, Hartemink AE, McBratney A. Individual, country, and journal self-citation in soil science. Geoderma. 2010;155(3–4):434–8.

52. Kellogg CE. More support for metric units. Soil Sci Soc Am J. 1968;32:605.

53. Kamprath EJ, McCants CB, Woodhouse W, Cox FR, McCollum RE, Woltz WG. Still more on metrics. Soil Sci Soc Am J. 1968;32:295.

54. Kellogg CE. Russian contributions to soil science. Land Policy Rev. 1946;9:9–14.
55. Faulkner EH. A second look. Norman: University of Oklahoma Press; 1947.
56. Faulkner EH. Plowman's folly. Norman: University of Oklahoma Press; 1943.
57. Picton L. Nutrition & the soil. Thoughts on feeding. New York: The Devin-Adair Company; 1949.
58. Bouma J, Hartemink AE. Soil science and society in the Dutch context. Neth J Agric Sci. 2002;50(2):133–40.
59. Anon. Successful farming's soils book. Des Moine, Iowa: Meredith Publishing Company; 1961.
60. Arno JR. Experiences in soil survey. Soil surveyor in Maine 1936–1976. Unpublished biography; undated.
61. Kellogg RL. Remembering Charles Kellogg. Soil Surv Horiz. 2003:86–92.
62. Bryson B. One summer: America 1927. London: Transworld Publishers; 2013.
63. Scholes R. A tribute to Robert Kellogg. James Joyce Quart. 2003;40:335.
64. Di J. In memory of a quiet Joycean: Robert L. Kellogg (2 September 1928–3 January 2004). James Joyce Quart. 2003;40(336–338).
65. Johannesson B. The soils of Iceland. Department of Agriculture Reports Series B—no. 13. Reykjavik: University Research Institute; 1960.
66. Kellogg CE. Our garden soils. New York: The MacMillan Company; 1952.
67. Anderson S. A Lament for B. NCSS Newslett. 1999;7:6–7.
68. Williams J. Stoner. Viking Press; 1965.

15

Epilogue

Soil science contributions are made in different ways by different kinds of scientists. Some make one or two significant contributions and, then, glide gradually into dormancy. Their contributions are recognized, cited for a while but, eventually, are surpassed by newer findings. One may hope that their contribution remains credited but the origin of their ideas may vanish in the generation that follows. Then there are those who make perhaps less significant contributions but keep going and generate results for decades. At some stage, the sheer volume of their work will be acknowledged and, if good, will be cited and create some history. Finally, there are those who unceasingly craft novelty and rejuvenate intellectually over the years. We have been blessed by only a few of these in soil science. I am not inclined to add names for these three categories of soil scientists but it is fair to say that both Charles Kellogg and Emil Truog made major contributions, and did that for many decades. Their legacy is not only scientific, they also built an institution and university department, trained a cohort of followers, and were part of the lifeblood of the global soil science community.

Emil Truog has been dead for 51 years, Charles Kellogg for 41 years. Their lives and works have been reviewed and their contributions celebrated in numerous publications and posthumous . Emil Truog's biography appeared in *Modern Americans in Science and Invention*, which also listed Leo Baekeland, who invented Bakelite, and Robert Williams, who isolated and synthesized vitamin B_1 [1, 2]. The pedigrees of Charles Kellogg and Emil Truog stretch across the world and their ideas have influenced the way we observe and manage soils today; they shaped the fields of pedology, soil survey, soil chemistry, and soil fertility. Emil Truog traveled, but resided his entire life in Wisconsin; Charles Kellogg traveled often and became a world citizen. The soil connected them, they knew its value for human survival and prosperity; both believed in the fruits of science and what it would bring to humanity. They were buoyantly optimistic about how good soil science increased agricultural production although in later years Charles Kellogg had more of an eye for the environmental impact of agriculture.

© The Author(s), under exclusive license to Springer Nature
Switzerland AG 2021
A. E. Hartemink, *Soil Science Americana*
https://doi.org/10.1007/978-3-030-71135-1_15

Gravestones of Emil Truog and Charles Kellogg. Charles Kellogg died on the 9th of March 1980 at Leland Memorial Hospital in Riverdale, Maryland. He had cardiovascular disease. Charles Kellogg was buried in Palo in Michigan. Lucille died in 2004. Emil Truog died 19th of December 1969 in Madison after he had spent his last years in a nursing home. He was buried on a drumlin of Forest Hill Cemetery in Madison, Wisconsin, and a few hundred meters apart from Franklin King and Andrew Whitson. Photos by the author

Like everyone else, the legacy of Emil Truog and Charles Kellogg will descend into the subsoil of oblivion and they will, eventually, be disremembered. Before that happens, what is their legacy beyond their scientific work? They both have grandchildren and great grandchildren who carry their surnames and family histories. Their professional archives are extensive, well-indexed, and accessible. Emil Truog has an award named after him: the *Soil Science Outstanding Dissertation Award* given annually by the *Soil Science Society of America*. Charles Kellogg's first position was at North Dakota State University which awards the *Dr. Charles Kellogg Scholarship* to an undergraduate or graduate student in Soil Science or Physical/Earth Science. In a barn west of the university campus, soil samples from the McKenzie County survey, as well as samples from Charles Kellogg's surveys in Wisconsin from the late 1920s are stored; the jars are labeled with the date, location, horizon, and the initials C.E.K.

Soil Series are almost invariably named according to the place where they are first described but, some 60 years after Charles Kellogg conducted soil surveys in northern Wisconsin, a soil series was named after him: The Kellogg series. It is a Spodosol (podzol) found on lake plains, mostly under forest; very deep, moderately well drained, and formed in sandy lacustrine deposits or outwash and the underlying clayey deposits. The series occurs solely in Northern Wisconsin and the Northern lower peninsula of Michigan and is similar to the soil on which Charles Kellogg was raised in Palo. In 2012, the soils laboratory at the *National Resource and Conservation Service* in Lincoln, Nebraska, was renamed the *Dr. Charles E. Kellogg Soil Survey Laboratory* at the instigation of the late Jon Hempel. The laboratory, originally started in Mandan in North Dakota in 1949, is part of the *National Soil Survey Center*. A plaque honoring Charles Kellogg was unveiled by his grandson, Steve Kellogg.

In June 2019, Jim Bockheim and I visited Palo and Ionia in Michigan. Jim grew up on a Spodosol in Grand Rapids and knows the soils, geomorphology and the geography of the Great Lakes Region [3]; he met Charles Kellogg at a soil judging contest at Cornell University in 1968. Jim and I took the ferry from Milwaukee across Lake Michigan, a body of water 1.6 times the size of the Netherlands. In Muskegon, on the Michigan side, we were greeted by sand dunes towering 80 m over the landscape. We drove through fields waiting to be sown with corn or soybeans but the wetness of spring and early summer, combined with the slow drainage, had delayed planting; weather like this is predicted to become common for much of the Midwest.

About 200 people live in Palo now. It has not changed much in size since Charles Kellogg left a century ago and he enrolled in a short course

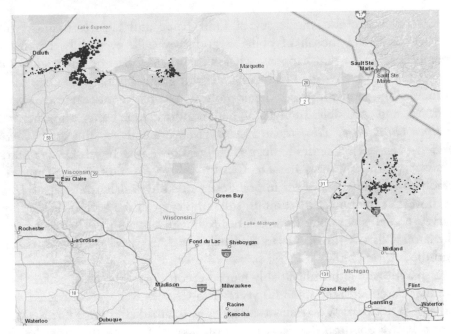

Kellogg soil series (in red) in Northern Wisconsin and the Northern lower peninsula of Michigan. The Kellogg soils are wet Spodosols, very deep, and formed in sandy lacustrine deposits or outwash and the underlying clayey lacustrine deposits. Its full name is: Sandy over clayey, mixed, active, frigid Alfic Oxyaquic Haplorthods. The series was established in 1993 in Michigan. Map from the Web Soil Survey

at *Michigan Agricultural College*. The main street was tarmac but side roads were unpaved. Some houses were well maintained, some trailers had been added, and Palo retained a small store and Methodist and Baptist churches. The school where Charles Kellogg attended had been expanded with some low buildings in the 1970s but was closed four years ago and rented out; several families now occupy the buildings. The cemetery on the south side of Palo was well maintained, and the graves of Charles and Lucille Kellogg were next to his parents, her parents, and his grandparents. The grave did not stand out—when we are dead, we are all equal.

We drove three kilometer south to the homestead on which Charles's grandfather had settled in the 1870s, and where he built the house in which Charles was born in 1902. We were reservedly but warmly welcomed by a woman with two small children. They rented and had lived in the house for only half a year. Her husband was deaf, and we engaged in a sign language conversation looking at Kellogg family photos on my laptop. The neighbors came over and told us that the grandfather's house had burned down many years ago, and that the tenants lived in the house where Charles had grown

Charles Kellogg (1902–1980) plaque in the laboratory of the *Natural Resources Conservation Services* in Lincoln, Nebraska. It was unveiled on the 4th of June 2012. The plaque reads: "Dr. Charles Kellogg achieved one of the most distinguished careers in the history of soil science, a career marked by a dedication to assist land users through the knowledge of soils. To accomplish that objective he redirected and refined the soil survey program of the Department of Agriculture, which he supervised from 1934 to 1971. His legacy continues in the work of soil scientists around the world." It includes his often-quoted saying from 1938: "Essentially, all life depends upon the soil... There can be no life without soil and no soil without life; they have evolved together." Left picture: Soil sample jars from his collection in Wisconsin (1928–1929) and North Dakota (1930–1933) stored in a shed of North Dakota State University. Photos by the author

up. The farmland had changed ownership four times since the 1950s and the house was separately sold from the land. The glossic horizon waved under our feet when we walked around the barn. A large sinkhole had filled up with water since the surrounding land was now artificially drained. The large red barn, once the source of continuous chores for a young Charles Kellogg, had lost its doors and some of its siding. Milkweed grew out of the cowshed. The sun set over the greening hills when we drove to Ionia. The day was over when we walked into a hotel built on an old terrace of the Grand River.

The Truog farm in Independence, Wisconsin was started in 1884 by Thomas and Magdalena Truog, in the same year Emil Truog was born. They cleared 72 hectares and brought it under the plough. Thomas Jr. took over the farm in the early 1900s and, in 1915, he and Emil Truog purchased the farm from the other heirs. The farm was operated by Emil's brother but sold in 1961 to the neighbor Don Woychik who paid $65,000 for the house, barn, and land. A cousin was also interested but could not afford it. The Woychiks built a small house where Emil's brother could live, and where he accidentally set himself on fire near the stove. He died a few days later.

The Truogs had hardy, calm Guernsey cattle. The Woychiks milked Holsteins, large cows that were too wide to fit the bays of the Truog barn and, so, they had to make adjustments to house their 60 cows. They built some smaller barns and large silos, and milked cows until the late 1990s when it became clear that their son, who had a degree in agriculture from the University of Wisconsin-River Falls, was not going to take over the farm. Their other son had tragically died in farm accident. Don Woychik sold the cows and constructed a chicken barn where they fatten 350,000 broilers per year.

Leo Walsh and I visited the old Truog farm in Independence in August 2019. Leo knew the area well. Raised on a farm in Iowa, he came in 1955 to the Department of Soils in Madison, obtained his PhD and, later, became chair of the department and Dean of the College of Agriculture. With a graduate student, Bob Hoeft, he studied the deficiency and excess of sulfur of alfalfa in the fields around Independence. Emil Truog had introduced alfalfa in Independence in the 1940s; prior to that farmers grew red and white clover. Alfalfa has high sulfur demands but the soils were low in sulfur and so deficiencies were common [4]. But there were also alfalfa fields that had high levels of sulfur in their foliar tissue. As Leo Walsh discovered, such excess was caused by the nearby coal plant in Alma where the sulfur in the air was 10 to 15 times higher than in other locations; they found that sulfur in the rainfall was as high as 168 kg per hectare [5]. It took another decade and dying forests before such deposition was termed 'acid rain.' In Alma, measures were taken to reduce the sulfur from the exhaust of the coal plant, and sulfur precipitation decreased sharply, to the extent that sulfur is now deficient in most soils of Independence.

The colossal three-story barn at the Truog farm was designed by Emil Truog and built in 1929. It was almost 40 m long and 11 m wide, constructed from solid wood and all bolted; the wood came from Emil Truog's in-laws, the Rayne lumber yards. The cow byre had a ceiling over three meters high which was much higher than in most barns built at that time and aimed to improve ventilation and airflow. Emil Truog had learned about ventilation from Franklin King, who had shown that it was critical for successful byres: "Not so very many years ago existed woeful ignorance of the importance—vital necessity of fresh air to the health of both man and beast" [6]. As Don Woychik told us, the high ceiling was good for the cows but so cold in winter that the drinking troughs would freeze. Since the Woychiks had stopped milking, the barn had slowly filled up with machinery, a caravan, recycled cans for the local football club, and random belongings that tend to accumulate in empty barns. A family room was created upstairs on the hay deck, with pictures, toys, old clothing, farm tools, and kitchen utensils from

The Truog farm in Independence, Wisconsin, in the 1950s some years before it was sold to their neighbors. In 1884, Thomas and Magdalena Truog bought 90 hectares of land in Independence, some 40 km of the Mississippi river. The same year Emil Truog was born. The house was built in 1898. Photo from the Truog family

the past. The wooden construction overarching the characteristic shape and angular bending majestically stood the test of time.

Only Place—the farm on which Roy Simonson grew up is no longer in the family. His parents had filed the homestead claims on quarter sections of land in the spring of 1897 and had built it up from very little. When Otto Simonson died, Bruce Simonson spent a summer farming and by that time it was already owned by the children of Roy's sister Ruth. But farm profits were slim, investments and interest rates incalculably high, and earnings minimal. So, the farm and land were sold. The house was taken apart and moved off the property some years ago.

The post office in Agate, which was established in 1907, had closed in 1964, but Bisbee, the nearby village where Roy Simonson had gone to high school in 1919, kept its post office open, as well as a chocolate shop and a Lutheran church. The train no longer stopped in Agate, and the railway track was overgrown. The grain elevators—those castle-like mansions of the prairie harvest—were demolished and in their last years, they had been occupied by pigeons. Now rock hunters occasionally traverse the landscape searching for petrified wood, jasper and agates. The wide-open lands of Agate are cultivated by large farm machinery, and people live somewhat evenly dispersed in modest houses along checkerboard roads. Those that cultivate the land do not seem to receive a fair share from their labor or the wealth of its soil; more and more is produced by fewer farmers that earn less year after year. In all that, the soil does what it does best: supportlife, lay low, and preserve its wonders broadly underfoot.

References

1. Yost E. Modern Americans in science and invention. New York and Toronto: Frederick A. Stokes Company; 1941.
2. Yost E. Modern Americans in science and technology. New York Dodd, Mead & Company; 1962.
3. Bockheim JG. Soils of the Laurentian Great Lakes, USA and Canada. Dordrecht: Springer; 2021.
4. Hoeft RG, Walsh LM. Effect of carrier, rate, and time of application of S on the yield, and S and N content of Alfalfa 1. Agron J. 1975;67(3):427–30.
5. Hoeft RG, Keeney DR, Walsh LM. Nitrogen and sulfur in precipitation and sulfur dioxide in the atmosphere in Wisconsin 1. J Environ Qual. 1972;1(2):203–8.
6. King FH. Ventilation for dwellings, rural schools and stables. Madison, WI: Published by the Author; 1908.

Correction to: Soil Science Americana

Correction to:
A. E. Hartemink, *Soil Science Americana*
https://doi.org/10.1007/978-3-030-71135-1

In the original version of this book, chapter-wise abstracts were omitted online. This has now been rectified and abstracts have been included in the online version. The book has been updated with the changes.

The updated version of the book can be found at
https://doi.org/10.1007/978-3-030-71135-1

A. E. Hartemink, *Soil Science Americana*
https://doi.org/10.1007/978-3-030-71135-1_16

Index

© The Editor(s) (if applicable) and The Author(s), under exclusive
license to Springer Nature Switzerland AG 2021
A. E. Hartemink, *Soil Science Americana*
https://doi.org/10.1007/978-3-030-71135-1

Person Index[1]

---

[1] Emil Truog (Chapter 2), Charles Kellogg (Chapter 3) and Roy Simonson (Chapter 4) are only listed where they are mentioned in any of the other chapters. Country of birth in parenthesis behind the name; second country is where the person worked. For example, Selman Waksman was born in Ukraine but worked his entire life in the USA.

Person Index by Country[2]

[2]People are listed by country of birth; country in parentheses where they emigrated to and worked